International Scientific and Practical Conference "Sustainable Development of Traditional and Organic Agriculture in the Concept of Green Economy" (SDGE 2021)

BIO Web of Conferences Volume 42 (2022)

Smolensk, Russia
27 November 2021

Editor:

Patlasov Oleg Yurievich

ISBN: 978-1-7138-4241-5

Printed from e-media with permission by:

Curran Associates, Inc.
57 Morehouse Lane
Red Hook, NY 12571

Some format issues inherent in the e-media version may also appear in this print version.

For additional information, please contact EDP Sciences – Web of Conferences
at the address below.

EDP Sciences – Web of Conferences
17, Avenue du Hoggar
Parc d'Activité de Courtabœuf
BP 112
F-91944 Les Ulis Cedex A
France

Phone: +33 (0) 1 69 18 75 75
Fax: +33 (0) 1 69 28 84 91

contact-edps@webofconferences.org

Additional copies of this publication are available from:

Curran Associates, Inc.
57 Morehouse Lane
Red Hook, NY 12571 USA
Phone: 845-758-0400
Fax: 845-758-2633
Email: curran@proceedings.com
Web: www.proceedings.com

TABLE OF CONTENTS

edp sciences

Web of Conferences

Statement of Peer review

In submitting conference proceedings to *Web of Conferences*, the editors of the proceedings certify to the Publisher that

1. They adhere to its **Policy on Publishing Integrity** in order to safeguard good scientific practice in publishing.
2. All articles have been subjected to peer review administered by the proceedings editors.
3. Reviews have been conducted by expert referees, who have been requested to provide unbiased and constructive comments aimed, whenever possible, at improving the work.
4. Proceedings editors have taken all reasonable steps to ensure the quality of the materials they publish and their decision to accept or reject a paper for publication has been based only on the merits of the work and the relevance to the journal.

Title, date and place of the conference

International Scientific and Practical Conference Sustainable Development of Traditional and Organic Agriculture in the Concept of Green Economy (SDGE 2021)

Smolensk, Russia, November 27, 2021

Proceedings editor(s):

Patlasov Oleg Yurievich

О.Ю. Патласов

Date and editor's signature

20/12/2021

17, avenue du Hoggar - PA de Courtabœuf - BP 112 - 91944 Les Ulis Cedex A (France)
Tél. : 33 (0)1 69 18 75 75 - Fax : 33(0)1 69 07 45 17 – www.edpsciences.org

Comparative analysis of rape seed production in Russia and Kemerovsk region

Ekaterina Kondratenko, *Olga* Soboleva, *Alexander* Vidyakin, *Tatiana* Miroshina, *Daria* Raushkina, and *A. V.* Zhdanov

Kuzbass State Agricultural Academy, 650056, st. Markovtseva, 5, Kemerovo, Russia

Abstract. The purpose of the research is to conduct a comparative analysis of rapeseed production in the territory of the Russian Federation and the Kemerovo region. During 2011-2019, on the territory of the Russian Federation (RF) and the Kemerovo region, a comparative analysis of the sown area, yield, gross harvest of spring rape was carried out. The research results showed that over the past 9 years, from 2011 to 2019, the sown area, both in the Russian Federation and in the Kemerovo Region, has doubled. At the same time, there is a strong variation in this indicator. The variation coefficient ranged from 43.3% to 54.1%. On the territory of the Russian Federation and in the Kemerovo region, an increase in the yield of spring rape is observed by 1.13 times and 1.04 times, respectively. It was revealed that an increase in the gross harvest of spring rape seeds occurs on the basis of an extensive development of production due to an increase in the sowing area and yield.

1 Introduction

Spring rape or colza *Brassica napus L.* is an important annual herbaceous oilseed crop of the cruciferous family. Not found in the wild. In culture, it is known for 4 thousand years BC [1]. According to D.N. Pryanishnikov (1963) began to cultivate rapeseed in Russia in the 18th century [2].

Rape is a cross between spring rape and cabbage (*Brassica oleracea*). S.V. Goncharov and L.A. Fedorchuk (2018) believe that rapeseed is superior to many agricultural crops in terms of nutritional and fodder merits. So, in the production of oilseeds, spring rape takes the 2nd place, second only to soybeans, ahead of sunflower [3].

According to S.K. Sandhu, G. Singh (2018), rape is the main oilseed crop and ranks third in the world in terms of production among oilseeds [4].

The largest world producers of rapeseed are: Canada - 14.2 million tons, China - 13 million tons and India - 6.5 million tons [5].

One of the important problems in fodder production is to increase the production of vegetable protein for animal husbandry. It can be solved by cultivating high-protein crops, including winter and spring rapeseed, which are used for the production of green fodder and silage, as well as in the form of cake and meal, grass meal and seeds for animal feed as high-energy protein supplements to concentrated feed.

Rape seeds accumulate 40-50% fat and 20-30% protein [6]. The meal obtained from rape seeds allows 1 ton of meal to balance 10 tonnes of compound feed in terms of protein

[7]. Its fats and proteins are of great biological importance. Rapeseed fats contain unsaturated fatty acids (oleic, linoleic and linolenic), which are not synthesized in the body of animals, but must be supplied with feed.

M.G. Dawood, M.S. Sadak (2014) believe that the cake remaining after the oil is extracted from the rapeseed is rich in protein and can be used as animal feed. As a fodder crop, rapeseed provides a good yield of nutrients. 1 kg of rapeseed flour obtained from rape seeds contains 400-450 g of fat and up to 380 g of protein, which is 1.9-4 times more than in pea, wheat and barley flour [8].

The fodder value of rapeseed is that its vegetative mass serves as fodder for cattle in early spring and late autumn, when the productivity of perennial grasses decreases. This allows the use of green forage to be extended by 25-30 days. Vegetative mass of rape is valuable forage that is not inferior in protein content to legumes. With a yield of 400 c/ha, each hectare gives 6400 c.u. and about 820 kg of digestible protein, which is significantly more than in the green mass of corn and sunflower. With good moisture supply, it is capable of producing 2-3 mows of green mass, because it has the ability to grow back, its crops can be used for grazing.

W. Long et al. (2018) confirm that rapeseed contains 25% protein and 40-42% oil, 60% oleic acid and 8.8% linoleic acid [9].

By subjecting proteins to heating with solutions of mineral acids, the researchers found all essential amino acids in the seed hydrolysis products, the main ones of which are lysine in an amount of 1.24%, methionine - 0.64%, cystine - 0.72% and methionine + cystine 1.32%. Most of the amino acids are derivatives of fatty acids, which have an amino group [10].

F.E. Açikgöz et al. (2014), based on scientific studies, found that rapeseed has a rich nutritional composition. It is a good source of vitamin C, crude protein, crude fiber and minerals [11].

According to Glazieli Marangoni de Oliveira et al (2015), rapeseed can also be used for human food purposes. Thus, rapeseed oil is used for salads and frying, as well as for the production of margarines and other food products [12].

The World Health Organization (WHO) estimates that cardiovascular disease causes 17.7 million deaths worldwide each year, or 31% of all deaths. At the same time, it is known that rapeseed products with a high content of unsaturated fatty acids have proven themselves well in the prevention of diseases of the cardiovascular system [13].

E.A. Weiss et al. (2000) confirm that rapeseed has the highest ratio of unsaturated fatty acids among vegetable oils. Rapeseed oil contains about 65% oleic acid, 20% linoleic acid, 9% linolenic acid and 2% stearic acid [14].

According to A.S. Tan et al. (2009), a high content of oleic and linoleic acids over 20% indicates a high quality of the oil. Rapeseed oil, rich in linoleic acid, lowers cholesterol and triglyceride levels, improves blood cell viscosity and thus prevents embolism [15].

A. Namvar, T. Khandan (2015) identified and claim that rapeseed oil helps to lower cholesterol levels in the human body, as it is rich in two important essential fatty acids: linoleic acid (ω-6 fatty acid) and α-linolenic acid (ω-3 fatty acid) in a ratio of 2:1, which is ideal from the point of view of human nutrition [16].

M.C. Martínez-Ballesta, R. Dominguez-Perles, D.A. Moreno et al. (2010) found that the concentration of minerals in rapeseed can vary depending on the culture of agricultural technology, region of production, species and various plant organs [17].

P. Malagoli, P. Laine, L. Rossato, A. Ourry (2005) believe that rape is also capable of absorbing high levels of nitrates from the soil. Therefore, rapeseed cultivation can be used to reduce nitrogen leaching in autumn and winter [18].

T. Emrebas (2010) argues that rape has a number of advantages over other oilseeds. It can be winter or spring; the growing season is shorter than that of other oilseeds; it has a higher yield per unit area; the seeds have a fairly high oil content; improved varieties do not contain erucic acid and glucocyanate, and each stage of growth from sowing to harvest is suitable for mechanization [19].

In the Kemerovo region in 2019, spring rape was cultivated on an area of 74.1 thousand hectares and was widely used not only for fodder purposes, but also for the production of seeds. Farms of the region use rape for green fodder for the preparation of silage and haylage. In the rations of feeding farm animals, rapeseed cake and meal are used.

The purpose of the research is a comparative analysis of rapeseed production in Russia and the Kemerovo region.

2 Objects and methods

The object of research was rapeseed cultivated on the territory of the Russian Federation and the Kemerovo region.

To characterize the dynamics of rapeseed production in the period 2011-2019 on the territory of the Russian Federation, the data of the official website of the Federal State Statistics Service of the Russian Federation [20] were used, on the territory of the Kemerovo Region - the official website of the Federal State Statistics Service of the Kemerovo Region [21].

The analysis of the dynamics of the cultivated areas, yield and gross harvest of rapeseed in the studied territories was carried out to identify the level of development of the production of this crop and the reserves of its development.

Mathematical processing was carried out according to B.A. Dospekhov. The variability of the studied indicators was assessed by the magnitude of the range of variation, defined as the ratio of the difference between the maximum value of the trait and the minimum to its maximum value, expressed as a percentage (V, %) [22].

3 Results and discussions

The analysis of the conducted scientific research revealed a significant increase in the acreage of rapeseed on the territory of the Russian Federation (RF). In 2019, according to the Federal State Statistics Service of the Russian Federation, it amounted up to 1,547.5 thousand hectares, with most of it accounted for spring rape, the area of which reached 1356.7 thousand hectares, while the winter crop was only 191 thousand hectares, which is 7.1 times more (Fig. 1).

On the territory of the Russian Federation for the period under review 2011-2019 the total growth of the area under crops of rape was 653.95 thousand ha (73.2%). According to our data, the sown area increased due to the expansion of spring rapeseed by 639 thousand hectares (89.1%), the increase in the area of winter rape was 15 thousand hectares (8.4%).

The sown area on average in the country in the studied series of years underwent significant changes from year to year. The maximum sowing areas of spring rape were in 2018-2019. - 1386.86 thousand hectares and 1356.70 thousand hectares, the minimum in 2011 - 717.58 thousand hectares. Its value varied under winter rape within 98.09-276.37 thousand hectares with an average value of 174.9 thousand hectares (V=64.5%), under spring rape - 717.58-1386.86 thousand hectares (V=48.2%) with the average value is 913.8 thousand hectares.

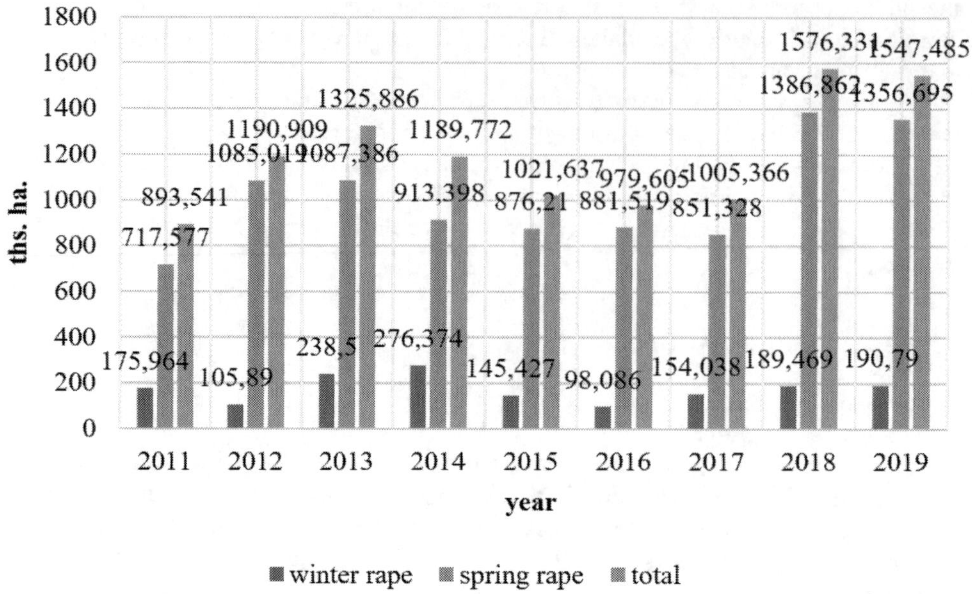

Fig. 1. Dynamics of the planting acreage of rapeseed in the Russian Federation for the period 2011-2019, ths.ha.

Analyzing the dynamics of sown areas by federal districts of the Russian Federation, it can be noted that in 2019 the largest share of all rapeseed crops in Russia fell on the Siberian Federal District (SFD) - 44.0% (680 thousand hectares). The Central Federal District (CFD) with an area of 317.3 thousand hectares ranked second in the country. Rape crops in the Siberian Federal District are 2.1 times higher than those in the Central Federal District.

Among the regions of the Siberian Federal District, the leading position is occupied by the Altai Territory, where 186 thousand hectares (12.0%) were sown in 2019. The Kemerovo region took sixth place with a sowing area of 74.1 thousand hectares, which is 4.8%. This is a fairly good result. Consequently, it is necessary to use the existing economic potential and new resources for the production of rapeseed in the study area.

Over the years of research on the territory of the Russian Federation (2011-2019), the gross harvest also changed significantly. Table 1 shows its dynamics. It has been established that the gross harvest of spring rape seeds is 3.7 times higher than that of winter rape.

An annual increase in the gross harvest of rapeseed was revealed. Its value varied from 854 thousand tons to 1662 thousand tons, with an average value of 1337.8 thousand hectares (V = 49.6%). The maximum increase in the gross harvest of rapeseed seeds was observed in 2018-2019, during this period there was an increase in the sown area and yield of this crop. The minimum is 2011, 2012 and 2015. Such fluctuations occurred due to a decrease in yields in these years and sown areas allotted for rapeseed.

Table 1. Dynamics of the rapeseed gross harvest in the Russian Federation, 2011-2019.

Indicator	2011	2012	2013	2014	2015	2016	2017	2018	2019	V, %
Gross harvest, thousand tons										
Winter Rape	261	140	348	396	235	138	284	290	351	64,6
Spring Rape	611	714	761	788	641	699	959	1313	1311	53,5
Total	872	854	1109	1184	876	837	1243	1603	1662	49,6

Comparative analysis of the yield of winter rape and spring rape cultivated in the fields of the Russian Federation for 2011-2019 shows that the highest average yield is formed when growing winter rape (Table 2). Its average value was 15.5 c/ha over the years of the study, with year-to-year variability from 13.21 c/ha to 18.38 c/ha. It was found that the yield was 1.8 times higher in winter rape compared to spring rape. On average, for 9 years, the yield of spring rape was 8.48 c/ha.

From the above, it follows that the highest average yield is formed during the cultivation of winter rape; on average for 2011-2019, it amounted to 15.48 c/ha.

Table 2. Variability of yield (c/ha) of spring rape and winter rape, t/ha, RF, 2011-2019

Year	Spring rape	Winter rape	V, % between rape species	Both rape species
	average	average		average
2011	8.51	14.83	42,6	9,75
2012	6.58	13.21	50,2	7,17
2013	7.00	14.56	51,9	8,36
2014	8.63	14.35	39,8	9,95
2015	7.32	16.21	54,8	8,57
2016	7.93	14.08	46,7	8,54
2017	11.27	18.44	38,9	12,37
2018	9.47	15.34	38,3	10,17
2019	9.66	18.38	47,4	10,74
среднее	8.48	15.48	–	9,51
V, %	45.5	27.8	–	33,0
НСР $_{05}$	0,2	0,3	–	0,4

On average, over 9 years, the range of variation in the yield of spring rape was 45.5%, winter rape - 27.8%. The variability of the yield of spring rape is higher in comparison with the yield of winter rape. On average for the period 2011-2019, winter rape plants formed a higher yield than spring rape plants. The positive features of winter rapeseed determine the desire to search for studies of the possibility of its cultivation in the territory of not only the country, but also in Siberia, to which the Kemerovo region belongs. The analysis shows that winter rapeseed is not spreading in the crops of farms due to insufficiently developed assortment of varieties and cultivation technology.

From the standpoint of adaptive crop production, the problem requires a solution based on the study of the specific growing conditions of this culture. Unfortunately, there is practically no scientific data on the features of the development and formation of the yield of winter rapeseed in this ecological niche (Kemerovo region).

As in the country as a whole, in recent years in the Kemerovo region there has been an increase in the area occupied by spring rape. So, if in 2011 the sown area was 34.6 thousand hectares, then in 2019 it was already 74.1 thousand hectares, that is, more by 39.5 thousand hectares (an increase of 114.2%) (Fig. 2). With an average sown area of 61.9 thousand hectares, the coefficient of variation was V = 53.3%. The maximum areas were allocated for this crop in 2019, the minimum - in 2011 and 2017.

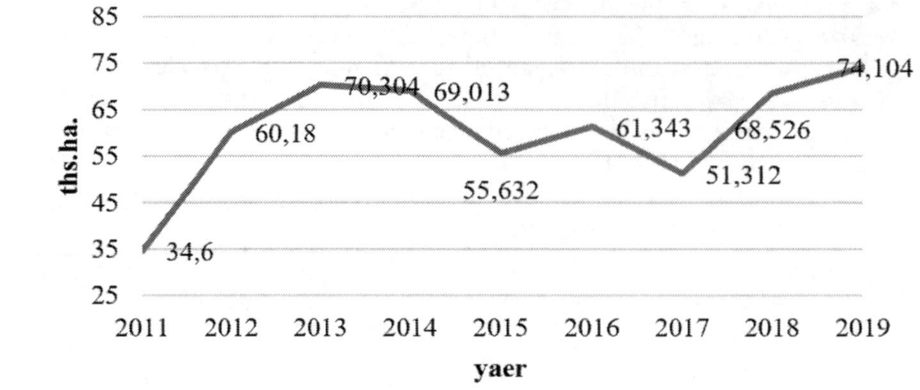

Fig. 2. Dynamics of rapeseed areas in the Kemerovo region, ths.ha.

On the territory of the Kemerovo region, along with the rapid expansion of the cultivated areas for rapeseed, the volume of its gross harvest is steadily growing. If in 2011 34.42 thousand tons of spring rapeseed were collected, then in 2019 it is already 85.05 thousand tons, an increase of more than 50.63 tons (147.1%) (Fig. 3).

From the data in Figure 3 it can be seen that an increase in gross harvests of rapeseed occurs due to an increase in the sown area and due to an increase in yield.

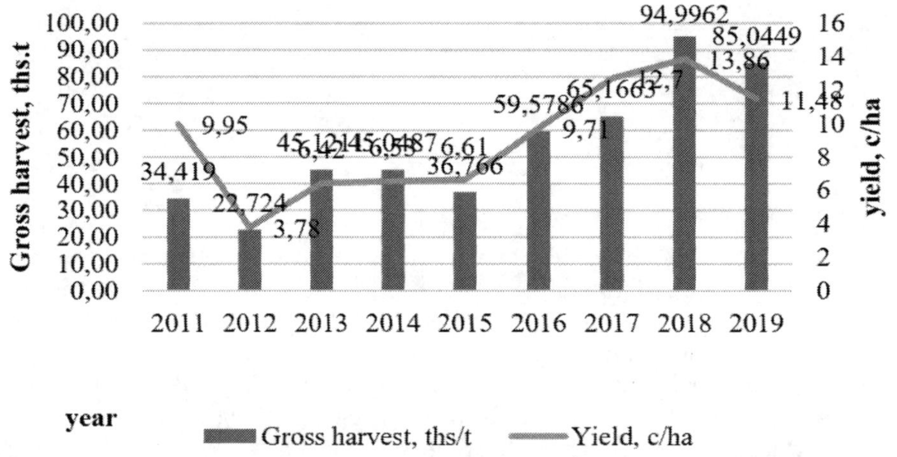

Fig. 3. Dynamics of gross harvest and yield of spring rapeseed in the Kemerovo region, 2011-2019.

It was found that the average yield for these years was 9.0 c/ha. In 2018, a record yield of seeds of this crop was obtained for the region - 13.86 c/ha.

Table 3. Comparative characteristics of rapeseed yield, RF, Kemerovo region, 2011-2019

Crop	Lim of yield, c/ha	Average yield, c/ha	V, %
the Russian Federation			
Spring rape	6.58 – 11.27	8.48	45.5
Winter rape	13.21 – 18.38	15,48	27.8
Kemerovo region			
Spring rape	3.78 – 13.86	9.00	72.6
HCP_{05}		0.17	

A comparative analysis of the yield of spring rapeseed cultivated in the Russian Federation and the Kemerovo region shows that the highest average yield is formed in the Kemerovo region, the excess was 1.12 times. The year-to-year variability of its value ranged from 6.58 c/ha to 11.27 c/ha (RF) and from 3.78 c/ha to 13.86 c/ha (Kemerovo region). A high degree of variation in the yield of spring rape was revealed during cultivation on the territory of the Kemerovo region V = 72.6%, this indicates an insufficiently developed technology of cultivation of rape in this ecological sphere.

4 Conclusions

Comparative analysis of the planting acreage on average for the period 2011-2019, allotted for rapeseed in the Russian Federation and the Kemerovo region, shows that the sown area increased by 682 thousand hectares and 39.5 thousand hectares, respectively. It was found that on the territory of the Russian Federation, winter rape plants formed a higher yield than spring rape plants. The average yield of spring rape was 8.48 c/ha, winter 15.48 c/ha, which is 1.8 times more than spring rape. The positive features of winter rapeseed determine the desire to search for studies of the possibility of its cultivation in the territory not only of the country, but also in Siberia. From the standpoint of adaptive crop production, the problem requires a solution based on the study of the specific growing conditions of this culture. Unfortunately, there are practically no scientific data on the features of the development and formation of the yield of winter rapeseed in this ecological niche.

The observed trend in the development of spring rapeseed production indicates that in the agro-climatic conditions of the Russian Federation and the Kemerovo region, this crop is cultivated successfully and can reasonably claim further growth and increase in the export of the obtained products, including outside the country.

References

1. E. N. Sinskaya, *Historical geography of cultural flora (at the dawn of agriculture)*, 480 (1969)
2. D. N. Pryanishnikov, *Selected plants*, 735 (1963)
3. S. V. Goncharov, Oilseeds: scientific and technical bulletin of the All-Russian Scientific Research Institute of Oilseeds, **1(173)**, 36 (2018)
4. S. K. Sandhu, G. Singh, Springer, Cham, 317 (2018)
5. A. L. Filimonov, Tendencies of agricultural production in modern Russia: materials of the XII international scientific and practical conference, 285 (2013)
6. V. V. Vydrin, *Varietal zoning of agricultural crops and the results of variety testing in the Tyumen region for 2011*, 15 (2011)
7. V. V. Karpachev, Feed production, **12**, 17 (2007)
8. M. G. Dawood, M. S. Sadak, Middle East J Agric Res., **3,** 943 (2014)
9. W. Long, et al., Frontiers in genetics, **9,** 399 (2018)
10. E. P. Kondratenko, O. M. Soboleva, A. S. Sukhikh, M. G. Kurbanova, IV International Scientific and Practical Conference "Modern S&T Equipments and Problems in Agriculture", 116 (2020)
11. F. E. Açikgöz, et al., Ziraat Fakültesi Dergisi, Uludağ Üniversitesi, **28(1),** 43 (2014)
12. G. M. de Oliveira, M. A. Stahl, A. P. B. Ribeiro, R. Grimaldi, L. P. Cardoso, T. G. Kieckbusch, Eur J Lipid Sci Technol., **117,** 1762 (2015)
13. WHO (World Health Organization), World health report (2017) www.who.int
14. E. A. Weiss, et al., Oilseed crops, Blackwell Science (2000)
15. A. Ş. Tan, et al., Anadolu, **19(2),** 1 (2009)

16. A. Namvar, T. Khandan, Arch Agron Soil Sci., **61(8),** 1137 (2015)
17. M. C. Martínez-Ballesta, R. Dominguez-Perles, D. A. Moreno, et al., Agronomy for Sustainable Development, **30,** 295 (2010)
18. P. Malagoli, P. Laine, L. Rossato, A. Ourry, Annals of Botany, **95(5),** 853 (2005)
19. T. Emrebas, *Determination of oil and seed yield with yield components of some canola varieties under Kahramanmaras ecological conditions* (2010)
20. Federal State Statistics Service of the Russian Federation, https://rosstat.gov.ru
21. Federal State Statistics Service for the Kemerovo Region, http://kemerovostat.gks.ru
22. B. A. Dospekhov, *Field experiment methodology (with the basics of statistical processing of research results): a textbook for students of agricultural universities in agronomic,* 352 (2011)

BIO Web of Conferences **42**, 01002 (2022) https://doi.org/10.1051/bioconf/20224201002
SDGE 2021

Microgreens - biologically complete product of the XXI century

E. P. Kondratenko, *S. N.* Vityaz*, *T. A.* Miroshina, and *A.S.* Kuznetsov

Kuzbass State Agricultural Academy, st. Markovtseva, 5, 650056 Kemerovo, Russia

Abstract. Microgreens are a new functional food crop that can facilitate adaptation to urbanization and global climate change, and improve human health. The research was carried out in 2021 at the Department of Landscape Architecture of the Kuzbass State Agricultural Academy. The aim of the research was to study the technology of cultivation of microgreens of the *Brassicaceae* family on an aqueous substrate. The objects of research were the seeds of cultivated plants of the *Brassicaceae* family: Brassica oleracea Broccoli Group broccoli or asparagus, Fortuna, Raphanus sativus radish, Violetta, Lepidium sativum, watercress, Dansky, and Eruca versicaria, arugula, Sicily. It was revealed that microgreens can be obtained from seeds of the *Brassicaceae* family in 6-12 days. Such a product does not have time to accumulate harmful substances from the atmosphere in a short period of time. When growing microgreens, it is not necessary to use mineral fertilizers, pesticides and, thus, it is possible to obtain environmentally friendly, biologically useful products with low material costs. It was found that, depending on the seeds of the studied crops and their genotype, the cycle of growing microgreens lasts from 6 to 10 days after germination. Depending on the type of culture, the sprouts reached a height of 5-10 cm. The laboratory germination rate was 96-98%.

1 Introduction

Studies by Kyriacou et al. (2016), Galieni et al. (2020), Turner et al. (2020) show that microgreens are a new class of foods that are gaining popularity. These are seedlings of edible plants, harvested 7-14 days after sowing, when the first true leaves begin to appear.

Verlinden (2019) defines microgreens as small lettuce greens grown from fully developed cotyledons that include one or two true plant leaves. There are many definitions of microgreening, but based on a review of the literature, it can be assumed that the above simple definition is quite broad. Microgreens are harvested without roots and seed coat and are therefore less likely to be exposed to two pathways of contamination during sprout production: the nutrient medium and the seeds themselves.

Microgreens are a new functional food crop that can sustainably diversify global food systems, facilitate adaptation to urbanization and global climate change, and improve human health. Microgreens are characterized by high quality, minimal environmental impact and wide consumer acceptance, according to Michell et al (2020).

* Corresponding author: svetlana_vityaz@mail.ru

Microgreens have become popular in the past two or three decades as a culinary aid to improve texture, color, flavor, aroma and visual appeal in a range of foods, and more recently they have been promoted as a beneficial dietary supplement. Microgreens are mainly used in the restaurant industry to decorate dishes and are most often consumed fresh in salads, soups and sandwiches.

A series of microgreen cultivation studies were conducted by Kyriacou et al. In 2016, scientists argue that interest in fresh, functional foods is growing, driven by growing consumer interest in health-promoting diets. The researchers emphasize that microgreens have tremendous potential to enhance the nutritional value of the human diet.

Chronic illness is a serious health problem. Accumulating evidence suggests that eating vegetables can significantly reduce the risk of many chronic diseases. Microgreens, despite their small size, have a delicate texture, characteristic flavor and a variety of nutrients. Because microgreens are rich in nutrients, smaller amounts can provide similar nutritional benefits compared to more mature vegetables (Choe U., Yu L.L., Wang T.T.Y., 2018).

According to Meyerowitz (2010), microgreens are nutritious functional foods. According to his research, soybean sprouts have the highest protein level (28%), followed by lentil and pea sprouts at 26% (Meyerowitz 2010). It has been found that soybean sprouts contain twice as much protein as chicken eggs.

Marton et. al. (2010) found that during seed germination, polysaccharides under the action of enzymes and water are hydrolyzed into oligo- and monosaccharides, fats into free fatty acids, proteins to oligopeptides and free amino acids. Therefore, sprouting can be considered as a type of pre-digestion that helps to break down high molecular weight complex organic substances into low molecular weight building blocks (Marton et. al. 2010).

Caracciolo et al. (2020) note that microgreens enrich the human diet with important nutrients. The nutritional composition primarily depends on the crop being grown, but also on the growing conditions. Young plants have a significantly higher vitamin content than mature plants. Due to the higher level of phytochemicals, microgreens are classified as functional foods (Sharma et al., 2012; Izhmulkina et al., 2020).).

Based on an analysis of the literature, Verlinden (2019) showed that most of the work on microgreens focuses on the effect of light during production to increase the level of mineral and bioactive compounds. Other publications focus on the microbial safety of this type of product. Most of the work indicates that microgreens are safer from a microbiological point of view than seedlings.

Microgreens are usually produced using soilless growing systems. The soil is replaced with a substrate or the plants are grown in an aquatic environment containing all the elements necessary for life. The production of microgreens can be carried out in a variety of environments and can be carried out commercially using modern techniques necessary to ensure continuity of production and good product quality or as a hobby, for self-consumption, using very simple methods and techniques and very limited spaces, e.g. windowsill. While growing microgreens for your own consumption can be fairly straightforward, commercial production of microgreens is not easy and requires specialized technology.

Scientists mainly investigate species belonging to the families *Brassicaceae, Asteraceae, Chenopodiaceae, Lamiaceae, Apiaceae, Amarillydaceae, Amaranthceae* and *Cucurbitaceae* for microgreens. Methods for improving bioactivity and principles of human health promotion, use, storage after harvest and microbial safety are considered.

The aim of our research was to study the technology of cultivation of microgreens of the *Brassicaceae* family on an aqueous substrate.

2 Material and research methods

The research was carried out in 2021 at the Department of Landscape Architecture of the Kuzbass State Agricultural Academy. The objects of research were the seeds of cultivated plants of the Brassicaceae family: broccoli variety *Fortuna*, radish variety *Violetta*, watercress variety *Dansky*, and arugula variety *Sicily*.

The experiment used calibrated seeds, without diseases and pests, which were not treated with chemical preparations. Sowing of seeds was carried out on March 10. The seeding rate was 5 g per 75 cm^2. Plants were grown at a temperature of +18 ... +24^{0}C without additional lighting on the windowsill. Mineral fertilizers were not applied. For the cultivation of microgreens, a hydroponic plant was used using a two-channel compressor on tap water, which did not contain heavy metals, pollutants and pathogenic microorganisms (Escherichia coli, Salmonella). To ensure good germination of the root system and optimal growth and development of seedlings, the level of aeration was 20-30% of the total volume. The microgreen medium had a pH of 5.5 to 6.5.

Germination energy was determined for broccoli and radish cabbage at 3 days, watercress - at 4 days and arugula - at 7 days; laboratory germination of broccoli and watercress seeds was determined at 10, radishes at 7 and arugula at 14 days in accordance with GOST 12038-84.

The color, smell, taste of microgreens were determined according to GOST R 54692-2011 (for broccoli), GOST 34216-2017 (radish), GOST 34215-2017 (watercress and arugula).

The sowing density (pcs/cm^2) was also determined.

The microgreens were harvested on day 10 and their yield was determined.

Dry matter in roots and shoots was determined according to GOST 33977 - 2016.

Mathematical processing was carried out according to B.A. Dospekhov.

3 Results and discussion

In the production of microgreens, the quality of the seeds is of great importance, since the speed and uniformity of germination, as well as the final result of the growing cycle, depend on their quality. To avoid weed infestation, which may be inedible or poisonous, the seeds were clean and 100% pure. We have studied the sowing qualities of the seeds of the studied crops. The research results are presented in table 1.

Table 1. Germination energy and laboratory germination of microgreen seeds, %

Crop	Variety	Germination energy	Laboratory germination
Broccoli	Fortuna	94±2,0	98±2,0
Radish	Violet	86±1,5	96±1,5
Watercress	Dansky	89±3,0	98±3,0
Arugula	Sicily	93±2,0	98±2,0

Analysis of the results of determining the quality of seeds of the studied crops made it possible to reveal that the indicators of germination energy and laboratory germination vary within 86-94% and 96-98%, respectively, which indicates a high bioenergetic potential of the seed. The highest values of laboratory seed germination were observed in broccoli and arugula, and the lowest in watercress.

In the course of analyzing the yield of plants used for growing microgreens, it was found that it depends on the type of crop. The maximum yield was obtained with broccoli.

Indicators of dry matter content in shoots and roots of this culture amounted to 3.3 g/m^2 (Table 2).

Table 2. Yield productivity of microgreens

Crop	Variety	Sowing density, pcs/cm^2	Phytomass, g/m^2		Dry matter content, g/m^2	
			Shoots	Roots	Shoots	Roots
Broccoli	Fortuna	12	5,6±0,01	1,1±0,005	2,8±0,01	0,5±0,005
Radish	Violet	6	3,7±0,01	0,5±0,005	1,9±0,01	0,3±0,005
Watercress	Dansky	11	2,4±0,01	0,5±0,005	1,2±0,01	0,3±0,005
Arugula	Sicily	12	1,3±0,01	0,3±0,005	0,7±0,01	0,1±0,005

The minimum yield was observed on the arugula cultivar Sicily (Fig. 1). The raw phytomass of this crops and dry matter were 1.6 g/m^2 and 0.8 g/m^2, respectively. It was revealed that in all variants of the experiment, the water content in the shoots and roots of microgreens is 2 times higher than in the dry phytomass in each test culture.

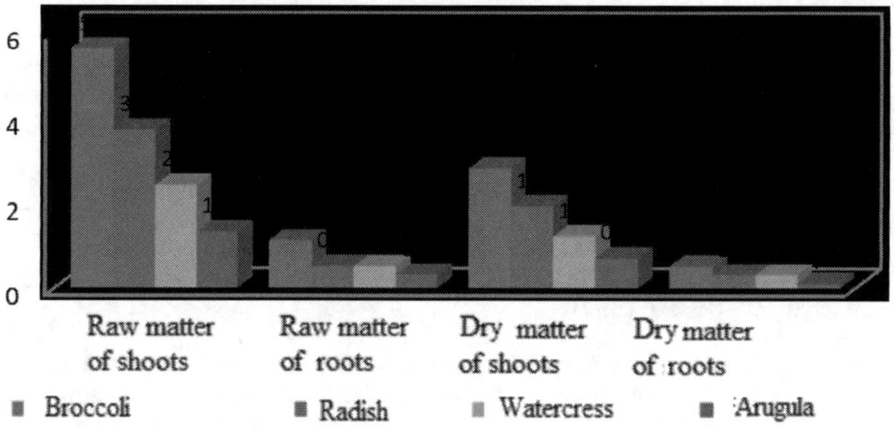

Fig. 1. Yield productivity of microgreens, g/m^2

Table 3. Coefficient of variation on average by types of crops

Indicators	max	min	среднее	V, %
Sowing density	12	6	10,25	50,0
Raw phytomass	6,7	1,6	3,85	76,1
Dry matter content	3,3	0,8	1,95	75,7

There was a strong variation in wet weight and dry matter content by crop. Thus, the studied parameters varied greatly depending on the genotypic characteristics.

It was found that the smell and taste of microgreens of all studied crops was characteristic of these botanical varieties, without any foreign smell and/or taste. The research results are presented in table 4.

Table 4. Organoleptic assessment, germination time, harvest time of microgreens of the *Brassicaceae* family

Crop	Color	Taste	Germination time, days	Harvesting, days
Broccoli	Leaves are bright green, stems are slightly pink	slight bitterness	1-2	7-10
Radish	Leaves are green, stems are white	spicy	1-2	6-7
Watercress	Leaves are green, stems are yellowish-white	peppery, spicy	3-5	5-7
Arugula	Leaves are pale green, stems are light purple	spicy	2-3	7-10

Previously studied indicators of the biological value of microgreens indicate that broccoli, radishes, watercress and arugula are rich in complete proteins, carbohydrates and have a low calorie content (Table 5).

Table 5. The content of proteins, fats and carbohydrates in microgreens

Crop	Content, g/100g			Energy value, kcal
	proteins	fats	carbohydrates	
Broccoli	3,0	0,4	7,0	34,0
Radish	1,1	0,1	2,4	14,9
Watercress	2,6	0,7	5,5	32,0
Arugula	2,6	0,7	2,1	25,0

During cultivation, the maximum accumulation of proteins and carbohydrates was noted in broccoli microgreens. Fat accumulates during the growing season, a small amount in the studied crops from 0.1 g to 0.7 g per 100 g of the product. The calorie content per 100 g of the product is, depending on the type of culture, from 14.9 kcal to 34 kcal. In addition, according to Xiao et. al. (2019), *Brassicaceae* microgreens are a good source of antioxidant phytochemicals, although there are significant differences within and between species. It is a valuable source of ascorbic acid, phylloquinone, carotenoids, tocopherols, glucosinolates and polyphenols. In general, microgreens contain more nutrients and trace elements beneficial to human health than their mature counterparts.

4 Conclusion

As a result of scientific research, it has been established that microgreens can be obtained from seeds of the Brassicaceae family in 6-12 days. Such a product does not have time to accumulate harmful substances from the atmosphere in a short period of time. When growing microgreens, it is not necessary to use mineral fertilizers, pesticides and, thus, it is possible to obtain environmentally friendly, biologically useful products with low material costs.

It was found that, depending on the seeds of the studied crops and their genotype, the cycle of growing microgreens lasts from 6 to 10 days after germination. Depending on the type of culture, the sprouts reached a height of 5-10 cm. The laboratory germination rate was 96-98%.

References

1. Marios Kyriacou, Youssef Rouphael, Francesco Di Gioia, Aggelos Kyratzis, Francesco Serio, Massimiliano Renna, Stefania De Pascale, Pietro Santamaria, Trends in Food Science & Technology, **57** (2016)
2. Angelica Galieni, Beatrice Falcinelli, Fabio Stagnari, Alessandro Datti, Paolo Benincasa, Sprouts and Microgreens: Trends, Opportunities, and Horizons for Novel Research. Agronomy, **10** (2020)
3. Francesco Caracciolo, Christophe El-Nakhel, Maria Raimondo, Marios C. Kyriacou, Luigi Cembalo, Stefania De Pascale, Youssef Rouphael, Agronomy, **10** (2020)
4. Z. Xiao, S. R. Rausch, Y. Luo, J. Sun, L. Yu, Q. Wang, J. R. Stommel, LWT – Food Science and Technology, **101**, 731-737 (2019)
5. Ellen Turner, Yaguang Luo, Robert Buchanan, Journal of Food Science, **85** (2018)
6. Sven Verlinden, Horticultural Reviews, **47** (2019)
7. P. Sharma, AK. Ghimerey, A. Gurung, CW Jin, HS. Rho, DH. Cho, African Journal of Biotechnology, **11(1)**, 184–190 (2018)
8. Kiri Michell, Hanan Isweiri, Steven Newman, Marisa Bunning, Laura Bellows, Michelle Dinges, Lauren Grabos, Sangeeta Rao, Michelle Foster, Adam Heuberger, Jessica Prenni, Henry Thompson, Mark Uchanski, Tiffany Weir, Sarah Johnson, Journal of Food Science, **85** (2020)
9. U. Choe, L.L. Yu, T.T.Y. Wang, J Agric Food Chem, **66(44)**, 11519-11530 (2018)
10. O.V. Ushakova, A.V. Molchanova, L.L. Bondareva, Vegetable crops of Russia, **1**, 96-104 (2021)
11. E.A. Izhmulkina, E.P. Kondratenko, O.M. Soboleva, O.A. Sartakova, XIX international scientific and practical conference "current trends of agricultural industry in global economy", 102-111 (2020)
12. GOST 12038-84, Agricultural seeds. Methods for determining germination, http://docs.cntd.ru/document/1200023365
13. GOST 33977-2016, Fruit and vegetable processing products

Influence of cow reproductive tract microbiota on formation of calf upper respiratory tract microbiota

Anastasiya Sergeevna Metleva*, *Oksana Vladimirovna* Smolovskaya, *Vladimir Aleksandrovich* Pleshkov, *Alexandr Nikolaevich* Mironov, and *Anastasia Leonidovna* Evstratenko

FSBEI HE Kuzbass State Agricultural Academy, 650056, Kemerovo, st. Markovtseva, 5, Kemerovo, Russia

Abstract. Diagnosis of latent endometritis and predicting the effect of the inflammatory process on the course of pregnancy and its outcome is a necessary measure when conducting clinical examination of animals. Lack of diagnostic criteria, including microbiological one, cause difficulties for veterinarians in assessing latent endometritis. Often, animals suffering from latent inflammation of the reproductive tract are infertile and cannot be treated. And from cows with dysbiosis of the genital tract, sick young animals are born, lagging behind in growth and development. The aim of the work was to establish the relationship between the genital tract microbiome, the number and species composition of commensals and opportunistic microorganisms on the ability to carry pregnancy. Washes from the oral and nasal cavities were taken from the young born to study the microbiome: lactic acid microorganisms in combination with opportunistic pathogens. As a result of the research, it was found that animals with a low number of lactic acid microorganisms and an increased titer of opportunistic microorganisms were infertile, which led to their culling. In young animals, they are present in the respiratory tract in all samples, together with coccal microorganisms. The study of lactic acid microorganisms is advisable in the diagnosis of inflammatory processes of the reproductive tract and upper respiratory tract.

1 Introduction

Inflammations of cattle pelvic organs cause economic damage to agricultural enterprises due to early culling of cows, shortage or receiving sick non-viable young animals. Microorganisms are transmitted from the parent stock to the offspring one in two ways: intrauterinely and/or exogenously through objects of the external environment. Commonly, absence of pronounced clinical manifestations, lead to the fact that latent inflammation of the pelvic organs is not provided with proper and timely treatment, and consequently to intrauterine infection of calves. Sick livestock releases pathogens into the environment with

* Corresponding author: zveryski@mail.ru

physiological and pathological secretions of the upper respiratory tract, gastrointestinal tract, urinary tract, thereby contributing to echogenic infection of healthy animals.

Timely diagnosis of latent endometritis can help to avoid culling of animals. Considering that this disease has a microbial etiology, it is necessary to take into account the species and biochemical properties of the microbial community when carrying out diagnostic measures. At the same time, the state of the indigenous microflora, its quantity and species composition is of importance as well.

The aim of the study is to establish the relationship between the composition and properties of the microflora of the cows' genital tract and the microflora biotypes' colonization of the calves obtained from these cows.

2 Materials and methods

A study of swabs from the cervical canal of cows of different ages was carried out in the amount of 20 samples before artificial insemination. Sampling is done with a sterile cotton swab from the cervical canal.

From 10 calves, at the age of 1-2 months, were washed with a sterile swab from the mucous membranes of the pharynx and nose. A visual examination of the animals revealed that the animals are clinically healthy, the appetite is preserved, the integrity of the mucous membranes of the pharynx and nose is not disturbed, and the color of the mucous membranes is pale pink.

After sampling in a laboratory, successive 10-fold dilutions from 101 to 1010 are prepared. Dilutions of 0.1 ml are inoculated into sectors of the following plate media: meat-peptone agar with the addition of 5% erythrocytes of ram, Endo, Ploskirev, yolk-milk-salt agar, Enterokokkagar, Saburo and Zeissler agar with incubation of the latter in an anaerostat. At the same time, inoculations are carried out on liquid, poured in a high column into test tubes, media: Blikfeldt and Bifidum medium, sugar broth. Crops are grown at 37 ° C for 24-72 hours (on Saburo the temperature is 24 °C - 5 days).

3 Literature review

Functional and inflammatory diseases of the genital organs of cows, such as cervicitis and endometritis, cause significant economic damage to farms. Taking into account the fact that every year more than 30% of cows are asymptomatic with endometritis, the inflammatory process in these animals turns into a chronic stage, leading to infertility. There is a hypothesis that the development of inflammatory diseases is caused by changes in the ratio of representatives of normal microorganisms and opportunistic pathogens, which together constitute the microbiome of the genital tract [2, 3].

In various literary sources, information on the number and species composition of representatives of the resident microflora differs. But the authors state that the normal microflora of the genital organs is diverse and is represented by aerobes, facultative and strict anaerobes. The basis of the normal microflora of the genital tract is microorganisms of the genus Lactobacillus spp. It is generally accepted that normally lactobacilli are sown in 80–100% of cases in an amount of more than 106 CFU/ml. Research Gordeeva I.V. (2005) show that normal microbiocenosis of the genital tract of clinically healthy cows contains: bifidobacteria (73.5%), coagulase-negative S. xylosus (14.6%), lactobacilli (4.2%), anaerobic microorganisms Bacteroides spp. (2.9%) and fungi of the genus Candida (2.1%) [1, 5, 15, 16, 18, 19].

Against the background of imbalance between the resident and opportunistic microflora of the genital tract, cervicitis and endometritis develop. With inflammation of the genital tract, a sharp decrease in the number of lactobacilli and bifidobacteria is observed, to their complete absence, and opportunistic microorganisms act in dominant positions with the formation of polymicrobial associations. Most often, when the microbocenosis is disturbed, bacteria of the genera Staphylococcus spp. and Candida spp. develop quickly. In clinically sick animals, several species were found in the reproductive tract: Staphylococcus aureus, Streptococcus pyogenes, Escherichia coli, Proteus vulgaris [5, 8, 12, 13, 14, 16].

L.N. Kosolovich (2013), examining cows with acute catarrhal-purulent endometritis, found that the main causative agents of the disease are: S. aureus (81%); S. pyogenes (12%), E. coli and associations of these microorganisms (5%), which is consistent with the data of many authors. For example, I. Gordeeva (2005) determined that acute endometritis occurs in the presence of polymicrobial associations consisting of S. pyogenes (40.2%), S. aureus (18.3%) and P. vulgaris (0.01%), with chronic form - P. aeruginosa (11.0%) and S. aureus (2.3%), with subclinical - P. aeruginosa (2.5%) and S. aureus (18.2%). Microorganisms of the species S. epidermidis, S. xylosus, fungi of the genus Candida, as well as anaerobic Bacteroides spp. and Fusobacterium spp. were found regardless of the form of the course of the disease. With the course of the inflammatory process, an increase in the number of opportunistic microflora and the proportion of gram-negative bacteria in its composition was observed [4, 7, 11, 17, 19, 22].

The research by Lyzikova Yu.A. (2005) confirms the hypothesis about the influence of dysbiotic states on the manifestation of inflammatory processes: in chronic endometritis, gram-positive coccal microorganisms prevail in the endometrial tissues, while in the body where chronic endometritis is not diagnosed; lactobacilli are the prevailing group of microorganisms [8, 21].

In this case, it is advisable to resort to laboratory diagnostics of the microbiome of the cervical canal and endometrium, which consists in studying the change in the ratio of the number of microbiological indicators of normal representatives of the microbiota to opportunistic microorganisms [6, 9, 24, 25].

It should be borne in mind that contamination with opportunistic strains of microorganisms occurs much easier with dysbiotic changes in the microbial community of biotopes. An increase in the number of transient microbiota and a decrease in the number of obligate microbiota indicate an equalization of the importance of members of the microbial community. Thus, inflammatory diseases of the reproductive tract are characterized by changes in the qualitative and quantitative characteristics of the microbial community. In this case, the study of the microbiome of biotopes shows an increase in the total number of bacteria along with an increase in the frequency of occurrence of various microbial associations [10, 26].

Fundamental importance for the development of infectious pathology, in addition to reducing the amount of obligate microflora, has the presence of pathogenic properties in transient microorganisms. Degtyareva S.S. (2008) showed that cultures with hemolytic activity (35.5% of cultures) gave a positive plasma coagulation reaction (18.3% of cultures), and 51% of cultures were pathogenic for laboratory animals [6].

When studying diseases of the reproductive organs of cows, it is necessary to take into account the likelihood of intrauterine infection of calves, with the subsequent long-term persistence of the pathogen and the development of diseases of the respiratory and gastrointestinal tracts. Perinatal transmission of pathogens of infectious diseases in farm animals is confirmed by many studies [23].

For example, S.A. Aksenenko. (2006) found that calves with gastrointestinal tract pathology are born from mothers with endometritis 2 times more often than from healthy cows [1].

In addition, the composition of the nasopharyngeal microbiota in dairy calves is directly proportional to the mother's vaginal microbiota. The resistance of animals to respiratory diseases depends on the composition and species diversity of microorganisms. In the presence of microorganisms of the genus Lactobacillus in the respiratory tract, animals have no signs of respiratory diseases, while the presence of bacterial pathogens such as Mycoplasma bovis, Mannheimia haemolytica or Pasteurella multocida causes upper respiratory tract disease (URT) in cattle. The respiratory tract of cattle is represented by a complex bacterial ecosystem known as the bacterial microbiome. The microbiome develops and forms under the influence of many factors: the mother's vaginal microbiota, environment, age, feeding, treatment used and stressful events [17, 23, 26].

The respiratory tract of cattle is colonized by various microorganisms immediately after birth. Airway colonization in dairy calves begins immediately after birth and develops rapidly during the first weeks of life. The number of bacteria in the nasopharynx increases from birth to 14 days of age, and then slightly decreases until the 35th day or remains unchanged until 42 days.

The microbiome of the nasopharynx of calves at the age of 3 and 14 days is characterized by the presence of proteobacteria up to 70% of the specific weight of the total number of microorganisms. After the 14th day, the species diversity increases, starting to include such divisions of microorganisms as Tenericutes, Firmicutes, Actinobacteria and Bacteroidetes. The most numerous genera of bacteria in the nasopharynx of dairy calves are Mannheimia, Moraxella, Mycoplasma, Psychrobacter and Pseudomonas. The relative abundance of these genera changes over time: the abundance of Moraxella decreases between the 14th and 35th day, and the relative abundance of Mannheimia and Mycoplasma simultaneously increases significantly. Thus, the nasopharyngeal microbiota of cattle changes significantly between weaning and the first weeks of feeding. This evolution may explain why cattle are more susceptible to respiratory tract disease during the first 40-60 days of feeding, as unstable microbiota is less resistant to colonization by pathogens [14].

When studying diseases of the upper respiratory tract and lungs of cattle, most scientists pay attention only to pathogenic biological agents, without studying the role of beneficial microorganisms in the immune response during the development of an infectious process. As a result, there is not enough information in the information field about the structure of the microbiota of the oropharynx and nasopharynx and its role in maintaining health.

Various studies show that pneumonia develops in cattle, when the number of commensal bacteria in the URT organs decreases. Normally, the types Proteobacteria and Firmicutes predominate, with smaller proportions of Actinobacter, Bacteroidetes and Tenericutes. Of the dominant genera, the most frequently identified are Mycoplasma, Moraxella, Pasteurella, Psychrobacter, and Mannheimia [15, 23].

It is becoming more and more obvious that the pathogenesis of infectious diseases of the reproductive and respiratory tract organs is influenced by several factors, including the interaction of pathogens with commensals. This interaction was the subject of this study.

4 Results

When examining the microflora of the cervical canal of cows, it was found that bifidobacteria are found in an amount of 10^1 in 7 samples, 10^2 in 2 samples, 10^3 in 5 samples; lactobacilli in the amount of 10^1 were not found, 10^2 - in 4 samples, 10^3 - in 2 samples, 10^4 - in 1 sample; microorganisms of the family Bacillus spp. cultivated in a dilution of 10^1 in 4 samples, in a dilution of 10^2 in 1 sample; E. coli without hemolytic properties was cultivated in 1 sample at a dilution of 10^2. The presence of coccal gram-positive microflora was also established: microorganisms of the family Staphylococcus spp. with pronounced hemolytic properties were found in 1 sample at a dilution of 10^2;

E.faecalis with hemolytic properties was found in 2 samples at dilutions 10^1 and 10^2; E.faecium with hemolytic properties - in 1 sample at a dilution of 10^1 (Fig. 1).

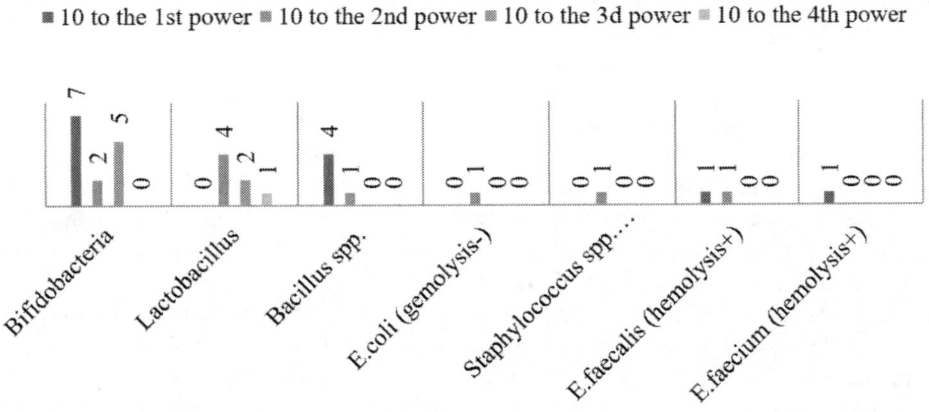

Fig. 1. Number of microorganisms isolated from samples of cow cervical canal.

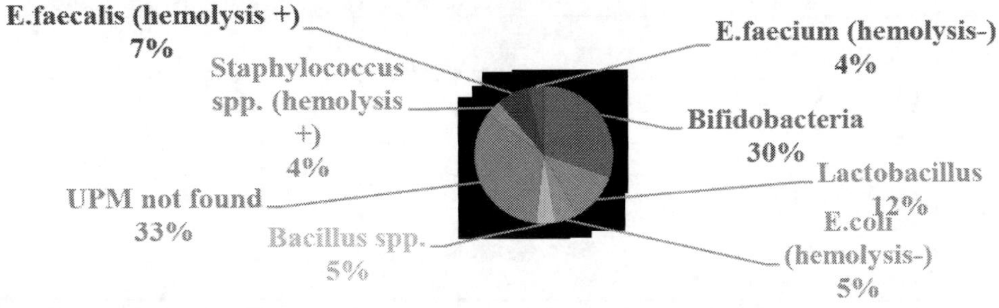

Fig. 2. Specific weight of microorganisms in samples of cow cervical canal.

The specific weight of microorganisms in the samples is: bifidobacteria - 30% of the samples; lactobacilli - 12% of samples; Bacillus spp. - in 5% of samples; E. coli - in 5% of samples; microorganisms of the family Staphylococcus spp. - in 4% of samples; E.faecalis - in 7% of samples; E.faecium - in 4% of samples; CPB was not found in 33% of samples (Fig. 2).

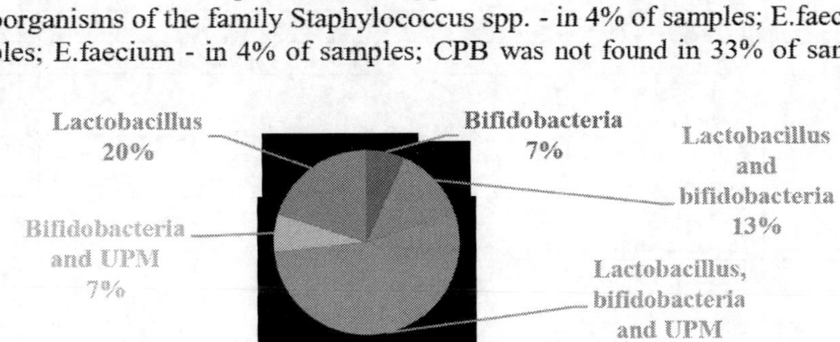

Fig. 3. Specific weght of samples containing associations of microorganisms and microorganisms in a monoculture.

https://doi.org/10.1051/bioconf/20224201003

Microorganisms were cultured in associations and monocultures. Representatives of the normal flora, together with CPB, were isolated in 53% of the samples; bifidobacteria and CPB were contained in 7% of samples; monocultures of lactobacilli are isolated from 20% of samples; monoculture of bifidobacteria - in 7% of samples; only representatives of normal flora - in 13% of samples (Fig. 3).

It was found that out of 4 cows, without clinical signs of inflammation, from the farm of the Prokopyevsky district, 2 heads were not inseminated and as of the reporting date they are not fertile. When analyzing the protocols of laboratory tests, it was found that in these heads in the cervical canal, the amount of CPB, namely microorganisms of the Bacillus spp. Family, exceeds the amount of normal flora, or there is no normal flora at all. At the same time, in 2 heads, which at the reporting date are pregnant, the CPB was either absent, or the amount of normal flora exceeded the number of CPB.

When sampling from the cervical canal of cows in the Belovsky region, 5 heads showed the presence of clinical signs of cervicitis and endometritis: hyperemia and swelling of the mucous membranes, the presence of purulent exudate with blood streaks. Analysis of these samples showed the presence of CPB: E. coli without hemolytic properties; E.faecium and E.faecalis with hemolytic properties (Table 1).

Table 1. Comparison of genital tract microbiome before artificial insemination and pregnancy

Sample number	Quantitative indicator of the normal flora representatives	Quantitative indicator of CPB representatives	Insemination performance
1	Bifidobacteria – 10^1,	Bacillus spp. - 10^2	No pregnancy
2	absent	Bacillus spp. - 10^1	No pregnancy
3	Bifidobacteria – 10^3	Bacillus spp. – 10^1	Pregnancy
4	Bifidobacteria – 10^1	absent	Pregnancy
5	Bifidobacteria – 10^2	absent	Pregnancy
6	Bifidobacteria – 10^1	absent	Pregnancy
7	Bifidobacteria – 10^1 Lactobacillus - 10^2	Bacillus spp. – 10^1 E.coli (gem-) - 10^1	Pregnancy
8	Bifidobacteria – 10^1 Lactobacillus - 10^2	Bacillus spp. – 10^1	Pregnancy
9	Bifidobacteria – 10^3 Lactobacillus - 10^4	E.faecium (gem+) - 10^1	Animal is culled
10	Bifidobacteria – 10^1 Lactobacillus - 10^3	Bacillus spp. – 10^1 S.cohnii – 10^2	Animal is culled
11	Bifidobacteria – 10^3	absent	Pregnancy
12	absent	absent	Pregnancy
13	Lactobacillus – 10^1	absent	Pregnancy
14	absent	Bacillus spp. – 10^1	No pregnancy
15	Bifidobacteria – 10^2 Lactobacillus – 10^2	absent	Pregnancy
16	Bifidobacteria – 10^2 Lactobacillus – 10^2	E.faecalis (gem+) - 10^1	No pregnancy
17	Bifidobacteria – 10^3 Lactobacillus – 10^3	absent	Pregnancy
18	Bifidobacteria – 10^1	absent	Pregnancy
19	Lactobacillus – 10^1	absent	Pregnancy
20	Bifidobacteria – 10^1 Lactobacillus – 10^1	E.faecalis (gem+) - 10^1	No pregnancy

In samples from the throat of calves, the presence of the following opportunistic microorganisms (CPB) was established:

-Enterobacteriaceae: E. cloacae - 60%, incl. 5 cultures (83.3%) with hemolytic properties; E.aerogenes 10%; E. coli - 10% with hemolytic properties (100%);

-Enterococcusceae: E.faecalis - 40%, where 4 cultures (100%) with hemolytic properties; E.faecium - 10% with hemolytic properties (100%);

- Staphylococcusceae: S. saprophyticus - 20%;
- microscopic fungi: Mucor - 40% (Fig. 4).

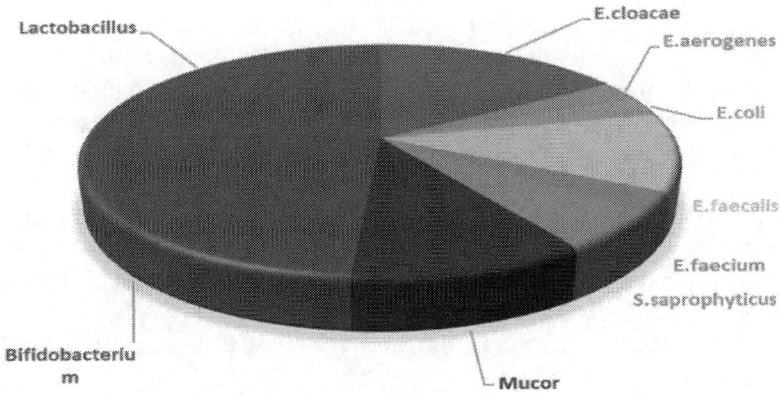

Fig. 4. Specific weight of microorganisms in samples from calf throat.

From the throat of some representatives of the CPB, the following quantitative characteristics were established::

- E.cloacae – 10^3 -20% of samples; 10^6 – 10%;
- S.saprophyticus – 10^1 -10%;
- Mucor – 10^1 – 30%, 10^2 – 10%, 10^3 – 10%.

In samples from the nasal cavity, CPB are presented in the following quantities:

- E.cloacae – 10^4 -10%;
- E.faecalis – 10^1 – 30%;
- Mucor – 10^1 – 20%, 10^2 – 10%, 10^3 – 30%;
- S.saprophyticus – 10^1 -10% (Fig.5).

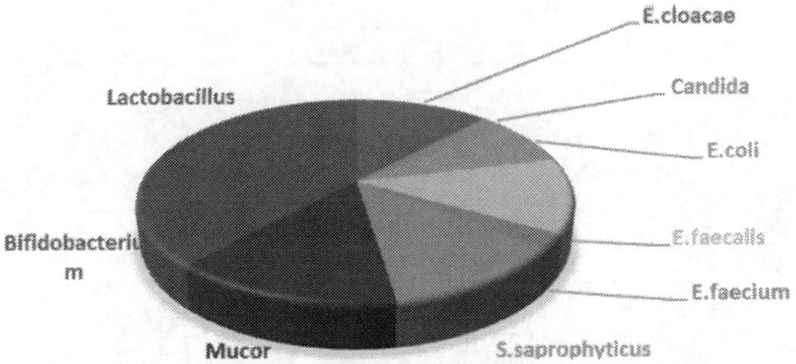

Fig. 5. Specific weight of microorganisms in samples from calf nasal cavity

In samples from the nasal mucosa, the presence of the following CPB was adopted:

-Enterobacteriaceae: E.cloacae – 50%, ncl. 4 cultures (80%) with hemolytic properties; E. coli - 30%, incl. 2 cultures with hemolytic properties (66.6%);

- Staphylococcusceae: S.saprophyticus – 50%;

-Enterococcusceae: E.faecalis – 50%, where 4 cultures (80%) with hemolytic properties; E.faecium - 10% with hemolytic properties (100%);

- microscopic fungi: Mucor – 60%, Candida – 10%.

Normal microflora in the pharynx is represented by lactic acid microorganisms of childbirth:

1. Bifidobacterium: in quantities of 10^1 – 10%; 10^3 – 80%.
2. Lactobacillus: in quantities of 10^1 – 10%, 10^3 – 70%, 10^6 – 10%.

In the nasal cavity, lactic acid microorganisms are contained in quantities:

1. Bifidobacterium: in quantities of 10^1 – 30%; 10^3 – 50%.
2. Lactobacillus: in quantities of 10^1 – 40%, 10^3 – 40%.

In most samples from the pharynx and nose (70% each), lacto- and bifidobacteria were found in a consortium with each other. In 10% of samples from the throat and 10% of samples from the nose, only representatives of the genus Bifidobacterium were found. The presence of Lactobacillus in 10% of samples was also established from the pharynx and nose.

Representatives of normal flora in 90% of samples from the pharynx and nasal cavity were in consortium with the UPM.

In 10% of samples from the pharynx, only bifidobacteria and lactobacilli with microscopic fungi of the genus Mucor were found.

In 10% of the nasal cavity, only the presence of UPM without representatives of indigenous microflora was found.

The most common association of microorganisms in samples from the pharynx (30%) and nasal cavity (20%) is represented by the following microorganisms: Mucor + E. cloacae + E. faecalis + Bifidobacterium + Lactobacillus.

Table 2. Microbial associations in throat samples from calves.

№	Microbial Association	Number of samples
1	E.cloacae + E.faecalis + Bifidobacterium + Lactobacillus	1
2	Mucor + E.cloacae + S.saprophyticus + Bifidobacterium + Lactobacillus	1
3	E.coli + Bifidobacterium + Lactobacillus	1
4	E.cloacae + S.saprophyticus + E.faecalis + Bifidobacterium + Lactobacillus	1
5	Mucor + E.cloacae + E.faecalis + Bifidobacterium + Lactobacillus	3
6	Mucor + Bifidobacterium + Lactobacillus	1
7	E.aerogenes + E.faecalis + E.faecium+ Lactobacillus	1
8	Bifidobacterium + Lactobacillus	1

Table 3. Microbial associations in nasal samples from calves.

№	Microbial Association	Number of samples
1	S.saprophyticus + E.faecalis + Bifidobacterium + Lactobacillus	1
2	Mucor + E.coli + E.faecium + Bifidobacterium	1
3	Mucor +E.coli + S.saprophyticus + E.faecalis	1
4	E.cloacae + S.saprophyticus + E.faecalis + Lactobacillus	1
5	Mucor + E.cloacae + E.faecalis + Bifidobacterium + Lactobacillus	2
6	Mucor + E.cloacae + Bifidobacterium + Lactobacillus	1
7	Candida+ Lactobacillus+ Bifidobacterium	1
8	Mucor + E.cloacae + S.saprophyticus+ Bifidobacterium + Lactobacillus	1
9	Mucor +E.coli + S.saprophyticus + Bifidobacterium + Lactobacillus	1

5 Discussion

1. The amount of normal microflora of the reproductive tract of cows does not exceed 10^4.

2. The number of CPB does not exceed the number of 10^2.

3. CPB is represented by microorganisms: Bacillus spp.; Staphylococcus spp.; E. coli; E.faecium; E.faecalis.

4. E.faecium; E.faecalis; Staphylococcus spp. possess hemolytic properties, which indicates the possibility of causing an infectious process.

5. A relationship has been established between the content of normal flora, CPB of the reproductive tract and the ability to reproduce: artificial insemination of animals with a quantitative content of CPB exceeding the amount of normal flora turned out to be ineffective.

6. The number of lacto- and bifidobacteria in the mouth and nasal cavity of calves reaches 10^3.

The dominant representative of CPB is E. cloacae, reaching 10^6 in the nasal cavity and 10^4 in the oral cavity.

The microbial composition of the oral and nasal cavities of healthy animals is similar in qualitative and quantitative characteristics.

7. In the oral and nasal cavities, the association of bifidobacteria, lactobacilli and opportunistic microorganisms prevails.

The most common association of microorganisms in samples from the pharynx (30%) and nasal cavity (20%) is represented by the following microorganisms: Mucor + E. cloacae + E. faecalis + Bifidobacterium + Lactobacillus.

6 Conclusion

The course of the infectious process of the reproductive tract is influenced not only by the presence of the pathogen itself, but also by the presence of lactic acid microorganisms that have a protective effect in the macroorganism. An excess of the amount of CPB with a simultaneous decrease in commensal microorganisms is an unfavorable indicator for the further fertilization of animals.

In young cattle, without clinical symptoms of respiratory tract diseases, the microbiome of the upper respiratory tract consists mainly of lactic acid microorganisms: Bifidobacterium spp., Lactobacillus spp. The specific gravity of CPB, represented mainly by coccal microflora, does not exceed the specific gravity of lactic acid microorganisms.

When diagnosing diseases of the reproductive tract and upper respiratory tract using a microbiological method, it is necessary to take into account the presence of lactic acid microorganisms, along with the determination of pathogenic microorganisms.

Acknowledgements

The paper was prepared based on the results of the implementation of a project aimed at conducting applied scientific research and experimental development funded from the funds of the FSBEI HE Kuzbass State Agricultural Academy - 2020.

References

1. S. A. Aksenenko, *On the role of the maternal factor in the etiology of neonatal diarrhea and its prevention in calves,* 30 (2006)

2. I. O. Bushtyreva, Ch. Doctor, **3(62),** 49 (2018)
3. L. G. Voitenko, V. Ya. Nikitin, Bulletin of Michurinsky State Agrarian University, **2(1),** 181 (2011)
4. I. V. Gordeeva, *Epizootological and microbiological screening of diseases of the reproductive organs of cows in the middle Volga region (microbiocenoses and their correction),* 20 (2005)
5. A. V. Danshina, N. I. Potaturkina-Nesterova, I. S. Nemova, Modern problems of science and education, **2,** 358 (2012)
6. S. S. Degtyareva, *Acute postpartum endometritis of bacterial-mycotic etiology in cows and its pharmacotherapy,* 27 (2008)
7. L. N. Kosolovich, S. N. Ivanova, Bulletin of the Ulyanovsk State Agricultural Academy, **1(21),** 83 (2013)
8. Yu. A. Lyzikova, Problems of health and ecology, **4(58),** 24 (2018)
9. A. Metleva, et al., Materials of the V National Scientific and Practical Conference, 108 (2020)
10. V. I. Pleshakova, T. V. Ermakova, A. S. Egorova, Scientific notes of the Kazan State Academy of Veterinary Medicine named after V.I. N.E. Bauman, **210,** 159 (2012)
11. V. A. Tovstanovskaya, Reproductive health. Eastern Europe, **6(2(44)),** 271 (2016)
12. Yu. Tulev, N. Tuleva, M. Levkovskaya, Dairy and beef cattle breeding, **5**, 29 (2009)
13. G. A. Khonin, Veterinary medicine of the Kuban, **5,** 11 (2010)
14. N. A. Castillo, et al., Food Res. Int., **45,** 831 (2012)
15. L. Castro, et al., Prebiotics and Probiotics Science and Technology, Springer, 1123 (2009)
16. H. Chen, Veterinary and Animal Science (2020)
17. L. S. Frizzo, et al., Anim. Feed Sci. Technol., **169,** 147 (2011)
18. L. M. Gorgos, Sex Transm Dis. **42(9),** 475 (2015)
19. R. Kasimanickam, Theriogenology, 9 (2004)
20. I. Lorenz, et al., Irish veterinary journal, **64(1),** 1 2011)
21. A. S. Metleva, O. V. Smolovskaya, V. A. Pleshkov, A. N. Mironov, XIX International Scientific and Practical Conference "Current Trends of Agricultural Industry in Global Economy", 166 (2020)
22. C. G. Moreno, *Characterization of native Escherichia coli populations from bovine vagina of healthy heifers and cows with postpartum uterine disease* (2020)
23. I. Moreno, J. M. Franasiak, Fertil Steril., **108(1),** 2 (2017)
24. O. B. Pascottini, Sci Rep., **10** (2020)
25. E. S. Ribeiro, J. Dairy Sci., **96,** 5682 (2013)
26. S. Sandes, et al., Microbiological Research, **200,** 1 (2017)

Index of epithelization rate of thermal burns in laboratory animals when application of the "acerbin" preparation

Alexander Nikolaevich Pereverzev*, and *Vladimir Aleksandrovich* Tolkachev

Kursk State Agricultural Academy named after I.I. Ivanov, Karl Marx, 70, 305021 Kursk, Russia

Abstract. Thermal burns in animals represent an urgent problem for veterinary medicine, since the anti-burn drugs available on the world pharmacological market are not always adapted for use on animals and are approved for use in veterinary medicine. In this regard, the aim of the work was to use the wound healing spray "Acerbin" for medical purposes in the treatment of thermal burns and to evaluate its therapeutic efficacy by the rate of epithelialization of burn defects in the skin in laboratory animals. The work was carried out at the Department of Surgery and Therapy of the Kursk State Agricultural Academy, on laboratory white mice in the amount of 60 individuals, in which thermal burns were simulated in the croup and treated with applications of the Acerbin spray. To assess the rate of growth of epithelialization, planimetric measurements of the area of burns were carried out before treatment on the 3rd, 5th, 7th, 10th and 14th days of treatment, as well as the corresponding mathematical calculations of the rate of reduction of burn defects. According to the research results, it was found that when applying the Acerbin spray, the burn rate index on the 3rd day was 0.25 ± 0.06; on the 5th day - 0.58 ± 0.03; on the 7th day - 0.74 ± 0.01; on the 10th day - 0.86 ± 0.02; on the 14th day - 0.95 ± 0.01. Thus, at the end of the therapeutic period, the index of the epithelialization rate of thermal burns approached the most physiologically justified numerical expression equal to "1.0", which allows it to be recommended for use in veterinary medicine as the preparation of choice for the treatment of thermal burns of the skin in animals.

1 Introduction

At present, about 600 thousand cases of burns, both in animals and in humans, are annually registered in Russia, which accounts for 50 to 60% of all diagnosed traumatic injuries. Thermal burns in animals most often occur when they are kept in the domestic conditions of apartment buildings when the body surface comes into contact with an open flame or electric heating devices. They cause charring of the coat and damage to the deep layers of the skin. Further damage causes significant damage to the conformation characteristics of companion pets. At the same time, thermal burns are difficult to treat, contribute to the development of multiple organ failure and, as a consequence, high mortality, which ranges

* Corresponding author: pereverzeff.sasha@yandex.ru

from 65 to 95%, depending on the severity and stage of the course. First, the available arsenal of veterinary anti-burn drugs is often not species-specific. Secondly, many veterinary anti-burn drugs are not able to have a wound-healing effect on damaged tissues in a short time and require long-term and repeated use. Thirdly, veterinary anti-burn pharmacological preparations ineffectively eliminate the phenomena of local intoxication, which affects the rate of epithelialization of damaged tissues.

Therefore, the problem of treating thermal burns remains the most urgent problem for medical and veterinary science and practice. Currently, a sufficient number of anti-burn drugs adapted for use in medical practice in the treatment of humans are presented on the world pharmacological market. However, for veterinary purposes, the same pharmacological market does not fully satisfy the needs of practicing veterinary specialists who are faced with the problem of treating thermal burns in animals. Multiple organ failure arising against the background of thermal damage not only contributes to the high mortality rate of combustive patients, but also slows down the regeneration processes. Due to the peculiarities of cardiac activity in the damaged tissues, the percentage of oxygenation decreases. In connection with the violation of the hepatetic and nephro-urological function of the corresponding internal organs, an imbalance of acid-base balance occurs in the tissues subjected to thermal action, which causes a slowdown in the processes of granulation and epithelialization.

In this regard, the relevance of the search and development of new innovative anti-burn pharmacological preparations adapted for use in veterinary medicine for the treatment of burnt animals increases.

2 Materials and methods

To solve the problem of developing new anti-burn agents adapted for use in veterinary medicine, it was considered necessary to test the wound healing spray "Acerbin" for medical purposes in the treatment of thermal burns in animals and evaluate its therapeutic efficacy in terms of the rate of epithelialization of burn defects. The drug, according to the current instructions, has a keratolytic, aseptic and wound-healing effect. It contains DL-malic acid, benzoic acid, salicylic acid, propylene glycol, purified water and is a solution for external use.

DL-malic acid in its pharmacological effect stimulates metabolism, normalizes cell metabolism, improves local and general blood circulation, strengthens the immune system and enhances the protective properties of the body as a whole. It has anti-inflammatory, anti-edema properties.

Benzoic acid is a monosyllabic carboxylic substance by its cyclic nature, it is a natural compound. In appearance, it is a white powder from elongated granules or crystals with a characteristic luster. Benzoic acid exhibits anti-inflammatory and antifungal effects, consisting in preventing the multiplication of lactic acid fermentation bacteria and yeast, as well as in suppressing the activity of enzymes of pathogenic microbial agents. These antiseptic and antibacterial properties of benzoic acid are widely used in the pharmaceutical industry for the production of antifungal drugs in the form of solutions, sprays or ointments.

Salicylic acid is clinically a dibasic acid, has keratolytic, antiseptic, anti-inflammatory, antifungal effects. The above pharmacological properties are confirmed by numerous studies that have revealed many factors of a positive effect on the state of the epidermis, namely, salicylic acid is well absorbed by the skin, dries it out, cleanses the skin pores from impurities, normalizes blood circulation, helps to reduce the inflammatory process, relieves irritation symptoms, eliminates itching, stimulates the regeneration of damaged tissues. Thus, due to its keratolytic, antiseptic, local healing properties, salicylic acid is widely used in the fight against dermatitis, seborrhea, psoriasis, keratosis, burns.

Propylene glycol is a clear viscous liquid with a density lower than that of ethylene glycol and glycerin, but higher than that of ethanol. Possessing hygroscopic properties, i.e. the ability to absorb and absorb water from surfaces and air, drying them, propylene glycol is widely used as filler in medicines for external use, as a bactericidal agent, as a base in the manufacture of creams, ointments, lotions, sprays, as well as soaps and toothpastes.

The work on approbation of the wound healing spray "Acerbin" for medical purposes in the treatment of thermal burns in animals was carried out at the Department of Surgery and Therapy of the Kursk State Agricultural Academy. To solve the formulated scientific and practical problem, thermal burns were simulated on 60 white laboratory mice. Simulation of burns was carried out in accordance with the "European Convention for the Protection of Vertebrate Animals Used for Experiments or Other Scientific Purposes", namely, all laboratory animals were deeply sedated by injections of the pharmacological preparation "Xila" at a dose of 0.1 ml per animal. In sedated experimental models, an operative field was prepared on the dorsal surface in the region of the croup in compliance with the rules of asepsis and antiseptics according to standard methods: they cut the hair with a razor, and treated the skin twice with an alcohol solution of iodine with an exposure of 2 minutes. Later, the third degree burn defects were clinically and experimentally simulated, namely, the tip of the working part of a red-hot electric soldering iron was applied to the previously prepared surgical field with an exposure of 5 seconds. After simulating a thermal burn, laboratory animals were taken out of the state of deep sedation by injections of sulfocamphocaine and the spray "Acerbin" was applied on the 3rd, 5th, 7th, 10th and 14th days. In the course of the experiment, planimetric studies were carried out before treatment and during the application of the "Acerbin" spray. Planimetric studies included: making cellophalograms of the affected skin on the 3rd, 5th, 7th, 10th, and 14th days, respectively. The contours of cellophalograms at the appropriate time of registration were transferred to graph paper and the area of thermal burns was obtained at the time of measurement by the number of whole squares included in the contours (mm^2), that is, Sn. Further planimetric studies included the calculation of ΔS (1) - the rate of epithelialization of the burn wound and Yt (2) - the value of relative healing according to the following formulas:

$$\Delta\mathrm{S} = \frac{(S - Sn) * 100}{\mathrm{S} * \mathrm{t}} \tag{1}$$

where S - the size of the wound area during the previous measurement; Sn - size of the wound area at the moment; t - number of days between the first and subsequent measurements.

$$Yt = \frac{S_0 \ - S_t}{S_0} \tag{2}$$

where S_0 - initial wound area; St - wound area on day t.

Based on the obtained planimetric data on the area of burn defects and the rate and epithelization index calculated according to the above formulas, there were made conclusions about the possibility of using the medical spray "Acerbin" in the treatment of thermal burns in animals.

3 Literature Review

Thermal burn is damage to integuments and deep-lying tissues caused by high temperatures [1, p. 198]. Burns patients present as one of the most challenging situations in both human and veterinary medicine [2, p. 232]. Thermal burns of small pets can occur as a result of exposure to flame, boiling water, steam or hot air, hot metal objects in the domestic conditions of maintenance of apartment buildings and private farmsteads [3, p. 1].

According to statistics from The British Small Animal Veterinary Association, among all surgical pathologies in small pets, burns are about 20% [4, p. 4]. According to the classification of Alden H., Harken, Ernst, burns are divided, depending on the cause that caused them, into thermal, chemical and radiation [3, p. 1]. Burns, even superficial, and all the more deep, quickly become life-threatening for the animal, since in addition to local changes they cause general disturbances in the body. At the same time, at the site of the burn defect, local coagulation tissue necrosis occurs, the strongest intoxication of the body of toxicological and microbial origin, a significant violation of the regional and peripheral blood flow, which complicates the healing process of defects. Healing of skin lesions causes inflammation of the tissues damaged by the burn, edema, hypertrophic and ugly scars [5, p. 475]. The initial assessment should include the evaluation of the general physical condition, systemic compromise, amount of body surface affected, plus degree of local injury [6, p. 17]. Thus, the choice of topical agent or type of coating to be used in the treatment of burns should be based on an assessment of the characteristics of the lesion and the evidence presented in the literature [7, p. 1]. These products should have characteristics such as antimicrobial or bacteriostatic activity, absence of toxicity and hypersensitivity, compliance, shortening of healing time, and cost / benefit ratio. However, many of the methods used to heal burn injuries are controversial. In this regard, there is an urgent need to develop new innovative means of adequate and highly effective pharmacotherapy [8, p. 5].

4 Results

The research results showed that initially the area of the simulated burns was 42.30±3.20 mm^2 ($p \le 0.05$). On the 3rd day of application, the burns had an area equal to 31.12±2.72 mm^2 ($p \le 0.05$), that is, they decreased by 8.35±2.12 mm^2 ($p \le 0.05$). On the 5th day of the spray application, burn defects decreased additionally by 8.54±0.87 mm^2 ($p \le 0.05$); an average wound area in the group was 17.50±1.70 mm^2 ($p \le 0.05$). After a week of appropriate treatment, the size of thermal burns on average in the group of supervised laboratory animals decreased to 10.60±0.73 mm^2 ($p \le 0.05$), that is, by 6.61±0.24 mm^2 ($p \le 0.05$). On the 10th day, the area of clinically - experimentally simulated thermal burns was 5.40±0.67 mm^2 ($p \le 0.05$) and was less than on the 7th day of treatment by 4.98±0.44 mm^2 ($p \le 0.05$). At the end of the clinical trial of the medical spray "Acerbin" in veterinary medicine for the treatment of thermal burns, it was found that on the 14th day their area was 2.14±0.03 mm^2 ($p \le 0.05$) and was less than the 10th day by 4.41±0.25 mm^2 ($p \le 0.05$).

5 Discussions

Analysis of the dynamics of healing of thermal burns in laboratory animals when applying the spray "Acerbin" showed that in the first three days of application to burn defects, their area decreased by 26.24% ($p \le 0.05$), and the epithelialization rate index was 0.25±0.06 ($p \le 0.05$). Subsequently, by the 5th day of supervision, thermal burns decreased by an additional 43.91% ($p \le 0.05$) from the dimensions of the previous planimetric measurements carried out on the 3rd day of the experiment, and the epithelialization index increased to 0.58±0.03 ($p \le 0.05$). On the 7th day, burns decreased by 39.43% ($p \le 0.05$) relative to the size recorded on the 5th day of observations, and on the 10th day by 49.05% ($p \le 0.05$) in comparison with the results of planimetric accounting of the 7th day of the experiment, that is, the epithelization index at the given time of registration was 0.74±0.01 ($p \le 0.05$) and 0.86±0.02 ($p \le 0.05$), respectively. On the 14th day, the burn epithelization index was 0.95±0.01 ($p \le 0.05$), and the area of burns decreased by 60.37% ($p \le 0.05$) relative to the measurement results on the 10th day of clinical observations. Thus, the results obtained testified to the positive effect of the "Acerbin" spray for medical purposes on the healing of third degree thermal burns of the skin in animals, since the epithelialization index by the 14th day approached the numerical value of "1.0", which is physiologically justified.

6 Conclusions

Scientific and applied research has shown that when applying the spray "Acerbin" the index of epithelialization of burns on the 3rd day is equal to 0.25±0.06 ($p \leq 0.05$), on the 5th day - 0.58±0.03 ($p \leq 0.05$), on the 7th day - 0.74±0.01 ($p \leq 0.05$), on the 10th day - 0.86±0.02 ($p \leq 0.05$), on the 14th day - 0.95±0.01 ($p \leq 0.05$), that is, at the time of completion of treatment it was the maximum close to the physiologically justified numerical expression in "1.0".

References

1. N. Salhi, et al., South African journal of botany, **121**, 128 (2019)
2. E. Yu Zakirova et al., BioNanoScience, **11(1),** 232 (2021)
3. D. S. Tavares Pereira, et al.. Journal of Biomedicine and Biotechnology, **2012**, 1 (2012)
4. W. Jean Dodds, Israel Journal of Veterinary Medicine, **73(2)**, 3 (2018)
5. C. B. Nielson, et al., Journal of Burn Care & Research, **38(1)**, e469 (2018)
6. L. H. Tello, Journal LAVECC, **5(1)**, 17 (2013)
7. E. V. Inzhevatkin, et al., Bulletin of Experimental Biology & Medicine, **169 (5),** 1 (2020)
8. M. L. Hossain, et al., Int J Complement Alt Med, **11(1)**, 1 (2018)

Stud-breeding work in sheep breeding based on cytogenetic monitoring

Vladimir Ivanovich Rossokha[1,*], *Ivan Andreevich* Pomitun[1], *Alexandr Vladimirovich* Tkachev[2,3], *Olga Leonidovna* Tkacheva[2], *Tatyana Vladimir* Zubova[4], *Vladimir Aleksandrovich* Pleshkov[4], and *Oksana Vladimirovna* Smolovskaya[4]

[1]National Academy of Agrarian Sciences of Ukraine, st. Livestock breeders, 1A, 61026 Kharkov, Ukraine
[2]Belgorod State Agrarian University, 308503, st. Vavilova, 1, vil. Maysky, Belgorodsky, Belgorod, Russia
[3]Federal State Budgetary Educational Institution of Higher Education Penza State Agrarian University, 440014, st. Botanicheskaya, 30, Penza, Russia
[4]Kuzbass State Agricultural Academy, 650056, st. Markovtseva, 5, Kemerovo, Russia

Abstract. The article presents the results of the cytogenetic monitoring in the breeding, selection and reproduction of sheep in the ecological conditions of Ukraine. A cytogenetic analysis of sheep with low and high vigor and different levels of fertility was carried out. In the ecological conditions of Ukraine, the individual level of chromosomal abnormalities in all the studied animals of the Tsigai breed and prekos is within the natural background. No translocations were found. Among the violations, such as single and paired fragments, hypo- and hyperploidy (mainly hypoploidy), polyploidy were encountered. The maximum average level of chromosomal abnormalities was found in local lambs (5.5 ±1.73%). The minimum average level of chromosomal abnormalities (2.0 ±1.41%) was recorded in lambs with high growth intensity. Among the structural changes, mutations of the chromosomal type prevailed in the 3-year-old group of rams - 0.46%, in the 8-year-old group - 0.59%. Chromatid disturbances were 0.37% and 0.34%, respectively. The average level of chromosomal abnormalities in rams by groups was 0.84 ±0.14 and 0.93 ±0.13, respectively. In the group of ewes with low fertility (n = 4), the level of chromosomal abnormalities was 3.5%, which is lower than ewes (n = 7) with high fertility by 0.79%.

1 Introduction

At the beginning of the XXI century, in the context of the European integration policy, on the agenda of livestock breeders was the question of harmonizing primary accounting in animal husbandry with the requirements of the International Committee for Animal Recording (ICAR), the mandatory requirement of which is the detection of carriers hereditary developmental anomalies based on genetic examination. In Ukraine, in 2004, the "Regulation on the procedure for carrying out genetic examination of the origin and

*Corresponding author: rossokha.v@ukr.net

anomalies of pedigree animals" was developed and introduced into practice, which regulates the genetic examination of the origin of pedigree animals, the identification of genetic and chromosomal anomalies [1-4].

2 Literature Review

In developed countries of the world, cytogenetic monitoring programs in breeding farms began to be applied without fail since the 70s of the last century. In Ukraine, cytogenetic studies of sheep were actively carried out in the 80s of the last century, which remained in the history of genetics as a cytogenetic boom. Most of the breeds of sheep and cattle were examined, the frequency and spectrum of chromosomal aberrations in blood cells were found, a number of carriers of chromosomal abnormalities characteristic of sheep were identified, which caused a decrease in vitality and reproductive function. In the scientific literature, many publications have appeared on the identification of various chromosomal aberrations and their influence on the phenotypic characteristics of animals of different species and breeds [5-9].

Based on the foregoing, cytogenetic monitoring is of great scientific and practical importance in all branches of animal husbandry (sheep breeding, horse breeding, cattle breeding, pig breeding) in order to eliminate animal carriers of chromosomal aberrations, in order to increase reproductive function, increase the efficiency of creating cryobanks of valuable producers [7-11]. Since modern methods of animal reproduction - artificial insemination, embryo transplantation, etc. lead to the risk of the spread of hereditary cytogenetic abnormalities. It has been established that various anomalies of the karyotype are often found in cattle and can have a significant effect on their vital activity, reproductive function and productivity. An economic analysis of the consequences of the use of carriers of hereditary chromosomal aberrations in the breeding process showed an impressive amount of losses for the livestock industry on farms [13-14].

3 Materials and methods

Cytogenetic monitoring was carried out in sheep-breeding farms of Ukraine in 2019-2020 at LLC "Kiseli", SPD "Tretyakova O.A." (Romanov breed), DPDG "Gontarovka" and CJSC AF "March 8" (prekos), "Volodarskoe" and LLC AF "Agrotis" (Askanian meat and wool), AF "Liya" and breeding plant LLC "Donagrolux" (Tsigai breed Azov type), LLC AF "Dobrobut" (Sokolskaya), SE AF "Shakhtar" (Merino Askanian Tavrichesky type).

Sheep chromosome preparations were prepared according to generally accepted methods [10-12] in a sterile box. The production of a culture of sheep peripheral blood lymphocytes included a number of stages: the cultivation of peripheral blood lymphocytes stimulated by phytohemaglutinin (PHA) (Sigma, USA) in a mixture of Igla nutrient medium for 48 hours in a thermostat at 37 ° C. Stopping mitosis was carried out 2 hours before the end of cultivation by adding colchicine at a concentration of 1 mg / ml. In sterile vials with blood samples (1 ml, about 5000 lymphocytes), 5-8 ml of culture medium 199 or Eagle was added, added to 2 ml of IST (inactivated calf serum) and PHA (6-12 µg / ml of the final volume of the culture medium), penicillin or streptomycin 100 mg / ml. The resulting suspension was thoroughly mixed and placed in a thermostat for cultivation for 48 hours at a temperature of 37 ° C. 30 minutes before the end of cultivation, colchicine was injected into each sample in an amount of 1 mg per 1 ml of medium. After the completion of the cultivation, centrifugation was carried out for 10 minutes at 1000 rpm. After centrifugation, the supernatant was aspirated and a hypotonic solution of 0.5% KCl warmed to 37 ° C was added. Then 3-4 drops of the cell suspension were dropped onto the glass slide from a height of 10-15 cm and dried. After hypotonic treatment with 0.5% KCl for 20 minutes, the cells were fixed with a mixture of ethanol and glacial acetic acid. The cell suspension was applied to wet cooled glass slides and dried. Analysis of chromosome preparations was carried out on preparations stained with Giemsa dye using a Jenaval microscope (Carl Zeiss, Germany) with oil immersion at a magnification of 1000 times. At the same time, the total chromosomal instability was taken into account as a percentage (percentage of metaphases with aberrations) in the structure of aberrations, single aberrations, paired aberrations, circular chromosomes or acentric rings, gaps and breaks of chromosomes, and others were taken into account [10-12].

Statistical processing of the data was carried out using generally accepted methods of variation statistics; the reliability of differences was assessed by the Student's t-test using a specialized software package SPSS for Windows ("IBM", USA). The tables show mean (M) and mean deviations (± m).

4 Results and Discussions

According to the results of the conducted studies of the sheep of the Tsigai breed of the Azov meat-wool type in the peripheral blood (Table 1-2), no characteristic anomalies were revealed for all aberrant cells. The structural variability of chromosomes was established, which was at the level of 0.89 ± 0.18.

Table 1. Structural mutations in peripheral blood cells of rams of different ages (M ± m; n = 991)

Age group (years)	Number of metaphases	Chromosomal	Chromatid	Average level of aberrations, %
up to 3	511	0,46	0,37	0,84 ±0,14
up to 8	480	0,59	0,34	0,93 ±0,13

Among the structural changes, mutations of the chromosomal type prevailed in the 3-year-old group of rams - 0.46%, in the 8-year-old group - 0.59%. Chromatid abnormalities were 0.37% and 0.34%, respectively. The average level of chromosomal abnormalities in rams by groups was 0.84 ± 0.14 and 0.93 ± 0.13, respectively.

Polyploidy was predominantly tetraploid and no significant difference was found between the two age groups.

Table 2. Genomic mutations in peripheral blood cells of stud rams of different ages (M ± m; n = 991)

Age group (years)	Number of metaphases	Aneuploidy, %	Polyploidy, %
up to 3	511	13,14 ±0,46	1,23 ±0,07
up to 8	480	20,40 ±0,9	1,38 ±0,13

A cytogenetic analysis of sheep with low and high vigor and different levels of fertility was carried out. The individual level of chromosomal abnormalities in all studied animals is within the natural norm. No translocations were found. Among the violations, there were such as single and paired fragments, hypo- and hyperploidy (mainly hypoploidy), polyploidy. The maximum average level of chromosomal abnormalities was determined in lambs of German origin (5.5 ± 1.73%). The minimum average level of chromosomal abnormalities (2.0 ± 1.41%) was recorded in lambs with a high growth rate. In animals with low growth intensity (n = 5), the average level of chromosomal abnormalities was 3.60 ± 2.05% (group 1) (Table 3).

Table 3. Chromosome profile of lambs with different growth rates

№ of heads	Aneuploidy, %	Polyploidy, %	Single fragments, %	Paired fragments, %	Level of chromo-somal abnormalities, %	Level of genomic abnormali-ties, %	Level of structural abnormali-ties, %
lambs with low growth rate (first group)							
201	4	2	4	0	10	6	4
344	0	0	0	0	0	0	0
334	0	0	0	0	0	0	0
347	0	4	0	0	4	4	0
248	0	2	2	0	4	2	2
M±m	0,8 ±0,89	1,6 ±0,84	1,2 ±0,89	0	3,6 ±2,05	2,4 ±1,30	1,2 ±0,89
lambs with high growth rate (second group)							
14	2	0	0	0	2	2	0
28	2	2	0	0	4	4	0
493	0	0	0	0	0	0	0
M±m	1,33 ±0,82	0,67 ±0,82	0	0	2,0 ±1,41	2,0 ±1,41	0
local lambs (third group)							
477	2	2	0	0	4	4	0
476	2	0	2	0	4	2	2
423	2	2	0	0	4	4	0
437	0	6	2	2	10	6	4
M±m	1,5 ±0,58	2,5 ±1,45	1,0 ±0,67	0,5 ±0,58	5,5 ±1,73	4,0 ±0,94	1,5 ±1,11

In the group of animals (n=3) with high growth intensity (second group), the average level of chromosomal abnormalities was 2.0 ± 1.41%. Structural disorders (1.20±0.89%) were present only in the first group of lambs. In the second group (with high growth intensity), only genomic abnormalities of 2.0±1.41% were recorded. In lambs of German selection (n = 4), the average level of chromosomal abnormalities was 5.5±0.73%, where genomic abnormalities accounted for 4.0 ± 0.94%, and structural abnormalities accounted for 5±1.11%, in this group of animals, the percentage of violations was higher in all parameters than in the previous two groups.

In the group of ewes (n = 11), the average level of chromosomal abnormalities is 4.0 ± 0.89%, which corresponds to the natural background. Among structural disorders (0.36±0.26%), both paired and single fragments were encountered with the same frequency. Among genomic disorders (3.27±0.71%), aneuploidy prevailed, mainly hypoploidy (Table 4).

Table 4. Chromosomal profile of ewes

№ of heads	Aneuploidy, %	Polyploidy, %	Single fragments, %	Paired fragments, %	Level of chromosomal abnormalities, %	Level of genomic abnormalities, %	Level of structural abnormalities,%
3584	0	0	0	0	0	0	0
4722	4	0	0	0	4	4	0
4661	0	0	0	0	0	0	0
2630	4	0	0	0	4	4	0
4572	2	0	0	2	4	2	2
3529	6	0	2	0	8	6	2
4657	4	2	2	0	8	6	2
3500	2	2	0	2	6	4	2
3489	2	4	0	0	6	6	0
3490	2	0	0	0	2	2	0
2582	2	0	0	0	2	2	0
M±m	2,55 ±0,57	0,73 ±0,43	0,36 ±0,26	0,36 ±0,26	4,0 ±0,89	3,27 ±0,71	0,73 ±0,32

In the group of ewes with low fertility (n = 4), the level of chromosomal abnormalities was 3.5%, which is lower than ewes (n = 7) with high fertility by 0.79%. Among the chromosomal abnormalities, genomic ones prevailed (Table 5).

Table 5. Chromosomal profile of ewes with different fertility

№ of heads	Aneuploidy, %	Polyploidy, %	Single fragments, %	Paired fragments, %	Level of chromosomal abnormalities, %	Level of genomic abnormalities, %	Level of structural abnormalities, %
ewes with low fertility							
3584	0	0	0	0	0	0	0
4722	4	0	0	0	4	4	0
3529	6	0	2	0	8	6	2
3490	2	0	0	0	2	2	0
M±m	3,0 ±1,49	0	0,5 ±0,58	0	3,5 ±1,97	3,0 ±1,49	0,5 ±0,58
ewes with high fertility							
4661	0	0	0	0	0	0	0
2630	4	0	0	0	4	4	0
4572	2	0	0	2	4	2	2
4657	4	2	2	0	8	6	2
3500	2	2	0	2	6	4	2
3489	2	4	0	0	6	6	0
2582	2	0	0	0	2	2	0
M±m	2,29 ±0,56	1,4 ±0,64	0,29 ±0,31	0,57 ±0,39	4,29 ±0,53	3,43 ±0,91	0,86 ±0,44

From the data in Table 5, it can be seen that in the environmental conditions of Ukraine, chromosomal instability increases in queens with low fertility.

5 Conclusions

It was revealed that in the ecological conditions of the south of Ukraine, the individual level of chromosomal abnormalities in all studied animals of the Tsigai breed and the prekos is within the natural background. No translocations were found. Among the violations, there were such as single and paired fragments, hypo- and hyperploidy (mainly hypoploidy), polyploidy. The maximum average level of chromosomal abnormalities was found in local lambs (5.5 ± 1.73%). The minimum average level of chromosomal abnormalities (2.0±1.41%) was recorded in lambs with high growth rates. Among the structural changes, mutations of the chromosomal type prevailed in the 3-year-old group of rams - 0.46%, in the 8-year-old group - 0.59%. Chromatid abnormalities were 0.37% and 0.34%, respectively. The average level of chromosomal abnormalities in rams by groups was 0.84±0.14 and 0.93±0.13, respectively. In the group of ewes with low fertility (n = 4), the level of chromosomal abnormalities was 3.5%, which is lower than ewes (n=7) with high fertility by 0.79%.

References

1. V.V. Aboneev, Collection of scientific papers of the Stavropol Scientific Research Institute of Livestock and Forage Production, **1-1**, 12(2003)
2. V.I. Glazko, Agricultural biology, **4,** 659 (2018)
3. A.V. Deikin, M.I. Selionova, A. Yu. Krivoruchko, D.V. Kovalenko, V.I. Trukhachev, Vavilov Journal of Genetics and Breeding, **5**, 576(2016)
4. T.E. Deniskova, A.V. Dotsev, E.A. Gladyr, A.A. Sermyagin, V.A. Bagirov, U. V. Khompodoeva, A.N. Ilyin, G. Brem, N.A. Zinovieva, Agricultural biology, **6**, 746 (2015)
5. L.G. Moiseykina, M. Rozikova, N.S. Marzanov, In the collection: Materials of the international scientific-practical conference on sheep and goat breeding, dedicated to the 65th anniversary of VNIIOK, 91(1997)
6. L.G. Moiseikina, N.S. Marzanov, In the collection: Topical issues of agricultural production in the Republic of Kalmykia Kalmyk Research Institute of Agriculture. Elista, 40(1997)
7. Yu.A. Stolpovsky, I.A. Zakharov-Gesekhus, Vavilov Journal of Genetics and Breeding, **4**, 477(2017)
8. A.V. Tkachev, V.I. Sheremet, O. L. Tkacheva, Scientific Bulletin of Lviv National University of Veterinary Medicine and Biotechnology imene S.Z. Ґzhitsky, **2-2(67),** 241(2016)
9. A.V. Tkachev, A.A. Evsyukova, A.D. Frundina, *Innovative solutions in agricultural science - a look into the future Proceedings of the XXIII international scientific and industrial conference*, 61(2019)
10. A.V. Tkachev, O. L. Tkachev, V.I. Rossokha, Agricultural biology, **2**, 302(2018)
11. A.V. Tkachev, Molecular and Applied Genetics, **19**, 79(2015)
12. A.V. Tkachev, O. L. Tkacheva, M.R. Shvetsova, N.V. Yavnikov, I.A. Koschaev, *Modern molecular genetic research methods in physiology, zoohygiene, veterinary medicine and biological safety: monograph* (Belgorod: Publishing house of the Belgorod GAU named after V.Ya. Gorin), 414 (2020)

13. V.I. Tyschenko, V.P. Terletskiy, Modern scientific research and innovations, **8(64)**,158(2016)
14. A.M. Kovalenko, A.V. Tkachev, O.L. Tkacheva, B.V. Gutyj, O.I. Prystupa, M.D. Kukhtyn, V.R. Dutka, Ye.M. Veres, O.O. Dashkovskyy, V.V. Senechyn, M.B. Riy, V.A. Kotelevych, Ukrainian Journal of Ecology, **10(1)**, 300 (2020)

Adaptive morphogenesis of lamb liver under Anthropobiocenosis conditions lamb liver morphogenesis

T.P. Skobelskaya*, *V.V.* Lemeshchenko, *N.V.* Saenko, *V.G.* Sokolov, and *E.V.* Nekhaichuk

Federal State Autonomous Educational Institution of Higher Education "Crimean Federal University named after V.I. Vernadskiy" of Agrotechnological Academy, 295492, Simferopol, Russia

Abstract. Investigated the histological features of the structure of hepar of one-day-old lambs (ODL) using a complex of morphological techniques. It was found that hepar in ODL is formed by stroma and parenchyma. The stromal elements of the organ are formed by loose lugs of hematopoietic connective tissue, which in turn forms a capsule (2.80-7.23 microns thick), as well as the septa of the lobules, which include the hepar triads and paravasal tissue of efferent blood vessels. As a result of the studies, the prevalence of the stroma in the hematopoietic foci around large afferent blood vessels, especially in the hepar hilus, was revealed. The hepar parenchyma in ODL has structural incompleteness. It is formed by hepar cells and numerous resinusoidal cells located in the subcapsular, peripheral, central zones in the lobus hepatis. Hepar cells located in the subcapsular zone (SZ) have a more optically cleared (OC) cytoplasm, and the practically cleared cytoplasm of the hepatocyte diameter (HD) decreases towards the central zone (CZ). In the hepar parenchyma of the ODL, hematopoiesis foci (HF) are established in each of the lobes, which tend to decrease from the periphery to the center of the lobes. Hemopoetic cells (HC) are located between hepatocytes, densely layering on top of each other, and there are also cells located sparsely between the liver cells (LC). The number of hepatocytes per 1 mm2 of the area of the histological specimen in the left lobe (LL) of the ODL reaches the highest value 5983.79 ± 90.40, in the right lobe (RL) of the liver it is 5358.80 ± 646.60, while in the middle it has a minimum value and reaches 5133.10 ± 205.75.

1 Introduction

The productivity of an adult animal depends on its adaptive abilities at the newborn stage of development, manifesting itself in the implementation of the functions of all organs, including hepar [4, 6, 9, 10, 11]. Hepar performs a number of vital functions, and in particular, one of the main functions in the prenatal period is hematopoietic and the formation of the liver tissue itself. In the postnatal period, hepar undergoes a significant increase in functions against the background of attenuation of hematopoiesis, which is certainly reflected in the histological structure of hepar [5, 7, 8]. The lack of information

*Corresponding author: www.150171986@mail.ru

about the structure of the hepar parenchyma and the degree of its incompleteness, as well as the performance of the main functions in newborn lambs, makes it difficult to clarify the features of pathomorphological changes in diseases of various etiologies.

There is practically no information in the literature on the structural incompleteness of hepar in lambs, which does not allow timely diagnosis and full prophylaxis of diseases in newborns, especially those occurring against the background of a general hematopoiesis inoculation of the body.

The aim of the study is to establish the hepar adaptive morphogenesis of male and female lamb under the conditions of an anthropically enlightened cytoplasmobiocenosis.

To achieve the goal, the following **tasks** were set:

- to establish the features of the structure of the parenchyma and stroma of the hepar in ODL;

- optically clarified cytoplasm to determine the peculiarities of intervascular and vascular-tissue relationships of hepar in male and female lambs.

2 Materials and methods

The morphological features of the parenchyma, stromal intervascular and vascular-tissue relationships of the hepar optically cleared cytoplasm were determined on histological organ preparations. For the preparation of histological sections, liver pieces with a thickness of 05-1 cm were taken, with an area of 1.5x0.5 cm in the right, left hepar lobes perpendicular to the gate of the organ, the middle (caudate) lobe parallel to the hepar gate. The liver of the studied animals was fixed in a 10% formalin solution. The preparation of histological preparations with a thickness of 25-30 μm was carried out on a microtome-cryostat MK-25-m at a temperature of -12°C...-16°C. Sections of hematopoietic foci stained hematopoietic foci with silin and eosin [2]. In order to obtain thin (8-12 μm) histological sections, the material was embedded in paraffin, which then stained the hematopoietic foci with Ehrlich's silin and eosin [3].

Morphometric parameters of hepar were determined using a Mikmed-5 binocular microscope. To establish the morphometric parameters of tissue components, the number of hepatocytes per 1mm^2 of the area of the hepar histological preparation, we used the Jeffreys method modified by S.A. Saltykov [1]. Longitudinal indicators of the structures were established using an eyepiece micrometer MOV-1-15^x.

$$X = z+0,5w+0,25u, \text{ where}$$

X – average number of hepatocytes in the cut area;
z – number of hepatocytes in the test system (ocular mesh insert);
w – number of hepatocytes crossed by the lines of the outer contour of the test system;
u – number of hepatocytes at the nodal points of the test system.

3 Results

In ODL, the tissue components of the hepar are represented by the stroma and parenchyma. The parenchyma is formed mainly by hepatocytes and a large number of resinusoid cells. In each lobes hepar of ODL, based on the morphological structure and its comparison with the stroma, it is possible to distribute: subcapsular (5.00-10.00 mm thick), peripheral (10.00-20.00 mm thick) and central (10.00- 28.00 mm) zone. The organ stroma in JSV is formed by loose fibrous connective tissue forming a capsule (2.80-7.23 μm thick), septa of lobules, which have hepar triads and paravasal tissue of efferent blood vessels. Stromal elements prevail around the large afferent blood vessels, predominantly in the hepar hilum. Inside the hepar, the amount of stroma is reduced; it is formed only by thin collagen fibers around the hepatic triads and hepatic veins. In ODL, the stroma does not form the characteristic boundaries of the outline of the hepar lobules.

In the SZ of RL in the ODL, liver cells (LC) are predominantly rounded, less often oval HD of 12.13-16.10 μm, the nuclei of the LC have a round or oval shape with HD from 3.97 μm to 7.23 μm. LC are close to each other, which creates a "honeycomb" look. But also insignificant areas were revealed in which LC form branched, short (3-7 hepatocytes) HD that anastomose with each other. The cytoplasm of hepatocytes (CH) is about 68.00-90.00% under the capsule in the RL in the of OC in the ODL, and about 10.00-32.00% is weakly eosinophilic, has a foamed appearance with different color intensity and different granularity (Fig. 1).

At the ODL LC in the SZ of RL, manifold branched beams are formed, the parameters of which are from 25.43 × 35.23 microns to 37.79 × 41.53 microns in diameter. Also distinctive is the formed single lobules (3-5 in the field of view of the microscope at a magnification of 10 × 10). LC, which ranges from 257.59×345.32 microns to 444.25×439.58 microns. LC are formed by branched that anastomose with each other, as well as groups of cells that form a radial order relative to the central veins (CV) only in insignificant areas of the histological specimen (Fig. 2).

HC are located between hepatocytes compactly overlapping each other, and also cells located weakly sparse between the LC were revealed. These cells form islets that are round, oval or ovoid 11.90 x 20.30 μm to 13.76 x 42.00 μm in diameter. The islets are represented by densely scattered 3-29 small cells 2.10 μm to 7.70 μm and with a richly stained basophilic nucleus. The diameter of the sinusoidal capillaries (DSC) in the SZ is 6.07-7.23 microns.

In the peripheral zone (PZ) in RL the ODL, LC with a regular rounded shape prevail, and the HD is slightly smaller, in contrast to the SZ (10.50-14.00 μm), while their nuclei are 4.67-6.30 μm. CZ is about 65.00-85.00% eosinophilic with various color saturation, and in 15.00-35.00% it is not completely OC, which is not inherent in such a ratio of cells under the hepar capsule.

In the PZ of the RL of the ODL, polymorphic branched beams are located, forming crowding with sizes from 22.16×22.30 μm to 76.52×77.92 μm in diameter, not ordered with respect to blood vessels. Thus, LC were identified that form anastomosing single beams, which have a radial orientation with respect to the CV, which were detected only in certain areas of the histological specimen. It should be noted that the formed lobules reach 148.39×129.73 microns - 275.32×301.46 microns, and the PSC - 4.67-6.77 microns. The hematopoietic islets (HI) in the PZ of the RL in the ODL have a polymorphic, oval or round shape. At the same time, LC are organized compactly or in a loose type. The number of cells in one islet in the PZ of the RL reaches 5-40, and that of the HD is 1.87-2.56 μm (Fig. 3, 4).

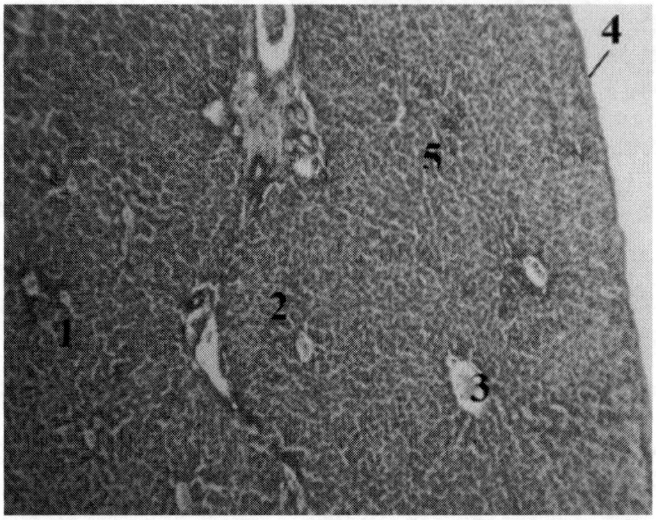

Fig. 1. Histological specimen of lamb liver. Age - day: 1 - hepatocytes; 2 - sinusoids, 3 - central vein; 4 - liver capsule; 5 - focus of hematopoiesis. Env. hematoxylin-eosin. Magnification - 100.

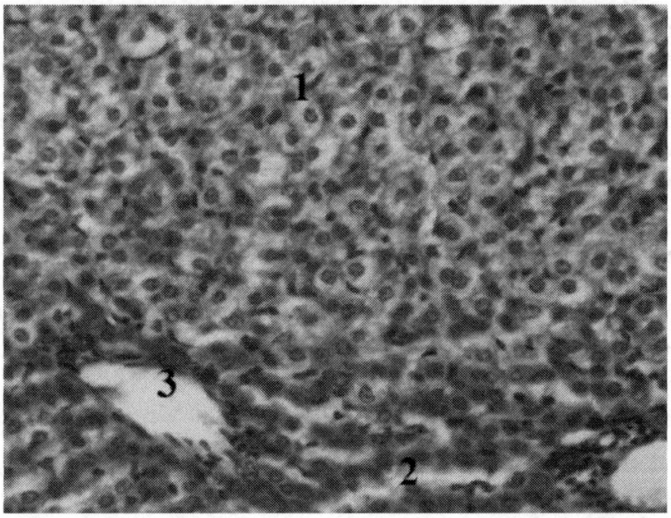

Fig. 2. Histological specimen of lamb liver. Age - day: 1 - hepatocytes; 2 - sinusoids, 3 - central vein. Env. hematoxylin-eosin. Magnification - 400.

The LC located in the central zone of the RL in the ODL have a rounded shape, and their diameter is 8.17-12.13 microns, while the nuclei (3.50-6.67 microns) are round. CZ of OC is about 10.00-25.00%. At the same time, LV with OC by the cytoplasm is much higher than the cells (about 75-95%), in which the cytoplasm is more stained. LC located near large II-III orders of blood vessels have a more strongly stained cytoplasm and a smaller diameter.

In the ODL LC, polymorphic branched clusters of beams are formed in the central zone of the RL, which are represented by an ovoid and oval shape (42.46×50.63−60.89×70.22 μm). Single lobules formed by hepatocytes, which do not have a radial location in relation to blood vessels, with rare beams of hepatocytes with 5-9 cells and have a radial orientation in relation to CV only in certain areas of the histological specimen were revealed. The parameters of such lobules are 129.73 × 215.59-195.06 × 291.19 μm. DSC is 3.27-5.83 microns. HI in the central zone are small, oval and rounded, predominantly located compactly and reach 15.86×19.83−23.80×22.39 μm in diameter.

In the ODL, the number of hepatocytes located in the RL per 1 m^{m2} of the area of the histological specimen is 5358.80±646.60 (V = 12.07%).

In the SZ of the middle lobe of hepar (ML) in the ODL, the LC have a round, oval shape, and the HD is 13.06-15.63 µm, while the CZ is about 75.00-96.00% of the LC, which is characteristic of significantly large LC 4.00-25.00% weakly eosinophilic cytoplasm with variable granularity. The shape of the LC nuclei in the NW is round and has a diameter of 6.30–7.47 µm.

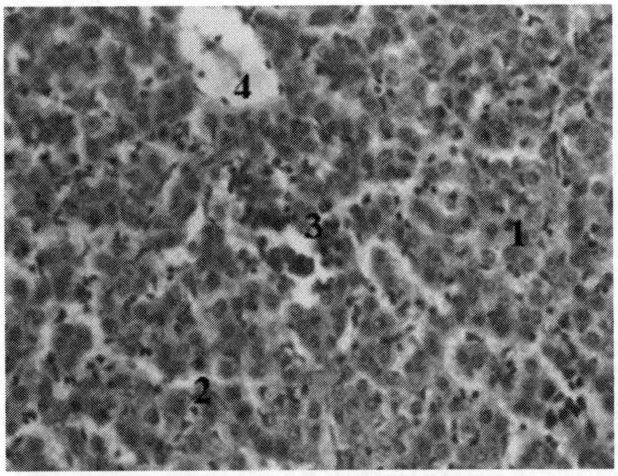

Fig. 3. Histological specimen of lamb liver. Age - day: 1 - hepatocytes; 2 - sinusoids, 3 - focus of hematopoiesis, 4 - central vein. Env. hematoxylin-eosin. Magnification - 400.

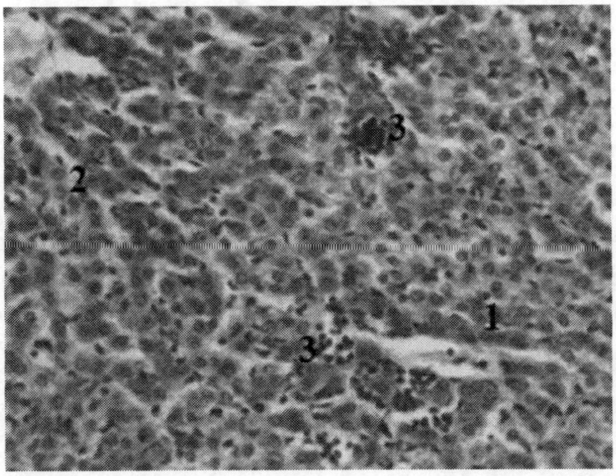

Fig. 4. Histological specimen of lamb liver. Age - day: 1 - hepatocytes; 2 - sinusoids, 3 - focus of hematopoiesis. Env. hematoxylin-eosin. Magnification - 400.

LC, which are located in the SZ of the ML at the ODL, are closely adjacent to each other, do not form an inherent radial beam structure. At the same time, they are combined into polymorphic groups of cells, the same pattern is also established in the RL, and their parameters reach from 15.16×22.86 µm to 26.13×36.63 µm. The lobular structure of hepar in the SZ of the ML is much less pronounced, but single lobules that do not have definite boundaries were revealed, their parameters are 216.53 × 169.86 µm - 293.06 × 188.53 µm. The DSC in the SZ of the ML is 3.97-6.53 microns, while no HI were detected in this zone.

The LC in the PZ of the ODL have a rounded shape with a diameter equal to 10.26-11.67 microns, and their nuclei from 6.30 microns to 7.23 microns. CZ is about 55.00-85.00% of OC, is 15.00-40.00% more strongly colored and 5.00-10.00% with slightly OD. In the PZ of the ML, the LC are not oriented in relation to the blood vessels and are organized into round, ovoid and oval multiforme groups of cells, and the diameter reaches from 30.79×37.09 microns to 47.36×45.73 microns. DSC in the PZ is 4.43-7.47 microns. In ODL, only a small amount of hepar lobulation was revealed in the vicinity of large blood vessels. At the same time, the LC combined into beams are determined, the thickness of which reaches 9.80-13.76 microns, and also have a radial orientation to the efferent blood vessels. The parameters of these lobules are larger than those in the RL and reach from 296.79×340.65 microns to 376.12×433.98 microns.

In the ML of PZ in the ODL, LC have the shape of an oval, round or polymorphic shape, which contains small cells, densely packed or scattered with a strongly stained basophilic nucleus, the number of which in one islet varies from 4 to 48; their diameter is of 0.93-5.83 microns. HI parameters reach from 11.43×26.83 μm to 17.26×29.63 μm.

The HD of the CZ ML in the ODL reaches 7.93-12.33 microns. Such LC are predominantly rounded, while their nuclei (3.97 μm to 5.83 μm) are rounded or oval, slightly basophilic. CZ is about 65.00-94.00% slightly eosinophilic, and the remaining LC are OC. CZ located near large blood vessels are much more stained, and in other areas LC with OC or slightly stained cytoplasm.

In the CZ of the ML in the ODL, LC are formed with branches that are anastomosed with each other, they have a radial placement in relation to the CZ only in insignificant areas of the histological specimen. Or they are defined in multiforme cell clusters from 28.70×41.76 μm to 49.69×50.86 μm. Hepar lobules have indistinct boundaries and are equal from 328.52×454.52 microns to 405.99×493.72 microns, and DSC located in the central zone is 2.80-6.07 microns.

The parameters of HI and CZ in the ML of the ODL range from 23.87×34.53 microns to 39.19×60.42 microns. The number of densely located LC in one focus is from 3 to 25, and their parameters reach 3.50-5.37 microns, and the number of hepatocytes in the ML per 1 mm^2 of the histological section area is slightly less than in the RL - 5133.10 ± 205.75 at V = 4.01%.

LC located in the LL of the SZ in the ODL have an irregular rounded shape and a slightly smaller diameter of 8.87-10.27 μm than in the right and middle lobes of the organ. However, these cells are well isolated from each other, their cytoplasm is about 40.00-70.00% slightly eosinophilic, while 30.00-55.00% have OC. LC have rounded nuclei, their diameter is 4.90-6.07 microns. LC is formed by anastomosing beams going in different directions, creating a radial placement relative to the CV and only in insignificant areas of the histological specimen. In many areas, cells form multiforme clusters of a group of cells that are not radially located and reach from 28.00 × 36.39 μm to 44.56 × 58.09 μm in diameter.

It is necessary to point out that in the LL of the CZ in the ODL, the form of HI is oval or round, with a predominantly compact structure. OC are organized by LC with a rather strongly stained basophilic nucleus, their number in one focus can reach from 5-32 cells, while cell parameters vary at the border from 1.87-6.30 μm. The HI index ranges from 22.86×26.60 microns to 30.56×43.63 microns in diameter. HC are located similarly to the right and middle lobes, that is, between hepatocytes.

The hepatocytes located in the LL of the PZ in the ODL are rounded with a diameter of 9.10-14.93 μm, their nuclei are round or oval and reach 4.90-6.53 μm. LC are compactly located to each other, and their cytoplasm is about 12.00-36.00% eosinophilic (with different color intensity), while about 68.00-88.00% of hepatocytes contain OC cytoplasm.

It was also found that LC are not distributed oriented with respect to blood vessels, they are formed in polymorphic cell clusters (from 37.09×49.23 μm to 69.52×72.32 μm).

In the LL of the PZ in the ODL, the formed lobules are detected only along the large blood vessels and range in diameter from 426.52×485.31 μm to 448.92×526.38 μm. LC are connected in short (4-9 cells) branched beams, with a radial orientation in relation to the parenchymal venules. DSC reaches 3.03-6.77 microns.

In the LL of the PZ in the ODL, the forming lobules are detected only along the large blood vessels and range in diameter from 426.52×485.31 microns to 448.92×526.38 microns. LC are connected in short (4-9 cells) branched beams, with a radial orientation in relation to the parenchymal venules. DSC reaches 3.03-6.77 microns.

In the LL of the PZ in the ODL, HI are round, oval or ovoid and reach from 29.86 × 28.69 μm to 34.76 × 48.76 μm in diameter. HC are round, 0.93-5.37 microns in diameter, while their abundance in one focus can reach up to 42 cells.

In the LL of the CZ in the ODL, LC are round or oval with a diameter of 8.40-13.30 μm, their nuclei are weakly basophilic, and their diameter is 3.50-6.77 μm. LC have eosinophilic cytoplasm with varying degrees of color (about 78.00-95.00%), and about 5.00-25.00% HI of the CZ. LC are formed into polymorphic groups of cells as in other parts of the hepar, but in diameter they range from 36.63×45.73 μm to 48.99×56.23 μm.

The forming lobules, the sizes of which range from 235.19×394.79 μm to 352.79×479.72 μm in diameter, It should be noted that LC in the central zone of the LL are formed into branched beams, among which, in isolated cases, radially directed beams are revealed, reaching a length of 5-10 cells, which form a hepar lobule with fuzzy boundaries.

HI are formed by small (1.87–4.67 μm) cells with an intensely stained nucleus, their number ranges from 5 to 39 cells in one focus. HI have not the same shape, but the most prevalent is oval, and the parameters of HI are from 23.80×28.70 μm to 34.76×64.86 μm. The number of hepatocytes in the LL per 1 mm^2 of the area of the histological section is the highest, reaching 5983.79±90.40 with a small variability, reaching only 1.51%.

4 Conclusion

Thus, in the parenchyma of the ODL, hepar has a structural incompleteness, which determines the metabolic activity of the organ. Located directly under the capsule in the SZ, especially the ML, the LC have predominantly OC cytoplasm, the beams in some areas are not differentiated. In the underlying PZ, the number of LC with cleared cytoplasm becomes less. They form branched beams, in rare cases uniting into forming lobules. The number of the latter is the greatest in the CZ of all hepar shares. In this case, the number of HI decreases from the periphery to the center of the lobes.

References

1. G. G. Avtandilov, Medical morphometry, 324 (1990)
2. N. P. Gavrilin, Actual nutrition of morphology, science practice of the II National Congress of Anatomists, Histologists, Embryologists and Topographers of Ukraine, 53 (1998)
3. A. P. Goralskiy, Fundamentals of histological technology and morphofunctional methods of preschooling in the norm with pathology, 288 (2005)
4. L. V. Kletikova, A. N. Martynov, N. P. Shishkina, D. I. Sinelshchikova, Bulletin of Agrarian Science, **1(82),** 73 (2020)
5. O. A. Matveev, A. A. Torshkov, Bulletin of the Orenburg State Agrarian University, **2,** 155 (2019)

6. P. Praphatsorn, Asian Biomedicine, **4(4),** 619 (2010)
7. S. M. Suleimanov, O. B. Pavlenko, V. S. Slobodyanik, L. P. Mironova, Izvestiya Orenburg State Agrarian University, **5,** 182 (2019)
8. C. Stephens, Curr. Opin. Allergy Clin. Immunol, **14(4),** 286 (2014)
9. P. S. Timkina, Agrarian science of Euro-North-East, **2(57),** 51 (2017)
10. M. D. TomoakiIkeda, M. D. YujiMurata, J. Edward, M. D. Quilligan, T. Julian, M. D. Parer, American Journal of Obstetrics and Gynecology, **182(2),** 449 (2000)
11. S. N. Chebakov, N. T. Silantyeva, Bulletin of the Altai State Agrarian University, 1(171), 110 (2019)

Synanthropic flora of Kuzbass

S.A. Sheremetova[1,2,*], *S.N.* Vityaz[1], *E.B.* Rotkina[1], and *S.I.* Mikhailova[3,4]

[1]Kuzbass State Agricultural Academy, Kemerovo, Russia
[2]Federal Research Center of Coal and Coal Chemistry SB RAS, Kemerovo, Russia
[3]Tomsk branch of the FSBU All-Russian Plant Quarantine Center, Tomsk, Russia
[4]National Research Tomsk State University, Tomsk, Russia

Abstract. The article presents the materials of studies of the adventive flora of the Kemerovo region. For the territory of the Kemerovo region, we identified 244 alien species belonging to 162 genera and 50 families. As a result of the increasing process of invasion of new species into the territory of the region, the need for constant monitoring of alien species for the degree of their naturalization, especially in natural cenoses, is urgent. The problem of synanthropic plant species is becoming increasingly important not only for economically developed territories, but also for the relatively preserved mountain regions of Kuzbass. As a result of the studies, it was found that the synanthropic fraction of the Kuzbass flora, consisting of adventive and apophyte species, accounts for about 18% of the total composition of the flora of vascular plants in the Kemerovo region (60 apophyte species, 244 – advent ones). The revealed heterogeneity in the chorological, ecological and biological terms of the species of the adventive fraction makes it possible to find suitable conditions in various types of ecotopes on the territory of the Kemerovo region. This type of work can serve as a basis for developing a strategy for the preservation of natural phytosystems of Kuzbass.

1 Introduction

Currently, the agro-industrial complex of the Kemerovo region is the most important sector of the regional economy, along with the main industrial complexes [10, p. 19]. According to the Passport of the Kemerovo Region - Kuzbass, in 2019, a little more than 14% (394 thousand people) of the population live in rural areas, and only 2.4% (29 thousand people) live in agriculture [10]. In the agricultural turnover there are 2651.3 thousand hectares of agricultural land, which is 27.7% of the total area of land resources of the Kemerovo region [1, p. 156]. In addition, the territory of the region belongs to areas with a high degree of soil disturbance, including as a result of plowing up areas suitable for this. This is facilitated by the high fertility of the soil and the needs of the region for crop production. In the XX century, when, along with the growing agricultural development of the country, there was a rapid development of coal enterprises, which led to the complete destruction of vegetation in large areas.

In 1950 A.V. Kuminova [5] pointed out that the most favorable natural conditions for agricultural production are in the central steppe and forest-steppe regions of the Kemerovo

* Corresponding author: ssheremetova@rambler.ru

region. Due to the ruggedness of the terrain, heterogeneous soil cover, they are concentrated in the Kuznetsk depression and the forest-steppe strip running along the northeastern end of the Kuznetsk Alatau.

Since the end of the 80s, the changes that began in the country, which led to privatization and the collapse of the collective economy, as a result of which the regional economy, like the economy of the entire country, entered a state of deep systemic crisis. As a result, a large number of abandoned agricultural lands was formed, and young birch forests were gradually formed on empty fields.

At the moment, the Kemerovo region - Kuzbass, is one of the most industrially developed and densely populated areas in the south of Western Siberia. The leading industries are coal and metallurgy. The extraction of minerals in the Kuznetsk coal basin is carried out not only by the mine, but also by the open method, as a result, the problem of lands disturbed by dumps arises. According to the Report "On the State of the Environment...", more than a billion tons of rock are extracted and stored on the surface every year. Thereof, the adjacent territories undergo strong and sometimes irreversible anthropogenic changes in the soil cover, animal and vegetation cover [1, p. 345]. In the process of such anthropogenic transformation, the indigenous flora of the Kemerovo region lost some of its species, while the invasion of adventive species took place [21].

All this ultimately led to the fact that some of the species of native flora adapt to living in changed conditions: arable land, fields, fallow lands, roadsides, railway embankments, landfills, etc.

Synanthropic plants in the flora of the Kemerovo region are of undoubted interest for study in light of the indicated problems, the constantly increasing anthropogenic pressure on natural ecosystems. The Kuznetsk Basin of the Kemerovo Region is a territory significant for the conservation of biological diversity in the Altai-Sayan region [3].

2 Materials and methods

The study area is located in the southeast of Western Siberia and is located in temperate latitudes between 52° 08' and 57° 54' north latitude and 83° 13'and 89° 28' east longitude, with a total area of 95.7 thousand km2. Located at the junction of the West Siberian Plain and the mountains of South Siberia, most of the Kemerovo region belongs to the Kuznetsk-Salair geomorphological province of the northwestern part of the Altai-Sayan mountainous region [5], which affects the distribution of the main types of vegetation. The mountain ranges of the Salair Ridge, Gornaya Shoria and Kuznetsk Alatau disturb the general distribution of the vegetation cover according to latitudinal zoning, making it possible to develop taiga formations at the geographical latitude of the forest-steppe and steppe zones of Western Siberia.

The first scientific information about the flora of the Kemerovo region dates back to 1721 and is associated with the name of D.G. Messerschmidt. He crossed the territory of Kuzbass from Tomsk to Kuznetsk and through the Kuznetsk Alatau proceeded to Abakan [4, p. 134]. Subsequently, the materials collected by Messerschmidt in the second half of the 18th century were used by I.G. Gmelin, P.S. Pallas when publishing works on the flora of Siberia and Russia. The study of the flora of the Kemerovo region by the authors of the article has been conducted since 2000, since the opening of the Kuzbass Botanical Garden. In general, the summary of the flora of the study area includes 1,700 species of higher vascular plants [9; 15].

The synanthropic flora of the Kemerovo region was identified according to the data of long-term (2000-2020) field studies, taking into account information on weeds [8; 20, p. 43] and adventive [14; 13; 16] plants, the samples are kept in the funds of the KUZ Herbarium.

3 Results and discussion

In the flora of Kuzbass, there are 60 apophytic species, which are subdivided, depending on the place of growth, into segetal - species growing in arable lands and agrocenoses: *Cirsium incanum* (S.G. Gmel.) Fisch., *Cirsium setosum* (Willd.) Besser, *Cirsium vulgare* (Savi) Ten., *Convolvulus arvensis* L., *Sagina procumbens* L., *Stellaria media* (L.) Vill., *Sonchus arvensis* L. and ruderal plants that prefer to grow near dwellings, garbage dumps, wastelands, along roads: *Capsella bursa-pastoris* (L.) Medikus, *Descurainia sophia* (L.) Webb ex Prantl, *Polygonum aviculare* L., *Potentilla anserine* L., *Potentilla norvegica* L., *Psammophiliella muralis* (L.) Ikonn. etc.

The maximum number of synanthropic species are noted in settlements and their environs. For the reasons described above, the leader in the representation of synanthropic species in the region is the Kuznetsk Basin. Since ancient times, these territories have been used as agricultural land, and now over 50% of all the region's land are allotted for arable land [1, p. 170].

In the course of the study, for the territory of the Kemerovo region, we identified 244 adventive species belonging to 162 genera and 50 families. The most impressive in terms of number of species is the Asteraceae family (46 species), followed by Poaceae (26 species), Brassicaceae (24 species), Fabaceae (18 species), Lamiacee (15 species) and Caryophyllaceae (12 species). The next group of families contains from 8 to 5 species: Rosaceae (8), Boraginaceae (7), Malvaceae (7), Scroplulariaceae (6) Solanaceae (5) Amaranthaceae (5), Onagraceae (5). From 4 to 2 adventive species include 10 families, 25 families - 1 species each.

The expansion of alien plants is gradually leading to the floristic pollution of the territory and is becoming a global environmental problem. As a result of this process, together with adventive species, new families are noted in the flora of Kuzbass: Aceraceae, Amaranthacea, Commelinaceae, Cucurbitaceae, Elaeagnaceae, Oleacae, Resedaceae, Ulmaceae, Zygophyllaceae and, accordingly, new genera (50 in total):

Acer, Alcea, Abutilon, Acalypha, Amelanchier, Anthyllis, Cannabis, Commelina, Conium, Daucus, Elaeagnus, Hippophaë, Hibiscus, Galega, Lupinus, Malus, Oenothera, Reseda, Sorbaria, Tribulus, Ulmus, Xanthoxalis, etc.

It should be noted that, despite certain changes in natural vegetation, they affect the entire territory of the Kemerovo region, but to the least extent, to date, the transformation processes have affected the mountainous regions of the Kuznetsk Alatau and Gornaya Shoria.

In the works of many authors, adventive species are classified according to three main separate criteria: the mode of immigration, the time of introduction and the degree of naturalization [8; 17; 18].

In Kuzbass, according to the time of introduction, it possible to distinguish 3 groups of species of the adventive fraction of flora:

Eukenophytes include 162 species. This is the largest group in terms of the number of species, the maximum number of alien species was recorded in the Kuzbass in the XX century. The active agricultural development of the Kemerovo region, the development of summer cottages, the construction of roads and, accordingly, the growth of traffic flows contributed to the introduction of such species as: *Aconogonon divaricatum* (L.) Nakai ex Mor, genus *Amelanchier, Arctium minus* (Hill) Bernh., *Aster novi-belgii* L., *Calendula officinalis* L., *Epilobium adenocaulon* Hausskn., *Lotus corniculatus* L., *Matricaria recutita* L., *Oenothera biennis* L., *Xanthium strumarium* L. It should be noted that the growth of invasions continues, and further penetration of other species into Kuzbass can be expected.

The species, that appeared in the XIX century on the territory of the Kemerovo region, belong to the group of kenophytes - 19 species. These are such weed-field and ruderal species as representatives of the genus *Chenopodium, Asperugo procumbens* L., *Acinos arvensis*

(Lam.) Dandy, *Cichorium intybus* L., *Camelina sativa* (L.) Crantz, and also traditionally cultivated in the farmsteads of *Armoracia rusticana* P.G. Gaertn., B. Mey. & Scherb. This group is so small in number due to the impossibility of determining the time of introduction of many species into the territory of Kuzbass.

The most ancient group - archaeophytes, also includes 19 species. These are species both deliberately introduced by humans: *Avena sativa* L., *Cannabis sativa* L., *Neslia paniculata* (L.) Desv., *Panicum miliaceum* L., and weed-field species accompanying cultivated plants: *Atriplex patula* L., *Bunias orientalis* L., *Echinochloa crus-galli* (L.) P. Beauv., *Lepidium ruderale* L. etc. Scientists believe that for the territory of the Kemerovo region, the construction of the Siberian tract and the activities of the Russian-American company served as the main frontier for the penetration of archaeophytes, which contributed to the migration of species to Siberia from the Far East (Ermolaev, 2013).

The classification of the advents by the method of introduction shows that for some species at the present time we cannot determine the scenario of their appearance on the territory of the region, this is a group of xenoergaziophytes: *Amaranthus cruentus L., Echinocystis lobata (Michx.) Torr. & A. Gray, Lolium perenne L.*, there are 9 species in total. Ergasiophyta species - 67 species, introduced into the territory of the region intentionally as decorative, food, etc. intentionally: genus *Amelanchier, Berberis vulgaris* L., *Eschscholzia californica* Cham., *Portulaca grandiflora* Hook., etc. The third, the most numerous group of xenophytes is 168 species. It includes, as a rule, weeds introduced by chance, together with cultivated species or traffic flows with various cargoes: *Agrostemma githago* L., *Vaccaria hispanica* (Mill.) Rauschert, виды родов *Amaranthus, Atriplex, Chenopodium*, etc.

The degree of naturalization of the species of immigrants is very different, but against the general background of Advent, growing mainly in disturbed habitats, a group of the most aggressive species, agriophytes, stands out. These species are capable of being part of natural phytocenoses in the Kemerovo region. Currently, there are relatively few such species - about 10: *Amoria hybrida* (L.) C. Presl, *Armoracia rusticana* P.G. Gaertn., *Malus baccata* (L.) Borkh., *Medicago sativa* L., *Calystegia spectabilis* (Brummitt) N.N.Tzvelev, *Saponaria officinalis* L.. The most aggressive of them require the closest attention, to which we can classify 3 species on the territory of Kuzbass: *Acer negundo* L., *Solidago canadensis* L., *Elodea canadensis* Michx. Colonophytes are a group of plants that have entrenched themselves in places of drift or run wild, but currently do not spread further and do not penetrate into natural communities. In Kuzbass, scientists classify about 30 species as colonophytes: *Aconogonon divaricatum* (L.) Nakai ex Mori, *Acer tataricum* L., *Padus maackii* (Rupr.) Kom., *Panicum miliaceum L. Papaver rhoeas L. Pastinaca sativa* L., *Tilia cordata* Mill., *Helianthus tuberosus* L., *Viola arvensis* Murray, species of the genera *Oenothera and Xanthium*, etc.

The third group in terms of the degree of naturalization is epecophytes, species that spread over disturbed habitats and penetrate only segetal or ruderal communities. This largest group includes 196 species: *Amelanchier alnifolia* (Nutt.) Nutt. ex M. Roem., *Elsholtzia ciliata* (Thunb.) Hylander, *Syringa vulgaris* L., *Lepidium densiflorum* Schrad., *Lupinus polyphyllus* Lindl., *Lycopsis arvensis L., Medicago varia* Martyn, *Medicago sativa L., Mentha piperita L., Mentha gracilis Soll, Panicum miliaceum* L., *Papaver rhoeas* L., *Pastinaca sativa* L., *Ulmus laevis* Pall., *Viola arvensis* Murray, representatives of the genera *Oenothera, Xanthium* and many others. The fourth group - ephemerophytes, includes 10 species that are known from individual finds, but currently do not naturalize on the territory of Kuzbass, in this regard, they disappear from the places of initial finds, but can appear as a result of repeated introduction again: *Abutilon theophrasti* Medikus, *Avena sativa* L., *Bromus mollis* L., *Eruca sativa* L., *Helianthus annuus* L., *Papaver rhoeas* L., *Portulaca oleracea* L., *Vaccaria hispanica* (Mill.) Rauschert.

The analysis of the ranges of representatives of the adventive flora of Kuzbass allows us to determine the ways of penetration of these species into the territory of the region. The heterogeneous habitats of the Advents indicate that the alien species that have now become part of the flora of the higher vascular plants of Kuzbass were brought in from the most diverse parts of the world. Some of the species entered the territory of Kuzbass from various regions of the Mediterranean: *Bromus mollis, Carduus acanthoides, Chorispora tenella (Pall.) DC., Erucastrum armoracioides, Grossularia reclinata, Reseda lutea, Syringa vulgaris*. There are representatives of North American origin: *Acer negundo, Ambrosia artemisiifolia, Aster novi-belgii* L., *Amelanchier spicata* (Lam.) K. Koch, *Cyclachaena xanthiifolia* (Nutt.) Fresen., *Helianthus annuus* L., *Solidago canadensis* L., *Lupinus polyphyllus* Lindl., etc. The smallest group consists of Central American and South American species: *Galinsoga ciliate* (Raf.) Blake, *Nicandra physalodes* (L.) Gaertn.. There are representatives from Central and Central Asia: *Elaeagnus angustifolia* L., *Malus domestica, Senecio dubitabilis*. However, it should be noted that, in general, Advent species have wide ranges (pluregional, Palaearctic, Holarctic); for many of them, it is difficult to establish the initial range: *Amaranthus blitoides, Armoracia rusticana, Buglossoides arvensis, Elodea canadensis, Echinochloa crusgalli, Epilobium pseudorubescens, Epilobium adenocaulon, Camelina microcarpa, Fumaria officinalis, Impatiens parviflora, Lepidium ruderale, Matricaria recutita, Malva pulchella, Medicago sativa, Mentha gracilis, Oenothera villosa, Sisymbrium loeselii*.

Analysis of the life forms of apophytes, according to the classification of life forms of Raunkier, showed that among the alien species of Kuzbass in the first place are annuals, which experience an unfavorable period in the form of seeds - therophytes. There are 156 species of therophytes in the adventive fraction of the flora, the second place is occupied by hemicryptophytes, which preserve the buds of renewal at a time unfavorable for the growing season at the soil level, there are 62 species. In third place are cryptophytes, in which the buds of excitation are located in the soil or at the bottom of the reservoir, there are 26 species of them: *Elodea canadensis, Potamogeton sarmaticus, Convallaria majalis, Iris pseudacorus*, etc.

The classification of life forms of angiosperms by I.G. Serebryakova, which is based on ecological and morphological principles, divides the adventive species of Kuzbass into the following groups: herbaceous summer-green one-biennial tap-root monocarps: *Ambrosia artemisiifolia, Chorispora tenella, Centaurea cyanus, Coreopsis tinctoria, Conyza canadensis*; herbaceous summer-green perennial short-rhizome-brush-root polycarpics: *Aster novi-belgii, Achillea setacea, Bellis perennis*; herbaceous summer-green perennial tap-root polycarpics: *Lupinus polyphyllus, Malva moschata, Nepeta cataria;* herbaceous summer-green perennial long-rhizome polycarpics: *Centaurea phrygia, Mentha gracilis, Solidago serotinoides*.

The ecological analysis of the adventive fraction of the flora of the Kemerovo region showed that among the invasive species in relation to the main ecological factors prevail: mesophytes - more than 78% (moisture regime), non-petrophytes - more than 90% (stony substrate), mesotrophs - more than 92% (soil trophicity) , heliophytes - about 90% (illumination).

4 Conclusion

As a result of the studies, it was found that the synanthropic fraction of the Kuzbass flora, consisting of adventive and apophyte species, accounts for about 18% of the total composition of the flora of vascular plants in the Kemerovo region (60 apophyte species, 244 - advent). Strengthening the process of invasion in modern phyto-systems is one of the major factors in changing the composition and structure of natural plant communities. And if the composition of apophytes is quite stable, then the species of Advent require close attention. As our research has shown, the composition of apophytes recorded in the territory of the Kemerovo region is rather heterogeneous, but we can note some peculiarities. By the time of introduction, eukenophytes predominate on the territory of the region - species introduced to Kuzbass in the XX century (more than 66% of the total composition of apophytes). According to the method of introduction, xenophytes predominate - alien species introduced by chance, together with cultural species or traffic flows with various cargoes (about 69%). By the degree of naturalization, epecophytes predominate - species that spread over disturbed habitats and penetrate only segetal or ruderal communities (more than 80%). The greatest concern is caused by agriophyte species that can be part of natural phytocenoses. Currently, their number is not high and amounts to 4%, but they require close attention in order to undertake measures to eliminate the irreversible transformation of natural communities. It was also established that the ways of penetration of apophytes are also varied, but the most numerous group are Adventists with wide areas. The analysis of life forms and attitudes towards the main ecological factors shows the ability of alien species to adapt to the conditions existing in the region and stay for a long time on its territory. At present, the phenomena of transformation of the communities of ruderal and weed plants that have formed by now have been noticed due to the additional introduction of more aggressive species. It was assumed that the intensification of the invasion process is inevitable, and the strategies of the behavior of alien species are insufficiently studied, which determines the need for constant monitoring of alien species on the territory of Kuzbass. It is necessary to control the degree of naturalization of adventive species, for the timely identification of aggressive transformer species, in order to develop measures to combat agriophyte species.

Acknowledgments

This study was partially supported by the RFBR grant No. 20-44-420007r_a, and the state assignment on the topic "Assessment of the state and protection of floristic diversity under the influence of anthropogenic and technogenic factors *in situ* and *ex situ*", registration No. AAAA-A17-117041410053.

References

1. Report on the state and protection of the environment of the Kemerovo region - Kuzbass in 2019, 474 (2019)
2. A.N. Ermolaev, *Russian-American company in Siberia and the Far East (1799-1971)*, 620 (2013)
3. I.E. Smelyansky, G.A. Pronkina, *Key botanical territories of the Altai-Sayan eco-region: the experience of isolation*, 260 (2009)
4. I.M. Krasnoborov, Bot. research on Siberia and Kazakhstan, **12**, 134 (2006)
5. A.V. Kuminova, *Vegetation of the Kemerovo region*, 167 (1950)
6. A.N. Kupriyanov, S.S. Kazmina, A.A. Zverev, Bulletin of the Tomsk State University. Biology, **43**, 66 (2018)
7. N.I. Naumenko, *Flora and vegetation of the Southern Trans-Urals*, 512 (2008)
8. V.V. Nikitin, *Weed plants of the flora of the USSR*, 452 (1983)
9. I.M. Krasnoborov, E. D. Krapivkina, M.N. Lomonosov, et al., *Guide of plants of the Kemerovo region*, 477 (2001)
10. Passport of the Kemerovo region - Kuzbass for 2019, http://ako.ru/
11. E.B. Rotkina, S.A. Sheremetova, Achievements of science and technology of AIC, **12**, 73 (2011)

12. M.M. Silantyeva, *Flora of the Altai Territory: Analysis and History of Formation: Author's abstract*, 35 (2008)
13. S.A. Sheremetova, R.T. Sheremetov, *Basin of the Tom River (floristic and physical-geographical features)*, 323 (2020)
14. A.L. Ebel, *Abstract of the flora of the northwestern part of the Altai-Sayan province*, 568 (2012)
15. S.A. Sheremetova, A.L. Ebel, T.E. Buko, Turczaninowia, **14**, 65 (2011)
16. A.L. Ebel, A.N. Kupriyanov, T.O. Strelnikova, E.S. Ankipovich, E.M. Antipova, E.M. Antipova, T.E. Buko and et al., *The Black Book of the Flora of Siberia*, 440 (2016)
17. M. Kleunen, P. Petr, D. Wayne, et al., Ecology, **100(1)**, e02542 (2019)
18. A. F. Gulyaeva, N. N. Lashchinsky, N. N. Revushkin, et al., International Journal of Environmental Studies, **72(3)**, 527 (2015)
19. Important Plant Areas, http://www.plantaeuropa.org/
20. E.P. Kondratenko, E.A. Egushova, A. Starovoĭtov, E. Starovoĭtova, I.A. Sergeeva, E.B. Rotkina, *Weed vegetation in spring wheat crops in the southeast of western siberia (on the example of Kemerovo region)*, 43 (2020)
21. S. A. Sheremetova, I. A. Khrustaleva, A. N. Kupriyanov, T. O. Strelnikova, G. I. Yakovleva, E. B. Rotkina, Botanical Journal, **106(7)**, 696 (2021)

Pathomorphological changes in the early postnatal period in a calf with abomasum rupture

Maria Sergeevna Mannova*, *Lyudmila Vladimirovna* Kletikova, and *Nina Nikolaevna* Yakimenko

Ivanovo State Agricultural Academy named after D. K. Belyaev, Ivanovo, Russia, 153012, st. Sovetskaya, 45, Ivanovo, Russia

Abstract. The main cause of the disease and death of calves in the early postembryonic period of development is a feeding disorder. The aim of the work was to analyze macro-and micromorphological changes in the digestive system of a calf with a rennet rupture. To achieve this goal, standard macro-and microscopic research methods were used. Macroscopic examination revealed thickening, erosion of the mucous membrane of the esophageal gutter rollers, accumulation of contents in the scar, mesh and book, hyperemia of the mucous membranes; thinning and rupture of the rennet wall; pronounced hyperemia of the mucous membrane of the small intestine, thickening of the mucous membrane of the thick section; an increase in mesenteric lymph nodes and a change in their consistency. Microscopically, the thickness of the epithelium, submucosal and muscle layer is most developed in the abomasum and was, respectively, 0.68-0.72 mm, 0.23-0.32 mm and 0.98-1.05 mm. Villi were found in the folds of the scar, numerous well – developed longitudinally oriented protrusions were found in the book, lymphocellular clusters and bottom glands were found in the rennet; folds with well – developed crypts were found in the colon. Thus, as a result of feeding coarse feed, the deceased calf has catarrhal erosive lesions of the esophageal trough; serous-catarrhal abomasitis with signs of hemorrhagic with dilation and perforation in the cardiac part; hemorrhagic omasitis against the background of a book blockage; catarrhal reticulitis; catarrhal ruminitis and catarrhal enterocolitis.

1 Introduction

The neonatal period in calves is the interval from birth to the 8th day after birth, which is determined by morphofunctional changes due to the change of living conditions and adaptation of newborns to environmental factors [3, 6, 11, 12]. During this period, significant changes occur in organs and systems. In the first days after the birth of a calf, its digestive system is insufficiently developed, where the abomasum is the only fully formed and functioning department. During the newborn period, the gastrointestinal tract of cattle undergoes changes due to the heterochronous development of membranes, tissues and their

* Corresponding author: mannova09@yandex.ru

cells, the predominance of degenerative processes over regenerative ones. Adaptive rearrangements of epithelial, connective, muscular and nervous tissues, the intestinal wall, its mucous membrane, villi, crypts and duodenal glands are expressed in intensive tissue destruction, an increase in "migrant" cells, asynchronous development of the intestinal wall membranes, and cytochemical differentiation of cell protoplasm [5].

In the early dairy period, the digestive juices of the calf contain few enzymes, while the mother's milk contains all the enzymes that promote the digestion of milk nutrients. Despite the accumulated knowledge, the main cause of diseases of the gastrointestinal tract of young animals is still a feeding disorder. As a result of errors in the diet, the incidence of calves during the newborn period can reach 50-65%, and mortality – 90 percent or more [4, 6]. With intensive animal husbandry and human intervention in the animal habitat, conditions for technological stress are created, leading to the death of young animals [1, 2].

The purpose of this study is to analyze macro – and microscopic changes in the rupture of the abomasum in a calf in the early postembryonic period of development.

2 Materials and methods

The study was performed at the Department of Obstetrics, Surgery and Non-Infectious Animal Diseases of the Ivanovo State Agricultural Academy in 2020. Sampling for microscopic examination was performed after a macroscopic analysis of the state of the digestive system of the fallen animal. The histological examination was carried out in accordance with the standard protocol.

3 Results

Examination of the digestive organs of the fallen animal found that the rollers of the esophageal gutter are thickened, pink in color with foci of redness; a bluish hue and more pronounced changes were noted in the bottom part and lips at the entrance to the book. On the right lip, the integrity of the mucous membrane with a size of 7.00×13.00 mm, oval in shape, is iolated (Fig.1).

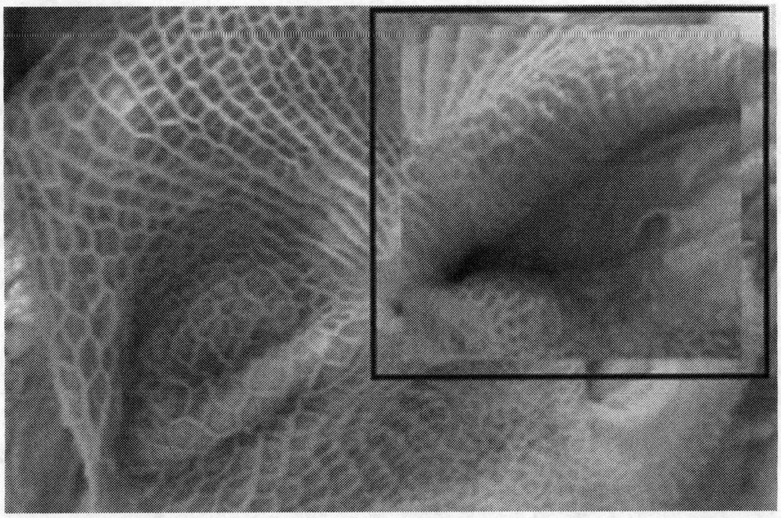

Fig. 1. The mucous membrane of the reticulum and oesophageal groove with hemorrhages and erosive lesions (macro-preparation).

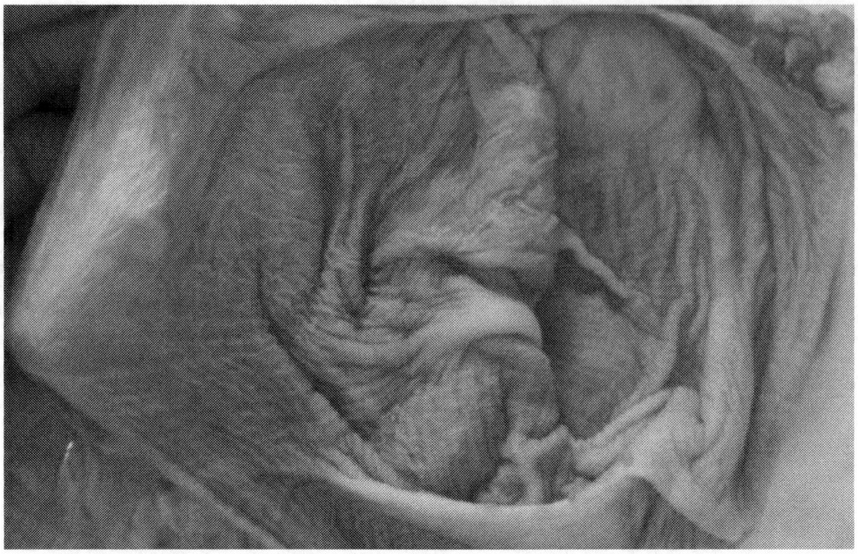

Fig. 2. The mucous membrane of the rumen with local redness (macropreparation).

In the scar, 3.00-4.00 kg of greenish-yellow contents of a dense consistency were found, the mucous membrane is yellow-gray in color with many papillae of different sizes and shapes (Fig. 2). The grid contains a small amount of dryish feed masses. The mucous membrane is pink in color with focal redness, collected in folds of various sizes, resembling network cells (Fig. 1).

The book is enlarged in volume, dense consistency, pink outside (Fig. 3). Along the serous surface of the book there is a cluster of hyperplastic, enlarged lymph nodes. In the area of the hole directed from the book to the abomasum, there are dry masses, the mucous membrane is dark cherry-red. The mucous membrane of the book is red-pink in color, collected in folds of different heights, between which there are dense dry feed masses (Fig. 4).

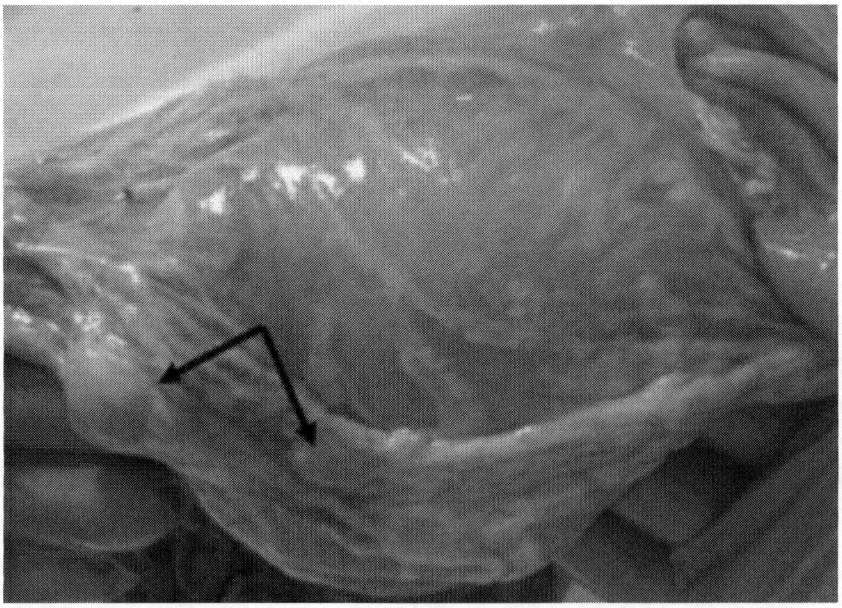

Fig. 3. Serous surface of the omasum (macropreparation) with hyperplastic lymph nodes.

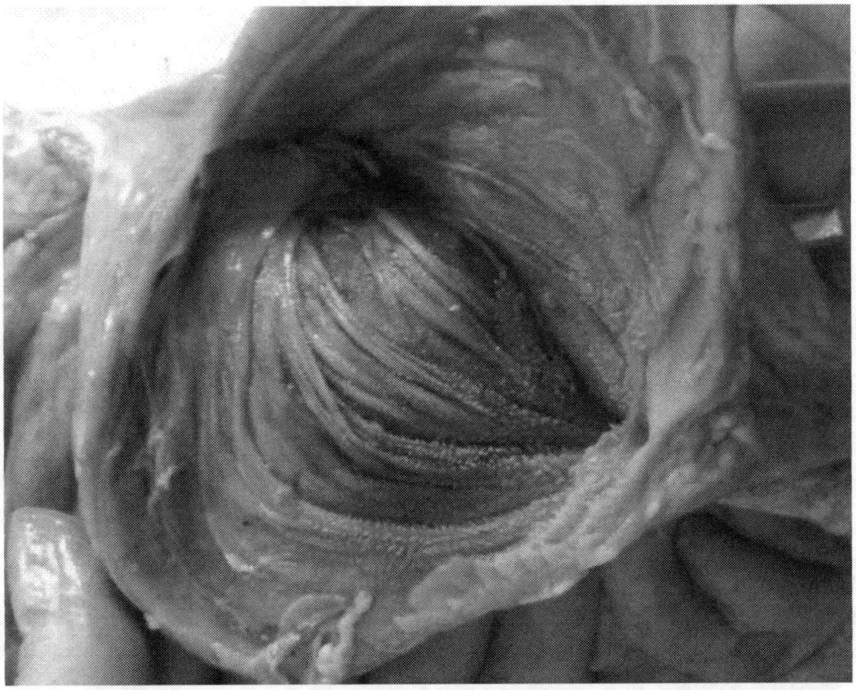

Fig. 4. The mucous membrane of the omasum (macropreparation).

On examination, the abomasum is enlarged in volume, it is colored dark cherry on the outside, due to the large curvature in the cardiac part, the shade becomes darker (Fig. 5). In the cardiac region of the rennet, stretching, thinning and violation of the integrity of the wall were revealed. The perforation hole is round-oval in shape with smooth edges, with a diameter of 38.00 mm (Fig. 6).

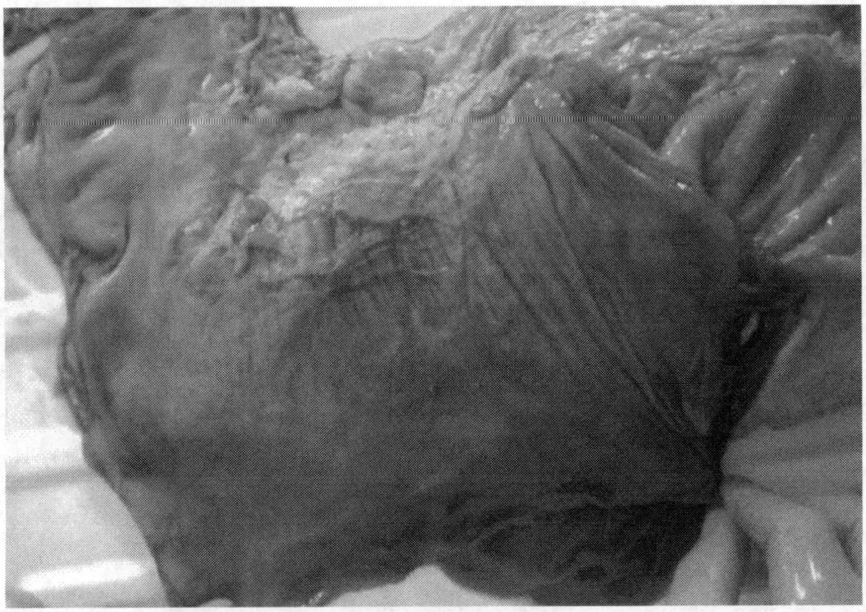

Fig.5. Serous surface of the abomasum (macropreparation).

Fig. 6. Perforation of the abomasum wall (macropreparation).

The mucous membrane in the fundal and cardiac regions of the rennet is dark cherry-red, gelatinous consistency with bloody contents. The spiral folds are thickened, jelly-like, in the area of perforation of dark cherry color.

In the pyloric part of the abomasum, the mucous membrane is slightly reddened, edematous, thickened, with jelly-like foci, serous-mucous exudate is present in the places of contact of the feed masses with the mucous membrane (Fig. 7). The longitudinal folds in the pyloric part are slightly thickened, yellowish in color, reddened in places.

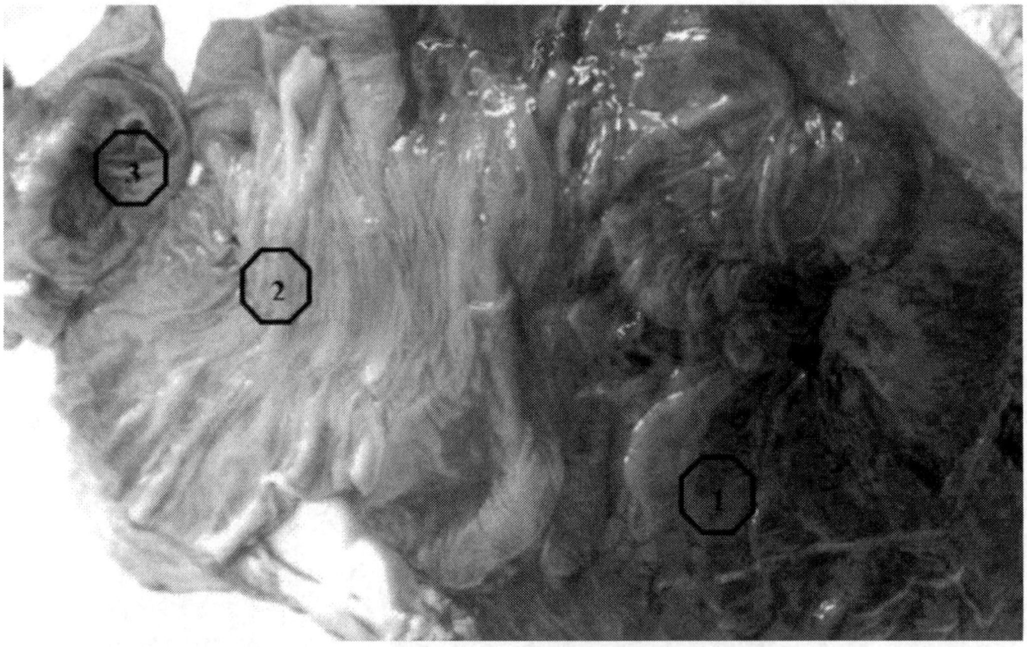

Fig. 7. The mucous membrane of the abomasum (macropreparation): 1 – gelatinous folds of the mucous membrane of the cardical part of the abomasum - hemorrhagic type of inflammation; 2 – fundus area of the abomasum pyloric part with signs of serous-catarrhal inflammation; 3 – sphincter area..

There is a small amount of yellow-brown content in the small intestine (12-duodenum, jejunum and ileum). Outside, the 12-digit and jejunum are pink-red in color. The mucous membrane is from red to dark cherry color, covered with a thick, viscous mucus with brown veins. Mesenteric lymph nodes are enlarged, dense consistency, elastic, gray in the section (Fig. 8).

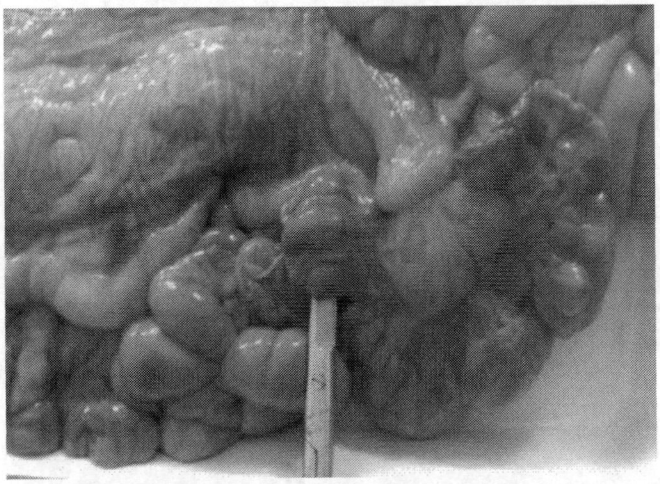

Fig. 8. Macropreparation of the calf's small intestine part with signs of catarrhal inflammation.

The large intestine: the blind, colon and rectum are moderately filled with a semi-liquid gray-brown mass with a specific smell. The mucous membrane is pale gray, shiny, in the area of the spiral labyrinth of the colon, the mucosa has a longitudinal folding, thickened, covered with a viscous gray mucus (Fig. 9).

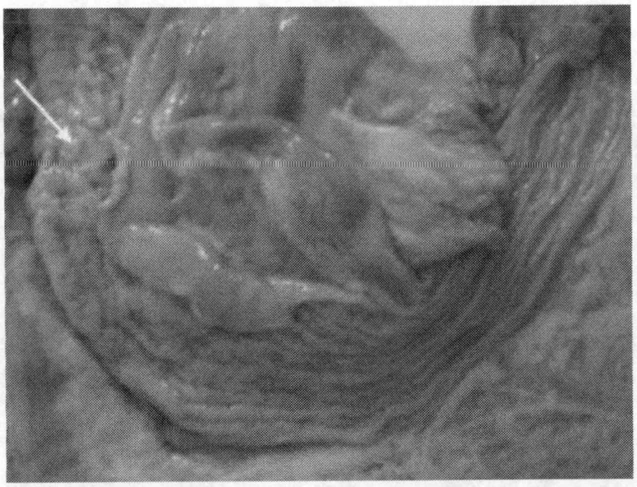

Fig. 9. The mucous membrane of the spiral colon of the middle intestine with signs of catarrhal inflammation (macropreparation).

Microscopically, it was found that the mucous membrane of the esophageal gutter was sent out by a multilayer flat non-corneating epithelium (MPNE), which together with the submucosal layer forms folds, the thickness of the epithelium was 0.18 mm, the submucosal layer was 0.27-0.55 mm. Below them is a well-developed muscle layer consisting of two ring-shaped and one longitudinal layer of smooth muscle tissue (Fig. 10).

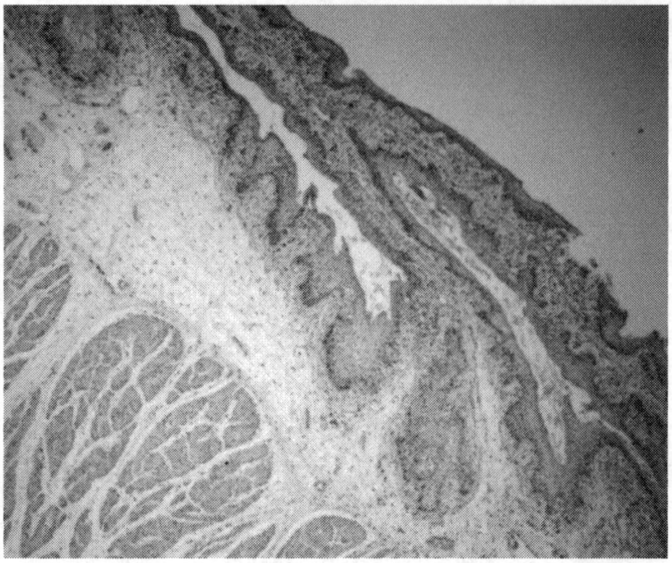

Fig. 10. Histological structure of the oesophageal groove wall Coloring: hematoxylin and eosin. Eyepiece ×10, objective×40.

The mucosa of the scar is sent out by MANE, which, together with the submucosal base of loose fibrous connective tissue (RVST), containing a large number of elastic fibers, forms protrusions (folds), inside which there is a cluster of lymphoid cells. The thickness of the mucosa varies from 0.09 to 0.13 mm, the submucosal base-from 0.17 to 0.60 mm, the height of the villi in the folds – 0.51-0.85 mm. The muscle membrane is represented by two layers: a well-developed internal circular and an external longitudinal one. The serous membrane is represented by RVST and mesothelium (Fig. 11).

The mucosa of the mesh is represented by the MPNE, which, together with its own plate of the mucosa with a thickness of 0.03-0.18 mm, forms a rimmed surface (Fig.12).

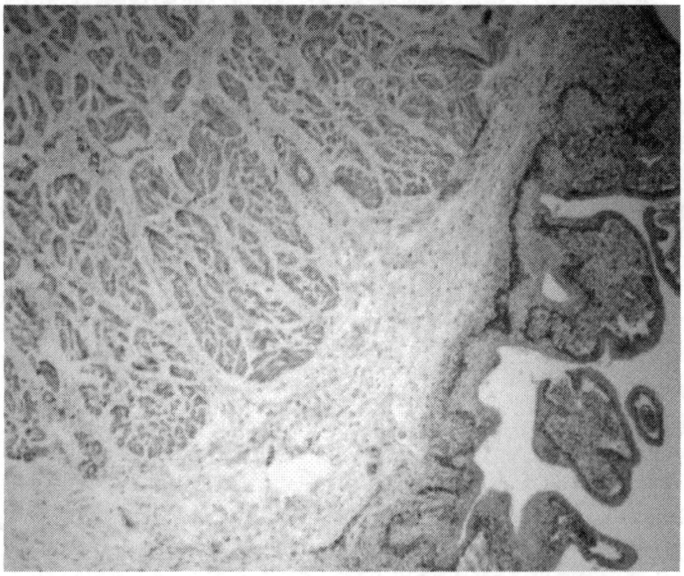

Fig. 11. Histological structure of the oesophageal groove wall Coloring: hematoxylin and eosin. Eyepiece ×10, objective×40.

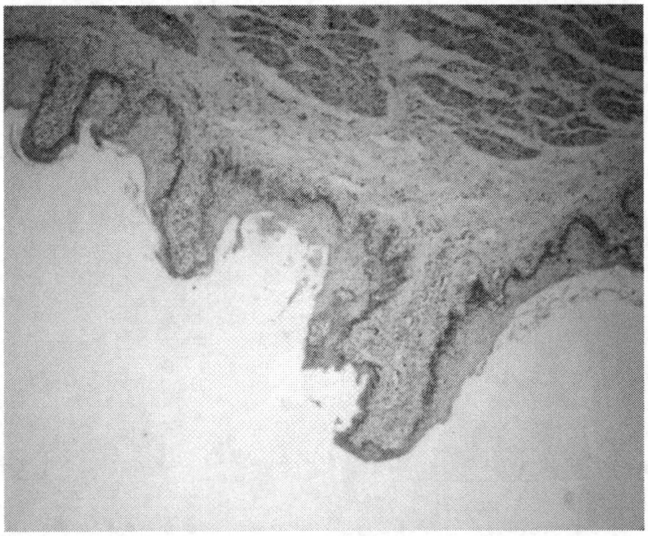

Fig. 12. Histological structure of the reticulum wall. Coloring: hematoxylin and eosin. Eyepiece ×10, objective×40.

The submucosal base is RVST with a thickness of 0.24-0.37 mm. Under them there is a well-developed muscle layer consisting of two annular and one longitudinal layer of smooth muscle tissue, outside there is a serous membrane consisting of RVST and mesothelium. At the same time, the epithelium in the area of the tip of the protrusion is exfoliated more intensively than at the base of the folds. The mucous membrane of the book, like other chambers of the pre-ventricles, was sent to ME, its thickness is from 0.08 to 0.11 mm (Fig. 13).

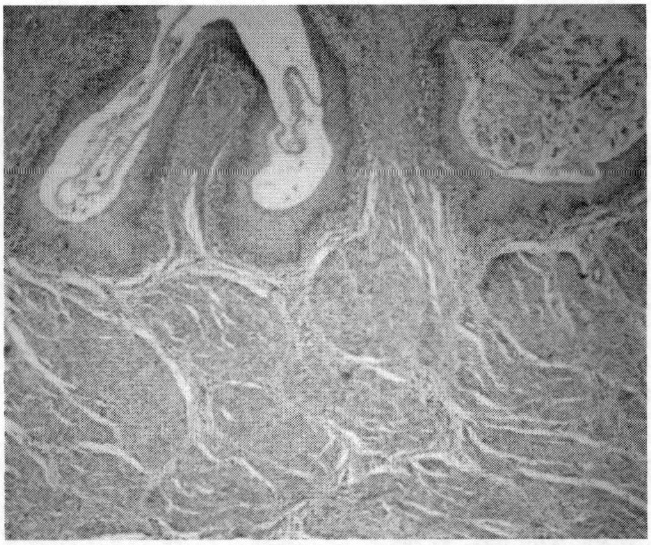

Fig. 13. Histological structure of the omasum wall. Coloring: hematoxylin and eosin. Eyepiece ×10, objective×40.

The epithelium and its own plate of the mucous membrane (SPZO) with a thickness of 0.07-0.13 mm form numerous, well-developed longitudinally oriented protrusions with a height of 0.43 to 0.75 mm. As in the grid, the epithelium on the tops of the folds is sloughed off more intensively. The height of the epithelium at the top is 0.04 mm, at the base-0.18 mm.

The muscle layer is represented by two ring-shaped and one longitudinal layer of smooth muscle tissue with a thickness of 0.89-1.08 mm, the outer, serous membrane, with a thickness of 0.05 mm, consists of RVST and mesothelium.

The mucous membrane of the abomasum is lined with a single-layer prismatic epithelium, in the SPSO there are lymphocellular clusters and bottom glands (fundal, pyloric and cardiac). The thickness of the mucous membrane (epithelium and MTR) was 0.68-0.72 mm, the submucosal base is represented by a loose connective tissue (PCT) with a thickness of 0.23-0.32 mm. The muscle membrane is 0.98-1.05 mm thick, consists of three layers of smooth muscle tissue: oblique, circular and longitudinal. The serous membrane is 0.05 mm. In the pyloric part of the abomasum in the area of the sphincter, the wall forms a powerful finger-like protrusion consisting of a mucous membrane 0.78-1.17 mm thick and a submucosal base 1.19 mm thick (Fig. 14-15).

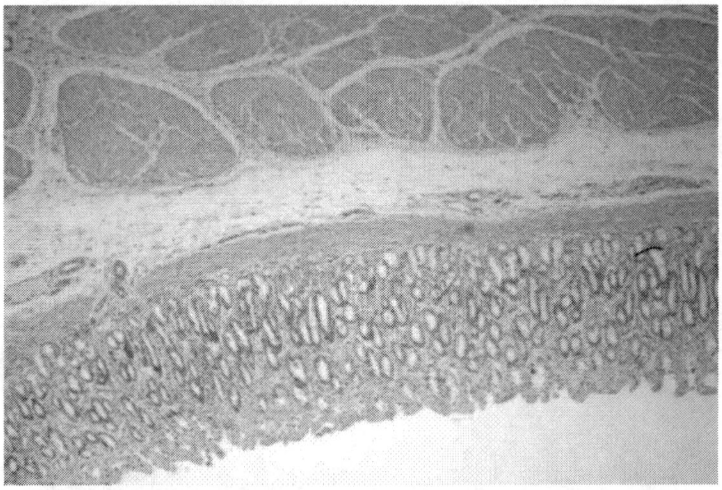

Fig. 14. Histological structure of the abomasum wall. Coloring: hematoxylin and eosin. Eyepiece ×10, objective×40.

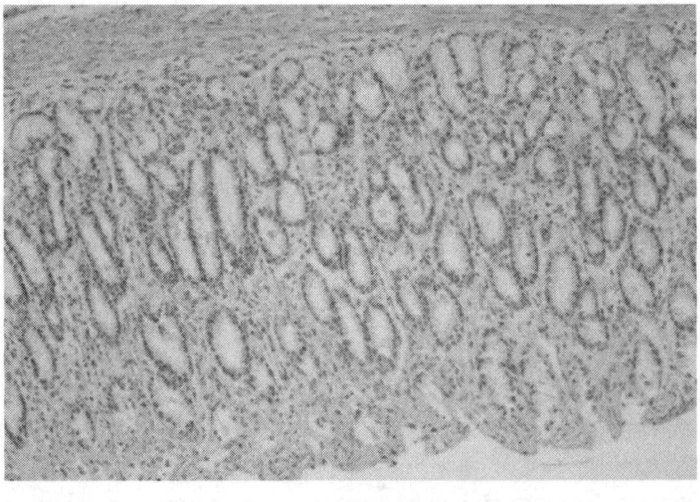

Fig. 15. Histological structure of the abomasum wall. Coloring: hematoxylin and eosin. Eyepiece ×10, objective×100.

The mucosa of the duodenum 12 forms villi lined with a high prismatic epithelium (Fig. 16).

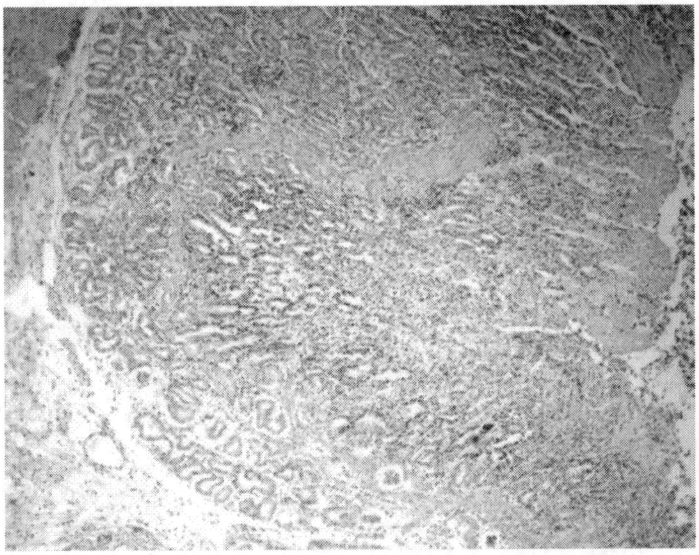

Fig. 16. Histological structure of the duodenal wall. Coloring: hematoxylin and eosin. Eyepiece ×10, objective×40.

The epithelium contains a large number of edged, goblet-shaped and endocrine cells. Going deeper into the SPSO, the epithelium forms crypts. The muscular plate of the mucous membrane (MPSO) is represented by several layers of smooth muscle cells. The thickness of the mucous membrane is 0.53-0.91 mm, the submucosal base is 0.72-0.78 mm. The latter is represented by PCT with well-developed duodenal glands of a complex tubular structure, which is typical only for the duodenum 12. The muscle membrane consists of circular and longitudinal layers with a thickness of 0.99-1.17 mm. The thickness of the serous membrane was 0.27-0.38 mm.

The mucous membrane of the jejunum is represented by a single-layer cylindrical epithelium, 0.03 mm thick (Fig. 17).

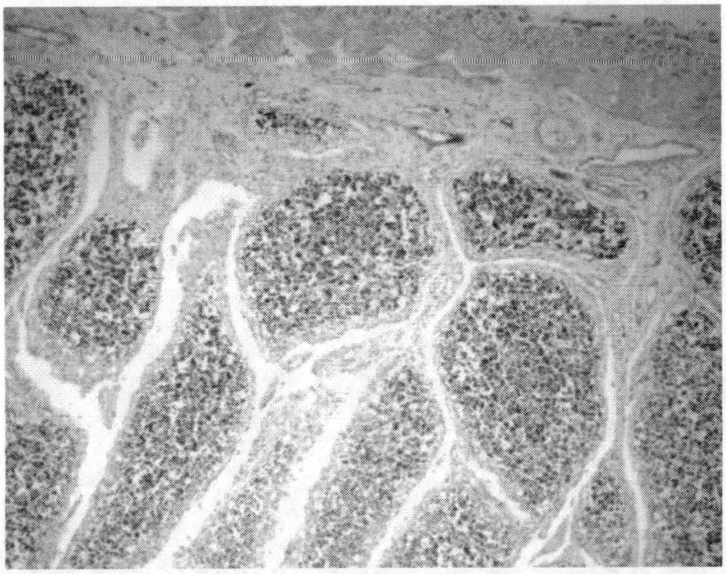

Fig. 17. Histological structure of the jejunum wall. Coloring: hematoxylin and eosin. Eyepiece ×10, objective×40.

The muscle membrane has a smaller thickness compared to the previous section and was 0.20 mm. The submucosal base is represented by PCT with numerous blood, lymphatic vessels and nerve plexuses with a thickness of 0.26-0.50 mm. The regional lymph node is covered with a capsule with a thickness of 0.045-0.055 microns from the surface, under the capsule is a cortical substance; the follicles are not morphologically formed, they are defined as denser clusters of lymphocytes without signs of a reactive center, reticular tissue is visualized. The border of the cortical and medulla is indistinct, the latter is blood-filled, the "loose" location of lymphocytes is noted.

The mucous membrane of the colon forms folds with well-developed simple unbranched tubular glands-crypts. The single-layer cylindrical epithelium of crypts is dominated by goblet-shaped cells. The height of the mucous membrane is 0.175-0.215 mm, the submucosa is 0.129-0.155 mm. In the submucosal base, there are lymphoid follicles with a size of 0.053×0.213 mm. The thickness of the muscle and serous membranes is 0.59-0.91 mm and 0.05 mm, respectively.

The mucous membrane of the rectum forms weakly pronounced folds, sent by a prismatic single-layer epithelium, in the crypts – goblet-shaped cells (Fig. 19).

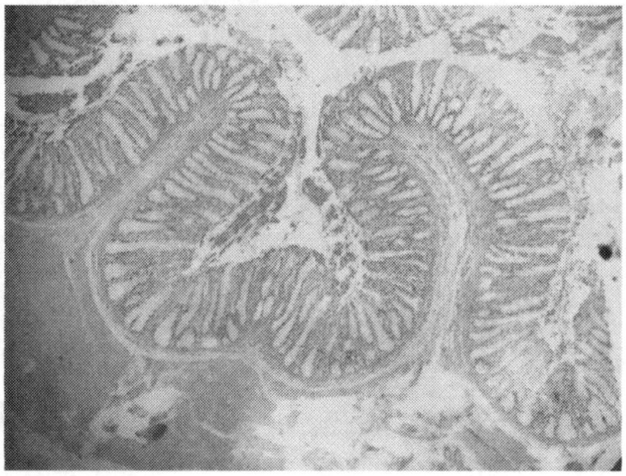

Fig. 18. Histological structure of the middle intestine wall. Coloring: hematoxylin and eosin. Eyepiece ×10, objective×40.

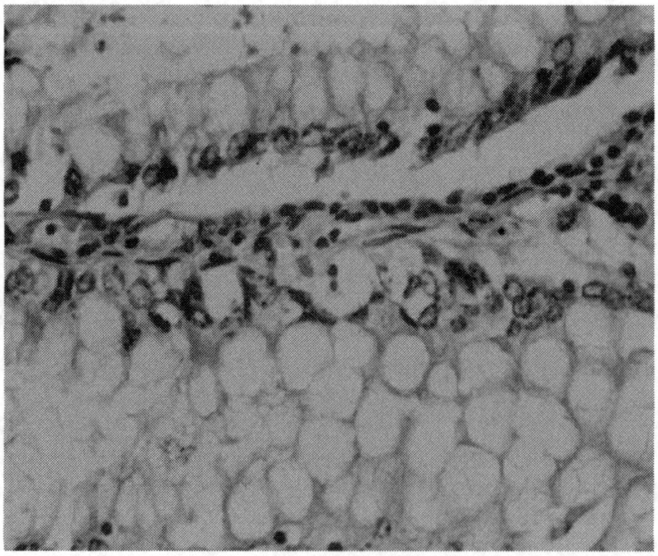

Fig. 19. Epithelium of the rectum. Coloring: hematoxylin and eosin. Eyepiece ×10, objective×400.

The height of the mucous, submucosal and muscular membranes is 0.33-0.54 mm, 0.06-0.13 mm and 0.50-0.85 mm, respectively. There are no lymphoid elements – sollitic follicles – in the submucosal layer. From the outer surface – Adventitia from the PCT.

4 Discussion

Individual development of an organism implies a complex of genetically programmed morphological, biochemical and physiological changes [8, 9, 10, 11, 12]. In seven-day-old calves, the gastrointestinal tract is not functionally and morphologically formed, there are no enzymes involved in the digestion of coarse and concentrated feeds, the "esophageal gutter reflex"is expressed. Only with age, as a result of unique transformations of the digestive system, as the scar develops, its colonization by microorganisms, the calf physiologically passes from a pseudomonogastric animal to a functioning ruminant animal [7].

Feeding the calf with concentrated feed caused thickening and erosion of the esophageal gutter rollers, accumulation of dense consistency contents in the rumen, mesh and book, hyperemia of the mucous membranes. In addition to redness, thinning and rupture of the abomasum wall, the calf has a red-cherry-colored mucous membrane of the small intestine, covered with a thick, viscous mucus with brown veins, the mucosa of the thick section is thickened, covered with a viscous mucus of gray color. Mesenteric lymph nodes are enlarged, of a dense consistency.

During microscopic examination, the mucous membranes of the digestive tract were sent out by a multilayer flat non-corneating epithelium. The thickness of the epithelium and submucosal layer, respectively, in the esophageal gut, scar, mesh, book, rennet, duodenum, colon, rectum was 0.18 mm and 0.27-0.55 mm; 0.09-0.13 mm and 0.17-0.60 mm; 0.03-0.18 mm and 0.24-0.37 mm; 0.68-0.72 mm and 0.23-0.32 mm; 0.08-0.11 mm and 0.07-0.13 mm; 0.53-0.91 mm and 0.72-0.78 mm; 0.33-0.54 mm and 0.06-0.13 mm. In the grid and the book, a more intense peeling of the epithelium at the tops of the folds was found.

Villi were found in the folds of the scar, numerous well – developed longitudinally oriented protrusions were found in the book, lymphocellular clusters and bottom glands were found in the rennet; folds with well – developed crypts were found in the colon. The muscle layer is well developed in all parts of the gastrointestinal tract.

5 Conclusion

As a result of the passage of coarse feed, the deceased calf has catarrhal erosive lesions in the esophageal trough; serous-catarrhal abomasitis with signs of hemorrhagic with dilation and perforation in the cardiac part; hemorrhagic omasitis against the background of a book blockage; catarrhal reticulitis; catarrhal ruminitis and catarrhal enterocolitis.

Acknowledgements

The study was carried out at the expense of the federal budget by order of the Ministry of Agriculture of Russia in 2020.

References

1. H. B. Baymishev, B. V. Krishtoforova, V. V. Lemeshchenko, *Biological foundations of veterinary neonatology: monograph,* 452 (2013)
2. V. P. Degtyarev, S. V. Fedorov, G. M. Udalov, *Neonatology and diseases of newborn calves: monograph*, 157 (2016)

3. A. A. Kataranov, *Clinical and immunological characteristics of newborn calves and non-drug methods of correction of their immunodeficiency,* 24 (2005)
4. Zh. N. Miholap, *Normalization of morphohistochemical abnormalities in the digestive organs of calves with dyspepsia with the proobiotic drug "Intenstevit",* 127 (2007)
5. L. P. Teltsov, T. A. Romanova, I. V. Dobrynina, E. O. Mikhailovskaya, A. G. Krasovskaya, I. G. Muzyka, A. I. Shadrina, Proceedings of the Orenburg State Agrarian University, **20-1(4),** 73 (2008)
6. D. N. Kharitonik, G. A. Tumilovich, *Morphofunctional changes in the body of young cattle and poultry against the background of the use of mineral-vitamin and probiotic drugs: monograph,* 219 (2019)
7. Q. Diao, R. Zhang, T. Fu, Animals (Basel), **9(8),** 490 (2019)
8. P. Gorka, Z.M. Kowalski, P. Pietrzak, A. Kotunia, W. Jagusiak, R. Zabielski, Journal of Dairy Science, **94(6),** 3002 (2011)
9. J. M. Vivo, A. Robina, S. Regodón et al, Histology and Histopathology, **5(4),** 461 (1990)
10. K. K. Knott, P. S. Barboza, R. Terrybowyer, Journal of Mammalogy, **86(1),** 121 (2005)
11. K. Dieho, A. Bannink, I.A.L. Geurts et al., Journal of Dairy Science, **99(3(1)),** 2339 (2016)
12. V. M. Shpygova, O.V. Dilekova, A.N. Kvochko, V.A. Meshcheryakov, V.V. Mikhaylenko, Dusunen Adam, **10(1),** 2063 (2019)

BIO Web of Conferences **42**, 01009 (2022) https://doi.org/10.1051/bioconf/20224201009
SDGE 2021

Effective use of innovative technologies in mixed sowing of annual crops

E. Yu. Podlasova*, and *S. V.* Lebedev

Federal Research Centre for Biological Systems and Agricultural Technologies of the Russian Academy of Sciences, Orenburg, Russia

Abstract. It is impossible to organize proper feeding of farm animals without a detailed study of the composition of the nutritional content of feed. Legume-grass mixtures occupy one of main places in solving the problem of high-quality feed production. Since mixed crops of legumes and cereals retain high fodder quality of green mass due to high protein content in legumes. To meet the need for a high-quality feed base, it is necessary to look for ways to increase the yield of fodder crops. The use of innovative technologies is one of the components of modern crop production. The aim of the study was to study the methods of seed treatment in conjunction with the technology of mixed crops; it allows obtaining a green mass with the desired productive properties. Thus, the article presents the materials of a field study on the development of a highly productive green feed. The results of the study of combined and mixed crops on the nutritional value and productivity of annual forages of peas, millet, and barley using pre-sowing seed treatment with ultrafine particles of SiO2; MoO2; Fe3O4 and growth regulators AgroVerm and RibavExtra were obtained. As a result of the experiment, it was shown that the use of pre-sowing treatment of seeds with ultradispersed SiO2 particles in combined crops showed the maximum leaf surface area of 85.5 thousand m2/ha. By the time of harvesting, the safety of plants on variants with the use of SiO2; Fe3O4; MoO2 was 88.4%; 87.5%; 86.5% respectively. According to the collection of digestible protein, variant with the use of ultrafine particles SiO2 was 20% in both sowing methods. Low protein content was 16.1%; 16.4% in the control variants respectively. According to the content of feed units in 1 kg of dry matter, SiO2 variants prevail with 0.91. Thus, the research is the basis for possible studying of combined and mixed crops with the use of pre-sowing seed treatment with ultrafine particles. It allows obtaining high-quality feed.

1 Introduction

To increase the production of high-quality complete feed, multi-species crops play the main role. The main source of high-quality green forage is the annual sowing of leguminous crops. We obtain different and the cheapest fodder from them. Sowing with a mixture of leguminous crops provides more stable yields, since the decrease in the yield of one crop compensates in another.

*Corresponding author: katerina.pryakhina@mail.ru

With the development of nanotechnology, science and technology are connected with a wide variety of areas of human activity. With the extensive development of innovative technologies, nanometals acquire new properties and biological effects that were not inherent in them, penetrate into cell and nucleus, increase bioavailability. [1]. In addition, the unique properties of ultrafine particles and their use in scientific and technical field have recently enriched this area of research. [2]. The development of agricultural production is an excellent task in agriculture with a minimum use of agrochemicals taking into account the conditions of growth and plant protection.

Ultrafine particles of iron oxide influence on growth and content of chlorophyll in corn plants [3]. Pretreating seeds with ZnO can alter the nutrient content of seeds; it affects plant growth, yield and quality. [4] In addition, ultrafine particles of copper and zinc helped to increase the resistance of two wheat varieties at the seedling stage by stabilizing photosynthetic pigments, thereby increasing the water content in the leaves. [5]

To some extent, high or low concentrations of ultrafine particles have opposite functions in plant development, they can inhibit or stimulate the plant. [6].

2 Study objective

Thus, the use of pre-sowing seed treatment in conjunction with the technology of mixed crops will allow to obtain green mass with the desired properties and productivity.

3 Materials and methods

Field test research was carried out in 2020 in the central zone of Orenburg region in village Nezhinka. The experiment studied the sowing of mixtures of annual forages (peas "Yamal" + barley "Natali" + millet "Orenburgskoe 27" in a ratio of 1: 2: 1) with combined and mixed sowing. The experiments were made in 4 repetitions, the site was randomized, the rigistered plot area was 1 m^2. In combined sowing, crops were in alternating rows or stripes. Before sowing, seeds were not mixed, but were sown separately with a row spacing of 15 cm. The following seed treatments were studied: SiO2 UFPs with a size of 30.7 ± 0.3 nm and a ζ-potential of 27 ± 0.12 mV, MoO2 NPs (100-120 nm) produced by Plasmoterm (Russia, Moscow, http://plasmotherm.ru), Fe3O4 NPs (80-100 nm, ζ-potential 20 ± 0.14 mV), purchased from Advanced Powder Technologies (Tomsk, Russia, www.nanosized-powders.com). Microfertilizer consumption was 100 g/ton. In the experiment, we used preparations based on biohumus AgroVerm ("Biokenetics" http://rusgumus.ru/) at a dose of 1 l per 1 ton; Ribav Extra (https://www.agroxxi.ru/goshandbook/prep/ribav-ekstra-p-2.html) at a dose of 10 ml per 1 t. The seeds of the mixture components were sown in rows with pre-mixed seeds. The components are mixed at the moment of treatment with ultrafine particles of SiO2 with a size of 30.7 ± 0.3 nm, MoO2 NPs (100-120 nm), Fe3O4 NPs (80-100 nm, ζ-potential 20 ± 0.14 mV). Concerning mixed seeds, the sowing rate is calculated for each component separately. Laboratory analysis was performed in 4 repetitions. Statistical processing of data was carried out in an Excel program. The data obtained are presented as arithmetic mean values for each option.

4 Results and discussion

The main condition for the formation of highly productive agrophytocenoses is the creation of an optimal plant density. It has a significant effect on growth processes, plant height, and yield structure [8]. The emergence of seedlings was registered 8-11 days after sowing. More even shoots were observed when ultrafine particles of SiO2 and MoO2 were used 8-9 days after sowing, the condition for the formation of high fodder productivity is the formation of optimal plant density, it has a significant effect on growth processes, plant height and weight, crop structure, timing of the phases growth. Pretreatment of seeds with ultrafine particles and growth regulators promoted an increase in the number of preserved plants by the time of harvest as compared to the control. Plant safety using SiO2; Fe3O4; MoO2 was 88.4%; 87.5%; 86.5% respectively.

Table 1. Density of standing and height of plants in mixed crops of annual forages.

Varant	Sowing date	Average number of plants at germination, pcs /m2	Density of standing of plants before harvesting, pcs / m2	Plant safety, %	Height of plants, cm
Combined sowing (peas + millet + barley)					
SiO_2	08.05	93+169+236	82+158+200	88,4	44+40+73
MoO_2	08.05	96+166+240	81+149+204	86,5	42+39+75
Fe_3O_4	08.05	92+170+242	78+145+218	87,5	40+37+72
Agroverm	08.05	93+162+233	76+147+212	89	40+34+72
Ribav Extra	08.05	90+153+229	71+128+211	86	29+33+65
Control	08.05	80+150+228	60+125+204	84	29+35+62
Mixed sowing (peas + millet + barley)					
SiO_2	08.05	57+61+172	40+52+144	81,4	43+41+74
MoO_2	08.05	61+50+169	48+41+139	81,4	41+40+70
Fe_3O_4	08.05	50+75+166	35+66+140	82,8	40+33+70
Control	08.05	45+75+162	33+46+124	71,9	29+33+60

Note: * – Differences with control are significant at $p \leq 0.05$

By the time of mowing ripeness at combined sowing, the average height of plants after seed treatment with SiO2 and MoO2 reached a height of 42-44 for peas, 39-40 for millet, and 73-75 for barley. The maximum height for mixed sowing was registered in variants with SiO2; pea was 43, millet was 41, barley was 74. Control both in combined and mixed sowing was characterized by the lowest height among the studied variants, it was 29 for peas; 33-35 for millet; 60-62 for barley, respectively.

Table 2. Leaf surface area of sowing according to the variants of the experiment, thousand m2 / ha.

Harvest	Yield		Harvest	
	Green weight t / ha	Dry substance t / ha	feed units	protein %
Combined sowing				
SiO_2	24,8±1,3	4,9±0,12	0,91±0,016	20,2±1,09
MoO_2	23,7±1,7	4,7±0,15	0,89±0,011	17,04±1,08
Fe_3O_4	22,7±1,6	5,1±0,12	0,88±0,011	18,8±0,75
Agroverm	23,9±2,9	4,3±0,09	0,88±0,059	18,9±0,85
Ribav Extra	21,6±1	4,0±0,21	0,77±0,023	17±0,89
Control	21,6±0,3	4,0±0,39	0,77±0,024	16,1±0,85
Mixed sowing				
SiO_2	20,3±1,7	5,1±0,12	0,91±0,017	20±1,29
MoO_2	19,8±0,9	4,8±0,16	0,88±0,017	18,2±1,11
Fe_3O_4	19,5 ±0,9	5,1±0,12	0,88±0,030	19,1±0,63
Control	18,1±0,9	3,9±0,15	0,79±0,028	16,4±0,74
Note: * - $P \leq 0,05$				

Leaf area is one of main factors in increasing the yield of green mass. The maximum leaf area was observed in combined crops in the SiO2 variant - 85.5 thousand m2/ha, and with mixed sowing the highest value was shown in the SiO2 variants 79.8 thousand m2 / ha and MoO2 - 78 thousand m2 / ha. Leaf area in the control variant of mixed sowing was minimal and amounted to 54.7 thousand m2 / ha.

Quality and nutritional characteristics of green fodder cannot be expressed by one factor. Such an assessment consists of the following indicators: mineral nutritional value, digestibility of nutrients, energy nutritional value, etc. It was established that the maximum yield of green mass at combined sowing was shown by the SiO2 variant - 24.8 t / ha, and by the yield of dry matter Fe3O4 - 5.1 t / ha. In mixed crops, the maximum value of the green mass yield is also registered in SiO2 variant -20.3 t / ha, and the highest dry matter content is noted in Fe3O4 variants; SiO2 and equals to 5.1 t / ha.

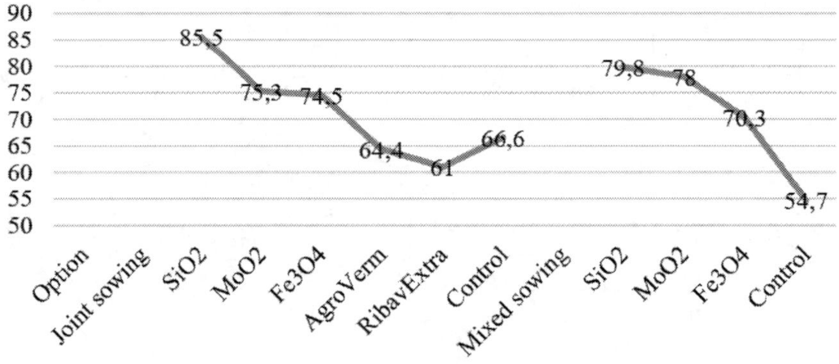

Fig. 1. Productivity of annual crops with combined and mixed crops.

The variant with the use of ultrafine particles SiO_2had 20% more digestible protein in both sowing methods, low protein content is registered in the control variants 16.1%; 16.4% respectively. According to the content of feed units per 1 kg of dry matter, SiO_2 variants with 0.91 prevail.

The main task for fodder production is to increase yields and improve the structure of sown areas by expanding multi-species sowing of protein crops with high-energy nutritional value and their mixture with cereals. It is better to cultivate peas in a mixture with cereals such as barley and millet, since they are located according to the vegetative structure and the location of the root system, thereby maximizing the use of soil fertility and environmental factors. [8]

It has been established that the pre-sowing treatment of seeds with ultrafine particles affects the nutritional value of feed, as well as the productivity of crops. Previously, the effect of ultradispersed Fe3O4 particles on wheat plants was shown; with their use, an increase in seed germination and leaf length was observed. [7] It was established that leaf area in combined sowing was at the highest level, and the largest area was observed with the use of ultrafine particles of SiO_2 - 85.5 thousand m2/ha. The cultivation of mixed and combined sowings of annual forages allows consistently obtaining the maximum yield of green mass with combined sowing, the SiO_2 option is 24.8 t / ha, and the dry matter yield Fe3O4 is 5.1 t / ha. In mixed crops, the maximum value of green mass yield was also recorded in SiO_2 variant -20.3 t / ha. The use of ultrafine particles of SiO_2 was registered to collect the digested protein in both sowing methods, it equals to 20% in the variant, the low protein content was registered in the control variants, 16.1%; 16.4%. According to the content of feed units per 1 kg of dry matter, variants with SiO_2 - 0.91 prevail.

5 Conclusion

In the conducted experiment, it was shown that the pre-sowing treatment of seeds with ultrafine particles had a stimulating effect on growth processes of peas, millet, barley, stimulating the germination energy and germination of seeds, productivity and the nutritional value of green forage.

Acknowledgements

The studies were carried out in accordance with the state task plan of the SSM2019-0004 NCB countries

References

1. L. Galaktionova, I. Gavrish, S. Lebedev, Toxicology and Environmental Health Sciences, **11(4),** 259 (2019)
2. N. A. Younes, M. F. A. Daud, A. A. Vardani, Chemosphere, **228,** 318 (2019)
3. N. Pariona, A. I. Martinez, H. M. Cdu-Garcia, L. A. Cruz, A. Hernandez-Valdez, Saudi J. Biol. Sci., **24,** 1547 (2017)
4. M. Rui, S. Ma, Ya. Bely, Yu. Hao, E. Yu. Wang, J. Tan H. Yang, F. Jiang, A. Ali, R. Yukui, W. Cao, Environ. Sci. Nano., **5,** 2088 (2018)
5. N. Taran, V. Storozhenko, N. Svetlova, L. Batsmanova, V. Shvartov, M. Kovalenko, Nanoscale Cut. Lett., **12,** 60 (2017)
6. G. Koo, J. Wang, Q. Zhang, Ch. Zhu, R. E. B. Shehab, L. Colvin, P. J. Alvarez, J. Braam, Environ Sci Technol., **49,** 626 (2014)
7. S. V. Lebedev, A. M. Korotkova, A. E. Osipova, Russ. J/ Plant Physoil., **61(4),** 564 (2014)
8. N. I. Kashevarov, D. Yu. Bashkaev, T. A. Sadokhina, Siberian Bulletin of Agricultural Science, **6,** 39 (2015)
9. A. S. Shpakov, N. M. Matveeva, M. I. Tarasenko, Adaptive forage production: problems and solutions, 170 (2002)

Prospective measures of control and prevention of Helminthosis among agricultural animals in animal complexes

G.R. Bayramgulova, *A.B.* Zulkarnaev, *S.M.* Muzafarov, *G.A.* Yagafarova, and *F.G.* Aminev

Federal State Budgetary Educational Institution of Higher Education Bashkir State University, Sibay Institute, Sibay, Russia

Abstract. Prospect of timely cleaning of livestock complexes from accumulated manure prevents environmental pollution. For this purpose, the most promising was the "cold" method of dehelminthization of a heap of manure, complete neutralization of chicken manure when heated and dried at temperatures above 100 °C, a "dry" method for neutralizing horse, pig, cow, sheep manure from helminth eggs using the Kranz biothermal method. For dehelminthization of humans and animals, science has recommended, and practice has successfully tested a fairly large, constantly growing and updated range of highly effective anthelmintics, mainly for gastrointestinal and partially for pulmonary helminthiasis.

1 Introduction

At present, untimely cleaning of accumulated manure and, as a result, environmental pollution is a huge threat to the fate of any livestock complex. In this regard, the issue of the need to accelerate development of scientifically based recommendations for its removal and neutralization arises. Development of animal husbandry at the present stage requires cultivation of farm animals free from helminthiases, and creation of appropriate sanitary and epizootic conditions on the farms, which can be achieved through disinvasion and modernization of livestock complexes (Bayramgulova, 2010).

2 Materials and Methods

The subject of the study is a review of effective methods of dehelminthization of farm animals, devastation of the environment of the livestock complex. To study the methods of devastation of the environment of the livestock complex, we studied the range of highly effective anthelmintics, well-known disinfectants, methods of dehelminthization of the external environment, heaps of manure in livestock complexes.

3 Results and Discussion

When solving the problems of cleaning, storing and using manure in livestock complexes, the specific biological characteristics of eggs and larvae of helminths shall also be considered. Being thrown out with feces from the body of animals into the external environment, eggs and larvae of a number of helminth species for several days (in spring and summer) remain non-invasive, not dangerous for infecting animals with helminthiases. The eggs and larvae of many helminths of farm animals have the ability to stay in feces, manure and water for several days and only after that reach the invasive stage. Having reached invasiveness, they take root in the body of animals, undergo further development in them until maturity. In the invasive stage, the larvae and eggs of such helminths are unusually resistant to desiccation, insolation, changes in heat and cold. Human roundworm eggs, for example, remain in the garden soil for up to 7 years or more without losing their viability during all this time (Vasilkova, 1950).

Strongylid larvae of horses remain alive for months in shallow water bodies, and in hay — for up to a year (Velichkin, 1952). The larvae of trichostrongylidae and dictyocaulus of sheep and cattle remain viable from three months to a year (Romanenko et al., 2000). Considering the great sanitary and epizootological significance of the ecological features of human and animal helminths, the method of dehelminthization offered by K.I. Scryabin (1946) in1925 distinguished from the methods of conventional therapy, namely, by the fact that it requires to neutralize not only the helminths released from the body, but at the same time their eggs. Later, the same author recommended to call dehelminthization also all methods of neutralizing eggs and larvae of helminths in manure, on pastures, in water bodies, soil, water, premises where animals are kept, even on fruits, vegetables and various household items, by any means.

For dehelminthization of humans and animals, science has recommended, and practice has successfully tested a fairly large, constantly growing and updated range of highly effective anthelmintics, mainly for gastrointestinal and partially for pulmonary helminthiasis. Unfortunately, anthelmintics for dehelminthization in tissue helminthiasis (cysticercosis of cattle and reindeer, coenurosis of sheep, larval echinococcosis) are still unknown. However, a scientific search in this direction is still underway. There is reason to believe that it will probably also succeed. The offensive and defensive methods of prophylaxis used against tissue helminthiasis of animals (Skryabin, 1945) are complex, in some places difficult to carry out and, as a result, are not very effective.

Considering the above, in a number of foreign countries — England, France, Yugoslavia, Germany, Poland, Australia and New Zealand, a number of researchers are very successfully conducting experiments on active immunization of cattle, sheep and rabbits with cysticercosis.

In Australia and New Zealand, active immunization of sheep against cysticercosis is already being widely adopted in sheep farming. Simultaneously with foreign authors, immunization of cattle with cysticercosis began in Russia (Romanenko et al., 2000). With our participation, experiments in vaccination of sheep against coenurosis were laid in Kazakhstan. As a vaccine, oncospheres, artificially freed from embryophores obtained from mature segments of taeniidaes of dogs and humans are mainly used.

According to the results of immunization of animals with tissue helminthiasis obtained in Russia and abroad, we believe that the time has come to significantly expand and deepen scientific studies on this issue in our country by the efforts of both central and peripheral research helminthological institutes and laboratories, departments of parasitology of some veterinary institutes and faculties. A further delay in implementation of this kind of study will undoubtedly aggravate the lag of Russian helminthology on this issue from world science.

Reliable chemicals for dehelminthization of the external environment, especially manure in pig and livestock complexes, are still unknown. Emulsions and solutions of well-known disinfectants widely used in practice — creolin, lysol, carbolic acid, bleach, mercuric chloride, and etc. — do not have a destructive effect on eggs and larvae of helminths (Skryabin, Petrov, 1964; Skryabin, Shultz, 1940; Velichkin, 1952). Even potassium cyanide does not kill Ascaris eggs (Zavadovsky, 1926). He explains this by presence of helminth eggs of an inner semipermeable lipoid membrane in the shell (Zavadovsky, 1915). It has been established that it is this that delays the permeability of solutions of disinfectants to embryos enclosed in eggs. The lipoid membrane is present in the shell, apparently, of helminth eggs of all types and classes. It was discovered by A. A. Skvortsov (1942) in the radial shell of oncospheres, A. A. Skvortsov (1936) — in an egg shell. R.S. Shultz and E.V. Gvozdev (1970), referring to the data of foreign authors, also indicate presence of a lipoid membrane in the structure of the embryophore of cestodes in general. According to M.M. Zavadovsky (1927), the lipoid membrane of the ascaris egg shell is destroyed by the action of fat-dissolving substances (gasoline, ether, acetone, chloroform, carbon tetrachloride), but alcohol does not penetrate through this membrane.

A.P. Markevich (1978) established that the lipoid membrane of the embryophore is shed from the oncosphere under the influence of intestinal juice on it with addition of bile, trypsin and some other agents.

Experimenting with eggs, K.I. Scryabin (1946) confirmed the data of other researchers that a 5 % solution of carbolic acid heated to 55-60 °C causes the death of horse roundworm eggs, and water heated to the same temperature has a detrimental effect on the eggs of the same nematodes. Exposure to water with a temperature of 100, 80 and 70 °C instantly kills the eggs of the horse roundworm. On this basis and the data of M.M. Zavidovsky and K.M. Sidorov (1926), VIGIS recommended a method for treating floors and walls in pigsties with boiling water, as well as cages for foxes and polar foxes (Skryabin, Petrov, 1964). However, in practice, this technique has not received wide recognition for a number of reasons: it is unsafe for performers, additionally humidifies the premises for pigs, and etc.

I.P. Gorshkov and E.E. Shumakovich (1935) used a "dry" method for neutralizing horse, pig, cow, sheep manure from helminth eggs using the Krantz biothermal method. With this method, manure is piled up on a layer of straw or dry neutralized manure in heaps, 1 m in size, loose to facilitate access to the manure for oxygen, and the top of the heap is covered with a layer of dry manure or soil. After a few days, horse or cow manure, pork manure mixed with horse manure, warms up under the influence of thermophilic microorganisms developing in it to a temperature of 55-60 °C. Under the influence of this temperature, the germ cells in the eggs of helminths die. However, aerobic microorganisms are still viable. In order to neutralize them in the warmed-up manure is additionally compacted, as a result of which oxygen access to it ceases and the aerobic microflora dies.

With the "cold" method of dehelminthization, heaps of manure with a volume of 1 m are also tamped tightly when laying and in this state are kept for 15-20 days in spring and summer, and in winter — up to 1.5-2 months. During these periods, anaerobic conditions are created and maintained in the thickness of the manure, under which helminth eggs die. Currently, the anaerobic process is used in practice: animal husbandry - when keeping chickens and cattle on a permanent manure bed, and sheep — on a dung mattress. On cattle сапры, bases and in shed, sheep feces are simultaneously pass dehelminthization under the influence of sheep urine (Gorshkov, Shumakovich, 1935).

However, the rapid neutralization of manure is most desirable for livestock complexes, and for this purpose no other factor other than high temperature can be recommended. The experiments of S.P. Gusev (1939) in this direction made it possible to achieve complete neutralization of chicken manure when heated and dried at temperatures above 100 °C. With this method of processing the chicken manure, a powdery mass that is safe for environmental pollution is formed, known as powder, a very valuable organic fertilizer and even a substance that can be used as feed for pigs and cattle. The powder contains 7-8 % nitrogen, 4-5 % phosphorus pentoxide, and 1-2 % potassium oxide. At one of the poultry farms near Moscow (Gorshkov, Shumakovich, 1935), the preparation of powder by drying the chicken droppings at high temperatures is used not only as a method of disinfection and dehelminthization, but also as an economically useful product, the production of which provides arrangement of an industrial enterprise without harmful waste.

For almost 20 years in our country, methods of removing, storing, neutralizing and using manure in pig farms have been studied. However, unfortunately, no complete solutions have yet been found that would satisfy livestock complexes in all respects. As a result, hundreds of pig and livestock complexes are forced to continue to use artificial concrete sedimentation tanks for manure (Romanenko et al., 2000).

Such sedimentation tanks do not stand up to criticism in sanitary, epizootic, nature conservation and economic terms. The manure in the settling ponds remains open. It poisons the surrounding atmosphere with fetid gases. By attracting insects and birds to itself, manure becomes a source of dispersal of infections and infestations. For a long time spent in the settling tanks, manure loses organics valuable for fertilization.

4 Conclusion

Therefore, the prospect of using timely cleaning of livestock complexes from accumulated manure prevents environmental pollution. For this purpose, the most promising was the "cold" method of dehelminthization of a heap of manure, complete neutralization of chicken manure when heated and dried at temperatures above 100 °C, a "dry" method for neutralizing horse, pig, cow, sheep manure from helminth eggs using the Kranz biothermal method. For dehelminthization of humans and animals, science has recommended, and practice has successfully tested a fairly large, constantly growing and updated range of highly effective anthelmintics, mainly for gastrointestinal and partially for pulmonary helminthiasis.

References

1. G. R. Bayramgulova, *Bioecological aspects of epidemiology, epizootology, prevention of intestinal invasions of humans and animals in the Republic of Bashkortostan,* 46 (2010)
2. Z. G. Vasilkova, *Fundamentals of sanitary helminthology*, 47 (1950)
3. P. A. Velichkin, The resistance of eggs and larvae of strongylids (delafondium, alfortium and trichonematine) to common disinfectants, **10,** 77 (1952)
4. I. P. Gorshkov, E. E. Shumakovich, Methods for dehelminthization of farm animal manure, **1,** 149 (1935)
5. M. M. Zavidovsky, K.M. Sidorov, Dependence of egg development on temperature. - Tr. Labor experimental biol. of Moscow zoo, **3,** 159 (1927)
6. N. A. Romanenko, I. K. Padchenko, N. V. Chebyshev, *Sanitary parasitology*, 354 (2000)
7. A. A. Skvortsov, Zool. magazine, **1-2,** 10 (1942)
8. K. I. Skryabin, *Total devastation — a new stage in improvement of humans and animals from helminthiases*, 6 (1945)
9. K. I. Skryabin, *Construction of Soviet helminthology*, 211 (1946)
10. K. I. Skryabin, A. M. Petrov, Fundamentals of Veterinary Nematodology, 566 (1964)
11. K. I. Skryabin, R. S. Shults, *Fundamentals of General Helminthology*, 470 (1940)
12. R. S. Shults, E. V. Gvozdev, Fundamentals of General Helminthology, **1,** 491 (1970)

New bacilli-based probiotic for aquaculture: efficacy study on *Macrobrachium rosenbergii*

Angelina Malkova[1,*], *Ivan* Evdokimov[1], *Maksim* Shirmanov [1], *Alena* Irkitova[1], and *Dmitry* Dementyev[2]

[1]EC "Prombiotech", Altai State University, 656049 Lenin Avenue 61, Barnaul, Russian Federation
[2]"Arsal" LLC, 658839 Gagarina Street 1C, Yarovoe, Russian Federation

Abstract. The data on the study of the effectiveness of a new probiotic based on strains *B. toyonensis* B-13249 and *B. pumilus* B-13250 when growing shrimps are presented. The experiments were carried out on a private farm in Kazakhstan. The control group of shrimps, which did not receive the probiotic, completed the molting process on the 28th day of the experiment, and the experimental group – 10 days earlier. It was also found that the probiotic has a positive effect on the hydrochemistry of water in recirculating aquaculture system (RAS). The concentration of ammonium (NH_3/NH_4^+), nitrite (NO_2) and nitrate (NO_3^-) nitrogen in the experimental group did not reach the limit value, in contrast to the control. Based on the obtained results, this probiotic can be recommended for greening aquaculture. The dosage of the probiotic when applied with brine shrimp as feed for shrimp is 1 g/100 g of dry cysts. The dosage of the probiotic when the RAS is added to the water is 5 g/600 l of circulating water once a week.

1 Introduction

Industrial breeding and rearing of aquatic animals (invertebrates and fish) on a large scale is accompanied by the spread of various infections. To prevent large losses during the reproduction of commercial objects, antibiotic substances have been used for many years, which led to the development of resistance to them in a large number of microorganisms. To prevent bacterial resistance to antibiotics from leading to environmental disaster, it is necessary to look for safer ways to fight microbes. In recent years, green agrobiotechnology and probiotics in particular have become increasingly popular. These preparations can be used prophylactically to prevent the development of infections [1].

Shrimp aquaculture is one of the fastest growing branches of aquatic animal husbandry, the total production of which is estimated at millions of tons per year [2]. The problems of breeding these crustaceans include water pollution, lack of quality feed and outbreaks of diseases of microbial etiology [3]. To address these latter challenges, researchers are actively exploring the effectiveness of beneficial microbial agents as a probiotic supplement for shrimp.

In the course of a series of experiments carried out by various scientists from Iran [4], Korea [5], China [6], India [7], Thailand [8], the island of Taiwan [9], the positive

* Corresponding author: gelishka96@mail.ru

dynamics of faster growth and digestibility of food in shrimp, as well as increased survival rates and immune response compared to control groups.

The probiotic preparations they added most often included various strains of the genus *Bacillus*. Modern scientists are trying to expand the range of bacillus species for use in aquaculture. Therefore, researchers are studying the safety and antagonistic activity of the species *B. pumilus* [10], *B. toyonensis* [11] and other microorganisms.

In view of all these facts, development of a multicomponent probiotic preparation based on antagonistically active strains of bacilli is relevant for development of organic agriculture, including aquaculture.

The aim of this paper is to study the effectiveness of use of a new probiotic preparation based on bacilli in the cultivation of *Macrobrachium rosenbergii* shrimp.

2 Materials and Methods

An experimental batch of probiotic was obtained on the basis of the Engineering Center Prombiotech, Altai State University. The multicomponent probiotic preparation is a powder consisting of lyophilized bacterial concentrates and maltodextrin. The active component of the biological product is a consortium based on patented antagonistically active strains *B. toyonensis* B-13249 and *B. pumilus B-13250* (Pat. No. 2693439 RF and Pat. No. 2694522 RF). The titer of bacilli in the finished preparation is not less than 1×10^{10} CFU/g [12].

The probiotic study was carried out on a private farm located in the Republic of Kazakhstan, in the vicinity of Karaganda. For the experiment, we used the offspring of a female Rosenberg shrimp (*M. rosenbergii*), which was divided into 2 groups: control (fed with *Artemia franciscana L.* incubated without the addition of probiotic) and experimental one (fed with brine shrimp incubated with addition of probiotic, i.e. "enriched" brine shrimp). Arsal LLC provided Artemia batch, and it was received from the Kuchuk Lake (Altai Krai, Yarovoe). Each group consisted of 200 shrimp larvae, which were raised in the same aquarium systems of 600 liters each.

Hydrochemical parameters (pH, temperature, O2, concentration of nitrogenous substances (NH_3/NH_4^+, NO_2, NO_3^-) were maintained within the normal range throughout the experiment [13].

Artemia nauplii were used as the only food. The probiotic was added at 20 hours of incubation of Artemia cysts, when the bulk of nauplii began to hatch, in an amount of 1 g per 100 g of dry cysts with a total incubation duration of 24 hours. In one pool about 20 g of cysts per day was added. When feeding shrimp at a later stage of development (about 15 days of rearing), nauplii were used for 48 hours of incubation. The probiotic was also added at 20 hours of incubation of cysts together with 0.5 ml of a suspension of live spirulina per 1 liter of incubation solution (5 ml per pool/day). Spirulina was introduced to increase the mobility of nauplii in order to attract the attention of shrimp larvae.

The shrimp molts were fixed visually so as not to injure the animals. If in the total mass was found at least one shrimp that had passed the molt, it was considered that the entire group had passed this molt. Because from the molt beginning, the whole group finishes it within 24 hours.

A homogeneous group of 1000 shrimp postlarvae, not involved in the previous experiment, was divided equally into experimental and control groups. Each group was grown for a month in a separate closed system with a volume of 600 liters. Wherein, a probiotic in a quantity of 5 g was added once a week to the water circulating in the RAS (circulating water). The shrimp were fed with dry food. During the month, measurements of the main nutrients – indicators of biological treatment - ammonium (NH_3/NH_{4+}), nitrite (NO_2) and nitrate (NO_3-) nitrogen were carried out on a regular basis. Measurements were carried out using drop tests (colorimetry for ammonium with Nessler's reagent, for nitrites and nitrates according to Bendschneider and Robinson [14].

3 Results and Discussion

According to the obtained data, in the experimental group, an earlier exit from the larval stage was observed – on the 18th day. In the control group – on the 28-th day (Fig. 1). The survival rate in both groups did not differ significantly and was more than 90 %.

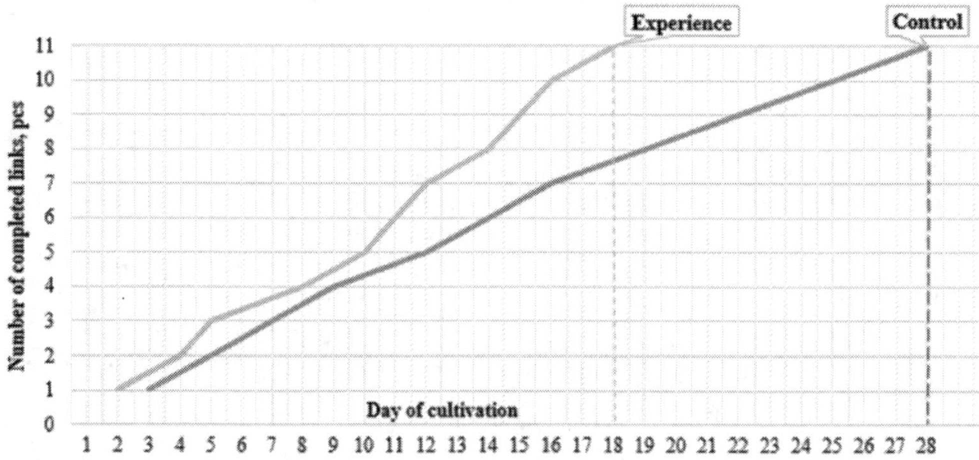

Fig. 1. The timing of the completion of the metamorphosis of *M. rosenbergii* larvae in the experimental and control groups

During the early stages of development, shrimp are demanding on the mobility of the food. In this regard, the technology of feeding shrimp prelarvae with live, newly hatched brine shrimp nauplii is widely used in aquaculture [15]. The incubation solution for *A. franciscana* is a very favorable environment for development of any, including unfavorable microflora that affects the quality of the final product. For this reason, breeders of brine shrimp are interested in adding probiotic and prebiotic preparations to their product to reduce the possible development of pathogens and transfer them to shrimp prelarvae [16]. The hatched Rosenberg shrimp larva goes through 11 different larval stages to become postlarva. According to the reference data, this process takes approximately 22–45 days [17]. According to the obtained data, in the experimental group of shrimps the molting process was completed in 18 days, which is 10 days faster than in the control group. This result indicates a positive effect of the probiotic preparation on the rate of passage of the larval stage in *M. rosenbergii*. This may be due to the antagonistic properties of bacilli in the probiotic in relation to shrimp pathogens, an increase in the efficiency of food assimilation by the larvae due to the synthesis of various digestive enzymes by microbes, and etc. However, more studies is needed to confirm these hypotheses.

In accordance with the data in Figure 2, the probiotic has a positive effect on the hydrochemical parameters of water. The concentration of ammonium (NH_4) and nitrite (NO_2) nitrogen in the control system by the end of the experiment reached the upper limit of the norm (0.5 mg/l), but not in the experiment. At a normal concentration of NO_3 nitrate up to 50 mg/l in the control unit on the 24th day of the experiment, a peak value of 60 mg/l was recorded. In the experimental installation this indicator did not exceed 40 mg/l.

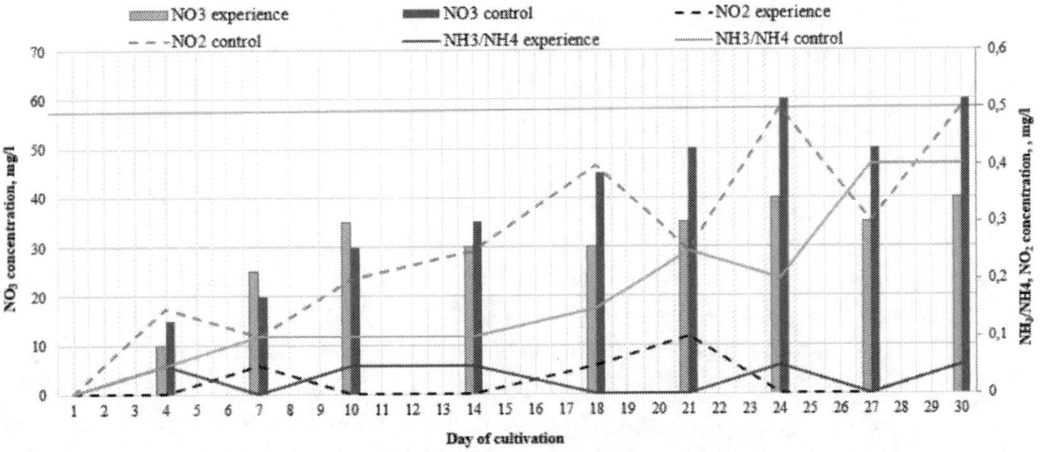

Fig. 2. Hydrochemical regime in experimental and control systems

Crustacean larvae are demanding on water quality and are classically controlled for ammonium, nitrites and nitrates. Since nitrogen-containing compounds in high concentrations have a detrimental effect on the development of shrimp [18]. The probiotic preparation under study reduces the concentration of toxic compounds in water as compared to the control. This contributes to improving the productivity of the system. Improvement of the hydrochemical parameters of water due to the addition of a probiotic to it indicates of the broader functional properties of the biological product.

4 Conclusions

Therefore, a new probiotic preparation based on strains *B. toyonensis* B-13249 and *B. pumilus* B-13250 promotes an earlier exit from the larval stage of *M. rosenbergii* (on day 18), as well as an improvement in the hydrochemical parameters of water in the RAS compared to the control. So, this biological product can be effectively used in agro-industrial complexes developing organic agriculture. The dosage of the probiotic when applied with brine shrimp as feed for shrimp is 1 gram per 100 grams of dry cysts. The dosage of the probiotic when the RAS is added to the water is 5 g per 600 liters of circulating water once a week.

References

1. L.V. Shulgina, E.V. Yakush, Yu.P. Shulgin, V.V. Shenderyuk, N.N. Chukalova, L.P. Baholdina, Izvestiya TINRO, **181(2)**, 216 (2015)
2. S. Tepaamorndech, K. Chantarasakha, Y. Kingcha, S. Chaiyapechara, M. Phromson, M. Sriariyanun, W. Visessanguan, Fish Shellfish Immunol., **86**, 4 (2018)
3. D. Srinivas, Ch. Venkatrayalu, Adv. Appl. Sci. Res., **7(2)**, 49 (2016)
4. S. Ziaei-Nejad, M.H. Rezaei, G.A. Takami, D.L. Lovett, A. Mirvaghefi, M. Shakouri, Aquac., **252(2-4)**, 516 (2006)
5. A. Sekar, M. Kim, H. Jeon, K. Kim, BB Rep., **19**, 100663 (2019)
6. K. Amoah, Q.C. Huang, B.P. Tan, S. Zhang, S.Y. Chi, Q.H. Yang, X.H. Dong, Fish Shellfish Immunol., **87**, 796 (2019)
7. S. Fernandes, S. Kerkar, J. Leitao, A. Mishra, Probiotics Antimicrob. Proteins, **11(4)**, 1309 (2019)

8. S. Tepaamorndech, K. Chantarasakha, Y. Kingcha, S. Chaiyapechara, M. Phromson, M. Sriariyanun, W. Visessanguan, Fish Shellfish Immunol., 86, 4 (2018)
9. Y.C. Wang, S.Y. Hu, C.S. Chiu, C.H. Liu, Fish Shellfish Immunol., **84**, 1050 (2019).
10. H.T, Truong, N.N. Tri, N. Ong, M. Quya, R. Fotedarc, K. Kannikad, S. Unajake, N. Areechon, Fish Shellfish Immunol., **60**, 391 (2017)
11. F. Feliatra, Nursyirwani, A. Tanjung, D.S. Adithiya, M. Susanna, I. Lukystyowati, IOP Conf. Ser.: Earth Environ. Sci., **116**, 012034 (2018)
12. A.V. Malkova, I.Yu. Evdokimov, M.V. Shirmanov, A.N. Irkitova, D.E. Dudnik, Proc Univ Appl Chem Biotechnol, **11(3)**, 393 (2021)
13. M.B. New, Farming freshwater prawns (Marlow, United Kingdom, 2002)
14. V.V. Sapozhnikov, A.I. Agatova, N.V. Arzhanova, N.V. Mordasova, N.M. Lapina, V.L. Zubarevich, O.N. Lukyanova, N.I. Torgunova, Guidelines for the chemical analysis of sea and fresh waters (RuFIR, Moscow, 2003)
15. T.D.N.J. Maldonado-Montiel, L.G. Rodriguez-Canche, Rev. biol. trop., **53**, 3 (2005)
16. H. Hamsah, W. Widanarni, A. Alimuddin, M. Yuhana, Jr. M. Zairin, AACL Bioflux, **10(1)**, 8 (2017)
17. M.B. New, W.C. Valenti, Freshwater Prawn Culture (Blackwell Science, Oxford, England, 2001)
18. W.J. Wasielesky, L.H. Poersch, T.G. Martins, K.C. Miranda-Filho, Braz. J. Biol., **77 (3)**, 558 (2017)

Application of photometry for diagnostics of nitrogen nutrition in spring barley

O. A. Shchuklina[1], and *R. A.* Afanasiev[2]

[1] Federal State Budgetary Institution of Sciences Tsitsin Main Botanical Garden of the Russian Academy of Sciences, Botanicheskaya st. 4, 127276 Moscow, Russian Federation
[2]All-Russian research Institute of Agrochemistry named after D.N Pryanishnikov, Pryanishnikova st. 31A, 127434 Moscow, Russian Federation

Abstract. The article discusses a method for diagnosing nitrogen nutrition of spring barley (*Hordeum vulgare* L.) in different phases of the growing season using a photometric device (N-tester) Yara. The studies were carried out in the conditions of the Moscow region (Russia) on the Mikhailovsky spring barley variety on sod-podzolic medium loamy soils with a high and medium content of phosphorus and potassium and a low content of humus (1.9). The agrometeorological conditions of the growing season of spring barley were characterized by sharp fluctuations in air temperature and the amount of precipitation over decades of the month and in comparison with average long-term observations. Abundant and prolonged precipitation during the grain ripening phase led to lodging in most of the experiments, which affected the yield. To diagnose crops, a model experiment was laid with the introduction of increasing doses of nitrogen fertilizers into pre-sowing cultivation with a step of 30, at which the dose of nitrogen ranged from 30 to 150 kg/ha. Diagnostics was carried out in three phases of vegetation: tube emergence ((Z42), earing (Z55), milk ripeness of grain (Z73) with the Yara N-tester and with the help of stem diagnostics according to the modified method of V. Zerling. The results of photometric diagnostics in the stemming phase (Z42) have a strong correlation with the yield of spring barley grain and with the results of stem diagnostics (R = 0.85). Wherein, the N-tester readings in the earing phase (Z55) and milk ripeness of grain (Z73) with yield R = 0.23 and R = 0.17, respectively, have a weak correlation. This is possibly due to lodging of crops and a change in yield, not as a result of poor plant nutrition, but with difficult mechanized harvesting.

1 Introduction

Modern agriculture, like many other industries, is moving along the path of automation. Devices that carry information about the relationship between the environment and the state of plants at a certain moment play an important role in this process. The most popular in recent years have received devices that allow quickly, without violating the integrity of agricultural plants to assess its physiological state by optical properties [1]. The measurement principle is associated with the photosynthetic activity of the plant, which correlates with provision of crops with nitrogen [2-3].

Plant chemical diagnostics makes it possible to determine the supply of plants with micro- or macroelements in the process of crop formation at one stage or another of development. It is tissue, in which fresh plant samples are analyzed without ashing in the laboratory, and leaf, according to which the total content of elements in the sample after ashing of leaves or other plant parts is determined [4, 5]. For leaf diagnostics, specialized laboratories are required, skills in working in them, and most importantly, a significant amount of time for analysis and decision-making. Rapid methods of tissue diagnostics have become widespread in the Soviet Union since the 30s of the last century [4]. However, application of plant diagnostics is complicated by the need for laboratory equipment or skills in working with acids in laboratory or field experiments.

The purpose of the study is to substantiate the efficiency of using the Yara photometric device for operational diagnostics of nitrogen nutrition in spring barley in the conditions of the Moscow region.

2 Materials and Methods

The studies were carried out in the field experiment of the Federal State-Owned Publicly-Funded Institution of Higher Education RGAU-Moscow Agricultural Academy named after K.A. Timiryazev (Moscow) in 2012-2015. The soils of the experimental site are well supplied with phosphorus (P_2O_5) - 361-483 mg/kg and potassium (K_2O) - 126-133 mg/kg (GOST 26207-91), an increased content of easily hydrolyzable nitrogen - 64-69 mg/kg (MU, 1985), the average humus content is 1.7-1.9 % (GOST 26213-91). A variety of spring barley *(Hordeum vulgare* L.) - Mikhailovsky was chosen as the object of study. The measurements were carried out with a portable optical instrument Yara (Konica Minolta, Japan). It measures the concentration of chlorophyll in leaves at two wavelengths using artificial light that is absorbed by chlorophyll at different intensities. The light not absorbed by chlorophyll is captured by the photodiode and converted into a measured value, which are conventionally referred to as "cu". To obtain a representative sample of one section of the field, it is necessary to take 30 measurements. For this, typical plants are taken, on which, respectively, measurements are taken on the youngest, well-developed leaves. In this case, sometimes individual values appear, which deviate significantly from the average. In this case, they are rejected by the device and are not taken into account in the calculation of the average. A defect in a single measurement may be due to incorrect positioning of the device on the sheet, damage to the sheet, or its strong contamination. N-tester readings are not affected by: time of day, measurements, pesticide treatment, moisture on the leaves. The Timiryazevskaya 42 variety is characterized by a small flag leaf, the area of which, however, is sufficient for taking readings with a Yara optical device. The first measurement was carried out in the phase of entering the tube (Z41-Z43), and the subsequent ones after reaching the next full phase of development (Z55, Z65, Z73). Simultaneously with photometric diagnostics, plant (stem, tissue) diagnostics were carried out according to the modified method of V.V. Zerling [4]. Evaluation takes place on a 3-point scale directly in the field. To detect the amount of NO_2, NO_3 ions in the juice of living plants, on a piece of which 0.5 g of diphenylamine dissolved in 100 ml of concentrated H_2SO_4 is applied.

The experiment scheme was developed to study the readings of photometric devices at different levels of fertilizing with nitrogen fertilizers. It included control – without the introduction of root fertilizing and the introduction of increasing doses of nitrogen fertilizers (NH_4 NO_3) from N_{30} to N_{150} kg r.a./ha for seedlings. Fertilizers were applied manually with an even distribution over the plot. The accounting area of the plots is 10 m^2, the replication is fourfold, the placement of options in all the years of the study is randomized, the agricultural technology adopted for the zone. Statistical processing of the results of the field experiment was carried out by methods of variance and correlation analysis [6].

Agroclimatic conditions in the year of study according to the V.A. Michelson (Moscow) were characterized by heterogeneity. Significant variation in the mean ten-day air temperature and precipitation was noted during the growing season of spring barley. In general, the air temperature was 1-2 °C higher than the average long-term data. Precipitation in the first ten days of May and June exceeded the mean annual values by more than two times. Abundant rainfall in the second and third ten days of August led to lodging of barley in most options of the experiment, which negatively affected the grain yield.

3 Results and Discussion

When carrying out stem diagnostics in the tube exit phase (Z42), fairly high results were obtained in options of the experiment with fertilizing in doses of 90 kg/ha nitrogen, which amounted to 2.5-2.6 points. These indicators indicate that at the moment the plants are sufficiently supplied with nitrogen and do not need additional feeding. The maximum indicator is 3 points – the plant is supplied with nitrogen. [7]. In options with nitrogen doses of 30-60 kg/ha, results were obtained showing the need for plants in feeding (1.3-2 points). In the control, the indicators of stem diagnostics determined the acute need for additional fertilizing – 0 points. In the earing phase (Z55), the stem diagnostics indices changed insignificantly in comparison with the first measurement in the direction of decreasing. In the phase of milky grain ripeness (Z73), when green leaves remained on the plants only in the upper tier, the indicators of stem diagnostics significantly decreased and averaged 0.48 points (Table 1). When assessing the results of stem diagnostics in the late phases of the growing season, such indicators are normal, since the amount of nitrates in the juice of plants, as well as the content of juice itself, decreases [4].

Table 1. Results of stem diagnosis for spring barley, points.

Option	Productivity, t/ha	Stem diagnostics, point		
		Z43	Z65	Z73
Control	3.31	0.1	0	0
N_{30}	3.50	1.3	0.3	0.7
N_{60}	4.41	2.0	1.5	0.9
N_{90}	3.51	2.5	2.5	0.7
N_{120}	3.77	2.5	2.7	0.3
N_{150}	3.80	2.6	2.1	0.3
NSR_{05}	0.54	0.6	1.1	1.1

The results of photometric diagnostics with the Yara N-tester in three phases of spring barley development showed the highest results in the earing phase (Z65) – on average 483 (**Fig. 1.**). In the phase of milk ripeness, the readings of the device decreased relative to the readings obtained in the phase of entering the tube (Z42) in options with nitrogen doses of 30-60 kg/ha and slightly increased in options with nitrogen doses greater than 90 kg /ha, which suggests that the additional fertilizing was reflected in the length of the growing season, as a result of which the plants stayed green longer.

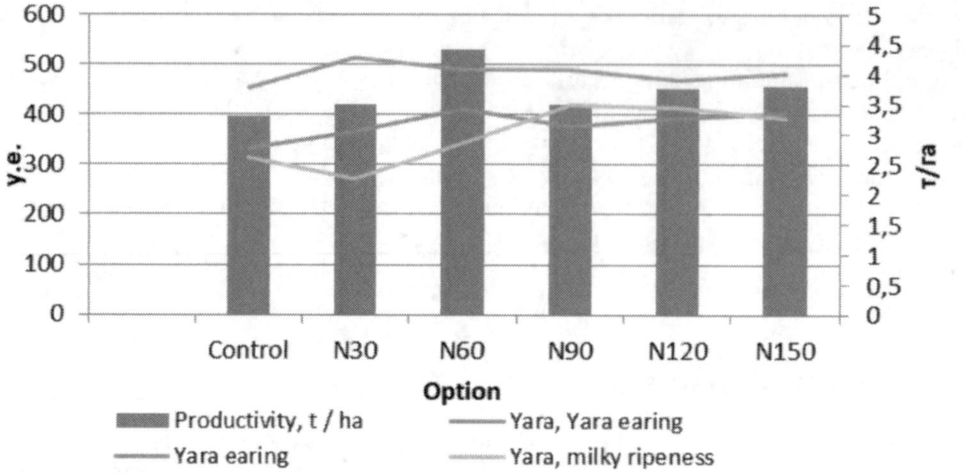

Fig. 1. Caption of the Figure 1. Below the figure.

Increasing doses of nitrogen fertilizers had a positive effect on the grain yield of spring barley. It ranged from 3.31 t/ha in the control, to 3.50-4.41 t/ha when applying top dressing. However, in spite of the actual increase in grain in all options of the experiment, only the increase in option N_{60} was statistically significant, which exceeded the grain yield in the control option by 33.2 %. The yield of barley grain was significantly affected by the lodging of crops.

4 Discussion

It is advisable to compare the indications of plant diagnostics during the growing season with the final yield [8, 9]. However, the value of the yield on the experimental plots was reduced due to lodging, which began during the period of abundant and prolonged precipitation in the phase of the beginning of milk ripeness of grain and in most options of the experiment amounted to 20 to 70 % of the crops. Lodging was not observed only on the plots of the first three options: control, N_{30} and N_{60}. When carrying out diagnostics in the stemming phase, the results of stem diagnostics had a weak correlation with the yield in the stemming phase (Z42) and in the earing phase (Z65) - R = 0.47 and 0.35, respectively. The average dependence in the phase of milk ripeness (Z73) - R = 0.57. The results of stem diagnostics during the exit into the tube (Z42) strongly correlated with the results of diagnostics by the Yara N-tester - R = 0.85. The same correlation coefficient was obtained between the Yara N-tester readings and the yield. Subsequently, the Yara N-tester readings had a very weak negative relationship with yield. This suggests that lodging in the crops did not affect the state of the plants, they continued to vegetate. However, it greatly affected the quality of mechanized harvesting.

5 Conclusion

Operational diagnostics using the Yara N-tester in a model experiment has shown its effectiveness at the early stages of plant development. The N-tester readings in the tube exit phase correlated with the readings carried out using the chemical method – stem diagnostics (R = 0.85). Lodging of crops makes its own adjustments to the level of yield, therefore, there was no correlation between the readings of the device and the yield in the later phases of the growing season. However, in the options of the experiment without fertilizers or with a low dose of nitrogen application (N_{30}), the N-tester shows the smallest indicators, which were 333 in the phase of entering the tube (Z42), in options N_{90} - N_{150} - 378-402; in the earing phase (Z65) - 453, in options N_{90} - N_{150} - 490-482; in the phase of milk ripeness (Z73) - 315, in options N_{90} - N_{150} - 418-394. To interpret the data of the photometric device, it is necessary to carry out test measurements in specific soil and climatic conditions.

Acknowledgments

The work was carried out within the framework of the State task of the MBG RAS (No. 19-119012390082-6).

References

1. F. A. Hamann, A. Fiebig, G. Noga, Canadian Journal of Agriculture and Crops, **5(2)**, 138-152 (2020)
2. I. Skudra, A. Ruza, Rural sustainability research, **37**, 29-37 (2017)
3. K.V. Belousova, M.O. Smirnov, V.V. Novikov, V.A. Litvinsky, O.A. Shchuklina, Fertility, **4**, 5-6 (2012)
4. O. Shchuklina, R. Afanasiev, I. Voronchikhina, I. Klimenkova, A.Komkova, Agronomy Research, **19(2)**, 595-600 (2021)
5. V. V. Tserling, *Diagnostics of nutrition of agricultural crops*, 236 (1990)
6. M. Almaliev, S. Kostadinova , G. Panayotova, Plant sciece, **50**, 10-14 (2013)
7. B. A. Dospekhov, Methods of field experience, 416 (1976)
8. K.A. Belousova, R.A.Afanasyev, E.V. Berezovsky, Agrochemistry, **7**, 78-87 (2015)
9. R. Hernández-Clemente, A. Hornero, M. Mottus, J. Penuelas, V. González-Dugo, P.J. ZarcoTejada, J.C. Jiménez, L. Suárez, L. Alonso, Current forestry reports, **5(3)**, 169-183 (2019)
10. O. Shchuklina, N. N. Langaeva, I. Voronchikhina, V. V. Voronchikhin, R. Afanasiev, IOP Conference Series: Earth and Environmental Science, 723, 022064 (2021)
11. A. Mecke, I. Lee, J.R. Baker jr., M.M. Banaszak Holl, B.G. Orr, Eur. Phys. J. E **14**, 7

Evaluation of Suppressiveness of chernozem upon the intake of organic fertilizers under the conditions of a model experiment

Lyudmila Galaktionova*, *Tatiana* Vasilieva, *Gulnur* Yamansarina, *Svyatoslav* Lebedev, and *Zhanara* Sulimenova

Federal Research Center of Biological Systems and Agrotechnologies of the Russian Academy of Sciences, Russia

Abstract. The problem of maintaining the sanitary function of soils of agroecosystems is one of the main ones, since it determines the danger for reproduction of various types of microorganisms, including human and animal pathogens. Within the study, the authors assessed the number of bacteria (Salmonella enterica and Escherichia coli) in samples of chernozem with introduction of organic fertilizers. The dynamics of the number of Salmonella enterica shows that organic matter in soils contributes to a sharp decrease in their number at the initial stages of biological pollution, and in the long term it contributes to total loss. By the 60th day, Salmonella was found only in the control option of the experiment, but continuation of the experiment showed their total loss on the 80th day. Introduction of organic ameliorants into the chernozem samples led to a decrease in the number of E. coli by the 80th day, but their total loss was not noted.

1 Introduction

It is known that microorganisms in agricultural soils have a significant effect on the level of soil fertility, especially in terms of the availability of plant nutrients and suppression of pathogenic microflora carried by the soil [1].

These key ecological soil functions will depend on critical constituents of the soil microbial community. In fact, the "health" of the soil can be defined in terms of its ability to counteract (suppress) pathogenic microflora – suppressiveness [2].

The topic of studying soil suppressiveness is actively discussed by scientists from all over the world. Among Russian scientists, the papers of E.Yu. Toropova, M.S. Sokolov, A.P. Glinushkin, who studied induction of soil suppressiveness, are highlighted, and among foreign scientists, the papers of G.A. Toranzos, R.P. Marcos, who studied bacteria pathogenic for humans, transmitted through the soil are highlighted.

Suppressiveness has been conceptually divided into "general" and "specific" one, where general suppressiveness is the result of the vital activity of many organisms. (for example,

* Corresponding author: anilova.osu@mail.ru

as a result of competition for habitat or nutrients), and specific suppressiveness is manifested in the antagonism of certain types of bacteria or fungi [3].

Approaches to increasing the suppressiveness of agricultural soils shall include the analysis and management of a wide range of factors [4].

Environmental factors that determine the "health" of a particular soil type include climatic variables such as rainfall, moisture and temperature, as well as physicochemical and biological activity indicators. Agronomic parameters can include types and varieties of crops, sowing dates, sequence of crop rotation and soil methods [5]. Other parameters may include the use of agrochemicals such as pesticides, growth regulators, fertilizers and ameliorants [6].

Microorganisms are another variable to consider when making agricultural decisions. The potential damage caused to plants by phytopathogenic microorganisms has a significant impact on the choice of crops and varieties of agricultural crops, crop rotations, planting density and timing, seed treatment and the list of agricultural chemicals used [7].

Studies aimed at identifying microorganisms involved in specific suppression face a number of methodological problems: first, the number of species inhabiting the soil is often very large, and estimates range from thousands to millions; secondly, most microorganisms are difficult to cultivate using standard methods and to assess their antagonistic interactions with pathogenic flora [8].

Alternatively, the grower may choose to knowingly use beneficial microorganisms when treating seeds or fertilizing the soil. For example, rhizobia and mycorrhiza can increase plant productivity, while many epiphytic microorganisms can enhance plant defenses [9].

While general suppression refers to suppression of pathogens resulting from the maximum amount of microbial activity, specific suppression refers to the antagonistic activity against a pathogen by individual microorganisms or a more narrowly defined group of organisms. One of the key points in implementing the potential of soils to reduce the number of pathogenic microorganisms is to determine their species, requirements for soil conditions and resistance [10].

Determining the number of both the pathogen and the organisms that suppress it will allow the development of more effective methods for predicting the potential danger of soils as a source of infection. The totality of the knowledge gained will make it possible to develop new strategies for combating pathogenic soil microflora by using various agricultural techniques, incl. application of organic and bacterial fertilizers. [11]. This determines the relevance of our paper.

The aim of the study was to assess the effect of organic fertilizers on the suppressiveness of chernozems in relation to *Salmonella enterica* and *Escherichia coli* under the conditions of a model experiment.

2 Materials and Methods

The objects of the study were 2 types of enterobacteria *Salmonella enterica* and *Escherichia coli*). The choice of study objects is determined by their high occurrence as agents of biological pollution of environmental components and well preserved in highly humic and mineral types of soils [12, 13].

Study materials: southern chernozem soils selected in the territory of the Park named after Perovsky, Orenburg (coordinates No:51.7718500, Eo:55.0934390). The following was used as organic fertilizers: soil conditioner Reasil Soil Conditioner, Super Gazon (NPO SILA ZHIZNI, Saratov) and biohumus AgroVerm (Manufacturer BioEraGroup, Moscow).

In the course of the work, we used classical microbiological methods: sampling method and preparation, soil treatment for bacteriological analysis; methods for isolating and accounting for bacteria in soil (R. Koch's method); method of limiting dilutions and sowing on solid nutrient media.

For the experiment, the selected soil was pre-dried, sieved and sterilized at a temperature of 120 °C within 40 minutes. Therefore, weighed portions of 27 g were taken and introduced into sterile Petri dishes. After that, solutions of organic ameliorants – soil conditioner and vermicompost at a dose recommended by the manufacturer were added to the soil. The soil without the addition of organic ameliorant served as a control.

Then the soils were kept for a week in a thermostat at a temperature of 25 °C, and then 10 ml of a suspension of bacteria (*Salmonella enterica* or *Escherichia coli*) at a concentration of 10^9 CFU/ml according to the turbidity standard (McFarland) was added to each Petri dish (Figure 1).

T H E S O I L	EXPERIENCE OPTIONS: 1. CONTROL (WITHOUT ADDING PREPARATIONS) . 2. BIOHUMUS. 3. SOIL REMOVER-SHITEL	ADD 10 ML (109 CFU / ML) BACTERIA SUSPENSION (SALMONELLA ENTERICA / ESCHERICHIA COLI) IN SIZE SOLUTION	SEEDING ON FOOD AND FOOD MEDIA FOR 10, 20, 40, 60 AND 80 DAYS

Fig. 1. Scheme of the experiment

After contamination, the soil was placed in a thermostat at a temperature of 25 °C and for the period of 10, 20, 40, 60, and 80 days, the number of bacteria was assessed by inoculation on elective media. BTN agar was used to isolate E. coli, and bismuth sulfite agar was used for Salmonella. Sowing was carried out in 7-10 fold repetition using standard identifiers.

3 Results and Discussions

3.1 Influence of organic fertilizers on some soil properties

The organic fertilizers taken for the experiments were characterized by a high content of organic matter, and the AgroVerm preparation was characterized by a sharply alkaline reaction of the medium (pH 11).

Organic fertilizers are an important factor in changing the content of soil organic matter, and also affect the complex of its agronomic properties. Changes in the chemical properties of soils with introduction of organic ameliorants affects the survival of pathogenic and opportunistic microorganisms [14, 15].

The results of determining the pH showed that introduction into the soil of biohumus AgroVerm and the soil conditioner Reasil Soil Conditioner, Super Gazon, did not lead to a change in the reaction of the soil solution (Table 1).

Table 1. Change in pH and humus content when adding ameliorants

Sample	pH_{H2O}	$C_{org.}$, %
Control	7.15±0.03	5.6±0.03
Biohumus	7.18±0.07	7.5±0.1
Soil conditioner	7.16±0.02	8.2±0.07

However, at the same time, an increase in the content of organic matter was noted with addition of vermicompost to 7.5 %, and a soil conditioner — to 8.2 %.

3.2 Salmonella enterica Population Dynamics in Chernozems

Enterobacteriaceae are among the most common microorganisms that cause acute intestinal diseases. They can serve as an indicator of biological pollution of lands of settlements and agricultural purposes. Bacteria actively enter the external environment with human and animal feces. They multiply in the soil in presence of a large amount of organic matter, but they can multiply even in clean water [16].

The results of determining the abundance of *S. enterica* show that after 10 days of the experiment, the option of the soil sample with a soil conditioner was the best suppressor of bacteria (Figure 2).

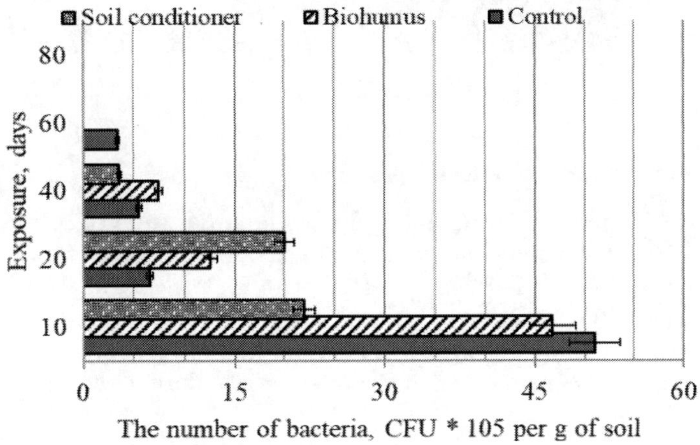

Fig. 2. Dynamics of the number of Salmonella enterica in samples of chernozem

The number of bacteria in the sample with addition of a soil improver was $22.2 \cdot 10^5$ CFU/g soil on the 10th day of the experiment and decreased to $20.1 \cdot 10^5$ on the 40th day, and on the 60th and 80th days of the experiment, bacteria in the soil were not detected

In the option with introduction of vermicompost, the number by 10 days was $47 \cdot 10^5$ CFU/g, and with an increase in the exposure of the experiment, it decreased to $7 \cdot 10^5$ CFU/g by 40 days, further studies showed the absence of viable *Salmonella enterica* cells in the samples of chernozem.

In the control option of the experiment, there was a sharp decrease in the number of Salmonella from $51 \cdot 10^5$ CFU/g to $7 \cdot 10^5$ CFU/g from 10 days to $5 \cdot 10^5$ CFU/g by 20 days. This was the only option of the experiment, where bacteria remained on the 60th day, and by the 80th day, Salmonella was not detected, as in other options of the experiment.

3.3 Changes in the Abundance of Escherichia coli in Soil Samples

Presence of microorganisms of the genus *Escherichia* in soils, including the type species *Escherichia coli*, is an indicator of fresh faecal contamination. To identify *Escherichia coli*, biochemical tests and elective media are used, which ensure the predominant growth of some organisms over others. E. coli is one of the indicators of the sanitary state of soils [17].

In control sample of soil on day 10 the number of bacteria was $46 \cdot 10^5$ CFU/g (Figure 3). Then, an increase in the duration of the experiment contributed to an increase in the number of bacteria to $136 \cdot 10^5$ CFU/g, and a further increase in the exposure time contributed to a decrease in the number with a minimum value of $15 \cdot 10^5$ CFU/g by 80 days.

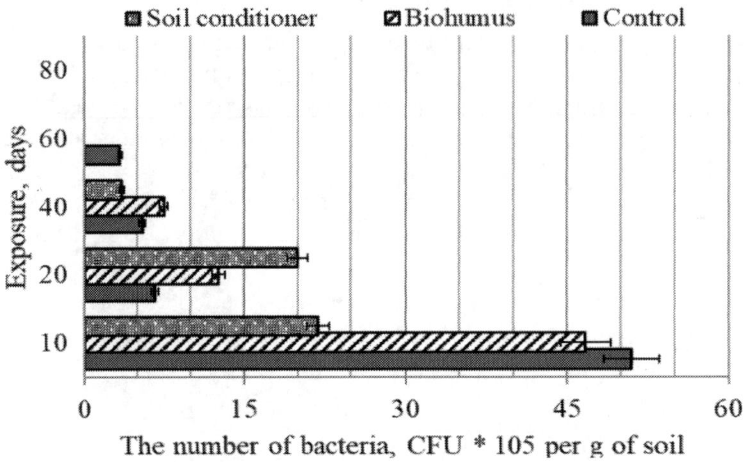

Fig. 3. Dynamics of survival of Escherichia coli in soils

The number of Escherichia coli in the option with introduction of vermicompost on the 10th day was $44 \cdot 10^5$ CFU/g soil, and by the 20th and 40th days it increased almost 2 times. Further dynamics of the number is associated with a decrease in their number with a minimum value of $14 \cdot 10^5$ CFU/g.

Analysis of the results of the experiment in the option of applying the soil improver showed a decrease in the number of bacteria by 73 % during the entire period of the experiment.

The main feature of soil contamination with *Escherichia coli* was preservation of viability by 80 days, i.e. total loss of bacteria is not observed.

3.4 Analysis of Results

According to studies by Sugiyama et al. (2010), Njira et al. (2013), and Silvina et al. (2019), a close relationship was found between the survival of pathogenic and opportunistic microflora representatives with the content of organic matter, soluble humus and salts in the soil. Therefore, we set the task of identifying the dependence of the number of bacteria on the content of organic matter and pH, which changed with introduction of organic ameliorants. The results of calculating the Spearman correlation coefficient are presented in Table 2.

The results of the correlation analysis showed that the indicator of the number of Salmonella is characterized by a high negative correlation with the exposure period of the experiment, which suggests that the older the pollution, the less likely the bacteria will retain their viability in the chernozem.

They also turned out to be more sensitive to the indicator of organic carbon content, it follows from this that the more organic matter in the soil, the less likely it is for bacteria to remain in the soil.

Table 2. Correlation analysis

Indicators	The number of *S. enterica*, CFU/g	The number of *E. coli*, CFU/g
Exposure time, days	-0.76	-0.23
pH_{H2O}	-0.54	-0.2
C_{org}, %	-0.73	-0.57

A similar relationship was noted for *E. coli* only between the indices of abundance and the content of organic matter, but it is weaker in comparison with the analogous values of the correlation coefficient for Salmonella.

4 Conclusion

The results of the study led to several conclusions:

1. Introduction of organic fertilizers into the soil practically did not affect the pH of the soil solution, but led to an increase in the content of organic matter with addition of vermicompost to 7.5 %, and the soil conditioner to 8.2 %.
2. The dynamics of the number of *Salmonella enterica* shows that organic matter in soils contributes to a sharp decrease in their number at the initial stages of biological pollution, and in the long term it contributes to total loss. By the 60th day, Salmonella was found only in the control option of the experiment, but continuation of the experiment showed their total loss on the 80th day.
3. Introduction of organic ameliorants into the chernozem samples led to a decrease in the number of *E. coli* by the 80th day, but their total loss was not observed.
4. The indicator of the number of *S. enterica* is characterized by presence of a high negative correlation (r^2 = - 0.76, at p = 0.038) with the indicator of the exposure period of the experiment and the content of organic carbon (r^2 = - 0.73, at p = 0.042), which indicates that the older the pollution, the less likely it is that bacteria will retain their viability in the chernozem, as well as the more rapid decrease in their number when organic ameliorants are introduced into the soil.
5. A weak correlation was found for the indicator of *E. coli* abundance with pH and organic carbon content, which confirms the thesis of its greater resistance to external environmental factors and the ability to survive in soils for a long time.

Author's contributions

Lyudmila Galaktionova, Tatyana Vasilyeva and Svyatoslav Lebedev conceived and planned the experiments. Gulnur Yamansarina carried out the experiments.

Zhanara Sulimenova contributed to the interpretation of the results, contributed to sample preparation.

Lyudmila Galaktionova and Gulnur Yamansarina took the lead in writing the manuscript. All authors provided critical feedback and helped shape the research, analysis and manuscript.

Acknowledgments

The study was conducted in accordance with the research plan for 2020–2021 of the Federal Scientific Center for Biological Systems and Agricultural Technologies of the Russian Academy of Sciences (No. 0761-2020-0003).

References

1. F. Magdoff, *Agric. Res. Educ.*, 27 (2009)
2. M. Strauss, *Hubei environmental sanitation study*, 54 (1994)
3. A. Cilimburg, Environ Manag., **25,** 587 (2000)
4. G. G. Feachem, D. J. Bradley, H. Garelick, D. D. Mara, *World bank studies in water supply and sanitation*, 196 (1983)
5. J. E. Scandura, Water Sci Technol. **35,** 141 (1947)
6. C.P. Gerba, *Groundwater pollution microbiology*, 39 (1984)
7. P. Payment, Water Sci Technol, **21,** 283 (1989)
8. P. Payment, Morbidity and mortality, **46**, 4 (1997)
9. C.P. Gerba, *Human viruses in sediments, sludges and soils,* 12 (1987)
10. A. Cilimburg, *Morbidity and mortality*, 1 (2000)
11. H. I. Shuval, A. Adin, B. Fattal, E. Rawitz, P. Yekutiel, UNDP (ed) UNDP project management report 6. (Integrated resource recovery series CLO/80/004) World Bank, Washington, 27 (1986)
12. G.A. Toranzos, *Soil biochemistry*, 461 (1997)
13. J.W. Doran, *In Advances in Agronomy*, 1 (1996)
14. G.W. Thomas, *Soil pH and soil acidity*, 475 (1966)
15. M. Y. Silvina, *Enterobacteria isolated from an agricultural soil,* 4 (2019)
16. M.G. Kibblewhite, *Soil health in agricultural systems,* 685 (2008)
17. K.O.W. Njira, J. Anim. Plant Sci., **18**, 2750 (2013)
18. A. Sugiyama, Plant Dis., **94**, 1329 (2010)

Strategies for Reducing Ruminant Methane Emissions

E. V. Sheida, *S. A.* Miroshnikov, *G. K.* Duskaev, *K.N.* Atlanderova and *V.V.* Grechkina

Federal Research Centre of Biological Systems and Agrotechnologies, RAS, Orenburg, Russia

Abstract. The paper studies the effect of additional administration of ultrafine particles on the cattle rumen microbiome composition. The in vitro method was used using the ANKOM Daisy II incubator according to a specialized method. Microflora analysis was performed using MiSeq (Illumina, USA) by a new generation sequencing method with a MiSeq reagent kit. After a detailed analysis of the structure and composition of the microbial community in the contents of the rumen sampled for different diets, it was found that no significant differences were observed in the bacterial communities, with the exception of a slight shift in the Bacteroidetes/Firmicutes ratio. However, we observed numerical differences in the abundance of some representatives, namely, with additional inclusion of Fe and Cr2O3, decrease in the abundance of the methane-forming species Methanobrevibacter, Methanobacterium, Methanosphaera, and Methnaomicrobium was noted regarding the control.

1 Introduction

Methane production by ruminants is of global concern due to accumulation of greenhouse gases in the atmosphere, as well as the loss of energy consumed by animals [1]. Methane is produced in the rumen and cecum of animals by a group of archaea collectively known as methanogens, which are of the Euryarcheota type. The greatest production of methane is observed in ruminants, since methanogens are able to freely produce methane within the normal process of food digestion [2]. Ruminants can produce from 250 to 500 liters of methane per day [3]. Methane emissions from cattle are influenced by many factors, for example, feed intake, type of carbohydrate in the diet, feed processing, additional ingredients [4, 5]. Manipulating these factors can reduce methane emissions. Digestion of food by microorganisms under anaerobic conditions leads to formation of acetate, propionate and butyrate, which are used by animals as an energy source, as well as for production of carbon dioxide (CO2) and CH4, which are eliminated by belching [6]. All gases formed in the rumen within methanogenesis, in addition to their negative impact on the environment, contribute to the loss of 2-15 % of the gross energy consumption [7] for the animal.

The feed composition or its quality affects the production of methane in ruminants. Digestion in the rumen depends on the activity of microorganisms, which require energy, nitrogen and minerals [8-10].

The aim of our study was to study the taxonomic composition of the rumen fluid with the additional inclusion of ultrafine preparations as a strategy for reducing formation of methane in the rumen.

2 Materials and Methods

2.1 Object of the Study

Ruminal fluid (CR), microbiome.

2.2 Experiment Scheme

The study of the rumen microbiome was performed on a model of bulls with a rumen fistula (n = 3). The studies were carried out by the Latin square method.

The studies used ultradispersed ferrum particles obtained by the method of electric explosion of a conductor in an argon atmosphere (Advanced Powder Technologies, Tomsk). Ultrafine particles (UFP) Fe (d= 90 nm, Z-potential 7.7±0.5 mV) contained 99.8 % of Fe. Before being included in the diet, ultrafine ferrum particles were dispersed in physiological solution using UZDN-2T (NPP Akadempribor, Russia) (35 kHz, 300 W, 10 μA, 30 min), at a dose of 1.4 mg/kg feed; Cr2O3 (d=91 nm, specific surface area – 9 m2/g, Z-potential – 93±0.53 mV), contained 99.8 % of Cr was obtained by plasma-chemical synthesis (Platina LLC, Moscow). Before being included in the diet, UFP was dispersed in physiological solution using an ultrasonic disperser UZDN-2 (NPP Akadempribor, Russia) (35 kHz, 300 W, 10 μA, 30 min), at a dose of 200 μg/kg of feed. Wheat bran was used as a control.

Investigations of the rumen microbiome were carried out by the in vitro method using the "artificial rumen" model using the device — ANKOM Daisy II incubator (modifications D200 and D200I) according to the specialized method. Distilled water was chosen as the dispersion medium.

In fistula animals, 3 hours after feeding, samples of rumen contents were taken, which were filtered through 4 layers of gauze and incubated in an artificial rumen at a constant T = 39 °C for 24 hours. After incubation, the ruminal fluid was sampled with a syringe dispenser OPA-2-20 ecochem manufactured by Ekroskhim LLC, St. Petersburg, into Eppendorf microtubes with a volume of 1.5 ml. For analysis, we took 1.5 ml of the substrate of the ruminal fluid, one sample for each sample. Keeping the animals and procedures within the experiments were in accordance with the instructions and recommendations of the Russian regulations (Order of the Ministry of Health of the USSR No. 755 dated August 12, 1977) and The Guide for Care and Use of Laboratory Animals (National Academy Press, Washington, D.C., 1996). Every effort has been made to minimize animal suffering and to reduce the number of samples used.

2.3 Metagenomic Analysis of Rumen Microorganisms

Microflora analysis was performed using MiSeq (Illumina, USA) by next generation sequencing (NGS) with MiSeq® Reagent Kit v3 (600 cycle). For bioinformatic processing of the results, the PEAR (Pair-End AssembeR, PEAR v0.9.8) program is used (Zhang J., Kobert K., Flouri T., Stamatakis A. PEAR: A fast and accurate Illumina Paired-End reAd merger. Bioinformatics, 2014, 30(5): 614-620 (doi: 10.1093/bioinformatics/btt593)). Filtration, de-replication, removal of chimeric sequences, clustering, sorting (cutting off singletons), removal of contamination is performed in the USEARCH program (usearch v8.0.1623_i86linux32, http://drive5.com/usearch). The fastq filter algorithm is used for filtering, derep prefix –for replication, cluster otus – for clustering and removal of chimeric sequences.

The sequencing results were processed using the Microsoft Excel 10 data analysis package, Microsoft Office software.

2.4 Statistical Processing

The numerical data were processed using the SPSS Statistics 20 software (IBM, USA), the mean (M), standard deviations (±σ), and standard deviation errors (±SE) were calculated. To compare the options, a nonparametric method of analysis was used. Differences were considered statistically significant at $p \leq 0.05$.

3 Study Results

Within the rumen microbiome, bacteria dominate and contribute most to the digestion and conversion of feed into VFA and microbial protein. The bacterial community of the rumen depends on the composition of the diet, types of feed, the age of the animals and the feeding strategy. Variations in these factors have a significant effect on rumen metabolism, which can affect both productivity and health of the animal.

Metagenomic sequencing the CR in the control group showed presence of 12 phyla, where 4 — dominant phyla, namely, the phylum Firmicutes — 17.28 %, Bacteroidetes — 50.42 %, Verrucomicrobia — 13.13 %, Proteobacteria — 10.18 %. The dominant bacteria found in the control sample are presumably responsible for transformation of the main components of the diet — cellulose, hemicellulose, starch, organic acids, proteins (Figure 1). With introduction of ferrum UFP into the diet, there is a predominance of the number of acid-fast bacteria of the phylum of Bacteroidetes (55.61 %), Firmicutes (16.26 %), Verrucomicrobia 11.06 %, and Proteobacteria 9.27 %. Additional inclusion of UFP Cr2O3 contributed to increase in the number of representatives of the Bacteroidetes phylum relative to control by 10.07 %, Firmicutes — by 0.43 %, as well as decrease in Verrucomicrobia — by 4.8 % and Proteobacteria — by 3.8 %.

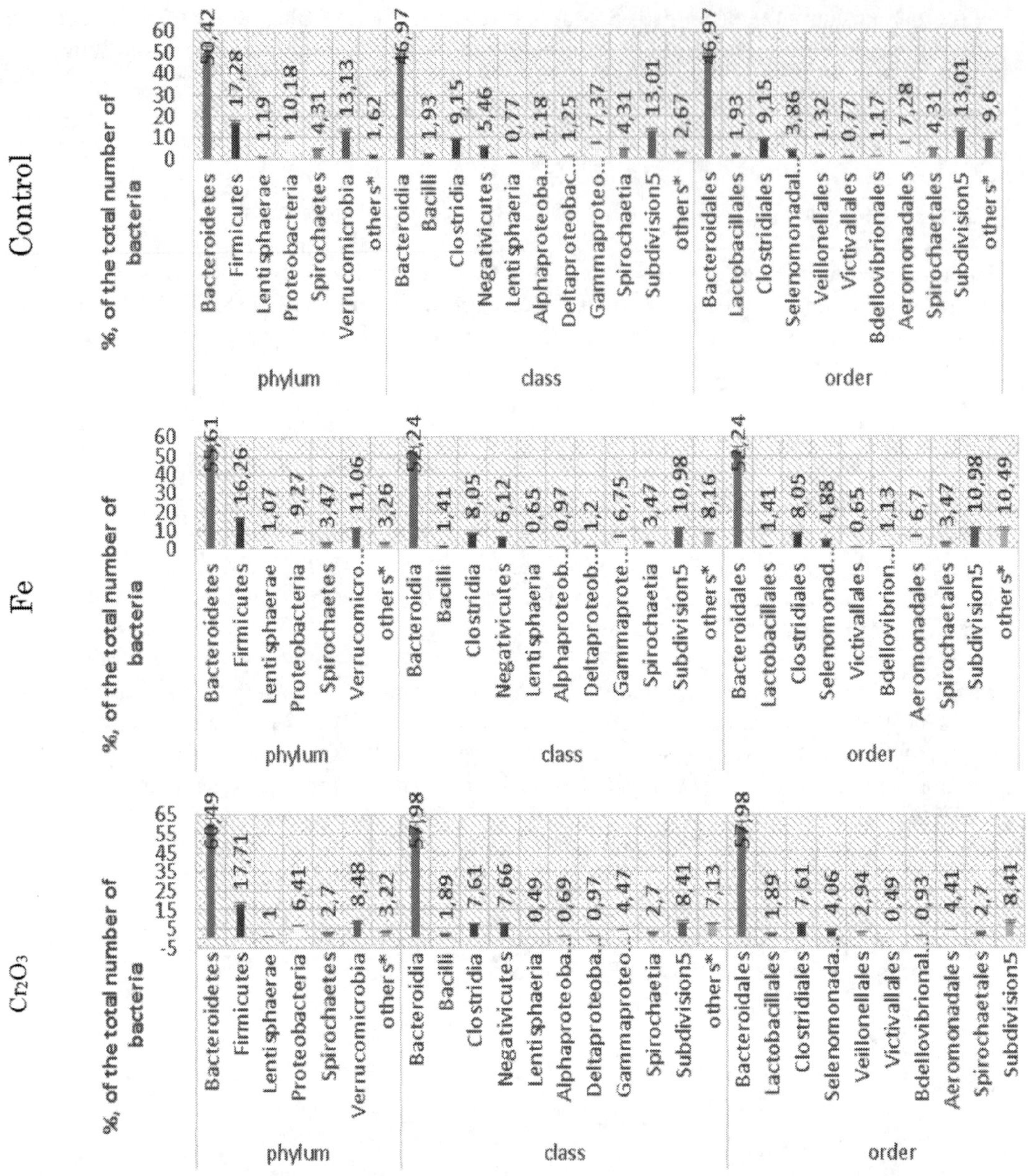

Other*: this group includes taxa, the number of which did not exceed 1.0 % of the total number of bacteria

Fig. 1. Taxonomic groups of the rumen microbiome, % of the total number of bacteria in the sample.

In both the control and experimental groups, the predominance of the Bacteroidia, Clostridia and Negativicutes classes was noted, however, it shall be noted that the total number of bacteria in the control group was significantly higher than in the experimental groups.

The additional inclusion of UFP promoted an increase in the number of bacteria of the Prevotellaceae families by 3.6 % (Fe) and 9.3 % (Cr2O3) relative to the control, as well as Selenomonadaceae, unclassified_Bacteroidales, and unclassified_Bacreria.

Also, in the study, when adding the UFP of ferrum and chromium, a decrease in the archaeal domain in the ruminal fluid was noted (Table 1). It was found, when ferrum and chromium UFP included, the number of representatives of the Methanobacteria methane-forming class decreased, the concentration in the group in presence of ferrum was lower by 11.6 % relative to the control, and with introduction of chromium UFP — by 6.4 % (p≤0.05). The number of representatives of the Methanomicrobia class in the experimental groups relative to the control also decreased significantly by 9.7 % in presence of Cr2O3 and by 2.8 % in presence of Fe (p≤0.01)

Table 2. Taxonomic diversity of the Archaea domain of the rumen of cattle when using various additives in the diet, n = 4

Group	class	order	family	genus
Control	Methanobacteria (9,02±0,77)	Methanobacteriales (9,02±0,77)	Methanobacteriaceae (9,02±0,77)	Methanobacterium (0,11±0,07)
				Methanobrevibacter (8,69±0,80)
				Methanosphaera (0,21±0,08)
	Methanomicrobia (0,72±0,10*)	Methanomicrobiales (0,72±0,10)	Methanomicrobiaceae (0,72±0,10)	Methanomicrobium (0,72±0,10)
Fe	Methanobacteria (7,97±0,45)	Methanobacteriales (7,97±0,45)	Methanobacteriaceae (7,97±0,45)	Methanobacterium (0,10±0,06)
				Methanobrevibacter (7,79±0,40)
				Methanosphaera (0,08±0,04)
	Methanomicrobia (0,70±0,08**)	Methanomicrobiales (0,70±0,08)	Methanomicrobiaceae (0,70±0,08)	Methanomicrobium (0,70±0,08)
Cr_2O_3	Methanobacteria (8,44±2,36*)	Methanobacteriales (8,44±2,36)	Methanobacteriaceae (8,44±2,36)	Methanobacterium (0,41±0,36)
				Methanobrevibacter (7,90±2,07)
				Methanosphaera (0,13±0,07)
	Methanomicrobia (0,65±0,06**)	Methanomicrobiales (0,65±0,06)	Methanomicrobiaceae (0,65±0,06)	Methanomicrobium (0,65±0,06)

* p≤0.05; ** p≤0.01 versus control versus control

4 Result Discussion

After analyzing the structure and composition of the microbial community in the contents of the rumen sampled for different diets, it was found that no significant differences were observed in the bacterial communities, with the exception of a slight shift in the Bacteroidetes/Firmicutes ratio, which is known to vary widely in individual animals. However, we observed numerical differences in abundance of some representatives. A decrease in the number of bacteria in the CR is possibly associated with the toxic effect of UFP additives on them; however, along with this, inhibition of methanogens in these groups was also noted [11]. In addition, the data obtained are contrary to the information obtained earlier on ferrum citrate [12]. The type of feed offered to ruminants, their ratio and presence of additional components in the diet affect rumen fermentation and, therefore, methane production. There is evidence of methane energy loss from 6 to 7 % of total energy consumption when feed was fed at the maintenance level , and a decrease to 2-3 % when giving concentrates against the background of a standard diet. It was found that a change in the diet for cattle led to a change in the nature of fermentation in the rumen and, as a consequence, the creation of a more hostile environment for methanogenic bacteria, in which the rate of passage increases, the pH of the content decreases and some populations of protozoa, mucosal cilia and methanogenic bacteria can be eliminated or suppressed [13, 14]. Similar effects were observed in our experiment.

Methanogens are a separate group of microorganisms that belong to the Archaea domain and the Euryarchaeota type [15]. In the rumen, methanogens use hydrogen and CO2 to produce CH4. Cultivated methanogens were assigned to seven species: Methanobrevibacter ruminantium, Methanobacterium formicicum, Methanobacterium bryantii, Methnaomicrobium mobile and Methanoculleus olentangyi [16]. With inclusion of Fe and Cr2O3 UFP, a decrease in the abundance of the Methanobrevibacter, Methanobacterium, Methanosphaera and Methnaomicrobium species was noted relative to the control. The data obtained differ from the results obtained earlier in relation to various forms of ferrum [17].

5 Conclusion

Reducing the production of methane in the digestive processes of ruminants is a complex problem, however, adjusting the diets by introducing additional ingredients into them, namely, ultradispersed mineral preparations (Fe and Cr2O3 UFP) can have a positive effect on reducing the production of methane, which is important in the ecology of the climate.

Acknowledgments

This study was performed with the financial support from the Russian Science Foundation (Project No.20-16-00088).

Conflict of Interest

Authors declares that they have no conflict of interest.

References

1. J. L. Black, T.M. Davison, I. Box, Animals (Basel), **11(4),** 951 (2021)
2. K. A. Beauchemin, E. M. Ungerfeld, R. J. Eckard, M. Wang, **14(S1),** 2 (2020)
3. D. Olijhoek, P. Lund, *Methane production by ruminants. Department of Animal science AU-Foulum.* Aarhus University, Denmark (2017)
4. I. Karimov, K. Kondrashova, G. Duskaev, O. Kvan, E3S Web of Conferences, **143,** 02034 (2020)
5. K. Logachev, I. Karimov, G. Duskaev, A. Frolov, S. Tulebaev, O. Zav'yalov, Asian Journal of Animal Sciences, **9(5),** 248 (2015)
6. C. Martin, D.P. Morgavi, M. Doreau, Animal, **4(3),** 351 (2009)
7. Y.Q. Guo, J.X. Liu, Y. Lu, W.Y. Zhu, S.E. Denman, Lett Appl Microbial, **47(5),** 421 (2008)
8. M. Ramin, P. Fant, P. Huhtanen, J. Dairy Sci., **104(5),** 5617 (2021)
9. J. C. Ku-Vera, O. A. Castelán-Ortega, F. A. Galindo-Maldonado, J. Arango, N. Chirinda, R. Jiménez-Ocampo, S. S. Valencia-Salazar, E. J. Flores-Santiago, M. D. Montoya-Flores, I. C. Molina-Botero, A. T. Piñeiro-Vázquez et. al, Animal, **14(S3),** 453 (2020)

10. P. K. Malik, S. Trivedi, A. Mohapatra, A. P. Kolte, V. Sejian, R. Bhatta, H. Rahman, PLoS One, **16(8),** e0256048 (2021)
11. M. Popova, J. Guyader, M. Silberberg, A. R. Seradj, C. Saro, A. Bernard, C. Gérard, C. Martin, D. P. Morgavi, Appl Environ Microbiol, **85(4),** e02657 (2019)
12. B. A. Clemmons, L. G. Schneider, E. A. Melchior, A. K. Lindholm-Perry, K. E. Hales, J. E. Wells, H. C. Freetly, S. L. Hansen, M. E. Drewnoski, S. J. Hartman, P. R. Myer Access Microbiol, **3(1),** acmi000180 (2020)
13. F. Dehareng, C. Delfosse, E. Froidmont, H. Soyeurt, C. Martin et al, Animal, **6(10),** 1694 (2012)
14. P. J. Van Soest, *Nutritional Ecology of the Rumi-nant*, 40 (1982)
15. A. K. Almeida, R. S. Hegarty, A. Cowie, Anim Nutr., **7(4),** 1219 (2021)
16. P. H. Janssen, M. Kirs, Applied and Environmental Microbiology, **74(12),** 3619 (2008)
17. H. Wu, Q. Meng, Z. Yu, Bioresour Technol., **211,** 603 (2016)

Changes in the concentration of methane in the ecosystem *in vitro* against the background of Asteraceae family plants biomass

E.V. Sheida, *V.A.* Ryazanov, *K.S.* Denisenko, and *O.V.* Shoshina

Federal Research Centre of Biological Systems and Agrotechnologies, RAS, Orenburg, Russia

Abstract. The assessment of Asteráceae Family Plants (rhizomes and roots of elecampane and wormwood) influence on the process of methane formation in the rumen ecosystem and metabolic processes was carried out. Studies (*in vitro*) were carried out using ANKOM Daisy II incubator (modifications D200 and D200I) according to a specialized method. Rumen contents were obtained from beef bulls with chronic rumen fistula. Gas analysis of air and volatile fatty acids samples was performed by gas chromatography. The results of the study showed that different dosages of phytobiotic preparations did not significantly affect the characteristics of fermentation *in vitro*. Phytobiotic preparations of elecampane and wormwood reduce the production of methane in the ruminal fluid, which may be associated with various active components or dosages of their administration.

1 Introduction

The problem of global warming requires immediate resolution, and a significant role in this problem is assigned to the issues of reducing methane emissions into the atmosphere. It has been established that the livestock industry accounts for approximately 25 % of the global anthropogenic methane emissions [1] and decrease in methane production by ruminants will greatly contribute to a decrease in it in the atmosphere. It will also reduce energy losses [2]. Various strategies are known to reduce the release of methane from the rumen, such as vaccination, inhibition by 2-bromoethanesulfonic acid, antibiotics, stimulation, defaunization, adaptation to diet, use of propionate precursors or alternative electron acceptors [3].

Due to reactivity phytobiotic preparations (plant extracts, pigments, antioxidants) can play the role of electron absorbers under anaerobic conditions in the rumen and contribute to decrease in methane production [4].

Food additives that modify rumen microbial fermentation can be used to improve the growth and health of animals. According to [5], most of the 13 million kg of antibiotics administered to animals in 2010 worldwide were used to stimulate the growth of livestock. However, the potential for bacterial resistance has sparked intense debate over use of antibiotics, highlighting the importance of natural and safe alternatives such as probiotics, prebiotics or phytobiotics [6]. It has been shown that addition of herbal anti-MON drugs had a similar effect on rumen fermentation and growth rates in young Holstein bulls consuming large amounts of concentrate [7]. Another study [8] showed a similar effect of neomycin and origanum leaves on the number of days of brushing, the severity of brushing, and mortality in newborn calves. Plants or their biologically active compounds with antimicrobial properties can improve feed utilization and animal performance by altering rumen microbial fermentation (for example, short-chain fatty acid profile, methane production, nitrogen metabolism or pH [9].

Herbs exhibit not only antibacterial effects that modulate rumen fermentation, but a wide range of health benefits such as antioxidant and anti-inflammatory effects [10]. In addition, plant products are currently used in the feed industry mainly as sensory additives, flavorings, and appetizing substances [11].

Purpose of the study: in our study, we looked for new natural substances to reduce production of methane in the rumen and assessed whether the selected plant materials affect metabolic processes in the rumen and methanogenesis.

2 Materials and Methods

The object of the study is the ruminal contents (RC) obtained from bulls of the Kazakh white-headed breed with a chronic fistula of the rumen, with an average weight of 220-225 kg, at the age of 9-10 months.

Maintenance of animals and experimental studies were carried out in accordance with the instructions and recommendations of Russian regulations (1987; Order of the Ministry of Health of the USSR No. 755 dated August 12, 1977 "On Measures to Further Improve Organizational Forms of Work using Experimental Animals") and Guide for the Carre and Use of Laboratjry Animals (National Academy Press, Washington, D.C., 1996). Within the studies, measures were taken to minimize the suffering of animals and reduce the number of tested samples.

For the study, we used Inulae rhizomata et radices (elecampane of rhizomes and roots), Artemisiae absinthil herba (wormwood, bitter herb).

Experiment Scheme. The studies were carried out by the *in vitro* method using the device — ANKOM Daisy ^{II} incubator (modifications D200 and D200I) according to the specialized method. Distilled water was chosen as the dispersion medium. Each experiment was carried out in three repetitions. After incubation, air samples were taken to determine the level of methane on Kristallux-2000M device by gas chromatography.

Laboratory studies were carried out at the Testing Center of the Center for Collective Use of the Federal Research Center of the BST of the RAS: the level of volatile fatty acids (VFA) in the contents of the rumen was determined by gas chromatography using gas chromatograph Crystallux-4000M, determination of nitrogen forms — according to GOST 26180-84.

Statistical processing. The numerical data were processed using the SPSS Statistics 20 software (IBM, USA), the mean (M), standard deviations (±σ), and standard deviation errors (±SE) were calculated. To compare the options, a nonparametric method of analysis was used. Differences were considered statistically significant at $p \leq 0.05$.

3 Study Results

Within *in vitro* studies, it was found that the VFA level in the samples with introduction of various dosages of phytobiotic preparations was different (Table 1). The rhizomes and roots of elecampane in all studied dosages and wormwood herb at a dose of 2.0 g and 5.0 g showed the same level of acetic acid in the ruminal fluid. The concentration of acetic acid in the sample with the presence of wormwood herb at a dosage of 10 g/kg DM was significantly higher than in other samples by 79.2 % (p≤0.05); it shall also be noted that the concentration of all detected VFAs in this group was higher than other samples. The same level of propionic acid was recorded in samples 2, 3, 4 and 5, its concentration was higher by 56.7 % in sample 1 and by 18.8% in sample 6. A low concentration of butyric acid was shown in samples containing elecampane 6.0 g and wormwood 5.0 g and 2.0 g, in the presence of wormwood at a dose of 10.0 g, the level of this acid increased by 90.5 % (p≤0.01).

The VFA concentration in the ruminal fluid was noted 64.6 % lower relative to sample 6 in the study of sample 1. The lowest VFA level was recorded in a sample with wormwood grass at a dosage of 5.0 g, the level of valeric and caproic acids in this group was minimal and amounted to 0.9 and 0.8 mmol/100 ml.

Table 1. Concentration of volatile fatty acids in ruminal fluid, mmol/100 ml

Sample No.	Name	Dose, g/kg DM	Volatile fatty acids, mmol/100 ml				
			Acetic	Propionic	Butyric	Valeric	Caproic
1	*Inulae rhizomata et radices*	3.0	1.6±0.002	3.0±0.001	2.9±0.004	2.1±0.006	1.1±0.0002
2		1.0	1.6±0.004	1.3±0.002	2.6±0.006	1.2±0.0004	0.8±0.0002**
3		6.0	1.6±0.004	1.3±0.001	1.1±0.001*	0.9±0.0004	2.9±0.006*
4	*Artemisiae absinthil herba*	5.0	1.6±0.012	1.3±0.002	1.1±0.001*	0.9±0.0002	0.8±0.0001**
5		2.0	1.6±0.006	1.3±0.001	1.1±0.002*	1.4±0.002	4.0±0.004*
6		10.0	7.7±0.004*	1.6±0.004	11.6±0.004 **	6.3±0.002	3.0±0.002*

Note: * - P≤0.05; ** - P≤0.01; *** - P≤0.001

The content of nitrogenous components of the ruminal fluid (RF) is one of the indicators of the degree of assimilation of nitrogen in the feed, as well as the general direction of the ruminal digestion processes. The highest concentration of total nitrogen in the ruminal fluid was noted in sample 4, and the lowest one in samples 1 and 2. The wormwood herb showed a higher content of total, non-protein and ammoniacal nitrogen in the ruminal fluid than the rhizomes and roots of elecampane, and regardless of the concentration (Figure 1). A direct relationship was found between the concentration of total and protein nitrogen. Protein nitrogen was approximately at the same level in samples 1, 3, 5 and 6, in sample 4 it increased by 45.1 % - 47.7 %, in sample 2 — by 15.7 % - 19.7 %.

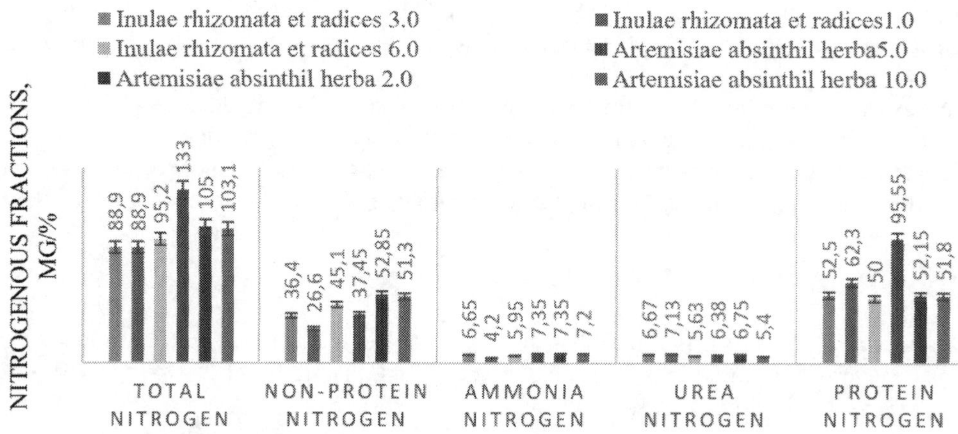

Fig. 1. Content of nitrogenous fractions in ruminal fluid, mg/%.

Wormwood herb at a dosage of 10.0 g per 1 kg of DM reduced methane production to a greater extent than its other dosages ($p \leq 0.05$), so a dose of 2.0 g showed an increase in the methane level in the ruminal fluid by 36.5 % ($p \leq 0.01$), and the dose of 5.0 g — 47.5 % ($p \leq 0.01$). Rhizomes and roots of elecampane in various dosages showed a high level of methane production, however, the *min* and *max* doses of this phytobiotic showed a 38.1 % - 38.6 % decrease in methane in the ruminal fluid (Table 2). In all groups, there was a direct relationship between the level of CH_4 and CO_2.

Table 2. Concentration of methane (CH_4), mg/m^3 and CO_2, e/g under *in vitro* conditions

Title	Dose, g / 1 kg of DM	Concentration CH_4 *in vitro*	
		CH_4, mg/m^3	Equivalent CO_2 *e/g*
Inulae rhizomata et radices	3,0	21500±10,26	537,5±4,36
	1,0	13300±9,74*	332,5±5,12*
	6,0	13200±9,16*	330±4,22*
Artemisiae absinthil herba	5,0	17900±10,54*	447,5±5,16*
	2,0	14800±10,12*	370±4,26*
	10,0	9400±8,58*	235±4,10*

4 Discussion

Herbal preparations or their active ingredients are characterized by antimicrobial activity against both gram-negative and gram-positive bacteria, as well as antifungal activity [12]. We assume that in our experiment, the probable mechanism of action is inhibition of deamination and methanogenesis, leading to decrease in the concentration of methane and acetate, and increase in the concentration of propionate and butyrate [10] (as a result of the addition of elecampane at a dose of 3.0 g/kg DM and wormwood in dose of 10.0 g/kg DM). Our findings are in part consistent with those reported in other sources [13, 14]. It has been shown that low doses of biologically active plant compounds increase the concentration of butyric acid [6]; herein, this was observed at low doses of Inulae rhizomata et radices.

In *in vitro* study, the effect of phytobiotics and their components on the VFA profile was observed, but it was noted that they did not affect the nitrogen concentration in the ruminal fluid [14, 15]. These authors suggested that herbal concentrations shall be higher than 35 mg/l in order to induce any significant changes in rumen nitrogen metabolism. In our study, we noted that the wormwood herb in all studied dosages contributed to increase in all nitrogenous fractions in the ruminal fluid, a similar level of nitrogen was recorded with introduction of the rhizome and roots of elecampane at the maximum dosage.

Previous studies have shown that saponins inhibit protozoa and also limit the availability of hydrogen for methanogenesis [16]. In a recent study by Holtshausen et al. [17] stated that they revealed a decrease in methane content at higher levels of inclusion of herbal preparations containing saponin (15 g/kg DM and more); in our experiment, we showed decrease in methane at low and high doses of roots and rhizomes of elecampane (1.0 and 6.0 g/kg DM of the diet).

Essential oils that make up phytobiotic preparations can also reduce the amount of available hydrogen for methanogenesis [18]; in the studies carried out, decrease in the level of methane was also noted with the additional inclusion of high doses (10.0 g/kg DM) of wormwood herb, which has essential oils in its composition.

5 Conclusion

Therefore, different dosages of dry plant biomass Inulae rhizomata et radices and Artemisiae absinthil herba did not have a significant effect on the fermentation characteristics in the rumen ecosystem, but they contribute to decrease in methane production in the ruminal fluid. Further study is necessary using the methods of metagenomic analysis of the rumen microbiome.

Acknowledgments

This study was performed with the financial support from the Russian Science Foundation (Project No.21-76-10014)

References

1. K.R. Lassey, Aust J Exp Agr, **48**, 114–118 (2008)
2. K.A. Johnson, D.E. Johnson, J Anim Sci, **73**, 2483–2492 (1995)
3. R. Bodas, S. López, M. Fernández, R. García-González, A.B. Rodríguez, R.J. Wallace, J.S. González, Anim Feed Sci Technol, **145**, 245–258 (2008)
4. I. Karimov, K. Kondrashova, G. Duskaev, O. Kvan, Evaluation of effects of rumen fluid in combination with probiotic preparations and vanillin on the luminescence of a recombinant strain E. coli., E3S Web of Conferences, **143**, 02034 (2020)
5. M. K. Chattopadhyay, Frontiers in Microbiology, **5**, 1–3 (2014)
6. R. Khiaosa-ard, Q. Zebeli, Journal of Animal Science, **91**, 1819–1830 (2013)
7. M. Devant, A. Anglada, A. Bach, Animal Feed Science and Technology, **137**, 46–57 (2007)
8. V. A. Bampidis, V. Christodoulou, P. Florou-Paneri, E. Christaki, Journal of Veterinary Medicine Series A, **53**, 154–156 (2006)
9. M. Akbarian-Tefaghi, E. Ghasemi, M. Khorvash, J Anim Physiol Anim Nutr (Berl). Jun, **102(3)**, 630-638 (2018)
10. S. Calsamiglia, M. Busquet, P. W. Cardozo, L. Castillejos, A. Ferret, Journal of Dairy Science, **90**, 2580–2595 (2007)
11. C. K. Franz, H. C. Baser, W. Windisch, Flavour and Fragrance Journal, **25**, 327–340 (2010)
12. K.S. Inchagova, G.K. Duskaev, D.G. Deryabin, Microbiology (Russian Federation) **88(1)**, 63-71 (2019)
13. D. Petrič, D. Mravčáková, K. Kucková, S. Kišidayová, A. Cieslak, M. Szumacher-Strabel, H. Huang, P. Kolodziejski, A. Lukomska, S. Slusarczyk, K. Čobanová, Z. Váradyová, Front Vet Sci., **8**, 630971(2021)
14. M. E. Abolfathi, S. A. Tabeidian, A. D. Foroozandeh Shahraki, S. N. Tabatabaei, M. Habibian, Arch Anim Nutr., **73(2)**, 88-110 (2019)
15. R. Temmar, M. Rodríguez-Prado, G. Forgeard, C. Rougier, S. Calsamiglia, Animals (Basel), **11(5)**, 1205 (2021)
16. F. Zhang, B. Li, Z. Ban, H. Liang, L. Li, W. Zhao, X. Yan, J Anim Physiol Anim Nutr (Berl), **105(4)**, 630-638 (2021)
17. K.A. Beauchemin, E.M. Ungerfeld, R.J. Eckard, M. Wang, Animal. Mar, **14(S1)**, 2-16 (2020)
18. K. A. Beauchemin, S. M. McGinn, Journal of Animal Science, **84(6)**, 1489–1496 (2006)

Dose-dependent effect of plants of the Lamiaceae family on the concentration of methane, fatty acids and nitrogen in the ecosystem in vitro

Vitaliy Ryazanov, *Galimzhan* Duskaev, and *Konstantin* Denisenko

Federal Research Centre of Biological Systems and Agrotechnologies, RAS, Orenburg, Russia

Abstract. Fermentation processes in the rumen of ruminants determine how much final metabolites and their derivatives will be formed, which are necessary for the full development of the organism, the level of productivity, and also affect the level of formation of endogenous substances, namely, greenhouse gas emissions. These criteria lead us to the search for new feed products that improve the metabolic processes of the rumen and the digestive system as a whole, so phyto-substances can serve as an alternative. The article presents the results of in vitro study of the influence of Salviae folia, Scutellaria baicalensis, Oríganum vulgáre on formation of methane, synthesis of volatile fatty acids and nitrogen, as the main indicator parameters of the enzymatic activity of the rumen of ruminants. It was found that when using phyto- substances: Salviae folia and Scutellaria baicalensis, more acetic and propionic acid was formed, Oríganum vulgare in various dosages shifted towards propionic and valeric acid. Formation of a larger amount of microbial protein ($P \leq 0.05$) with use of Salviae folia, Scutellaria baicalensis, Oríganum vulgáre in various dosages was established. Methane production decreased with use of Oríganum vulgáre.

1 Introduction

Fermentation processes in the rumen of ruminants determines how much final metabolites and their derivatives will be formed, which are necessary for full development of the body, the level of productivity, and also affect the level of formation of endogenous substances, namely, greenhouse gas emissions.

Use of phyto-substances in ruminant feeding can regulate enzymatic processes in the rumen [1,2]. Wherein, use of phytobiotics can serve as the basis for alternatives to use of antibiotics in animal husbandry [3]. It is known that feeding of phytobiotics or preparations based on them to cattle promotes better feed conversion, improves nitrogen metabolism, namely, formation of microbial protein, synthesis of volatile fatty acids and the general state of the animal's body [4]. Currently, impact of plant extracts and phyto-substances on formation of methane in the rumen is being assessed [5], a positive effect is noted from their use in terms of reducing the concentration.

A wide range of properties of phyto-substances opens up a wide range of possibilities for studying their effect on metabolism in the rumen of ruminants. From the above, it follows that the issue of use of phyto-substances in feeding cattle remains open and requires further study.

2 Materials and Methods

2.1 Object of the Study

Ruminal fluid, phyto-substances

2.2 Experiment Scheme

The study of the influence of phyto-substances was carried out on a model of Kazakh white-headed bulls with chronic rumen fistula (n = 3), (at the age of 9-10 months, with an average weight of 225-230 kg, diet: hay – 70 %; grain feed – 30 %, free access to the water). The studies were carried out by the Latin square method. Keeping the animals and procedures within the experiments were in accordance with the instructions and recommendations of the Russian regulations (Order of the Ministry of Health of the USSR No. 755 dated August 12, 1977) and The Guide for Care and Use of Laboratory Animals (National Academy Press, Washington, D.C., 1996). Every effort has been made to minimize animal suffering and to reduce the number of samples used. The studies were carried out by the *in vitro* method using the device — ANKOM Daisy ^{II} incubator (modifications D200 and D200I) according to the specialized method. Each experiment was carried out in triplicate. After incubation, air samples were taken to determine the level of methane on Kristallux-2000M device by gas chromatography. Laboratory studies were carried out at the Testing Center of the Center for Collective Use of the Federal Research Center of the BST of the RAS: the level of volatile fatty acids (VFA) in the contents of the rumen was determined by gas chromatography using gas chromatograph Crystallux-4000M, determination of nitrogen forms — according to GOST 26180-84. The study used *phyto*-substances of the Lamiaceae family: Salviae folia, *Scutellaria baicalensis, Oríganum vulgáre.*

3 Study Results

When assessing the efficiency of using phyto-substances it is possible to judge by the amount of volatile fatty acids formed in the rumen, which are directly a direct source of energy for the body.

Therefore, when using *Salviae folia,* the highest concentration values for volatile fatty acids were 1.6 g/kg DM for the dosage: acetic (7.3±0.004 mmol/100 ml) ($P \leq 0.01$); propionic (4.6±0.007 mmol/100 ml) ($P \leq 0.01$); butyric (5.8±0.002 mmol/100 ml) ($P \leq 0.05$); valeric (4.1±0.004 mmol/100 ml) ($P \leq 0.01$); caproic (1.4±0.002 mmol/100 ml) ($P \leq 0.05$) acid.

The concentration of volatile fatty acids using *Scutellaria baicalensis* was lower for all dosages, except for caproic acid, its content was significantly different from the others and was 61.5 % higher than that of *Salviae folia* and 42.3 % of *Oríganum vulgáre.* Further study of phyto-substances, namely, *Oríganum vulgáre* at a dosage of 0.2 g/kg DM, was noted by increase in concentration for propionic, butyric and valeric acids. The content of propionic acid was higher by 7.7 %, butyric acid — by 2.7 %, and valeric acid — by 42.5 % in comparison with *Salviae folia.* The volume of acetic acid was at the same level as for *Scutellaria baicalensis* and amounted to (1.6±0.003) ($P \leq 0.05$) mmol/100 ml (Fig. 1).

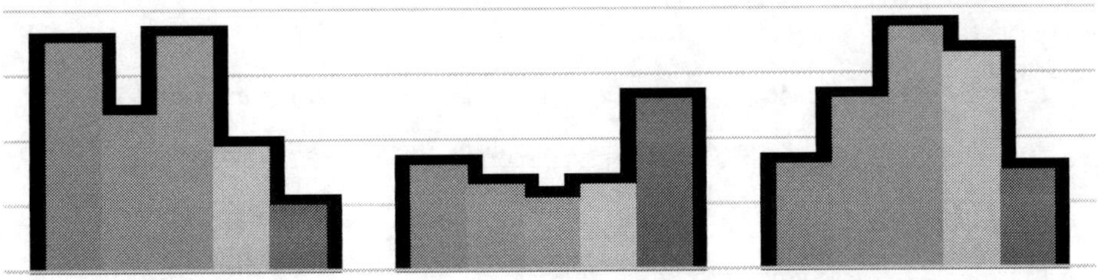

Fig. 1. Concentration of isolated volatile fatty acids in rumen fluid when using phyto-substances, mmol/100 ml

Supply of the body with amino acids and the synthesis of microbial protein determines the intake and formation of nitrogenous substances in the rumen. Studying the effect of phyto-substances in our experiment, the highest value for the content of total and protein nitrogen was found for a dosage of 0.8 g/kg DM *Salviae folia*, 0.15 g/kg DM *Scutellaria baicalensis*, and 0.3 g/kg DM *Oríganum vulgáre*. An insignificant decrease in the ammoniacal form of nitrogen was revealed for dosages of 1.6 and 2.4 g/kg DM of *Salviae folia,* in contrast to the dosage of 0.8 g /kg DM, the difference was 5.5 %. For a dosage of 0.2 and 0.3 g/kg DM *Oríganum vulgáre, the* content of urea nitrogen was 5.3 % and 10.0 % higher, respectively, than for a dosage of 0.6 g/kg DM (Table 1).

Table 1. Nitrogen content in rumen fluid in vitro at various dosages of phyto-substances, mg/%

Name	Dose,	Forms of nitrogen, mg/%				
Lamiaceae family	g/kg DM	Total	Non-protein	Ammonia	Urea	Protein
Salviae folia	0.8	86.1±1.4*	28.7±2.1	6.3±1.3*	6.38±1.0	57.4±0.9*
	1.6	72.1±1.8	37.1±1.3	5.95±0.5	6.0±0.8	35.0±1.2
	2.4	80.5±1.7*	31.5±1.4	5.95±1.1	4.88±1.6	49.0±1.1*
Scutellaria baicalensis	0.15	85.9±1.5*	36.6±1.2	6.1±1.4	6.2±1.4	49.3±1.7*
	0.2	83.0±1.6	36.2±2.5	5.9±1.0	4.7±1.3	46.8±1.4*
	0.3	84.3±1.5*	39.6±2.3*	6.23±1.2	5.8±1.6	44.7±1.3
Oríganum vulgáre	0.2	53.2±2.3	26.2±1.9	5.6±1.6	7.13±0.9	26.95±1.1
	0.3	89.6±1.4*	36.4±1.1	5.95±1.5	7.5±1.7	53.2±0.7*
	0.6	74.2±1.9*	23.1±0.9	6.65±1.8*	6.75±1.2	51.1±0.5*

Note:* - P≤0.05; DM – dry matter

Use of phyto-substances in feeding the cattle can influence the processes of methanogenesis and contribute to lower energy losses in animals with its conversion into the body's products, reducing the effect of greenhouse gas emissions. An in vitro study of phyto-substances of the Lamiaceae family made it possible to obtain the following data on formation of methane, the highest concentration value was noted for a dosage of 1.6 g/kg DM of *Salviae folia;* it was higher than for a dosage of 0.8 and 2.4 g/kg DM in the same group by 20.5 % and 29.7 %, respectively.

It is worth noting that in the *Salviae folia* group, the average methane concentration was higher than in the *Scutellaria baicalensis* and *Oríganum vulgáre groups.* With introduction of *Scutellaria baicalensis* at a dosage of 0.3 g/kg DM, the lowest methane concentration was noted in *Oríganum vulgáre* as well as at a dosage of 0.6 g/kg DM, the difference was 23.1 %. Wherein, the average value of methane concentration in the *Oríganum vulgáre* group in comparison with *Salviae folia* and *Scutellaria baicalensis* was lower by 32.4 % and 26.5 %, respectively (Table 2).

Table 2. Methane concentration using phyto-substances in vitro, g/m³, CO_2 *e*/g

Name	Dose, g/kg DM	Concentration of methane CH_4 *in vitro*	
Lamiaceae family		CH_4, g/m³	Equivalent of CO_2 *e*/g
Salviae folia	0.8	11.2±2.6*	280±1.2*
	1.6	14.1±3.2*	352.5±2.3
	2.4	9.9±2.3	247.5±1.9
Scutellaria baicalensis	0.15	10.4±2.4	260±3.4
	0.2	10.1±2.3*	252.5±1.6*
	0.3	5.3±1.2*	132.5±1.7*
Oríganum vulgáre	0.2	8.5±1.3**	213.8±2.4
	0.3	8.4±3.1*	210.8±1.1
	0.6	6.9±2.4*	174.8±1.8*

Note: * - P≤0.05; ** - P≤0.01; *** - P≤0.001

4 Discussion of Results

Increasing the volume of production of raising and fattening the cattle is associated with many factors leading to development of a number of diseases (acidosis of the salivary glands, liver abscess, and etc.) and, as a consequence, a decrease in animal productivity [6]. In this regard, the search for new substances that can reduce the load on the digestive tract of animals without disrupting the enzymatic processes in the rumen and without losing productivity is relevant.

In our in vitro study using phyto-substances, the main volatile fatty acids formed during fermentation of feed in the rumen (acetic, propionic, butyric, valeric, caproic acids) were determined. It was found that when using *phyto*-substances: Salviae folia and *Scutellaria baicalensis*, more acetic and propionic acid was formed, introduction of *Oríganum vulgáre* in various dosages led to a shift towards propionic and valeric acid. Similar results were noted in studies [7,8] when using various phyto-substances.

The ability of plant substances to shift the ratio of acetate to butyrate is described in [9], similar results are confirmed in a study [10], which indicates the ability of phytonutrients and phytobiotics to shift fermentation towards formation of propionate and butyrate with inhibition of methanogenesis. In our case, the methane concentration for the *Oríganum vulgáre* sample was lower than for *Salviae folia* and *Scutellaria baicalensis.* Studying the phyto-substances according to their ability to form methanogenesis, the authors of [11] found that in different dosages they are able to reduce the amount of methane produced and can be used in feeding. In contrast to our earlier study [12], no decrease in rumen methane production was observed for Oríganum vulgáre. Assessment of the impact of *Salviae folia, Scutellaria baicalensis, Oríganum vulgare* on the metabolism of various forms of nitrogen in the rumen fluid, namely, formation of a protein form, indicates normal microbiological processes. In study [13], the authors established similar results; it is shown that absorption of non-ammonium nitrogen in the gastrointestinal tract of animals increased when using a diet with addition of plant substances.

Therefore, use of the biomass of plants from the *Lamiaceae family: Salviae folia* and *Scutellaria baicalensis* contributes to formation of acetic and propionic acids, *Oríganum vulgáre* — propionic and valeric acids under in vitro ecosystem conditions, an increase in the synthesis of protein fractions, against the background of a decrease in methane values (*Oríganum vulgáre*).

Acknowledgements

This study was performed with the financial support from the Russian Science Foundation (Project No.21-76-10014).

Conflict of Interest

Authors declares that they have no conflict of interest.

References

1. I. Karimov, K. Kondrashova, G. Duskaev, O. Kvan, E3S Web of Conferences, **143**, 02034 (2020)
2. K. S. Kondrashova, D. B. Kosyan, K. N. Atlanderova, Animal Husbandry and Fodder Production, **103(4),** 128 (2020)
3. K. S. Inchagova, G. K. Duskaev, D. G. Deryabin, Microbiology, **88(1)**, 63 (2019)
4. P. Kuralkar, S. V. Kuralkar, J Ethnopharmacol, **278**, 114246 (2021)
5. T. C. de Jesus Pereira, M. L. Pereira, J. V. Moreira, J. A. Azevêdo, R. Batista, V. F. de Paula, B. S. Oliveira, E. de Jesus Dos Santos, Environ Sci Pollut Res Int, **24(5)**, 4301 (2017)
6. K. M. Benedict, S. P. Gow, T. A. McAllister, C. W. Booker, S. J. Hannon, S. L. Checkley, N. R. Noyes, P. S. Morley, PLoS One., **10(12)**, e0143995 (2015)
7. G. J. Kolling, S. C. B. Stivanin, A. M. Gabbi, F. S. Machado, A. L. Ferreira, M. M. Campos, T. R. Tomich, C. S. Cunha, S. W. Dill, L. G. R. Pereira, V. Fischer, J Dairy Sci., **101(5),** 4221 (2018)
8. P. Løvendahl, G. F. Difford, B. Li, M. G. G. Chagunda, P. Huhtanen, M. H. Lidauer, J. Lassen, P. Lund, Animal, **12(s2)**, s336 (2018)
9. F. Hashemzadeh-Cigari, M. Khorvash, G. R. Ghorbani, M. Kadivar, A. Riasi, Q. Zebeli, J. Dairy Sci., **97(12),** 7487 (2014)
10. J. Oh, E. H. Wall, D. M. Bravo, A. N. Hristov, J Dairy Sci., **100(7)**, 5974 (2017)
11. A. M. Akanmu, A. Hassen, F. A. Adejoro, Antibiotics (Basel), **9(9)**, 601 (2020)
12. D. W. Olijhoek, A. L. F. Hellwing, K. Grevsen, L. S. Haveman, M. R. Chowdhury, P. Løvendahl, M. R. Weisbjerg, S. J. Noel, O. Højberg, L. Wiking, P. Lund, J Dairy Sci., **102(11),** 9902 (2019)
13. J. A. Aguilar-Hernández, J. D. Urías-Estrada, M. A. López-Soto, A. Barreras, A. Plascencia, M. Montaño, V. M. González-Vizcarra, A. Estrada-Angulo, B. I. Castro-Pérez, R. Barajas, H. I. Rogge, R. A. Zinn, J. Anim Sci. **94(1),** 267 (2016)

The effectiveness of agrotextile cover with organic photoluminophore in rooting cuttings of Hungarian lilac (Syringa josikaea J. Jacq. ex Rchb.)

Robert Khramov[1], *Natalia* Martynova[2,*], *Natalia* Besschetnova[2], and *Vladimir* Besschetnov[2], and *Yuriy* Luponosov[3,4]

[1]Institute of Theoretical and Experimental Biophysics of the Russian Academy of Sciences, 142290 Pushchino, Russia
[2]Nizhny Novgorod State Agricultural Academy, 603107 Nizhny Novgorod, Russia
[3]Enikolopov Institute of Synthetic Polymeric Materials of the Russian Academy of Sciences, Moscow, 117393, Russia
[4]Moscow State University, Chemistry Department, 1/3 Leninskie Gory, Moscow, 119991, Russia

Abstract. The effectiveness of the use of a light-transforming shelter of vegetation structures consisting of a polypropylene spunbond with an organic photoluminophore integrated into its structure during the rooting of physiologically active stem cuttings of Hungarian lilac was studied. The object of the study was reproductively mature Hungarian lilac plants located in the arboretum of the Nizhny Novgorod State Agricultural Academy with geographical coordinates 56°14'32.7" N 43°57'20.7"E. The unequal reaction of the tested samples of Hungarian lilac to the use of luminophore in the shelters of vegetation structures during the rooting of cuttings was established, which manifested itself in all characteristics of regeneration processes. High rates of callus formation were in the shelter variants with a higher concentration of luminophore: 73.16 ± 5.95% and 65.25± 4.80%. Lower than in other shelters, the result was recorded in the variant with the lowest luminophore density: 47.00 ± 3.62%.

1 Introduction

The active introduction of woody and shrubby plants into the composition of landscaping facilities and depositing carbon polygons requires the mobilization of a variety of technologies for replicating planting material, including rooting cuttings in vegetation structures [1, 2]. The effectiveness of greenhouse farms is largely determined by the technical properties and optical characteristics of synthetic coatings used in their creation [3, 4]. The use of light-transforming and light-correcting films and plastics for these purposes makes it possible to regulate the spectral and other parameters of the luminous flux in quantitative and qualitative terms [4, 5]. When at the same time, the influence of the parameters of the luminous flux on vital processes in plant organisms is being discussed

*Corresponding author: martynova-natasha94@yandex.ru

quite actively [6-9]. Hungarian lilac (*Syringa josikaea* J. Jacq. ex Rchb.) undoubtedly belongs to the number of plants that successfully perform sanitary-hygienic, decorative-aesthetic and recreational-balneological functions, [10]. The purpose of the research is to evaluate the effectiveness of using a light-transforming shelter of vegetation structures consisting of a spunbond with a photoluminophore integrated into its structure when rooting physiologically active stem cuttings of Hungarian lilac.

2 Materials and methods

The object of the study was reproductively mature Hungarian lilac plants located in the arboretum of the Nizhny Novgorod State Agricultural Academy with geographical coordinates 56°14'32.7"N 43°57'20.7"E and a height above sea level of 178 m. Stem cuttings were simultaneously separated from the peripheral zone of the middle tier of a well-lit section of the crown of plants. They were planted according to uniform placement schemes (5 × 10 cm) in vegetation structures identical in design and metric parameters. The number of cuttings in each variant is 40 pcs. Environmental parameters were monitored using an automatic METEOSCAN PRO 929 RST02929 weather station. Illumination conditions were recorded under cover and without it using a TKA-PKM 08 luxmeter. Light-transforming agrotextile was used as covering materials: polypropylene nonwoven spunbond 30 ± 3 g/m2, coated with polylactide varnish containing a new organic photoluminophore (PL) with different concentrations with absorption mainly in the 460-560nm band, and re-emitting with a spectral maximum of 660 nm with a half-width of the spectrum of 610-730nm. Fluorescent molecules of the donor-acceptor type based on triphenylamine with effective radiation in the red range were used as PL [11, 12]. Mass concentration of PL in the spunbond was varied: 0.25% (background 1); 0.10% (background 2); 0.05% (background 3) 0.025% (background 4). In addition, a variant of the spunbond having 10 cm wide alternating strips uncovered / covered with PL (concentration of 0.25%) (background 5) was introduced into the scheme of the experiment. To assess the potential impact on the regeneration processes occurring on the lower cut of cuttings, a shelter made of the same spunbond with a lacquer film applied without luminophore (standard) was tested. The control was an unmodified shelter – polypropylene nonwoven spunbond (control). It was a non-woven thermally bonded polypropylene fabric with a density of 30 g/m^2. The effect of the optical characteristics of the covering material was tested by the reaction to its use of regeneration indicators according to the following signs: sign 1 - callus formation; sign 2 – the number of adnexal roots formed, pcs.; sign 3 – the length of the leading root, cm; sign 4 – the total total length of the adnexal roots, cm; sign 5 – the total average length of the adnexal roots, cm; sign 6 – the total length of the lateral roots, cm; sign 7 – the average length of the lateral roots, cm; sign 8 – the proportion of the length of the leading root in the total length of the root system; sign 9 – the proportion of the length of the lateral roots in the total length of the root system; sign 10 – the ratio of the length of the leading root to the average length of the lateral roots; sign 11 is the ratio of the length of the leading root to the total average length of the roots; sign 12 – the height of the aboveground part of the stalk; sign 13 – the diameter of the root neck; sign 14 – the ratio of the height of the aboveground part of the stalk to the diameter of its root neck.

The statistical analysis was carried out taking into account the existing methods [13, 14]. Derived features were used, which is customary in forestry and biological studies [15-21].

3 Results and discussion

The reaction of the tested samples of Hungarian lilac to the use of agrotextile with FL in the shelters of vegetation structures during rooting of cuttings was not the same, which was manifested in all characteristics of regeneration processes (Tables 1-4).

Cuttings showed a selective response to covering materials with different optical properties (see Table 1).

Table 1. Callusogenesis on cuttings of Hungarian lilac (%)

Variant	M	CKO	max.	min.	Δ_{lim}	± m	Cv, %	t	P, %
Background -1	73,16	36,62	100,00	10,00	90,00	5,94	50,06	12,31	8,12
Background -2	65,25	30,36	100,00	15,00	85,00	4,80	46,53	13,59	7,36
Background -3	53,50	39,26	100,00	10,00	90,00	6,21	73,38	8,62	11,60
Background -4	53,16	25,27	100,00	10,00	90,00	4,10	47,53	12,97	7,71
Background -5	47,00	22,89	100,00	20,00	80,00	3,62	48,71	12,98	7,70
Standard	45,80	22,31	80,00	10,00	70,00	3,53	48,71	12,99	7,70
Control	46,00	30,62	100,00	10,00	90,00	4,84	66,56	9,50	10,52
Total	54,72	31,44	100,00	10,00	90,00	1,89	57,46	28,91	3,46

[1]The values of indicators: M – arithmetic mean; SKO - standard deviation; max. - maximum value; min. - minimum value; Δlim - range of values; ± m - error of representativeness of the sample mean; Cv, - coefficient of variation, %; t - Student's criterion; P - accuracy of experience or relative error, %.

The highest rates of callus formation were noted in the shelter variants with the highest concentration of luminophore: background 1 - 73.16± 5.95% and background 2 - 65.25± 4.80%. Less than in the other tested shelters, the result was recorded in the variant with the lowest luminophore density: background 5 - 47.00 ± 3.62%. This is close to the results of the control variant (46.00 ±4.84%) and the reference sample (45.80± 3.53%). The generalized value for all variants of the experiment was 54.72 ± 1.89%. In the variants with a light-transforming shelter, the best result exceeded the corresponding minimum by 1.56 times or by 26.15 percentage points.

The influence of lighting conditions on the activity of rhizogenesis on cuttings of Hungarian lilac (see Table. 2) caused, as in the previous case, an ambiguous response of the tested samples to the introduction of a light-transforming substance into the composition of the covering materials of different volumes.

Table 2. Number of adventitious roots on Hungarian lilac cuttings (pcs.)

Variant	M	CKO	max.	min.	Δ_{lim}	± m	Cv, %	t	P, %
Background -1	12,89	5,93	22,00	2,00	20,00	0,96	45,96	13,41	7,46
Background -2	12,65	5,53	20,00	2,00	18,00	0,87	43,72	14,46	6,91
Background -3	10,85	4,70	21,00	5,00	16,00	0,74	43,31	14,60	6,85
Background -4	9,32	3,89	19,00	3,00	16,00	0,63	41,77	14,76	6,78
Background -5	9,50	3,21	15,00	3,00	12,00	0,51	33,80	18,71	5,34
Standard	9,68	3,07	19,00	4,00	15,00	0,48	31,70	19,95	5,01
Control	9,45	4,02	15,00	2,00	13,00	0,64	42,53	14,87	6,72
Total	10,61	4,62	22,00	2,00	20,00	0,28	43,58	38,12	2,62

The main trends in the nature of reactions to photoluminophore established for callus formation (see Table. 1), were preserved for this indicator (see Table 2). A greater number of subordinate roots arose in the case of shelters with increased luminophore density: background 1 - 12.89 ± 0.96 pcs. (background 1) and background 2 – 12.65±0.87 pcs. (background 2). A smaller number of adventitious roots (9.50 ± 0.51 pcs.) was formed in the variant with a reduced luminophore density (background 5). This ratio of estimates formed the excess of their maximum over the minimum by 1.38 times or by 3.58 pcs.

Against the background of the generalized values for all variants of the experiment (10.61 ± 0.28 pcs.), the standard and control had noticeably smaller values: 9.68 ± 0.48 pcs. and 9.45 ± 0.64 pcs., respectively. The best result in the variants with photoluminophore exceeded the control by 1.37 times or by 3.20 pcs., and the standard - by 1.33 times or by 3.22 pcs.

The growth rates of the leading root of the systems depended on the density of the luminophore applied to the surface of the greenhouse covering materials (see Table 3). Despite the fact that the indicators of most variants of the experiment differed little from the control (8.54± 0.58 cm) and the standard (8.57± 0.41 cm), the variant with an increased density of luminophore (background 1), having reached an estimate of 10.53 ± 0.74 cm, demonstrated a noticeable superiority over them: 1.23 times or 1.99 cm (relative to the control) and 1.23 twice or by 1.96 cm (relative to the standard). The generalized average value was 8.84±0.23 cm.

Table 3. Length of the leading root on cuttings of Hungarian lilac (cm)

Variant	M	CKO	max.	min.	Δ_{lim}	± m	Cv, %	t	P, %
Background -1	10,53	4,55	15,00	0,90	14,10	0,74	43,27	14,25	7,02
Background -2	8,62	3,80	13,00	0,80	12,20	0,60	44,10	14,34	6,97
Background -3	8,59	3,25	11,50	1,00	10,50	0,51	37,85	16,71	5,98
Background -4	8,25	3,32	13,50	1,30	12,20	0,54	40,17	15,35	6,52
Background -5	8,84	5,07	15,00	1,00	14,00	0,80	57,39	11,02	9,07
Standard	8,57	2,59	11,00	0,10	10,90	0,41	30,22	20,93	4,78
Control	8,54	3,69	13,50	0,40	13,10	0,58	43,22	14,63	6,83
Total	8,84	3,85	15,00	0,10	14,90	0,23	43,56	38,13	2,62

One of the most significant indicators of regeneration occurring on cuttings is the total length of root systems (see Table 4). There is clearly an increase in the growth of adventitious roots in variants in which covering materials containing a higher amount of luminophore were involved: 92.88 ± 9.24 cm (background 1) and 90.53±9.09 cm (background 2).

Table 4. Total length of root systems on cuttings (cm)

Variant	M	CKO	max.	min.	Δ_{lim}	± m	Cv, %	t	P, %
Background -1	92,88	56,95	177,00	1,90	175,10	9,24	61,31	10,05	9,95
Background -2	90,53	57,49	162,40	1,50	160,90	9,09	63,51	9,96	10,04
Background -3	65,81	38,48	136,90	2,20	134,70	6,08	58,47	10,82	9,24
Background -4	55,22	29,19	99,80	3,80	96,00	4,74	52,87	11,66	8,58
Background -5	54,70	37,97	130,50	2,50	128,00	6,00	69,42	9,11	10,98
Standard	54,81	38,21	184,10	7,00	177,10	6,04	69,71	9,07	11,02
Control	54,86	34,44	107,00	0,60	106,40	5,44	62,78	10,07	9,93
Total	66,87	45,46	184,10	0,60	183,50	2,74	67,99	24,43	4,09

To a lesser extent, the accessory roots grew in variants in which the luminophore concentration was lower: 55.22 ± 4.74 cm (background 4) and 54.70± 6.00 cm (background 5). These estimates were close to the control indicators (54.86±5.44 cm) and the standard (54.81±6.04 cm). The introduction of photoluminophore into the structure of the covering materials created a positive effect: the control and standard indicators were 1.69 times less than in the best version of the experiment, or 38.07 cm and 38.03 cm, respectively.

The recorded differences manifested themselves against a leveled background of environmental conditions, which gave reason to recognize the effect of the optical properties of the covering materials used as the cause of their occurrence. Confirmation of this was obtained during the analysis of variance (Table 5). For a number of indicators of the regenerative ability of cuttings, the differences were significant – Fisher's experimental criteria significantly exceeded the corresponding limit values for a given number of degrees of freedom at both 5 percent and 1 percent significance levels, taking values from 2.85 (sign 10) to 6.38 (sign 4) and 6.84 (sign 6) at $F_{05} = 2.13$ and $F_{01} = 2.87$.

Table 5. The significance of the differences between the application options [1]

Signs of regenerative ability	$F_{оп}$	The share of the influence of the factor ($h^2 \pm s_h^2$)						Criteria for differences	
		by Plokhinsky			by Snedekor				
		h^2	$\pm s_h^2$	F_h^2	h^2	$\pm s_h^2$	F_h^2	HCP_{05}	D_{05}
Sign 1	4,76	0,0961	0,0202	4,76	0,0872	0,0204	4,28	13,342	20,116
Sign 2	4,80	0,0967	0,0201	4,80	0,0879	0,0203	4,32	1,962	2,973
Sign 3	1,51	0,0325	0,0216	1,51	0,0127	0,0220	0,57	1,690	2,561
Sign 4	6,38	0,1246	0,0195	6,38	0,1201	0,0196	6,12	18,987	28769
Sign 5	1,14	0,0248	0,0218	1,14	0,0035	0,0222	0,16	1,179	1,787
Sign 6	6,84	0,1324	0,0194	6,84	0,1290	0,0194	6,64	17,829	27,014
Sign 7	1,46	0,0316	0,0216	1,46	0,0116	0,0220	0,52	1,176	1,782
Sign 8	1,94	0,0415	0,0214	1,94	0,0233	0,0218	1,07	0,058	0,088
Sign 9	1,94	0,0415	0,0214	1,94	0,0233	0,0218	1,07	0,058	0,088
Sign 10	2,85	0,0597	0,0210	2,85	0,0448	0,0213	2,10	0,471	0,714
Sign 11	1,98	0,0424	0,0214	1,98	0,0243	0,0218	1,12	0,249	0,378
Sign 12	4,71	0,0938	0,0199	4,71	0,0848	0,0201	4,22	1,576	2,388
Sign 13	1,35	0,0288	0,0213	1,35	0,0087	0,0218	0,40	0,355	0,538
Sign 14	1,42	0,0302	0,0213	1,42	0,0103	0,0218	0,48	0,385	0,583

[1] Indicators: F_{op} – the experimental value of the Fisher criterion; F_{05} – the tabular value of the Fisher criterion at a 5% significance level ($F_{05} = 2.13$); F_{01} - the tabular value of the Fisher criterion at a 1% significance level ($F_{05} = 2.87$); h^2 – the share of the influence of the organized factor; $\pm$ sh^2 - the error of the share of the influence of the organized factor; F_h^2 – the Fisher criterion in assessing the reliability of the share of the influence of the organized factor; NCP_{05} – the smallest significant difference at a 5% significance level; D_{05} – the Tukey criterion is at a 5% significance level; signs according to previously accepted designations.

At the same time, in most of the signs (signs 3, 5, 7, etc.), the significance of the differences between the variants of light-transforming materials, the significance of the differences has not been confirmed. In variants with confirmed significance of differences, the analysis of variance confidently refutes the null hypothesis that there are no differences between the variants of rooting cuttings using different categories of modified covering material of vegetation structures (see Table 5). This made it possible to continue its implementation on such grounds in terms of calculating estimates of the share of the influence of an organized factor (in our case, the specifics of the regenerative ability of biological samples due to the influence of a particular shelter on them) on the formation of a general phenotypic variance.

In calculations using the Plokhinsky algorithm, estimates ranged from 5.97±2.10% ($F_{h2} = 2.85$; $F_{05}/_{01} = 2.13/2.87$) in terms of balanced development of root systems (sign 10) to 12.46±1.95% ($F_h^2 = 6.38$; $F_{05/01} = 2.13/2.87$) in terms of the total length of root systems (sign 4) and 13.24±1.94% ($F_h^2 = 6.84$; $F_{05/01} = 2.13/2.87$) in terms of the total length of lateral roots (sign 6). The calculation of the same values using the Snedekor algorithm gave quite comparable estimates. It can be stated that there is a tendency in the increase of regeneration activity and the processes of root system growth and development of the aboveground part of Hungarian lilac cuttings occurring after it against the background of an increase in the amount of luminophore in the covering material.

4 Conclusions

Agrotextile covering materials containing organic photoluminophore emitting in the red spectral range cause an increase in the activity of callusogenesis and rhizogenesis, as well as an increase in the rate of post-regenerative development of root systems and the aboveground part of stem cuttings of Hungarian lilac. An increase in the concentration of photoluminophore in the covering material of summer vegetation structures leads to an increase in the regenerative ability of summer stem cuttings of Hungarian lilac.

Acknowledgements

The authors are grateful to D.O. Balakirev (Enikolopov Institute of Synthetic Polymeric Materials, of the Russian Academy of Sciences) for the production and provision of PL-containing agrotextile for experiments. The work was financially supported by Ministry of Science and Higher Education of Russian Federation [the State Assignments Numbers 075-00381-21-00] with the partial financial support from the Russian Foundation for Basic Research [Grant Number 18-29-17073].

References

1. J.B. Scarratt, For. Chron. **61**(4) (1985)
2. V.P. Besschetnov, E.Zh. Kentbaev, Russian Forestry journal, **4** (2018)
3. C. Edser, Plast. addit. compd. **4(3)** (2002)
4. R.P. Brown, Rapra Rev. Rep. **15(2)** (2004)
5. R. N. Khramov, V. D. Kreslavski, E. A. Svidchenko, N. M. Surin, A. A. Kosobryukhov, Opt. Express. **27(22)** (2019)
6. M.A.K. Jansen, Physiol Plant.**116**, 3 (2002)
7. M.L. Delprato, J. Photochem. Photobiol. **91(5)** (2015)
8. Y. Guo, J. Tan, J. Photochem. Photobiol. **91(1)** (2015)
9. A. Losi, C. Mandalari, W. Gärtner, J. Photochem. Photobiol. **91(5)** (2015)
10. S. Wierdak Acta Soc. Bot. Pol. **1(2)** (1923)
11. Yu. N. Luponosov, A. N. Solodukhin, D.O. Balakirev, N. M. Surin, E. A. Svidchenko, S. A. Pisarev, Yu. V. Fedorov, S. A. Ponomarenko, Dyes and Pigments **179**, 108397 (2020)
12. Yu. N. Luponosov, A. N. Solodukhin, A. L. Mannanov, P. S. Savchenko, B. A. L. Raul, N. M. Surin, S. M. Peregudova, A. V. Bakirov, M. A. Shcherbina, S. N. Chvalun, M. S. Pshenichnikov, D. Yu. Paraschuk, S. A. Ponomarenko, Materials Today Energy. **22**, 100863 (2021)
13. J.H. Zar, Biostatistical Analysis: Fifth Edition. Edinburg Gate: Pearson New International edition - Pearson Education Limited, 756 (2014)
14. A. Dean, D. Voss, D. Draguljić, Design and Analysis of Experiments (Springer Texts in Statistics). 2nd Edition, Kindle Edition. Heidelberg, Germany: Springer-Verlag GmbH. 865 (2017)
15. F. Noll, For. Chron. **86(5)** (2010)
16. N.N. Besschetnova, V. P. Besschetnov, Vavilov Journal of Genetics and Breeding. **21(2)** (2017)

17. N.N. Besschetnova, V.P. Besschetnov, A.V. Kul'kova, I.V. Mishukova, Russian Forestry journal, **4** (2017)
18. N.N. Besschetnova, V.P. Besschetnov, P.V. Ershov, Russian Forestry Journal, **1** (2019)
19. V.P. Besschetnov, N.N. Besschetnova, A.O. Esichev, Russian Forestry journal, **1** (2018)
20. A.V. Kul'kova, N.N.Besschetnova, V.P.Besschetnov, Russian Forestry Journal, **6** (2018)
21. M.M. Ulitin, V.P. Besschetnov, Russian Forestry Journal, **6** (2020)

Performance improvement of aquaponics systems when using microbiological strains in them

Olga Krotova[1,*], *Sergey* Chelbin[2], *Maria* Krotova[3], *Olga* Sangadzhieva [4], and *Kermen* Khalgaeva[4]

[1] Don State Technical University, Gagarin Sq., 1, 344003 Rostov-on-Don, Russia
[2]Branch of the Federal State Budgetary Institution "Russian Agricultural Center" in the Rostov region, 344029 Rostov-on-Don, Russia
[3] Volgograd State Agricultural University, 26 Universitetskiy ave, 400002 Volgograd, Russia
[4] Kalmyk State University named after B.B. Gorogovikov, Pushkin St., 11, 358000 Elista, Russia

Abstract. The essence of the method of operation of the aquaponic system is to use the vital waste of aquatic animals as food for plants. At the first stage, the fish produces waste products. Then, in the second stage, microbes and worms convert the waste into fertilizersfor plants. During the process, at the third stage, plants consume the necessary products of excretion of living organisms. Aquaponics is an artificial ecosystem in which three types of living organisms are key: aquatic animals (usually fish or shrimp), plants and bacteria. This technology works on the principle of an ecosystem of fish and plants: fish provides foodto plants, and plants purify water and are environmentally friendly. The essence of this method is to use the vital waste of aquatic animals (fish, shrimp) as a nutrient medium for plants. These are chemicals such as nitrogenous, potassium, phosphorus compounds, carbon dioxide. Thanks to aquaponic farms, the costs of water and other natural resources, soil pollution and the use of insecticides are already being reduced in different places of our planet. Such systems are a sustainable ecosystem capable of producing organic matter that does not contain toxic substances often present in vegetables grown by traditional methods.

1 Introduction

The progress of the food industry is connected with the progress of science. Competition in the food market and the deterioration of the environmental situation are one of the important factors of development. All this leads to the creation and improvement of agricultural practices.

Currently, there are global environmental problems, including food production, which directly concern agriculture. This is the impoverishment and destruction of arable land, the absence and pollution of fresh water. In addition, scientific progress does not always bring benefits. Putting high yields in the first place in a short time, manufacturers use fertilizers

* Corresponding author: Alb9652@yandex.ru

that negatively affect the health of consumers in the future. Therefore, new approaches to agriculture are already needed to obtain high-quality products without harming the surrounding world and the consumer.

The purpose of the research work is to study a high – tech method of farming-aquaponic.The relevance and importance of the chosen topic lies in the fact that aquaponics today is the most promising direction in terms of production technologies and scientific research. This system can be used all year round, regardless of climatic and soil conditions. aquaponics allows you to look at agriculture through the prism of the latest technologies in agriculture. This technology will take a very important place in the development of agriculture in the future, since the production process excludes the use of chemical fertilizers. The process of chemicalization and purification occurs naturally in a closed cycle. Aquaponics is considered the most environmentally friendly way of growing plants due to the lack of chemical fertilizers [1-3]. The practical significance of this method is quite large. For example, the development of this technology of growing plants can significantly reduce the cost of obtaining fresh vegetables in closed conditions in the off-season, which will have a positive impact on the health of consumers.

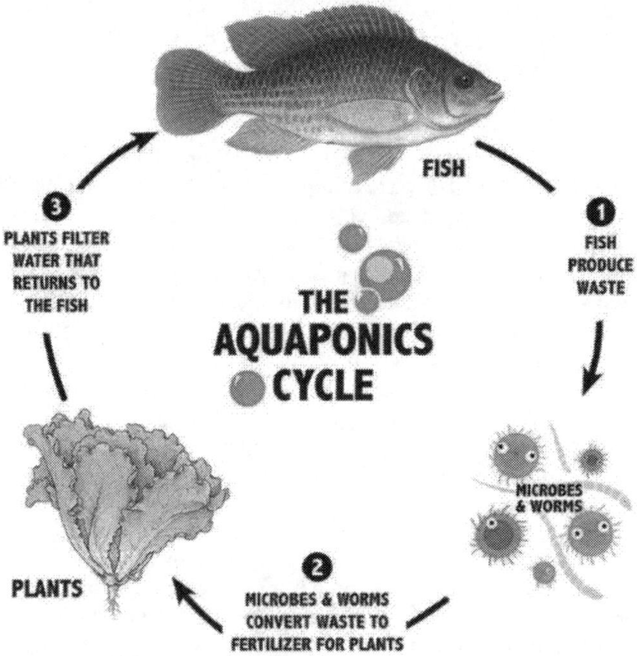

Fig. 1. Aquaponics cycle

The essence of the method of operation of the aquaponic system is to use the vital waste of aquatic animals as food for plants. At the first stage, the fish produces waste products. Then, in the second stage, microbes and worms convert the waste into fertilizers for plants. During the process, at the third stage, plants consume the necessary productsof excretion of living organisms. These are chemicals such as nitrogenous, potassium, phosphorus compounds, carbon dioxide.

Aquaponics is an artificial ecosystem in which three types of living organisms are key: aquatic animals (usually fish or shrimp), plants and bacteria. This technology works on the principle of an ecosystem of fish and plants: fish provides food to plants, and plants purify water and are environmentally friendly. The essence of this method is to use the vital waste of aquatic animals (fish, shrimp) as a nutrient medium for plants [4-6].

Wastewater generated as a result of malnutrition of feed or the vital activity of fish accumulates in the water due to closed circulation in most aquaculture systems. Water containing effluents becomes toxic and dangerous to animals in high concentrations, but at the same time contains nutrients that are necessary for plant nutrition and growth. Aquatic animals emit toxic waste products for themselves: chemical compounds.

The accumulation of these substances in the water is a serious problem, both in closed industrial aquaculture and in a conventional aquarium. These substances are absolutely necessary in hydroponics and are added to water to obtain nutrient solutions for plants. In aquaponics, this problem is solved in this way: the vital products of fish are removed by bacteria and plant [7-9].

Although aquaponics mainly consists of these two parts, aquaponic systems are usually grouped into several components or subsystems responsible for effective waste disposal, adding a base for acid neutralization or enriching water with oxygen.

The main product of microbiological decomposition of fish waste and other representatives of aquatic fauna is ammonia-NH3, released and dissolved in water, which, in turn, during the vital activity of aerobic bacteria, as well as oxygen dissolved in water – O2, oxidizes ammonia and its gas derivatives – amine, forming nitrites and nitrates — "saltpeter". This reduces the chemical toxicity of water in the vital activity of animals and allows plants to remove the resulting nitrate compounds, consuming the necessary elements for their life support.

Plants can also absorb ammonia directly from water, but dissolved salts are absorbed much more readily and more easily. Colonies of bacteria living in the soil and the root system of plants in a closed cycle of aquaculture purify water from toxic substances, and plants consume salts, gases and chemical elements dissolved in water – nitrates, nitrogen – N, phosphorus – P, carbon dioxide – CO2 and, to a certain extent, enrich oxygen into water, which is also returned to animals in the form of soda.

Thus, this system simulates the natural circulation of substances in nature, where cultivated plants act as producers, aquatic animals, as consumers – consults, and bacteria – as reducers [9-11].

Typical elements of an aquaponics system include: a fish breeding tank; a sump: an aggregate powered to absorb feed residues, lost biofilm and to separate small particles; a biofilter: a place where nitrifying bacteria grow and turn ammonia into nitrates necessary for plant nutrition; hydroponics subsystems: a part of the system in which plants are grown through the absorption of excess nutrients from water; a tray: the lowest point in the system in which water enters and rises back into tanks.

Aquaponics is studied all over the world. Figure 2 shows an aquaponics system developed by students in the Nature Laboratory of the Rhode Island School of Design in the United States of America.

Fig. 2. Aquaponics system developed by students in the Nature Laboratory of the Rhode Island School of Design

The aquaponics system developed in the Nature Laboratory of the Rhode Island School of Design was built using a one hundred and thirty gallon aquarium, PVC pipelines, two LED lamps for growing, two separate beds for growing, various pumps and filters. This project illustrates why aquaponics is an important event in the future of food production. This system is installed in the student canteen as a reminder of the importance of fresh and nutritious food in the human diet [12].

Depending on the cost and complexity of the aquaponic system, waste tanks, biofilters or subsystems of hydroponics can be combined into one component or subsystem, which does not allow the flow from the aquaculture part of the system to the part of the hydroponics system. Aquaponics makes it possible to significantly reduce, but in some cases even reduce to zero, the discharge of wastewater.

In the valve system, it is desirable to grow green-leaved plants, such as beans, peas, cauliflower, radishes, tomatoes, melon, strawberries, pumpkin, onion, pepper. In these plants, at the early stages of development, there is a need for nitrogen, which is just released into the water as a result of the vital processes of fish. Within this system, the number of aquatic animals directly depends on the number of cultivated plants [13, 14].

In Russia, the first aquaponic farm is being built in the Krasnodar Territory in the village of Maryina Roscha (near Gelendzhik). Here, using the unique aquaponics technology, tilapia fish and green crops, including lettuce and mint, will be grown. At the moment, the construction is at the initial stage [15-16].

The area of the farm is 1.3 hectares, two modules will be located here, the design capacity of each is 40 tons of fish per year. Aquaponics is a very organic method of growing plants and fish in a closed cycle. Fish saturates water with

nutrients necessary for optimal plant growth. In the future, it is possible to grow such fish species as trout, baramundi and sturgeon. Aquaponics is an interconnected hybrid system of pond agriculture and hydroponics. The system for growing vegetables and herbs can be both large and small, depending on the wishes, and is potentially able to provide fish and vegetables in huge quantities [17].

Changing the concept requires not only opportunities in agriculture – at least in the development of small farms, but also in the way of harvesting and its further consumption.

The application of aquaponics on an industrial scale is shown in Figure 3.

Fig. 3. Application of aquaponics on an industrial scale

In the first section, the relevance of scientific research in the field of aquaponics was considered. The environmental friendliness of this method is considered.

Fish are kept in a tank with filtered water from all harmful impurities. No human waste gets into the water, for example, garbage, polyethylene, petroleum products. In addition, the absence of mud in the water excludes the appearance of a grassy smell in the fish. Fish production takes place without the use of synthetic mineral fertilizers that regulate growth, artificial food additives, as well as without genetically modified products

Thanks to aquaponic farms, the costs of water and other natural resources, soil pollution and the use of insecticides are already being reduced in different places of our planet. Such systems are a sustainable ecosystem capable of producing organic matter that does not contain toxic substances often present in vegetables grown by traditional methods.

2 Materials and methods

Criteria for evaluating the effectiveness of the studied object: the experiment was carried out in two versions: with the addition of a bacterial strain (experiment) and without the introduction of a bacterial strain (control).

For statistical data processing, we used the Microsoft Office Excel program. Statistical processing of the results of the study was carried out using the Student's criterion. The changes at $p \leq 0.95$ were considered to be significantly significant.

Parameters controlled in the studies: Table 1 shows data on the cultivation of tilapia in an aquaponic installation for 30 days in the experiment.

Table 1. Dynamics of mass characteristics of tilapia in an aquaponic installation

Indicator	Initial state	After 30 days
Weight, g	246,35 ± 11,07	389,81 ± 11,73
Absolute increase, g	-	143,46
Average daily increase, g/day	-	4,74
Average daily growth rate, %	-	1,52
Mass accumulation coefficient, units.	-	0,10
Fulton fatness coefficient, units.	0,56 ± 0,02	0,68 ± 0,01

At the end of the experiment, a study of the physiological parameters of tilapia in both versions of the experiment was conducted, presented in Table 2.

Table 2.Dynamics of physiological and biochemical parameters of tilapia blood when grown in an aquaponic installation

Indicator	Experience	Control
ESR, mm/h	3,83 ± 0,14	3,82 ± 0,15
Hemoglobin, g / l	41,92 ± 4,271	43,51 ± 1,210
Total protein, g/l	45,30 ± 0,86	46,70 ± 0,34

The values of hemoglobin, ESR, and total protein were stable in both versions of the experiment, and did not changesignificantly. In the control, compared with the experiment, the hemoglobin index of tilapia was higher by 3.7%.

Thus, it can be concluded that the culture fluid Serratia ficaria TP3 does not significantly affect the blood parametersof tilapia.

Strawberries were treated with the same solution of the bacterial strain. Previously, before planting the plants in the system, the tubers were soaked in a solution of a bacterial strain. Then, during the entire experiment, the plants were treated in the experiment by spraying every 10 days. The length of the stem at the end of the experiment in the experiment is 69 mm, and in the control 46.2 mm. Figure 4 shows the change in the length of the stem.

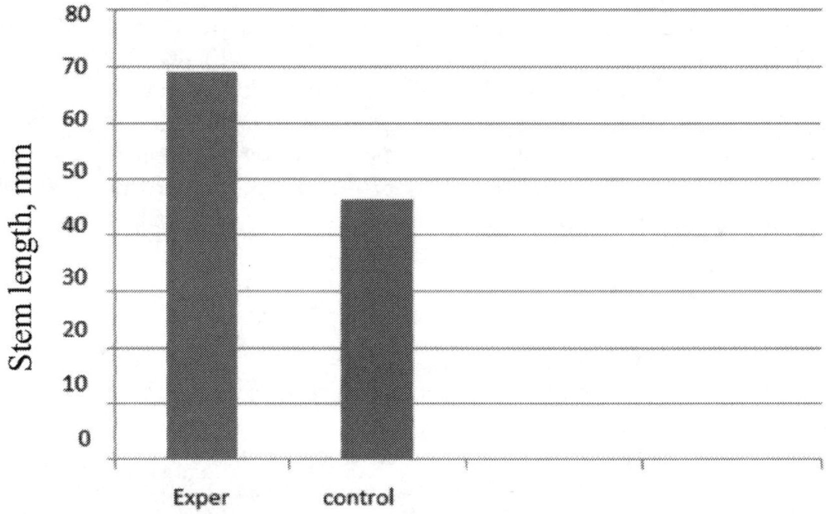

Fig. 4. The length of the strawberry stem

When harvesting, the level of nitrates in strawberries was much lower in the experiment than in the control, which proves the effectiveness of the biological product.

Figure 5 shows the level of nitrates in strawberries.

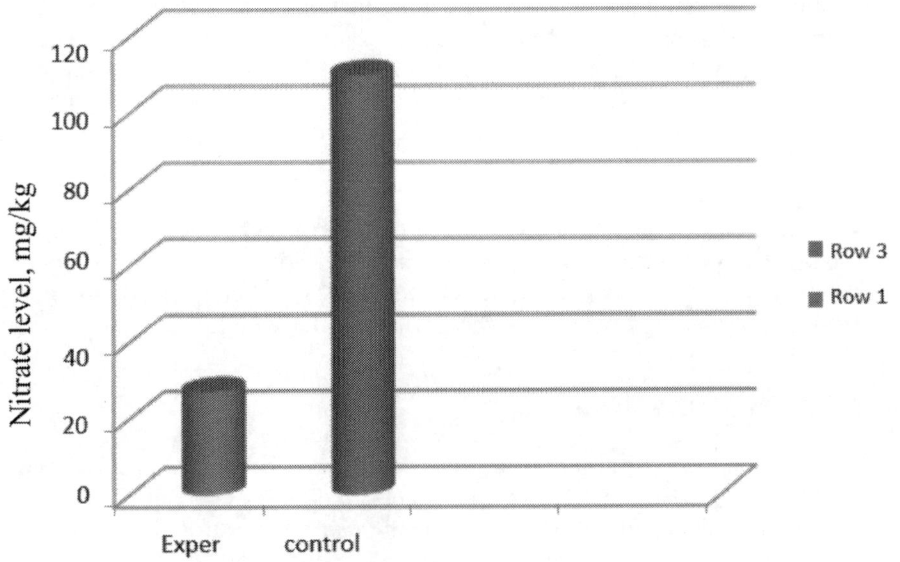

Fig. 5. The level of nitrates in strawberries

Therefore, the treatment of strawberries with a solution of a bacterial strain by spraying plants at the stage of creating the first brush has the greatest positive effect on the growth indicators of strawberries in the flowering phaseand in the fruiting phase. The accumulation of strawberry nitrates was significantly below the level of the maximum permissible concentration.

Equipment, experimental installations, instruments, equipment, accessories:

Fig. 6. Shows an aquaponic plant for growing strawberries.

Conditions and procedure for conducting experiments, the composition of experiments.

Tilapia was grown in modular systems with a volume of 500 liters, with a water temperature of 28 ± 1 °C, an oxygen content of 5 ± 1 mg/l, a pH of 7 ± 1, nitrites-0.1-0.2 mg/l, ammonia-ammonium-0.5-1 mg/l. The density of planting tilapia in growing pools was 35 copies. per 1 m3 in experience and control. Together with tilapia, the method of aquaponics was used to grow strawberries of the "Mashenka" variety. The experiment was carried out in two versions: with the addition of a bacterial strain (experiment) and without the introduction of a bacterial strain (control).

In the experiment, a microbiological biological product based on a strain of bacteria was used to accelerate plant growth and reduce the proportion of nitrates in the production. The culture liquid of the isolate has been previously tested and is a solution that is safe for fish and the environment. To prepare a suspension of the Serratia ficaria SP3 isolate, it was cultivated on potato agar with continuous stirring on a rocking machine (120 rpm) at a temperature of 280

°C [7, 8].

A sterile scalpel was scraped from the surface of the fish body, the resulting material was placed on solid nutrientmedia and rubbed with a spatula. At the same time or at the end of the crops, smears were made from the mucus of the body surface, gill sections, as well as smears-prints from internal organs, staining them with Leffler's methylene blue, by Gram.

3 Results and Discussion

Processing of research results and their analysis. Studies have been conducted to study the microflora of the tilapia surface during the joint cultivation of tilapia and various agricultural plants in an aquaponic installation. The number of cells from the tilapia surface in a pure sample was 4.2 • 107 CFU/ml. In the sample without the addition of a bacterial strain, the development of a similar microflora was observed, but less intense. The number of cells was 5.3 • 107 CFU/ml. The working solution of the isolate Serratia ficaria TP3 activated the physiological processes of the strawberry variety "Mashenka", had a favorable effect on the development of plants, contributed to a decrease in the level of nitrates and an increase in yield: the yield of strawberries in the experiment with the culture liquid Serratia ficaria TP3 was 5.2 kg/m3, and in the control – 3.8 kg/m3. The effect of the culture fluid Serratia ficaria TP3 on tilapia was not observed. Aquaponics is a progressive direction of agricultural development. It has many advantages, both for the consumer and for the manufacturer. But the most important thing is that aquaponics solves the problems of the impact of traditional agriculture on the environment. Aquaponics is widely used in the world, but the distribution of its use is uneven. Aquaponics has become stronger in the countries of Asia, Europe and North America. In Russia, the method of joint cultivation of fish and plants is not so popular yet, but it is already used by some companies.

This method of farming is based on the interaction of several groups of living organisms that mutually affect each other, as in natural ecosystems. It is the use of the closed system principle that makes aquaponics a unique, efficient and environmentally friendly type of agriculture. The main processes in aquaponic systems are nitrification, phosphorylation, photosynthesis and root nutrition of plants, the processes of nutrition and excretion in fish. All theseprocesses are absolutely natural and occur in nature.

Aquaponics has several methods of management, so you can choose the best conditions for growing plants and fish.

4 Conclusions

Many plant crops can be grown in aquaponics. Leaf lettuce, strawberries, onions of various varieties, spicy herbs, tomatoes, cucumbers, peppers have become widespread.

Studies were conducted to study the microflora of the tilapia surface during the joint cultivation of tilapia and various agricultural plants in an aquaponic installation.

Aquaponics in Russia is a promising direction. This is an eco-friendly and compact type of production that can exist within the city. High productivity allows us to provide environmentally friendly products to the consumer in a fairly short time with proper organization of production.

The transition of agriculture from soil and groundwater pollution to "clean" aquaponic production is a big step towards an environmentally friendly future.

References

1. Z. Schmautz, C.A. Espinal, A.M. Bohny, F. Rezzonico, R. Junge, E. Frossard, T.H.M., BMC Microbiology, **21(1)**, 12 (2021)
2. G.L. Lenz, A.Loss, C.R. Lourenzi, D. Luiz de Alcantara Lopes, Aquaculture Research, **52(10)**, pp. 5008-5021(2021)
3. A.A. Bracino, J.L. Espanola, A.A. Bandala, E.P. Dadios, E. Sybingco, R.R.P. Vicerra, Journal of Advanced Computational Intelligence and Intelligent Informatics, **25(5)**, 632-638 (2021)
4. R.S.Concepcion, S.C. Ii, Lauguico, J.D. Alejandrino, A.A. Bandala, E. Sybingco, R.R.P. Vicerra, E.P. Dadios, J.L., Journal of Advanced Computational Intelligence and Intelligent Informatics, **25(5)**, 610-617 (2021)
5. M. Eck, I. Szekely, S. Massart, M.H. Jijakl, Acta Horticulturae, **1321**, 213-219 (2021)
6. P.Derikvand, B. Sauter, L.Y. Stein, Applied Microbiology and Biotechnology, **105(18)**, 7009-7021(2021)
7. W.A. Lennard, Aquaculture and Fisheries, **6(5)**, 495-505 (2021)
8. John, J., Mahalingam P., M.P.R., *Automated Fish Feed Detection in IoT Based Aquaponics System*, 8th International Conference on Smart Computing and Communications: Artificial Intelligence, AI Driven Applications for a Smart World, ICSCC 2021, 286-290 (2021)
9. Z.J. Ong, A.K. Ng, T.Y. Kyaw, *Intelligent Outdoor Aquaponics with Automated Grow Lights and Internet of Things*, Proceedings of 2019 IEEE International Conference on Mechatronics and Auto mation, ICMA 2019, 8816577, 1778-1783 (2019)
10. D. Karimanzira, T. Rauschenbach, Information Processing in Agriculture, **6(3)**, 375-385 (2019)
11. O. Witzel, S. Wilm, D. Karimanzira, Baganz, Information Processing in Agriculture, **6(3)**, 326-334(2019)
12. V. Orobets, E. Lisovets, S. Zabashta, A. Ermakov, *Control of fish parasites in aquaculture* , IOP Conference Series: Earth and Environmental Science. 12th International Scientific Conference on Agricultural Machinery Industry, INTERAGROMASH 2019, 012065 (2019)
13. A.L.J. Rico, International Journal of Recent Technology and Engineering, **8(3)**, 7650-7653 (2019)
14. G. G. Matishov, E. N. Ponomareva, *State and prospects of development of aquaculture biotechnologies in the south of Russia*, Collection of joint publications of the staff of the UNC RAS and DSTU. Under the general editorship of G. G. Matishov, B. Ch. Meskhi, I. V. Karmanov (ed.). Rostov-on-Don, 18-20 (2020)
15. M.F. Mohd Pu'Ad, K.A. Sidek, M. Mel, International Journal of Innovative Technology and Exploring Engineering, **9(1)**, 4181-4184(2019)
16. H.H. Bennouri, A.A. Berqia, N.K. Patrick, Journal of Communications, **14(12)**, 1191-1197 (2019)
17. K. Al-Kodmany, Modeling and Optimization in Science and Technologies, **17**, 203-232 (2020)

Influence of the concentrations of essential and toxic elements in the blood serum on the indicators of sperm quality in Arabian purebred stallions

V. V. Kalashnikov[1], *A. M.* Zaitsev[1], *M. M.* Atroschenko[1], *O. A.* Zavyalov[2], *A. N.* Frolov[2], and *M. Ya.* Kurilkina[2]

[1]All-Russian Research Institute of Horse Breeding, Russia
[2]Federal Research Center for Biological Systems and Agricultural Technologies of the Russian Academy of Sciences, Russia

Abstract. Analysis of trace elements in blood serum can be an important tool for monitoring micronutrient supply in stallions in order to maintain reproductive health. In this regard, the purpose of this study was to determine the relationship between the concentrations of macro- and microelements in the blood serum of stud stallions with the qualitative characteristics of fresh sperm and sperm after cryopreservation and thawing. The studies were carried out on purebred Arabian stallions (n=50). As a biomaterial for studies, we used blood serum, fresh and thawed sperm after cryopreservation. The elemental composition of seminal plasma was determined by 25 chemical elements Co, Cr, Cu, Fe, I, Mn, Se, Zn Ca, K, Mg, P, B, Li, Si, V Na, Ni, Sn, Hg, Sr, Al, As, Cd, Pb) by ICP-DRC-MS methods. It was found that the concentrations of Mg and Co in the blood serum positively correlate with the qualitative characteristics of the sperm of Arabian purebred stallions. Wherein, the optimal range of concentrations of Mg in the blood serum, typical for stallions with relatively high indicators of sperm quality, is 22.64 to 27.05 μg/g; Co – 0.001 to 0.0015 μg/g.

1 Introduction

Currently, along with development of new reproductive technologies in animal husbandry, a search is underway for methods to recover the natural fertility of animals and humans. Moreover, chemical elements represent an important group of ecophysiological factors that determine the morphological and functional characteristics of the reproductive system [1,2]. Namely, it was found that the concentration of magnesium (Mg) significantly correlates with the concentration of spermatozoa and the level of Ca in the seminal plasma of stallions [3]. Copper (Cu) is an essential component of many metalloenzymes such as superoxide dismutase, ceruloplasmin and lysine oxidase, which are critical for antioxidant processes in sperm. It affects the progressive motility, viability, membrane integrity and prevents DNA damage after sperm dilution and cryopreservation [4, 5, 6]. Zinc (Zn) plays an important role in sperm motility and directly affects its morphology [7, 8], and is also an important factor in the normal functioning of the prostate gland and the reproductive system as a whole [9,

10, 11]. Ferrum (Fe) deficiency reduces the activity of iron-containing and iron-dependent enzymes in seminal plasma [12]. Selenium (Se) has antioxidant properties and is essential for spermatogenesis and the maintenance of male fertility [13]. A high level of Lead (Pb) is associated with development of infertility caused by induction of a spontaneous premature acrosome reaction, and an increase in the concentration of Pb in seminal plasma may adversely affect the rates of in vitro fertilization [14, 15]. Exposure to Cadmium (Cd) can have a detrimental effect on reproductive function, causing atresia of ovarian follicles [16], uterine edema [17] and, in some cases, degeneration of the testes [18]. Previous studies have shown that administration of a certain dose of Cadmium decreases sperm motility [19]. Nickel (Ni) can impair the function of the testes, seminal vesicles and prostate [20]. In general, it can be stated that despite the large number of scientific papers published in recent years on the study of the effect of the metabolism of chemical elements in relation to reproductive functions, all of them mainly relate to humans, some agricultural species, and less often to stallions. Wherein, seminal plasma is more often used as a biosubstrate for assessing the exchange pool of chemical elements in the body. Wherein, the analysis of microelements in blood serum can be an important tool for monitoring the micronutrient supply of the body in order to maintain the reproductive health of farm animals [21, 22, 23, 24] and, namely, stallions [25].

In this regard, the purpose of this study was to determine the relationship between the concentrations of macro- and microelements in the blood serum of stud stallions with the qualitative characteristics of fresh sperm and sperm after cryopreservation and thawing.

2 Materials and Methods

2.1 Study object

The studies were carried out on stallions of the Arabian purebred horse breed. The live weight of stallions within sampling period was 410.5±25.3 kg, age — 4-5 years.

Animal care and experimental studies were performed in accordance with the instructions and recommendations of Russian Regulations, 1987 (Order No.755 on 12.08.1977 the USSR Ministry of Health) and "The Guide for Care and Use of Laboratory Animals (National Academy Press Washington, D.C. 1996)". Efforts have been made in the studies to minimize animal suffering and reduce the number of samples used.

2.2 Experiment scheme

Clinically healthy stallions of the Arabian purebred breed (n=50) bred under the conditions of one biogeochemical province (Tersk pedigree stud farm No. 169, Stavropol Territory) were selected for the study. Blood serum, fresh sperm, and thawed sperm after cryopreservation were used as biosubstrates for the study.

At the first stage of the experiment, the correlation coefficients were calculated for concentrations of chemical elements in the blood serum and indicators of the quality of the sperm of stallions. At the second stage, based on the data on the levels of concentrations of chemical marker elements (Co, Mg) of reproductive qualities established within the first stage of the study, the stallions were divided into groups by the percentile method [26]: I — up to the 25th percentile, II — within the 25-75th percentile, III — above the 75th percentile.

The conditions of feeding and keeping for all examined animals were identical. The consumed daily ration of feeding of experimental animals in the two-month period preceding sampling contained: Ca – 77.3-97.5 g, P – 54.8-68.3 g, Fe – 740-930 mg, Cu – 128-169 mg, Zn – 440-550 mg, Mn – 601-769 mg, Co – 8.6-11.9 mg, I – 8.1-12.3 mg.

2.3 Selection and analysis of sperm parameters

Sperm intended for the study was obtained from stallions for a mare in season using an artificial vagina sample from the Horse Breeding VNII, with an interval of 48 hours. We received at least 5 ejaculates from each stallion. After a period of sexual rest in stallions, the sperm values of the first two ejaculates were not used in the treatment. After receiving sperm, each ejaculate was initially assessed according to the following indicators: volume, concentration, total sperm count in the ejaculate, progressive sperm motility, total sperm count with progressive motility in the ejaculate, sperm survival at a temperature (T) of 2-4 °C. For dilution of sperm, we used lactose-chelate-citrate-yolk (LCCY) medium in a volume ratio of 1:3. Sperm were frozen in liquid nitrogen vapor using the technology of the Horse Breeding VNII. The frozen sperm was stored in liquid nitrogen at a temperature of –196 °C. The cryopreserved sperm was thawed in a water bath at a temperature of 40 °C for 60 seconds. After thawing the cryopreserved sperm, the progressive motility and survival of spermatozoa were also determined at T 2-4 °C. Sperm concentration (billion) was estimated using a digital photometer (IMV Technologies). Sperm activity (score) was studied using a phase contrast microscope (Nikon ECLIPSE E400, Tokyo, Japan).

2.4 Selection and assessment of the elemental composition of blood serum

Blood samples (6 ml) were taken from the jugular vein in the upper third of the neck into vacuum tubes. Separation of serum was carried out by centrifuging the samples for 10 minutes at a speed of 1,000 g. The content of elements was estimated using Elan 9000 mass spectrometer (Perkin Elmer, USA) and an Optima 2000 V atomic emission spectrometer (Perkin Elmer, USA). The elemental composition of seminal plasma was determined by 25 chemical elements (Co, Cr, Cu, Fe, I, Mn, Se, Zn Ca, K, Mg, P, B, Li, Si, V Na, Ni, Sn, Hg, Sr, Al, As, Cd, Pb).

2.5 Statistical analysis

Correlation coefficients were calculated according to Spearman (K_s). To test the hypothesis about the normal distribution of other quantitative traits, the Shapiro-Wilk test was used. Significance of differences was checked using the Mann-Whitney U test. Significance level (P) was taken to be less than or equal to 0.05. The data were processed using Statistica 10.0 software package (Stat Soft Inc., USA).

3 Result

Correlation analysis showed that the survival rate of spermatozoa in fresh sperm at T 2-4 °C is reliably associated with the level of Mg in the blood serum, the concentration of spermatozoa is positively correlated with the level of Co (Table 1).

Table 1. Correlation coefficients for the concentrations of chemical elements in the blood serum and indicators of sperm quality in Arabian purebred stallions, K_s

Element	Sperm quality indicators					
	Ejaculate volume after filtration, ml	Sperm concentration, mln/ml	Progressive sperm motility in fresh sperm, %	Sperm survival in fresh sperm at T 2-4 ^{0}C, hour	Progressive sperm motility after cryopreservation, %	Sperm survival after cryopreservation at T 2-4 ^{0}C, hour
Macroelements						
Ca	-0.09	0.14	-0.13	0.25	0.16	0.22
K	0.20	0.02	-0.07	0.00	-0.27	-0.15
Mg	-0.13	-0.02	-0.02	0.46*	0.27	0.28
P	0.21	-0.13	-0.03	-0.14	-0.33	-0.27
Essential microelements						
Co	-0.09	0.41*	-0.25	0.11	0.10	0.09
Cr	0.24	-0.01	0.02	-0.12	-0.33	-0.26
Cu	0.13	0.19	-0.10	0.04	-0.31	-0.08
Fe	-0.17	0.05	0.06	0.23	0.07	0.06
I	0.21	-0.16	0.25	0.17	0.19	0.04
Mn	0.06	0.06	-0.19	-0.07	-0.17	-0.13
Se	0.19	0.10	-0.12	-0.26	-0.06	-0.02
Zn	-0.04	-0.03	-0.21	0.00	0.00	-0.06
Conditionally essential microelements						
Li	-0.14	0.35	0.19	-0.03	0.21	0.21
Ni	-0.06	0.24	-0.08	-0.06	-0.39	-0.19
V	0.21	-0.08	0.14	0.05	-0.21	-0.15
B	-0.10	0.28	0.29	-0.01	0.10	0.19
Toxic elements						
Cd	0.14	-0.17	0.24	-0.10	-0.15	-0.14
Hg	0.20	-0.16	0.14	-0.26	-0.31	-0.25
Pb	-0.02	0.08	0.11	-0.24	-0.18	-0.19
Sn	-0.28	0.14	-0.01	-0.18	-0.25	-0.07
Sr	0.26	-0.19	0.05	-0.12	-0.21	-0.18
Al	-0.08	0.14	-0.27	-0.19	-0.18	-0.14
As	0.03	-0.08	0.15	-0.10	-0.25	-0.20

* *significant correlation ($P \leq 0.05$)*

Due to the fact that a significant correlation was established only for Mg and Co, these elements were selected for further analysis.

Cobalt

Comparative analysis of data on the concentration of Co in the blood serum of stallions in the context of formed groups showed that the blood serum of stallions of group III contained 0.0012±0.000213 μg/g, which is 39.7 (P≤0.001) and 103.7 % (P≤0.001) is higher in comparison with groups I and II (Fig. 1). In this case, the actual range of variation of Co concentration in the blood serum of stallions of group I was from 0.00032 to 0.00067 μg/g, II — from 0.00068 to 0.00099 μg/g, group III — from 0.001 to 0.0015 μg/g.

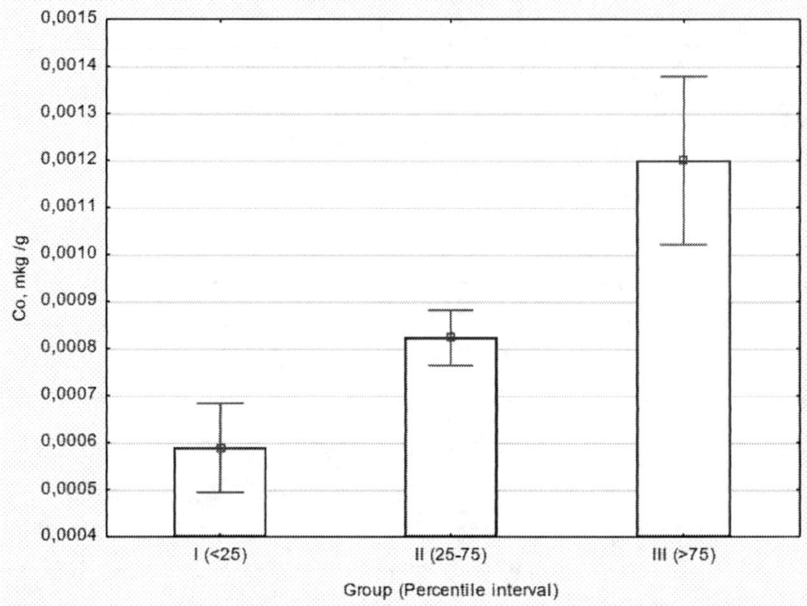

Fig. 1. Co concentration in the blood serum of stallions depending on the selected percentile interval, μg/g

The results obtained in our study demonstrated a positive trend in the indicators of sperm quality as the concentration of Co in the blood serum increased (Table 2).

Table 2. Indicators of sperm quality of Arabian purebred horse stallions depending on the concentration of Co in the blood serum

Indicators	Group (percentile interval)		
	(<25)	II (25-75)	III (>75)
Ejaculate volume after filtration, ml	29.63±7.78	27.21±11.83	33.63±11.26
Sperm concentration, mln/ml	299.1±77.75	211.9±96.38	228.1±110.1
Progressive sperm motility in fresh sperm, %	53.38±8.19	59.71±12.34	63.13±9.23*
Sperm survival in fresh sperm at T 2-4 ^{0}C, hour	123.0±44.32	136.3±49.95	147.0±123.0
Progressive sperm motility after cryopreservation, %	20.06±10.22	29.24±14.02	21.42±13.58
Sperm survival after cryopreservation at T 2-4 ^{0}C, hour	53.00±42.80	74.57±37.11	60.75±65.60

** P≤0.05 compared to group I*

The lowest rate of progressive sperm motility in fresh sperm was observed in animals with serum Co content below the 25th percentile. As the Co content increased from minimum to maximum in percentile intervals of 25-75 and more than 75 percentile, this indicator increased by 11.9 and 18.3 % ($P \leq 0.05$).

Magnesium

The actual differences between groups of stallions in terms of serum Mg concentration are shown in Figure 2.

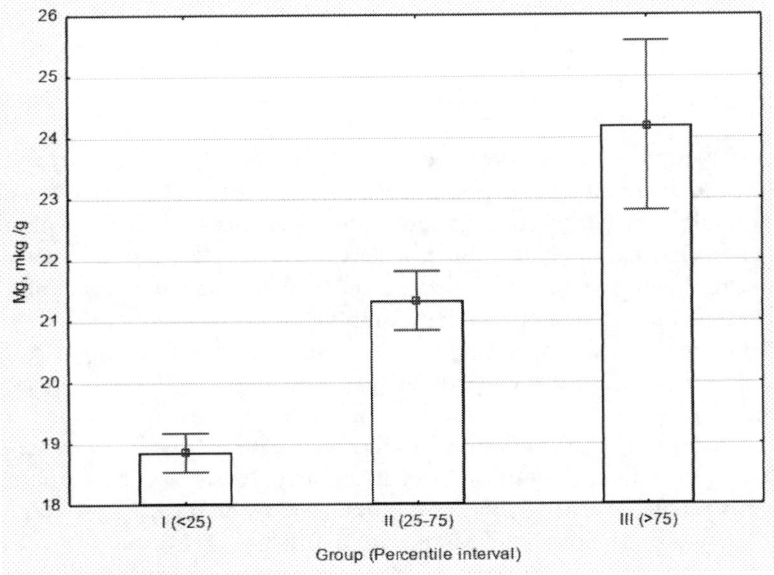

Fig. 2. Concentration of Mg in the blood serum of stallions depending on the selected percentile interval, µg/g

The blood serum of animals of group I contained 18.85±0.384 µg/g Mg, which is 11.63 ($P \leq 0.001$) and 22.08 % ($P \leq 0.001$) less in comparison with groups II and III. Wherein, the range of Mg concentrations in the blood serum of group I stallions was from 18.16 to 19.43 µg/g, from 19.79 to 22.37 µg/g in group II stallions, from 22.64 to 27.05 µg/g in group III stallions.

The qualitative indicators of sperm in the compared groups, in our study, increased as the concentration of Mg in the blood serum increased (Table 3).

Table 3. Indicators of sperm quality of Arabian purebred horse stallions depending on the concentration of Mg in the blood serum

Indicators	Group (percentile interval)		
	I (<25)	II (25-75)	III (>75)
Ejaculate volume after filtration, ml	30.50±11.03	29.00±12.69	29.71±6.05
Sperm concentration, mln/ml	232.4±91.34	256.7±108.11	210.7±96.07
Progressive sperm motility in fresh sperm, %	58.38±11.58	57.67±8.37	62.29±15.38
Sperm survival in fresh sperm at T 2-4 ^{0}C, hour	90.00±32.71	132.0±41.32*	195.4±116.3*
Progressive sperm motility after cryopreservation, %	18.67±16.86	25.19±12.56	30.57±8.06
Sperm survival after cryopreservation at T 2-4 ^{0}C, hour	40.25±41.16	64.80±45.99	94.29±42.95*

* *$P \leq 0.05$ compared to group I*

Stallions with a serum Mg concentration of less than the 25th percentile (group I) were characterized by minimal values of sperm survival rates in fresh and thawed sperm after cryopreservation. As the Mg content increased from minimum to medium (group II) and maximum (group III), there was an increase in sperm survival in fresh sperm by 46.7 ($P \leq 0.05$) and 117.0 % ($P \leq 0.05$); in sperm thawed after cryopreservation by 60.8 and 134.0 ($P \leq 0.05$), respectively.

4 Discussion

Comparison of sperm parameters obtained from stallions with different levels of chemical elements in the blood serum revealed a number of significant differences. An explanation of this fact is possible considering information on the concentration of elements in the blood serum as a characteristic reflecting the exposure of elements from the diet [27] and, in general, associated with the size of the exchange pool of these substances in the animal's body [28]. On the other hand, a close relationship is known between the elemental status of stallions and their reproductive qualities [29, 30], which in turn is determined by the influence of elements on spermatogenesis and sperm quality in mammals [31, 32, 33]. For this reason, the significant influence of the concentrations of microelements in the blood serum on the qualitative characteristics of the sperm of stallions revealed by us is quite understandable. However, despite the fact that in our experiment an increase in progressive sperm motility in fresh sperm was found by 11.9-18.3 % as the content of Co in the blood serum increased, in the vast majority of previous studies, negative effects of Co on male fertility were demonstrated. Namely, it was found that Co ions in extremely low concentrations exhibit powerful immobilizing properties of spermatozoa, leading to the destruction of thiol groups on the surface of spermatozoa, which provokes an increased formation of superoxide anion radicals and makes them the most likely sources of oxidative stress [34, 35]. Spermatozoa are sensitive to oxidative stress due to the lack of cytoplasmic protection [36, 37, 38]. In some cases, an excess of Co can manifest itself in the form of a decrease in motility and an increase in the number of abnormal spermatozoa [39, 40, 41] and disruption of the mechanisms of testosterone synthesis [42].

Another important element of cell physiology is Mg, which is present in high concentration in the sperm of stallions [43]. Previous studies have shown that the concentration of Mg in sperm is closely related to progressive motility, general motility and sperm viability [44, 45, 46]. In addition, Mg is considered a marker of seminal vesicle secretion and acts as an intracellular calcium antagonist [47]. In our experiment, results demonstrating a positive relationship between the level of Mg in the blood serum and the survival of sperm in fresh and thawed sperm after cryopreservation were obtained (Tables 1, 3). It shall be noted that a similar nature of the interaction of Mg with the quality characteristics of sperm has been reported in other publications. Namely, a relationship was found between the concentration of Mg in the blood serum and the survival of spermatozoa within hypothermic storage of diluted sperm ($r = 0.46$) in Arabian purebred stallions [25]. It was also reported that Mg levels were positively correlated with sperm viability in rams [48]. As one of the probable reasons for this nature of the interaction, it is possible to consider the physiological role of Mg within implementation of the function of maintaining the osmotic balance necessary for the function and fertility of spermatozoa, as well as its interaction with enzymes such as Ca^{2+}-dependent Mg^{2+} – ATPase [49]. In addition, it is believed that Mg can reduce the intensity of lipid peroxidation in the seminal fluid of males by inhibiting the metabolism of Cd [50]. For this reason, Mg supplements are often viewed as antidotes to the toxic effects of Cd on the reproductive system in laboratory animals [50]. Wherein, in some studies, no positive effect of Mg on the indicators of the quality of sperm of stallions was found [51,52], which determines the prospects for further study.

5 Conclusions

Therefore, it was found that the concentrations of Mg and Co in the blood serum are positively correlated with the qualitative characteristics of the sperm of Arabian purebred stallions. Wherein, the optimal range of concentrations of Mg in the blood serum, typical for stallions with relatively high indicators of sperm quality, is 22.64 to 27.05 μg/g; Co – 0.001 to 0.0015 μg/g.

Acknowledgments

The research was carried out with the support of the Russian Science Foundation (project No. 17-16-01109 P).

References

1. V.A. Bagirov, V.V. Kalaschnikov, A.M. Zaitsev, O.A. Zavialov, A.N. Frolov, Sel'skokhozyaistvennaya Biologiya, **52(6)**, 1184–1193 (2017)
2. H. Bertelsmann, S. Keppler, M. Höltershinken, H. Bollwein, D. Behne, D. Alber, G. Bukalis, A. Kyriakopoulos, H. Sieme // Reprod Fertil Dev. 2010;22(5). P. 886-91.
3. Lu Sy., Huang ZM., Huang WK., Liu XY., Chen YY., Shi T., et al. Proteins, **81(5)**, 740-53 (2013)
4. Vyvial, Michal, et al. Acta Veterinaria Brno, **88.4**, 377-384 (2020)
5. ElSaid, Halla, et al. Benha Veterinary Medical Journal, **37.1**, 210-215 (2019)
6. Tabassomi, Mehdi, and Sayed Mortaza Alavi-Shoushtari, Veterinary research forum: an international quarterly journal, **4(1)** (2013)
7. Z. Liu, L. Chen, Y. Shang, P. Huang, L. Miao, Development, **140(10)**, 2103-2107 (2013)
8. AIS. Villaverde, EG. Fioratti, RS. Ramos, RC. Neves, JCP. Ferreira, GS. Cardoso, et al. Blood and seminal plasma concentrations of selenium, zinc and testosterone and their relationship to sperm quality and testicular biometry in domestic cats. Anim Reprod Sci. 2014; 150(1):50-5.
9. W.Y. Wong, G. Flik, P.M. Groenen, D.W. Swinkels, C.M. Thomas, J.H. Copius-Peereboom, H.M. Merkus, R.P. Steegers-Theunissen, Reprod Toxicol, **15(2)**, 131-136 (2001)
10. B. Eskenazi, S.A. Kidd, A.R. Marks, E. Sloter, G. Block, A.J. Wyrobek, Hum Reprod, **20**, 10 (2005)
11. V. Cigankova, P. Mesaros, J. Bires, V. Ledecky, J. Ciganek, E. Tomajkova, Slov Vet J, **23**, 97-100 (1998)
12. P. Mudron, W. Baumgartner, G. Kovac, P. Bartko, I. Rosival, I. Zezula, Dtsch Tieraztl Wschr, 103, 131-133 (1996)
13. Mahdiyeh Mirnamniha, Fereshteh Faroughi, Eisa Tahmasbpour, Pirooz Ebrahimi and Asghar Beigi Harchegani, From the journal Reviews on Environmental Health (2019)
14. S. Benoff, G.M. Centola, C. Millan, B. Napolitano, J. L. Marmar, I. R. Hurley Hum, Reprod, **18**, 374-383 (2003)
15. J.P. Bonde, M. Joffe, P. Apostoli, A. Dale, P. Kiss, M. Spano, et al. Occup Environ Med, **59**, 234-242 (2002)
16. P. Massanyi, V. Uhrin, Reprod Dom Anim, **31**, 629-632 (1996)
17. P. Massanyi, V. Uhrin, J Environ Sci Health, **32**, 1459-1466 (1997)

18. R. Toman, P. Massanyi, Trace Elem Electrolytes, **19**, 114-117 (2002)
19. P. Massanyi, K. Massanyiova, F. Pizzi, J. Trandzik, R. Toman, Agriculture, **45**, 287-295 (1999)
20. Zs Forgacs, Zs Nemethy, Cs Revesz, P. Lazar, J Toxicol Environ Health, **A62**, 349-358 (2001)
21. J.F. Vázquezarmijo, R. Rojo, D. López, J.L. Tinoco, A. González, N. Pescador, et al., Tropical and Subtropical Agroecosystems, **14**, 1-13 (2016)
22. E. Humann-Ziehank, M. Ganter, I. Hennig-Pauka, A. Binder, Small Rumin Res, 75, 185-191 (2008)
23. Y.H. Wu, W. Lai, Z.H. Liu, H.K. Wei, Y.F. Zhou, J.J. Tan, et al., Biol Trace Elem Res, 10 (2018)
24. M. Zakošek Pipan, P. Zrimšek, B. et al. Macro- and microelements in serum and seminal plasma as biomarkers for bull sperm cryotolerance. Acta Vet Scand **63**, 25 (2021)
25. V.V. Kalashnikov, M.M. Atroshchenko, A.M. Zaitsev, A.A. Nezalenova, A.A. Datsyshin, XIX International Scientific and Practical Conference "Current Trends of Agricultural Industry in Global Economy", 112-117 (2020)
26. M. G. Skalnaya, V. A. Demidov, A. V. Skalny, Micro-elements in medicine, **4(2)**, 5-10 (2003)
27. K. Chojnacka, A. Zielińska, I. Michalak, H.Górecki, Environ Toxicol Pharmacol, **30(2)**, 188-194 (2010)
28. S.V. Lebedev, G.K. Duskaev, O.V. Kvan, E.V. Sheida, I. Alijanova, S.G. Rakhmatullin, S.A. Miroshnikov, Biol Med (Aligarh), **7(4)**, 1-6 (2015)
29. S.M. Ghallab, A.M. Shahat, A.M. Fadl, M.M. Ayoub, A.R. Moawad, Cryobiology, **S0011-2240(17)**, 30213-30214 (2010)
30. A. Contri, I. De Amicis, A. Molinari, M. Faustini, A. Gramenzi, D. Robbe, A. Carluccio, Theriogenology, **75(7)**, 1319-1326 (2019)
31. S. Hamameh, J-L. Gatti, Hum Reprod, **13(4)**, 20–30 (1998)
32. S.E. Chia, C.N. Ong, L.H. Chua, L.M. Ho, S.K, Tay. J Androl., **21**, 53–57 (2000)
33. P. Massanyi, J. Trandzik, P. Nad, R. Toman, M. Skalick, B. Kornekov, Asian J Androl, **5**, 101–104 (2003)
34. G. Pradeep Kumar, Malini Laloraya, Manmohan M. Laloraya, Contraception, **41(6)**, 633-639
35. S.J. Stohs, D. Bagchi, Free Radic Biol Med, **18**, 321 (1995)
36. E.T. Donnelly, N. McClure, S.E. Lewis, Fertil Steril, **72**, 484 (1999)
37. R.A. Saleh, A. Agarwal, J Androl, **23**, 737 (2002)
38. E Kawakami, A Tekemura, M Sakuma, M Takano, T Hirano, T Hori, T Tsutsui, J Vet Med Sci, **69(2)**, 133–136 (2007)
39. J.R. Bucher, M.R. Elwell, M.B. Thompson, et al. Fundam Appl Toxicol, **15**, 357 (1990)
40. U. Marzec-Wróblewska, P. Kamiński, P. Łakota, et al. Biol Trace Elem Res, **188**, 251–260 (2019)
41. Kocabaş, Mehmet, Filiz Kutluyer, Bulletin of environmental contamination and toxicology, **99.6**, 690-694 (2017)
42. N.G. Pedigo, W.J. George, M.B. Anderson, Reproductive Toxicology, **2(1)**, 45-53 (1988)
43. Halo Jr, Marko, et al., The Journal of Microbiology, Biotechnology and Food Sciences, **7.5**, 516 (2018)

44. M. Eghbali, S.M. Alavi Shoushtari, S. Asri Rezaei, M.H. Khadem Ansari, Veterinary research forum, **1(1)**, 12-20
45. A. Kasperczyk, et al., Magnesium Research, **28(1)**, 14-22 (2015)
46. Zhonghua nan ke xue, National Journal of Andrology, **16(11)**, 1019-1022 (2010)
47. M. Eghbali, S.M. Alavi Shoushtari, S. Asri Rezaei, M.H. Khadem Ansari, Veterinary research forum, **1(1)**, 12-20 (2020)
48. H Abdel-Rahman, MS El-Belely, AA Al-Qarawi, et al., Small Rumin Res, **38(1)**, 45–49 (2000)
49. NS Juyena, C. Stelletta, J Androl, **33(4)**, 536–551 (2012)
50. N. Babaknejad, S. Bahrami, A.A. Moshtaghie, Biol Trace Elem Res, **185**, 106–115 (2018)
51. M. Mrackova, M. Zavadilova, M. Sedlinska, Mac Vet Rev, **38(1)**, 91–96 (2015)
52. A. Usuga, B. Rojano, G. Restrepo, J Equine Vet Sci, **58**, 103–111 (2017)

The expediency of using a plant protection system using the microbiological fungicide BisolbiSan, W in the cultivation of winter wheat in the Southern natural and agricultural zone of the Rostov region

Gennady Urban[1], *Olga* Krotova[2,*], *Denis* Efimov[3], *Konstantin* Savenkov[4], and *Maria* Savenkova[4]

[1] Branch of the Federal State Budgetary Institution "Russian Agricultural Center" in the Rostov region, 1st Cavalry Army Street, 15, 344029 Rostov-on-Don, Russia
[2] Don State Technical University, Gagarin Sq., 1, 344003 Rostov-on-Don, Russia
[3] Novocherkassk Engineering and Reclamation Institute named after A.K. Kortunov branch of the Don State Agrarian University, Pushkinskaya Street, 111, 346428 Novocherkassk, Rostov region, Russia
[4] St. Petersburg State Agrarian University, Peterburgskoe highway, 2, G. Pushkin, 196601 St. Petersburg, Russia

Abstract. The authors analyzed the biological effectiveness of the integrated plant protection system formed when the microbiological fungicide BisolbiSan, Zh was included in the classical protection scheme adopted in the farms of the Rostov region. The biological effectiveness of the integrated protection system, including the fungicide BisolbiSan, was analyzed, the impact of the studied protection system on the yield, the quality of the grain obtained, the phytosanitary condition of the crop during the growing season on winter wheat crops of the variety Nakhodka with the use of a plant protection system, including the fungicide BisolbiSan, Zh. The fungicide BisolbiSan, containing a culture of rhizospheric bacteria Bacillus subtilis strain H-13, suppresses the germination of spores and mycelium growth of phytopathogenic fungi due to the multilateral action of bacterial metabolites, without causing the formation of resistance in pathogens. It was found that the protection system used practically does not cause a decrease in grain quality — in terms of the average protein and gluten content, the grain at the experimental site even slightly exceeds the grain obtained from the control site by 16.88% and 27.8%. With the cost of winter wheat grain of 1100 rubles/kg, an increase in the yield in the experiment of 3.8 kg/ha in relation to control and the cost of an experimental protection system in the amount of 3,357 rubles/ha, revenue amounted to 823 rubles/ha.

* Corresponding author: Alb9652@yandex.ru

1 Introduction

Biologization and ecologization of plant protection, the introduction of integrated protection systems based on the preventive role of the use of entomophages, microbiological rodenticides and fungicides, microbiological destructors of stubble, microbiological fertilizers and fertilizers based on humic acids have been given special importance in recent years [1-3]. Carrying out measures to restrain the number of pests is assumed only taking into account the assessment of the physiological and phytosanitary condition of crops, the forecast of the development of harmful organisms and economic thresholds of harmfulness [4]. The greatest effectiveness in maintaining high yields and environmental safety is shown by the use of integrated systems aimed at obtaining well-developed plants, carrying out preventive treatments with biological products, destroying the wintering stock of harmful objects and pest breeding stations (weeds and others), as well as growing varieties resistant to pests and diseases, preserving and activating the activity of natural entomophages [5- 7].

The use of intensive crop cultivation technologies, in which the main emphasis is on chemicalization, without taking into account the actual phytosanitary situation on the land, the state of plant development, the number of harmful organisms and stages of their development, is associated with the depletion of natural soil fertility, degradation of useful soil microbiota, accumulation of harmful substances in the soil (residual amount of pesticides, nitrites and nitrates), the development of resistance to pesticides in pests, pathogens and weeds. And the expansion of the use of integrated systems, their maintenance for a number of years, contributes to the transition to organic farming, which ensures the highest quality of agricultural products [8-11].

The purpose of our work was to analyze the biological effectiveness of the integrated plant protection system formed when the microbiological fungicide BisolbiSan, Zh was included in the classical protection scheme adopted in the farms of the Rostov region [12]. As a result of the successful completion of the research work, in accordance with the terms of reference, the biological effectiveness of the integrated protection system, including the fungicide BisolbiSan, was analyzed, the impact of the studied protection system on the yield, the quality of the grain obtained, the phytosanitary condition of the crop during the growing season was assessed [13-14]. The object of the study is winter wheat crops of the Nakhodka variety using a plant protection system including the fungicide BisolbiSan, Zh [15].

2 Materials and methods

To determine the effectiveness of the use of drugs and their combinations in the cultivation of agricultural crops, the branch of the Federal State Budgetary Institution "Russian Agricultural Center" in the Rostov region conducted a field demonstration test of a plant protection system, including the fungicide BisolbiSan, Zh., produced by Bisolbi Plus LLC, containing a live culture of rhizosphere bacteria Bacillus subtilis strain H–13. The venue of the experiment is located on the territory of the FGBNU "ANC "Donskoy". The sites where the experiment was conducted are located in the Southern natural and agricultural zone. Winter wheat crops of the Nakhodka variety were studied. The type of soils on which the cultivation of the crop was carried out are ordinary chernozems, the granulometric composition of the soils is heavy loamy. The area of the control and experimental plots is 0.024 hectares each.

On the plots (experimental and control) on 08/27/2020, ammophos 10:48 was introduced at a dose of 100 kg /ha. The fertilizer distributed over the surface of the field was embedded in the soil by cultivation. Sowing was carried out on 27.09.2020 with the

SS-11 "Alpha" seeder. The depth of seeding was 5 cm. Germination took place at air temperatures from 10 °C to 21 ° C.

A plant protection system based on the use of three chemical pesticides in tank mixtures (insecticide, fungicide, herbicide), liquid organomineral fertilizer and microbiological fungicide BisolbiSan, Zh. No treatments were carried out at the control site. A plant protection system based on the use of three chemical pesticides in tank mixtures (insecticide, fungicide, herbicide), liquid organomineral fertilizer and microbiological fungicide BisolbiSan, Zh. No treatments were carried out at the control site.

3 Results and Discussion

Pesticide treatments at the experimental site were carried out according to the approved scheme of the experiment (Table 1)

Table 1. The scheme of the treatments carried out

Date of processing	Culture phase	Working fluid consumption l/t, l/ha.	The applied drug	Consumption rate, l/t, l/ha, t/ha
24.09.2020	Seeds	10	Imidor Pro, CS	1,2
			Benefit, ME	0,8
			Biostim start	1,0
			BisolbiSan, W	1,0
08.04.2021	Exit into the tube	200	Diva, SE	0,9
			BisolbiSan, W	2,0
07.05.2021	The beginning of earing	200	BisolbiSan, W	2,0

Winter wheat treatments with preparations were carried out in three stages, consisting of pre-sowing seed treatment and two treatments of crops during the growing season.

Stage I. Pre-sowing seed treatment

The composition of the tank mixture for seed treatment included the following preparations: chemical insecticide Imidor Pro, CS in a dose of 1.2 l / t to protect seedlings from damage by larvae of bread beetle and cereal flies; chemical fungicide Benefit, ME in a dose of 0.8 l / t to protect germinating seeds and seedlings from pathogens of fusarium and helminthosporiosis root rot detected during phytopathological examination, seed mold (including alternarium seed infection), as well as to prevent infection with powdery mildew (in the early stages of development), snow mold, dusty and hard smut; liquid organomineral fertilizer Biostim brand Start at a dose of 1.0 l/t to stimulate seed germination and root system development; microbiological fungicide BisolbiSan, Zh in a dose of 1.0 l / t for the protection of germinating seeds and seedlings from pathogens of mold formation, fusarium and helminthosporiosis root rot detected during phytopathological examination, as well as to strengthen plant immunity, overcome the effects of chemical stress caused by chemical pesticides, increase germination and amity of seed germination, stimulate intensive growth and development of plants, increase plant resistance to autumn frosts.

Stage II. The first treatment during the growing season

Fig. 1. Carrying out the first processing of crops 08.04.2021

The first treatment of crops during the growing season was carried out on 08.04.2021 at an air temperature of 3 ° C in the phase of entering the culture tube. The composition of the tank mixture for wrapping included two drugs — herbicide Prima Donna, SE at a dose of 0.9 l / ha to protect against clogging of crops with dicotyledonous weeds in the early growth phases of the latter and microbiological fungicide BisolbiSan, W at a dose of 2.0 l / ha to enhance immunity and resistance of plants to leaf diseases such as powdery mildew, septoria, pyrenophorosis and to overcome the effects of chemical stress caused by herbicide, stimulation intensive growth and development of plants, increasing the resistance of plants to morning spring frosts and lack of moisture.

Stage III. The second treatment during the growing season

Fig. 2. Carrying out the second processing of crops 07.05.2021

The second treatment of crops was carried out on 07.05.2021 at an air temperature of 14 ° C during the earing phase of the crop. The treatment was carried out with the microbiological fungicide BisolbiSan, Zh at a dose of 2.0 l/ha to strengthen the immunity and resistance of plants to leaf diseases (such as powdery mildew, septoria, pyrenophorosis), stimulate intensive growth and development of plants, increase plant resistance to moisture deficiency, increase yield.

The results of the phytopathological examination of seeds carried out before sowing showed a total infection rate of 32%. Infection with fusarium was 3%, helminthosporiosis - 2%, which indicated the likelihood of the development of fusarium and helminthosporiosis root rot on crops. Infection with mainly saprophytic fungi - alternaria and mold fungi was mainly 18% and 9%.

In all areas, crops developed normally. The phytosanitary condition before wintering was good, there were no diseases. The number of winter wheat plants that reached the tillering phase by 10/23/2020 was 64 per 1 running meter at the control site, and 67 specimens at the experimental site.

By 01.11.2020, the height of winter wheat plants on both control and experimental plots averaged 19.5 cm.

The height of winter wheat plants at the control site in the tillering phase on 03/19/2021 was 27-32 cm, in the tube exit phase 08.04.2021 - 40-44 cm, at the beginning of the earing phase 07.05.2021 - 47-62 cm.

The height of winter wheat plants at the experimental site in the tillering phase on 19.03.2021 was 27-32 cm, in the tube exit phase on 08.04.2021 was 40-44 cm, at the beginning of the earing phase on 07.05.2021 - 56-68 cm.

Phytosanitary surveys of crops in order to determine the effectiveness of the tested protection system were carried out 23.10.2020, 01.11.2020, 19.03.2021, 08.04.2021, 07.05.2021, 29.05.2021 (Table 2, Table 3).

Table 2. Results of phytosanitary monitoring at test sites

Date of the survey	Culture phase	Region	Disease	Distribution %	Development, %
19.03.2020	End of tillering	Control	Septoria	9	0,5
			Root rot	12	1,5
		Experience	Septoria	7	0,4
			Root rot	3	0,1
08.04.2020	Exit into the tube	Control	Septoria	20	2,0
			Root rot	15	1,5
		Experience	Septoria	13	1,3
			Root rot	3	0,1
17.04.2020	Flag sheet	Control	Septoria	14	1,4
			Root rot	15	0,7
		Experience	Septoria	5	0,3
			Root rot	3	0,2
07.05.2020	The beginning of earing	Control	Septoria	5	0,3
			Root rot	15	0,3
		Experience	Septoria	2	0,1
			Root rot	3	0,1
29.05.2020	Waxy ripeness	Control	Septoria	0	0
			Root rot	13	0,2
		Experience	Septoria	0	0
			Root rot	2	0,1

Infection with root rot and septoria was manifested both on the experimental plot and on the control one. At the experimental site, the development and spread of diseases was lower than at the control site. Weed vegetation was not observed on the crops. The pests common piavica and thrips were noted with insignificant numbers.

Table 3. Results of phytosanitary monitoring at test sites

Date of the survey	Culture phase	Region	Pest	Number, copies/plant
19.03.2021	End of tillering	Control	–	0
		Experience	–	0
08.04.2021	Exit into the tube	Control	Piavica	2
		Experience	Piavica	2
17.04.2021	Flag sheet	Control	Piavica	3
			Thrips	1
		Experience	Piavica	2
			Thrips	1
07.05.2021	The beginning of earing	Control	Piavica	0
			Thrips	3
		Experience	Piavica	0
			Thrips	3
29.05.2021	Waxy ripeness	Control	Piavica	0
			Thrips	0
		Experience	Piavica	0
			Thrips	0

Fig. 3. Winter wheat at the control site on 03/19/2021

Fig. 4. Winter wheat at the experimental site on 03/19/2021

Fig. 5. Winter wheat at the control site 08.04.2021

Fig. 6. Winter wheat at the experimental site 08.04.2021

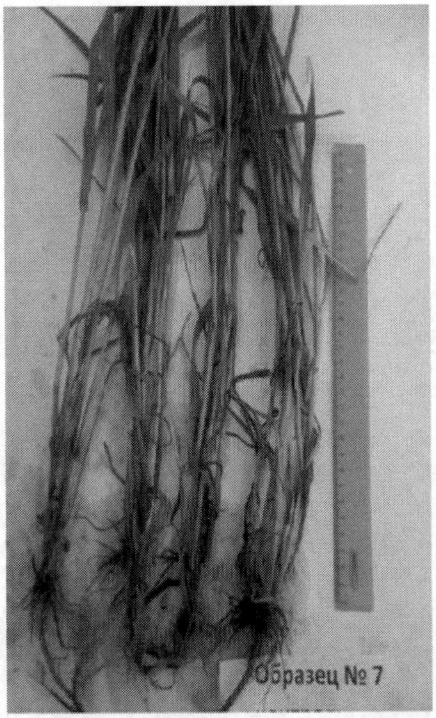

Fig. 7. Winter wheat at the control site 07.05.2021

Fig. 8. Winter wheat at the experimental site 07.05.2021

Fig. 9. Crops in the control area in the ripening phase

Fig. 10. Crops at the experimental site in the ripening phase

Table 4. Results of determining yield and profitability

№ n/a	Option	Yield, c/ha	Grain moisture, %	Content	
				squirrel, %	gluten, %
1	Control	39,8	5,6	16,88	27,80
2	Experience	43,6	5,5	17,41	28,80

Fig. 11. Harvesting of winter wheat 12.07.2021

4 Conclusions

The results of the study provide the basis for a number of conclusions about the effectiveness of the use of the studied plant protection system in the agro-climatic conditions of the Southern natural and agricultural zone of the Rostov region.

Pre-sowing seed treatment with the microbiological fungicide BisolbiSan, in addition to protecting seeds and seedlings from the harmful activity of root rot pathogens and seed mold, had an impact on increasing field germination, activating growth and shaping processes, increasing the resistance of seedlings to adverse factors such as pest damage, frost, lack of moisture, as well as stress caused by the use of chemical pesticides, stimulating seed germination and root system development against the background of the protective action of chemical protectants - insecticide Imidor Pro, CS and fungicide Benefit, ME. This made it possible to obtain healthy developed seedlings that quickly reached the stage of tillering before wintering.

The treatment of crops in the phase of entering the tube and in the phase of earing the culture with the microbiological fungicide BisolbiSan, had a beneficial effect on increasing resistance to adverse environmental factors, improving the phytosanitary condition and grain quality, increasing yields.

It should be noted the positive effect of the studied protection system on the improvement of plant development, which allowed overcoming the adverse consequences of the appearance of root rot and septoria.

Over time, in the control area, the proportion of the spread and development of these diseases gradually increased throughout all phases of wheat development up to the "earing" phase. The magnitude of the spread of root rot in the control area increased from 12% to 15%, while in the experimental area the spread and development of the disease stopped at the level of 3%. By the onset of the flag leaf phase, the spread of septoria at the experimental site was 5%, which is 3 times lower than in the control.

It was found that the protection system used practically does not cause a decrease in grain quality — in terms of the average protein and gluten content, the grain at the experimental site (17.41% and 28.8%, respectively) even slightly exceeds the grain obtained from the control site (16.88% and 27.8%, respectively).

When the cost of winter wheat grain is 1100 rubles/kg, the yield increase in the experiment is 3.8 kg/ha in relation to the control (Table. 5) and the cost of an experimental protection system in the amount of 3,357 rubles/ha, revenue amounted to 823 rubles/ha.

Table 5. Profitability of the experimental plant protection scheme

№ n/a	Option	Yield at natural grain moisture, c/ha	The cost of the crop excluding costs, RUB/ha	The cost of the crop, including costs, RUB/ha
1	Control	39,8	43 780	43 780
2	Experience	43,6	47 960	44 603

It is worth noting that if the optimal crop rotation is observed and the results of phytopathological examination of the seed material are used, it is possible to optimize the classical protection scheme by adding biologics of Bisolbi Plus LLC in order to reduce costs during pre-sowing etching.

The fungicide BisolbiSan, containing a culture of rhizospheric bacteria Bacillus subtilis strain H-13, suppresses the germination of spores and mycelium growth of phytopathogenic fungi due to the multilateral effects of bacterial metabolites (lytic enzymes and antibiotics), without causing the formation of resistance in pathogens. Strengthens the immunity and resistance of plants, stimulating the production of phenolic compounds and phytoalexins. Relieves chemical stress caused by pesticides, increases germination and friendly germination of seeds, stimulates intensive growth and development of plants, increases plant resistance to stresses caused by infection with pathogens, damage by pests, frost and lack of moisture. At the same time, the drug is not phytotoxic in recommended doses and is compatible in tank mixtures with chemical fungicides, insecticides, herbicides and fertilizers, it is not dangerous for fish, earthworms and wild fauna. It is especially important that the use of an integrated preventive protection system based on the use of a combination and alternation of traditional chemicals with biological products contributes to the maintenance of natural soil fertility, reduces the pesticide load of agricultural land, prevents the threat of environmental pollution, makes it possible to obtain an environmentally friendly crop that does not contain a residual amount of pesticides. Expanding the use of integrated systems, maintaining them for a number of years, contributes to the transition to organic farming, which ensures the highest quality of agricultural products.

References

1. C. Meunier, L. Alletto, L. Bedoussac, J.-E. Bergez, P. Casadebaig,, J. Constantin, N. Gaudio, R. Mahmoud, J.-N. Aubertot, F. Celette, M. Guinet, M.-H. Jeuffroy, M.-H.Robin, S. Médiène, L. Fontaine, B. Nicolardot, E. Pelzer, V. Souchère, A.-S. Voisin, B. Rosiès, M. Casagrande, G. Martin, European Journal of Agronomy, **132**, 126412 (2022)
2. V.E. Reutov, D.D. Burkaltseva, S.S. Zmiyak, K.V. Kuzmina, A.V. Osipova, Lecture Notes in Networks and Systems, **198**, 870-885 (2021)
3. L.A. Alexandrovskaya, A.S. Cheshev, Terra Economicus, **12(2-3)**, 84-88 (2014)
4. N.N. Glazunova, J.A. Bezgina, A.N. Shipulya, E.V. Volosova, E.V. Pashkova, *The effect of herbicides on the infestation of winter wheat crops,* IOP Conference Series: Earth and Environmental Science, **839(2)**, 022041 (2021)
5. Y. Cohen, A.E. Rubin, M. Galperin, PLoS ONE, 16 (10 October), 0258280 (2021)
6. T.W. Kahn, N.B. Duck, M.T. McCarville, L.C. Schouten, K. Schweri, J. Zaitseva, J. Daum, Nature Communications, **12(1)**, e3380 (2021)
7. L. Lyu, H. Wang, R. Liu, W. Xing, J. Li, Y.B. Man, F. Wu, Journal of Hazardous Materials, **424**, 127323 (2022)
8. R. Jevtić, N. Skenderović, V. Župunski, M. Lalošević, B. Orbović, S. Maširević, Scientia Agricola, **79(3)**, e20200046 (2022)
9. M. Bayat, M. Zargar, Acta Physiologiae Plantarum, **43(12)**, 150 (2021)
10. H. Li, H. Wang, B. Jia, D. Li, Q. Fang, R. Li, Scientific Reports, 11(1), 16901 (2021)
11. A.I. Klimenko, A.V. Grinko, M.A. Kholodova, *Principles of development of agricultural consumer cooperation in the agricultural sector of Russia,* E3S Web of Conferences, Rostov-on-Don, 08059 (2021)
12. G. Cruppe, E. DeWolf, B.R. Jaenisch, K. Andersen Onofre, B. Valent, A.K. Fritz, R.P. Lollato, Field Crops Research, **273**, 108300 (2021)
13. B. Brozović, I. Jug, D. Jug, B. Stipešević, M. Ravlić, B. Đurđević, Agronomy, **11(10)**, 2028 (2021)
14. S. Liang, L. Li, P. An, S. Chen, L. Shao, X. Zhang, Agricultural Water Management, **256**, 107114 (2021)
15. Bakaeva, N.P., *Efficiency of growth regulators with anti-stress properties in agricultural technology of winter wheat in the Middle Volga region,* IOP Conference Series: Earth and Envi ronmental Science, 848(1), 012121 (2021)

Analysis of the genus thymus of the southeastern part of the northern caucasus

Lidia Khashieva[1,*], *Maret* Dakieva[1], and *Zazu* Iriskhanova[2]

[1]FSBEI HE "Ingush State University", Department of Biology, 386001 Magas, Russia
[2]FSBEI HE "Chechen State University named after A.A. Kadyrov", Department of Botany, Zoology and Bioecology, 364024 Grozny, Russia

Abstract. The paper presents the analysis of the genus *Thymus* of the flora of the South-East of the North Caucasus. Revealed the total species, compiled an aspect of the genus *Thymus*, and conducted the analysis's bioecological, geographical composition. In the investigated territory, we have registered eight *thymus* species, distributed mainly in xerophilic types of vegetation: steppe, friganoid, alpine, meadow-steppe, and meadow societies.

1 Introduction

The flora of the Caucasus is rich and original in the presence of endemic, relict elements, which is due to the islet effect and the presence of refugia in high mountain conditions, which always arouses great interest in terms of floristic research. The study area is characterized by high altitudes, a significant length of mountain ranges, orography, various soils, and complex dissected relief. A significant variety of species formed the conditions for uneven distribution of heat and moisture on its territory. Five zones of vegetation are distinguished here: steppe, forest, mid-mountain, subalpine and alpine.

One of the exciting groups for floristic research is the genus *Thymus L.* - the most significant and taxonomically complex genus in the *Lamiaceae* family. One of the critical characteristics of the *Lamiaceae* is its floral architecture, in which the petals merge into the upper lip and lower lip. The family includes 236 genera and about 7200 species, distributed almost all over the world [1].

The genus *Thymus* includes about 350 perennial aromatic grasses and shrubs found in Europe and North Africa. The center of diversity of the genus *Thymus L.* is located in the Mediterranean [2]. On the territory of Russia, there are about 140 species of the genus *Thymus* [3]. This study aimed to conduct a floristic analysis of species of the genus *Thymus* in the southeastern part of the Northern Caucasus.

2 Material and Methods

The genus Thymus served as the object of floristic research. The studies were carried out in the altitude range from 200 to 3500 m above sea level, within the whole South-Eastern part

* Corresponding author: Lkhashieva@yandex.ru

of the North Caucasus. The results of this work are based on our field studies, on the study of herbarium specimens (ISU, CSU), literature data.

The collection of the actual herbarium material, field observations, the study of the ecological and biological characteristics of the species of the studied genus were carried out using classical keys [4, 5].

3 Results and Discussion

On the South-Eastern part of the North Caucasus, we have registered eight genus *Thymus: Th. Collinus Beb., Th. marschanus Willd., Th. dagestanicus Klok., Th. saucasicus Willd., Th. nummularius Bieb., Th. pastoralis Iljin ex Klok., Th. pseudopulegioids Klok. et Shost., Th. transcausicus Ronn.,* distributed mainly in xerophilic types of vegetation: steppe, friganoid, as well as high-mountain meadow-steppe and meadow communities. In the study area, species of the genus *Thymus* are combined into groups with similar ecology, coenotic relationships and are confined to specific plant communities.

The distribution of species of the genus *Thymus* over the altitudinal belts of the study area showed a wide amplitude of the altitudinal distribution of the taxon under consideration, covering the range from the steppe to the alpine belt: *Th. marschallianus and Th. pastoralis* are found in the steppe cenoses both in the lowlands and in the distribution of upland xerophytes. The mid-mountain belt includes 7 out of 8 species (*Th. Collinus, Th.dagestanicus, Th. Marschallianus, Th.nummularius, Th. Pastoralis, Th. pseudopulegioides, Th.dagestanicus*). Three different zonation - mid-mountain, alpine and subalpine, cover two species - *Th.nummularius and Th. transcausicus. Th. caucasicus* (Table 1).

Table 1. Distribution of species of the genus *Thymus* in altitudinal zonation

№	Species name	Altitudinal zonation			
		Alpine	**subalpine**	**steppe**	**mid-mountain**
1.	*Th. caucasicus* Willd. TxRonn.	+	-	-	-
2.	*Th. collinus* Bieb.	-	-	-	+
3.	*Th. dagestanicus* Klok.	-	-	-	+
4.	*Th. marschallianus* Willd.	-	-	+	+
5.	*Th. nummularius* Bieb.	+	+	-	+
6.	*Th. pastoralis* Iljin ex Klok.	-	-	+	+
7.	*Th. pseudopulegioides Klok. etShost.*	-	-	-	+
8.	*Th. transcaucasicus* Ronn.	+	+	-	+

The genus *Thymus* is characterized by a wide range of habitats - talus, stony and gravelly slopes, dry meadow-steppe, meadows, which, as we have already noted, is a characteristic feature of the taxon under study. The largest number of species of the genus *Thymus* in the study area are confined to rocky and gravelly slopes. There are 7 types here: (*Th.nummularius, Th. pastoralis, Th. caucasicus, Th. marschallianus, Th. dagestanicus, Th. pseudopulegioides, Th. transcaucasicus),* however, only one species (*Th. transcausicus*) exhibits strict stenotopicity to this floracenoelement (Table 2).

Species of the genus Thymus play a significant role in the composition of the vegetation cover of the slopes of the southern, southeastern exposures of arid basins, characteristic of the mid-mountain zone, both for the territory under consideration and for the entire Caucasus, being part of the herbage with the dominance of cereals, legumes, or herb-cereal cenoses forming in places thinned thickets.

Table 2. Placement of species of the genus *Thymus* by ecotopes

№	Species name	Ecotopes					
		Stony and gravelly slopes	Dry slopes	Scree	Forest glades	Meadow	Meadow-steppe
1.	*Th. caucasicus* Willd. TxRonn.	+	-	+	-	-	-
2.	*Th. collinus* Bieb.	-	+	-	-	-	-
3.	*Th. dagestanicus* Klok.	+	-	-	-	-	+
4.	*Th. marschallianus* Willd.	+	-	-	+	-	+
5.	*Th. nummularius* Bieb.	+	-	-	-	+	-
6.	*Th. pastoralis* Iljin ex Klok.	+	-	-	-		+
7.	*Th. pseudopulegioides Klok. etShost.*	+	-	-	-	+	+
8.	*Th. transcaucasicus* Ronn.	+	-	-	-	-	-

When analyzing the geographical structure developed by N.N. Portenier [7], two geographical elements were identified - Caucasian, Pontic-South Siberian.

1. Caucasian element. Species, the central area of which is limited by the Caucasian province with frequent irradiation to the adjacent provinces [8]. These are the following types: *Thymus nummularius, Th. dagestanicus, Th. collinus, Th. caucasicus., Th. pastoralis, Th. pseudopulegioides u Th. transcaucasicus.* An analysis of the relevance of the species of the Caucasian element to various ecotopes shows that the majority of Caucasian species are found on gravelly and stony slopes.

2. Pontico-South Siberian element. The Pontic Proposition limits the types of this element in the understanding of A. L. Takhtadzhyan [9]. It includes one species - *Thymus marschallianus.*

The analysis of life forms carried out based on the system of I.G. Serebryakova [10] showed that due to adaptation to the ecological conditions of an unstable petrophytic substrate, the genus *Thymus* developed similar adaptations, expressed in the form of semiprostatic - four species (*Th. Collinus, Th. Nummularius, Th. Dagestanicus, Th. Pseudopulegioides*) and steles half-shrubs - two species (*Th. caucacicus, Th. transcaucacicus*). Herbaceous polycarpic - two species (*Th.marschallianus, Th. Pastoralis*), confined to the conditions of meadow and meadow-steppe cenoses.

For a more visual picture of the distribution of species of the genus Thymus within the Greater Caucasus, in a comparative aspect, it was presented data for individual regions, covering the territory from the Western to the Eastern Caucasus (Table 3).

Table 3. The taxonomic spectrum of the genus Thymus of the southeastern part of the North Caucasus and adjacent territories.

№	Species	Western part of the Caucasus (Karachay-Cherkess Republic)	(Central Caucasus) Kabardino-Balkarian Republic	South-Eastern part of the North Caucasus (Republic Ingushetia and Chechen Republic)	Eastern Caucasus (Republic Dagestan)
1.	*Thymus borysthenicus* Klok.	-	+	-	-
2.	*Th. caucasicus* Willd. ex Ronn.	+	+	+	+
3.	*Th. collinus* Bieb.	+	+	+	+
4.	*Th. dagestanicus* Klok.	+	+	+	+
5.	*Th. elisabethae* Klok	+	+	-	-
6.	*Th. fedtschenkoi* Ronn.	-	-	-	+
7.	*Th. karamarianicus* Klok.	-	-	-	+
8.	*Th. liaculatus* Klok.	-	+	-	-
9.	*Th. majkopensis* Klok. etShost.	+	-		-
10.	*Th. marschallianus* Willd.	+	+	+	+
11.	*Th. nummularius* Bieb.	+	+	+	+
12.	*Th. pallasianus* H. Br.	-	-	-	+
13.	*Th. pastoralis* Iljin ex Klok.	-	+	+	+
14.	*Th. pulchellus* C.A. Mey	+	-	-	-
15.	*Th. pseudopulegioides Klok. et Shost.*	+	+	+	+
16.	*Th. transcaucasicus* Ronn.	-	+	+	+

The high polymorphism of the genus Thymus [11, 12, 13, 14, 15] is most likely the result of a wide amplitude of plasticity to environmental conditions, which allows them to develop particular adaptations to environmental conditions to inhabit both xerophilic types of vegetation: steppe, friganoid, and conditions of alpine, meadow-steppe and meadow habitats with a high degree of humidity and rockiness of the relief.

4 Conclusion

On the territory of the South-Eastern part of the North Caucasus, we have registered eight species of the genus *Thymus: Th.collinus* Bieb., *Th. marschalianus* Willd., *Th. dagestanicus* Klok., *Th. caucasicus* Willd., *Th. nummularius* Bieb., *Th. pastoralis* Iljin ex Klok., *Th.*

pseudopulegioides Klok. et Shost., Th. transcaucasicus Ronn., distributed mainly in xerophilic types of vegetation: steppe, friganoid, as well as high-mountain meadow-steppe and meadow communities. Seven of the eight species in the study area are confined to rocky and gravelly slopes. Besides, they play a significant role in the composition of the vegetation cover of the slopes of the southern, southeastern exposures of the study area, being part of the herbage with the dominance of cereals, legumes, or herb-cereal cenoses. According to the geographical structure, the species of the genus under study belong to two geographical elements - Caucasian (seven species) and Pontic-South Siberian (one species).

References

1. M.K. Dakieva, Materials of the scientific international conference of students, graduate students and young scientists Perspective, volume VII. Pp. 123 (2009)
2. A.I. Galushko, Flora of the North Caucasus. - Rostov: Russian State University, **1-3,** 351 (1978-1980)
3. R.M. Harley, S. Atkinson, A.L. Budantsev, P.D. Cantino, B.J. Conn, R. Grayer, M.M. Harley, R. De Kok, T. Krestovskaja, R. Morales, A.J. Paton, O. Ryding, T. Upson, The Families and Genera of Vascular Plants, **7,** 167 (2004)
4. F.Yu. Kasumov, Bulletin of the Botanical Garden-Institute of the Far Eastern Branch of the Russian Academy of Sciences, **2,** 34 (2008)
5. Yu. L. Menitskiy, Bot. Journal, **77(6),** 63 (1992)
6. R. Morales, Thyme The Genus Thymus, 1 (2012)
7. R.A. Murtazaliev, Conspect of the flora of Dagestan, **3,** 227 (2009)
8. N. N. Portenier, Bot. Journal. T., **85(9),** 26 (2000)
9. V. M. Prima, *Flora and vegetation of the Eastern Caucasus,* 46 (1974)
10. I.G. Serebryakova, *Ecological morphology of plants,* 380 (1962)
11. S.Kh. Shkhagapsoev, *Vegetation cover of Kabardino-Balkaria,* 218 (2015)
12. D.S. Shilnikov, *Flora survey of Karachay-Cherkessia,* 279 (2010)
13. A. L. Takhtadzhyan, *Floristic regions of the Earth. – L. Science,* 248 (1978)
14. M. U. Umarov, M. A. Taisumov, *Abstract of the flora of the Czech Republic,* 100 (2011)

Systematic and Ecological-Cenotic Analysis of the *Cyperaceae* Family of the Chechen Republic

Zazu Iriskhanova, *Madina* Takaeva, and *Luiza* Molochaeva

Chechen State University named after A.A. Kadyrov, 364024 Grozny, Russia

Abstract. In this communication, a systematic and ecology-coenotical analysis of representatives of the *Cyperaceae* family of the Chechen Republic is given. A complete list of genera of the studied family of the Chechen Republic is specified. These studies were carried out based on processing herbarium materials and field observations of the authors. Representatives of *Cyperaceae* family of the Chechen Republic for the first time will be well studied.

1 Introduction

The Chechen Republic is located on a territory where six altitudinal zones (from semi-desert to alpine) are successively replaced from north to south; therefore, the composition of plants is peculiar and original and remains deficiency studied.

Plants belonging to the *Cyperaceae* family have been recognized since antiquity, and this suggests that at least some family species have long been considered worthy of human attention. Even though the species of this family are sometimes considered plants of minimal practical importance, especially in comparison with *Poaceae*, a more detailed examination showed that this point of view is erroneous. According to a review by Simpson (2008), there is a wide range of historical or current uses for Sedge by humans. *Cyperus papyrus L.* is a prime example of food for domestic animals and even for humans (for example, the corms of the edible rhizome *Eleocharis dulcis* (Burm. f.).

Some species are interesting for horticulture as ornamental plants, and others are used to compact soils at risk of erosion or landscaping remarkably infertile areas. In addition, several family species (e.g., *Cyperus rotundus L.* or *C. esculentus L.*) are agricultural weeds.

In addition to purely practical aspects, the critical role of sedges in plant communities and entire ecosystems is undeniable. The species of this family inhabit various habitats from the tropics to the arctic regions, especially in wetlands. However, on the other hand, many sedge species are inferior in competitiveness, are limited to vulnerable habitats, and, therefore, are rare, endangered, and essential from a conservation point of view.

2 Material and Methods

These studies are based on the analysis of field studies and observations of the authors. Field research is a qualitative research method. Route-geobotanical, route-floristic methods were used. An integral part of the work is the inventory of the flora and the study of its current state, identifying the taxonomic composition of phytocenoses. Fieldwork was carried out using the route-floristic method in different growing seasons of 2020-2021. Based on the research results, a list of the studied family was compiled, numbering 70 species. The collected material was processed at the Department of Botany, Zoology, and Bioecology of the Chechen State University, named after A.A. Kadyrov.

3 Results and Discussion

In the course of the study, 70 species of plants belonging to the Sedge family, united in 14 genera, grow in the wild on the territory of the Chechen Republic.

Table 1. Generic composition of the *Cyperaceae* family in the flora of the Chechen Republic

№	**Genus name**	**Number of species**	
		quantity	%
	Latin name		
1.	*Pycreus*	1	1,43
2.	*Juncellus*	2	2,86
3.	*Cladium*	1	1,43
4.	*Cyperus*	4	5,71
5.	*Schoenoplectus*	1	1,43
6.	*Bolboschoenus*	2	2,86
7.	*Blymus*	1	1,43
8.	*Eleocharis*	5	7,14
9.	*Kobresia*	3	4,29
10.	*Carex*	43	61,42
11.	*Schoenoplectus*	4	5,71
12.	*Schoenus*	1	1,43
13.	*Scirpus*	1	1,43
14.	*Scirpoides*	1	1,43
Total		70	100

The genus *Carex* (*Cyperaceae*) in the flora of the Chechen Republic is represented by the most significant number of species - 43, which is (61.43%) of the total number of studied species. The genus *Eleocharis* (Bolotnitsa) is represented by five species (7.14%); genus *Cyperus* (Syt), *Schoenoplectus* (Shenoplektus) - four species (5.71%); the genus *Kobresia* (Kobresia) includes three species (4.29%), the genus *Juncellus* (Sitnichek), *Bolboschoenus* (Klubnekamysh) (2.86%) are represented by two species, the genus *Pycreus* (Sitovnik), *Cladium* (Sword-grass), *Isolepis* (Shenoplectus), *Blymus* (Blimus), *Schoenus* (Schinus), *Scirpus* (Reed), *Scirpoides* (Goloshenus), represented by one species (1.43%).

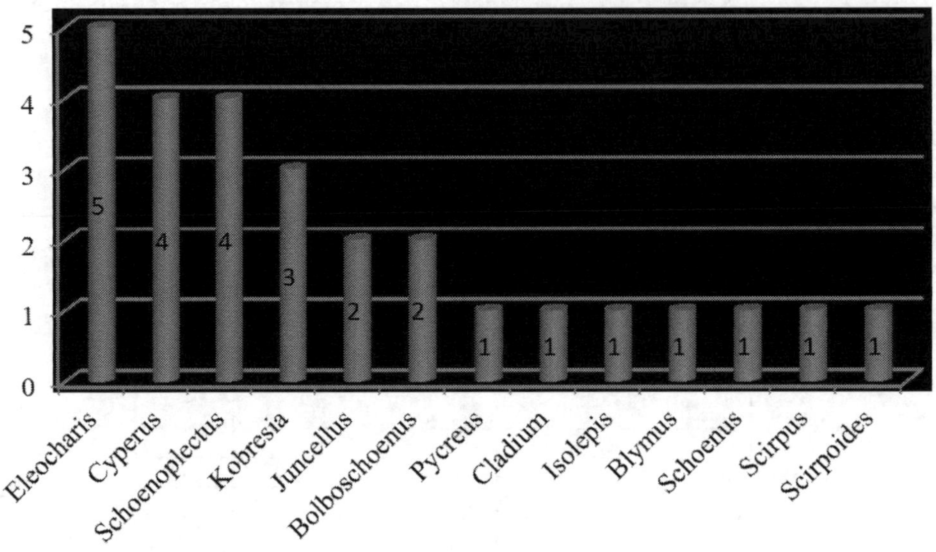

Fig. 1. Generic structure of the *Cyperaceae* family in the flora of the Chechen Republic.

In the ecological- coenotical analysis of the Sedge family, we have identified 105 floracenoelements, the spectrum of which is shown in Table 2. The number of florocenoelements in the flora is always more significant than the number of species. This position is confirmed by many researchers of the flora of the North Caucasus, who analyzed regional flora [1, 4, 5, 6, 7, 8].

Table 2. The phytoene-ecological spectrum of *Cyperaceae* of the Chechen Republic.

FLOROCENO ELEMENT	quantity flora-ceno-elements	% of the total number of species
Forest	8	7,62
Flat	10	9,52
Subalpine	7	6,67
Alpine	16	15,24
Steppe	6	5,71
Calcepetrophilic	7	6,67
Psammophilic	3	2,86
Halophilic	2	1,9
Hygrophilic	33	31,43
Hydrophilic	13	12,38
TOTAL	105	100

There are eight species of forest floracenoelements (7.62%). These are such species as *Carex Transsilvanica* Schur (*C. depressa* Link), *C. Transsilvanica* Schur (*C. depressa* Link), *C. digitata* L., *C. divulsa* Stokes, *C. polyphylla* Kar.et Kir., *C. contigua* Hoppe , *C. remota* L., *C. leporine* L., *Scirpus sylvaticus* L. [10].

There are ten types of flat floracenoelements (9.52%). These are species such as Carex songorica Rfr. et Kir., C. Michelii Host., C. Transsilvanica Schur (C. depressa Link), C. Transsilvanica Schur (C. depressa Link), C. caryophyllea Latourr. C. humilis Leyss. (C. Buschiorum V. Crecz.), C. pallescens L., C. tomentosa L., C. contigua Hoppe, C. leporine L.

Subalpine floracenoelements, seven species in total (6.67%). This is *Carex songorica* Rfr. et Kir., *C. Michelii* Host., *C. Transsilvanica* Schur(*C. depressa* Link), *C. Transsilvanica*

Schur(*C. depressa* Link), *C. caryophyllea* Latourr. –, *C. humilis* Leyss. (*C. Buschiorum* V. Crecz.), *C.pallescens* L., *C. tomentosa* L., *C. contigua* Hoppe, *C. leporine* L..

Subalpine floracenoelements, seven species in total (6.67%). These are types such as *Carex tristis* Bieb. (C. *meinshauseniana* V.Krecz.), *C. rostrata* Stokes, *C. capillaries* L., *C. cancasica* Stev., *C. transcaucasica* Egor., *C. canescens* L., *C. brunnescens* (Pers.) Poir. [6].

There are 16 species of Alpine floracenoelements (15.24%). These are types such as *Blymus compressus* (L.) Panz. ex Link, *Eleocharis quinqueflora* (F.X. Hartm.) O.Schwarz (*E. pauciflora (*Lightf) Link), *Kobresia schoenoides* (C.A. Mey.) Steud., *K. macrolepis* Meinsh. (*K. Capillifolia* (Decne) Clarke), *K. persica* Kuk. et Bornm. (*K. Humilis (C.A. Mey. ex Trautv.) Serg.), Carex tristis* Bieb. (C. *meinshauseniana* V.Krecz.), *C. dichroa* Freyn (C. *pamirica (*O.Fedtsch.) O. et B. Fedtsch.), *C. Huetiana* Boiss., *C.capillaries* L., *C. oligantha* Steud., *C. cespitosa* L., *C. cespitosa* L., *C. kotschiana* Boiss. et Hohen. (C. *orbicularis* Boiss. et Hohen.), *C. canescens* L., *C. brunnescens* (Pers.) Poir., *C. heleonastes Ehrh*. [10].

There are six species of steppe florocenosis of elements (5.71%). These are types such as *Carex caryophyllea* Latourr., *C. tomentosa* L., *C. Bordzilowskii* V. Krecz. (*C. Schkuhrii* Willd; *C.liparocarpos* Gaudiri), *C. supine* Willd. ex Wahlenb., *C. praecox* Schreb., *C. stenophylla* Wahlenb. (C. *uralensis* Clarke; *C. dimorphotheca* Stschegl.; *C. Stenophylloides* (V.Crecz.) Egor.) [10].

Calcepetrophilic floracenoelements there are one species (6.67%) - *Carex alba* Scop. [10].

There are three types of psammophilic elements (2.86%). Of these, only on sands are inhabited by *Carex colchica* J.Gay, *C. colchica* J.Gay, *C. praecox* Schreb. [10].

There are two types of halophilic floracenoelements (1.9%). They are all coenotypically correct. These are species such as *Eleocharis uniglumis* (Link) Schult. (*E. multiseta* Zinserl.), *Carex divisa* Huds. [10].

There are 33 types of hygrophilic floracenoelements (31,43%): *Pycreus flavescens* (L.) Beauv. ex Reichenb., *Juncellus pannonicus* (Jacq.) Clarke (*Acorellus pannonicus* (Jacq.) Palla), *J. serotinus* (Rottb.) Clarke (*Cyperus serotinus* Rottb.), *Cyperus fuscus* L., *C. glaber* L., *C. glomeratus* L., *C. longus* L., *Isolepis setacea* (L.) R.Br. (*Schoenoplectus setaceus* (L.) Palla; (*Scyrpus setaceus* L.), *Bolboschoenus glaucus* (Lam.) S.G.Sm., *B. maritimus* (L.) Palla *(B. compactus (Hoffm.) Drob.)*, *B. compressus* (L.) Panz. ex Link, *Eleocharis acicularis* (L.) Roem.et Schult., *E. quinqueflora* (F.X. Hartm.) O.Schwarz (*E. pauciflora (*Lightf) Link), *E. parvula* (Roem. et Schult.) Bluff. Neeset Schauer, *E. palustris* (L.) Roem.et Schult. (*E. eupalustris* Linld.fil., *E. crassa* Fisch. et C.A. Mey. ex Zinserl.; *E. intersita* Zinserl.), *E. uniglumis* (Link) Schult. (*E. multiseta* Zinserl.), *Carex hirta* L., *C. rostrata* Stokes –, *C.dichroa* Freyn (C. *pamirica* (O.Fedtsch.) O. et B. Fedtsch.), *C. melanostachya* Bieb. ex Willd., *C. songorica* Rfr. et Kir., *C. omskiana* Meinsh. (C. *Elata* All.), *C. cespitosa* L., *C. cespitosa* L., *C. remota* L., *C. leporine* L., *C. canescens* L., *C. heleonastes* Ehrh., *Schoenoplectus litoralis* (Schrad.) Palla (*S. litoralis* Schrad.), *Sch.s triqueter* (L.) Palla (*S. triqueter* L.), *Schoenus nigricans* L., *Scirpus sylvaticus* L., *Scirpoides holoschoenus* (L.) Sojak (*Holoschoenus vulgaris* Link incl. *H. romanus* (L.) Fritsch) [10].

There are 13 species of hydrophilic floracenoelements (12.38%). These are types such as *Cladium mariscus* (L.) Pohl, *Bolboschoenus maritimus* (L.) Palla *(B. compactus (Hoffm.) Drob.)*, *Eleocharis acicularis* (L.) Roem.et Schult., *E. palustris* (L.) Roem.et Schult. (*E. eupalustris* Linld.fil., *E. crassa* Fisch. et C.A. Mey. ex Zinserl.; *E. intersita* Zinserl.), *Carex vesicaria* L. (C. *inflata Huds.)*, *C. acutiformis* Ehrh., *C. riparia* Curt., *C.pseudocyperus* L., *C. hordeistichos* Vill., *C. diluta* Bieb., *C. Otrubae* Podp., *Schoenoplectus lacustris* (L.) Palla

(*Scyrpus lacustris* L.), *Sch. tabememontanii* (C.C. Gmel.) Palla (*S. tabernemontanii* C.C.Gmel.) [10].

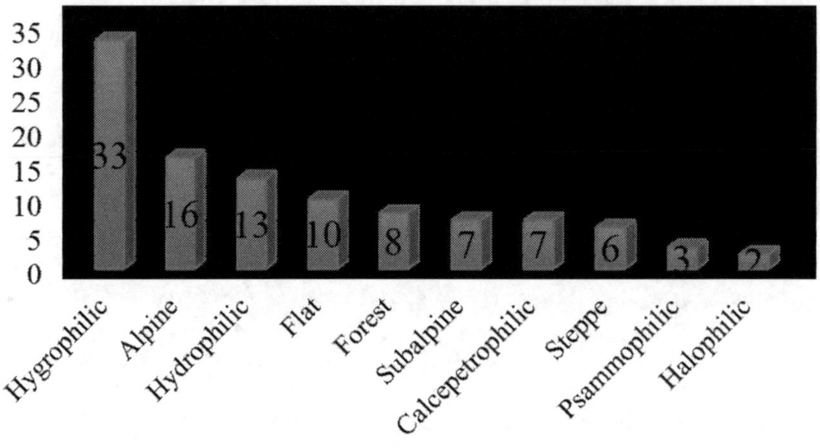

Fig. 2. Spectrum of floracenoelements of the *Cyperaceae* family of the Chechen Republic.

The spectrum of floracenoelements of the Sedge family of the Chechen Republic is shown in Table 3. Hygrophilic floracenoelements occupy first place, and the Alpine floracenoelements take second place. Lowland floracenoelements occupy third place. The smallest is the halophilic florocenoelement.

Occurrence of species of the family *Cyperaceae.*

A cosmopolitan family with a wide variety of habitats, with a preference for open countryside and wetlands. The third most prominent family of monocots and the seventh-largest among the angiosperms.

Usually (26 species) there are representatives of the family such as *Bolboschoenus maritimus* (L.) Palla *(B. compactus (Hoffm.) Drob.)*, *Blymus compressus* (L.) Panz. ex Link, *Kobresia schoenoides* (C.A. Mey.) Steud., *K. macrolepis* Meinsh. (*K. Capillifolia* (Decne) Clarke), *Carex tristis* Bieb. (C. *meinshauseniana* V.Krecz.), *C. rostrata* Stokes, *C.acutiformis* Ehrh., *C.riparia* Curt., *C. melanostachya* Bieb. ex Willd., *C. Michelii* Host., *C. Huetiana* Boiss., *C. caryophyllea* Latourr., *C. diluta* Bieb., *C. digitata* L., *C. pallescens* L., *C. tomentosa* L., *C. alba* Scop., *C. cancasica* Stev., *C. oligantha* Steud., *C. omskiana* Meinsh. (C. *Elata* All.), *C. Otrubae* Podp., *C. praecox* Schreb., *C.leporine* L., *C.canescens* L., *C.brunnescens* (Pers.) Poir., *Schoenoplectus lacustris* (L.) Palla (*Scyrpus lacustris* L.).

There are 24 scattered species: *Pycreus flavescens* (L.) Beauv. ex Reichenb., *Juncellus serotinus* (Rottb.) Clarke (*Cyperus serotinus* Rottb.), *Cyperus fuscus* L., *Isolepis setacea* (L.) R.Br. (*Schoenoplectus setaceus* (L.) Palla; (*Scyrpus setaceus* L.), *Bolboschoenus glaucus* (Lam.) S.G.Sm. –, *Eleocharis acicularis* (L.) Roem.et Schult., *E. palustris* (L.) Roem.et Schult. (*E. eupalustris* Linld.fil., *E. crassa* Fisch. et C.A. Mey. ex Zinserl.; *E. intersita* Zinserl.), *Kobresia persica* Kuk. et Bornm. (*K. Humilis (C.A. Mey. ex Trautv.) Serg.)*, *Carex hirta* L. – Осока волосистая, *C. vesicaria* L. (C. *inflata Huds.)*, *C. hordeistichos* Vill., *C.humilis* Leyss. (*C. Buschiorum* V. Crecz.), *C. Bordzilowskii* V. Krecz. (*C. Schkuhrii* Willd; *C. liparocarpos* Gaudiri), *C. supine* Willd. ex Wahlenb., *C. capillaries* L., *C. divulsa* Stokes, *C. polyphylla* Kar.et Kir., *C.contigua* Hoppe, *C. colchica* J.Gay, *C. divisa* Huds., *C. remota* L., *Schoenoplectus tabememontanii* (C.C. Gmel.) Palla (*S. tabernemontanii* C.C.Gmel.), *Sch. triqueter* (L.) Palla (*S. triqueter* L.), *Scirpus sylvaticus* L., *Scirpoides holoschoenus* (L.) Sojak (*Holoschoenus vulgaris* Link incl. *H. romanus* (L.) Fritsch).

17 species are rare: *Juncellus pannonicus* (Jacq.) Clarke (*Acorellus pannonicus* (Jacq.) Palla), *Cyperus glaber* L., *C.Glomeratus* L., *C. Longus* L., *Eleocharis quinqueflora* (F.X.

Hartm.) O.Schwarz (*E. pauciflora (*Lightf) Link), *E.s parvula* (Roem. et Schult.) Bluff. Neeset Schauer, *E. uniglumis* (Link) Schult. (*E. multiseta* Zinserl.), *Carex dichroa* Freyn (C. *pamirica (*O.Fedtsch.) O. et B. Fedtsch.), *C. songorica* Rfr. et Kir., *C. pseudocyperus* L., *C. Transsilvanica* Schur(*C. depressa* Link), *C. transcaucasica* Egor., *C.kotschiana* Boiss. et Hohen. (C. *orbicularis* Boiss. et Hohen.), *C. stenophylla* Wahlenb. (C. *uralensis* Clarke; *C. dimorphotheca* Stschegl.; *C. Stenophylloides* (V.Crecz.) Egor.), *C.heleonastes* Ehrh., *Schoenoplectus litoralis* (Schrad.) Palla (*S. litoralis* Schrad.), *Schoenus nigricans* L..

Very rare 1 species – *Cladium mariscus* (L.) Pohl.

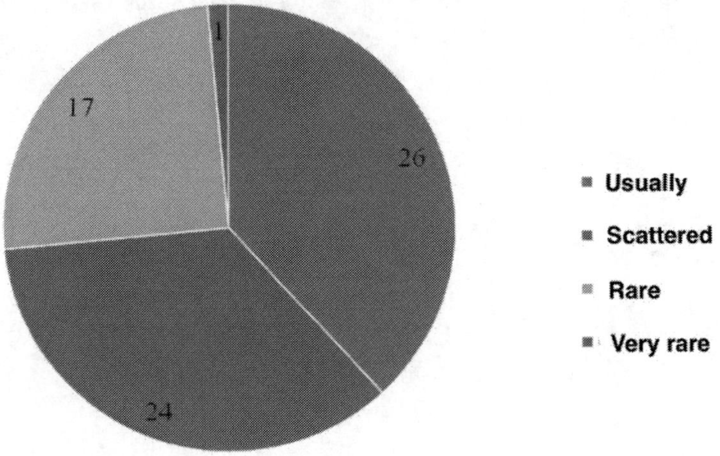

Fig. 3. Frequency of occurrence of species of the family *Cyperaceae* in the Chechen Republic.

4 Conclusions

The Sedge family in the flora of the Chechen Republic has 70 species belonging to 14 genera. The genus *Carex* (sedge) has the most significant number of species - 43; in second place in the genus *Eleocharis* (Bolotnitsa), which is represented by five species (7.14%); the genus *Cyperus* (Syt) *Cyperaceae in* the third place in terms of the number of species; one species (1.43%) includes the genus *Pycreus* (Sitovnik), *Cladium* (Sword-grass), *Isolepis* (Shenoplektus), *Blymus* (Blimus), *Schoenus* (Schenus), *Scirpus* (Reed), and *Scirpoides* (Goloshenus).

When carrying out the ecology-coenotical, it was revealed that the most significant number of species is confined to hygrophilic, alpine, hydrophilic floracenoelements. The smallest are halophilic floracenoelements [10]. In terms of prevalence, 26 species are common, 24 are scattered, 17 are rare, and one is very rare.

References

1. E.V. Wulf, *The experience of dividing the globe into vegetation areas based on the quantitative distribution of species*, 66 (1934)
2. A.I. Galushko, *Vegetation cover of Checheno-Ingushetia*, 118 (1976)
3. A.I. Galushko, Flora of the North Caucasus and questions of its history, **1**, 5-130 (1976)
4. A.L. Ivanov, Bulletin of the Stavropol State University, **6**, 47-57 (1996)
5. A.L. Ivanov, *Flora of the Pre-Caucasus and its genesis*. Stavropol: Publishing House of SSU, 204 (1998)

6. A.A. Golovlev, *Zapovednoe delo: Problems of protection and ecological restoration of steppe ecosystems*, 35-41 (2004)
7. A.A. Golovlev, News of the universities of the North Caucasus region. Natural sciences, **8**, 30-36 (2005)
8. A.I. Tolmachev, *Comparative floristic methods and problems of florogenesis*. Novosibirsk: Science, 195 (1986)
9. S.H. Shkhagapsoev, N.V. Starikova, *Analysis of the natural flora of Kabardino-Balkaria*. -Nalchik, 113 (2002)
10. M.U. Umarov, M.A. Taisumov, *Abstract of the flora of the Chechen Republic*. Grozny, 152 (2021)

Influence of iron nanoparticles (Fe_3O_4 and Fe_2O_3) on the growth, photosynthesis and antioxidant balance of wheat plants (*Triticum aestivum*)

Vladimir Kreslavski*, *Anatoly* Ivanov, *Alexander* Shmarev, *Alexandra* Khudyakova, and *Anatoly* Kosobryukhov

Institute of Basic Biological Problems of the Russian Academy of Sciences - a separate subdivision of the FRC PSCBI RAS, Pushchino, Russia

Abstract. More and more attention is paid to the development of technologies using iron nanoparticles in agriculture. In this regard, the effect of treatment of wheat seeds with various concentrations of iron nanoparticles Fe_3O_4 and Fe_2O_3 on the accumulation of biomass, the rate of photosynthesis and respiration, as well as on photochemical activity and antioxidant balance was studied. The seeds were treated for 3 h, germinated for 2 days in Petri dishes, transplanted into sand and grown under light for 18 days without mineral nutrition until the third leaf appeared. At a Fe_3O_4 concentration of 200 mg L^{-1} a significant increase in the dry biomass of the second leaf by 45% and the rate of photosynthesis by 16% was observed. At a concentration of nanoparticles in the form of Fe_2O_3 of 200 and 500 mg L^{-1}, an increase in the rate of photosynthesis in the second leaf was also observed, but not in the biomass of the leaves. The activity of photosystem 2, estimated from the F_v/F_m value, also increased in experiments with nanoiron. However, the activity of antioxidant enzymes, guaiacol-dependent peroxidase and superoxide dismutase, decreased. It is assumed that the acceleration of growth at an early stage of wheat development is associated with an increase in photosynthetic processes.

1 Introduction

In the last decade, nanoparticles (NPs), that is, particles less than 100 nm in size, have been widely used in various areas of the national economy, including medicine, for water purification and in the food industry. However, in agriculture, the use of nanoparticles has not been studied enough so far. Recently, more and more attention has been paid to the study of the ways of action of NPs of various metals, in particular, in the form of iron oxide in order to increase the productivity of plants, as well as reduce the environmental risk of using fertilizers [1–3]. The effect of nanomaterials largely depends on the method of their introduction. Thus, [4] showed that plant growth depends on the method of using nanomaterials. Foliar application of nano-Fe_3O_4 (leaf spraying) significantly improved the

* Corresponding author: vkreslav@rambler.ru

growth and a number of other physiological parameters of *Ocimum basilicum* plants compared to soil application. The use of Fe_3O_4 nanoparticles significantly increased the growth and yield of pea plants by increasing the leaf area. In work [5], an increase in dry biomass, protein content and the number of pods was shown in comparison with the control.

Iron (Fe) is an essential trace element for plants and plays a key role in the regulation of many cellular processes, including chlorophyll biosynthesis, photosynthesis and mitochondrial respiration, as well as redox balance [6]. However, the effect of iron in nanoform on plants is still insufficiently studied. Studies have shown different results on the effect of iron oxide NPs on plants [7–10].

It is assumed that the use of iron NPs increases productivity by enhancing various physiological processes, such as seed germination, photosynthetic activity, plant growth, synthesis of various metabolites, including proteins and nitrogen-containing compounds [2, 11]. It is also assumed that iron NPs can enhance growth by regulating the content of phytohormones, changing the activity of antioxidant enzymes and enzymes of the Calvin cycle [2, 12]. However, at high concentrations, iron NPs can exhibit toxicity [3].

In the presented work, the effect of iron nanoparticles Fe_3O_4 and Fe_2O_3 on growth parameters, rates of photosynthesis and respiration, primary processes of photosynthesis and changes in the activity of antioxidant enzymes in 20-d-old wheat plants under conditions of mineral nutrition deficiency was studied.

2 Materials and methods

2.1 Seeds treatment and plant growth

Wheat seedlings (*Triticum aestivum* L.) were used as the object of the study. The seeds were soaked for 2 h in an aqueous suspension of iron NPs obtained as a result of ultrasonic treatment of Fe_3O_4 (Advanced Powder Technology, Tomsk, Russia, 99%) or Fe_2O_3 (99%, China) for 20 min at 22 kHz using an UZG13 ultrasonic generator (0.1/22 (Russia). The size of the used Fe_3O_4 nanoparticles was 95 ± 10 nm, and the size of Fe_2O_3 nanoparticles – 80 ± 10 nm (China). After treatment, the seeds were germinated in Petri dishes on filter paper. Two days later, germinated seeds of the same size were transferred into vessels (13x13x8 cm), filled with well-washed river sand. Plants were grown without adding nutrients in the soil at a 12 h photoperiod, at a temperature of 24/20 °C (day/night), a relative air humidity of 70–80%, and a light intensity of 300 ± 20 μmol (photons) m^{-2} s^{-1}. During cultivation, the plants were watered with distilled water 3 times a week. 20-day-old plants were used in the experiment. The growth parameters of plants were determined for the 1st and 2nd leaves of seedlings. Plants without nanoiron treatment were used as control.

Fresh and dry leaf weights were determined using an analytical balance (Scout Pro SPU123, Ohaus Corporation, USA). The relative water content (RWC) in the leaves was calculated using the formula: $RWC = 100\times(FW - DW)/DW$, where FW is the wet weight, DW is the dry weight.

2.2 CO_2 gas exchange and photochemical activity

The photosynthesis rate (P_n) was determined using a portable infrared gas analyzer LCPro + (ADC BioScientific Ltd., United Kingdom) in a leaf chamber with an area of 6.25 cm^2 and a saturating light intensity of 1000 μmol (photons) m^{-2} s^{-1}. The rate of dark respiration (R) was determined after turning off the light.

The parameters of photochemical activity were determined using a fluorometer by a JIP test based on the setup described in [13]. Chlorophyll a fluorescence induction curves (OJIP) were recorded using the data of the JIP test for photosystem 2 (PSII) under blue light illumination with an intensity of 6000 µmol (photons) m^{-2} s^{-1}. Fluorescence parameters (F_V/F_m, Y(II), NPQ, Y(NO), and Y(NPQ) were calculated according to [14, 15]. F_V/F_m is the maximum quantum yield of PSII, where F_V is the variable fluorescence equal to the difference between F_m and F_0, F_0 is the minimum fluorescence amplitude (F), and F_m is the maximum fluorescence amplitude. Y(II) is the effective quantum yield of PSII fluorescence was calculated using the formula: $Y(II) = (F'_m - F_t)/F'_m$. NPQ – non-photochemical quenching of fluorescence was calculated as $NPQ = (F_m - F'_m)/F'_m$. Y(NO) – quantum yield of non-photochemical quenching of fluorescence unregulated by light was calculated according to $Y(NO) = 1/(NPQ + 1 + qL \cdot (F_m/F_0 - 1))$. Y(NPQ) – the quantum yield of controlled light-induced non-photochemical fluorescence quenching was calculated as $Y(NPQ) = 1\text{-}Y(II)\text{ -}Y(NO)$.

2.3 The evaluation of activity of antioxidant system

The content of malonic dialdehyde (MDA) was determined according to [16]. The concentration of MDA was measured at 532 and 600 nm using a spectrophotometer (Genesis 10UV, ThermoSpectronic, USA). For comparison, a mixture without plant extract was used. The MDA content was calculated using an extinction coefficient of 155 µmol^{-1} cm^{-1} and is expressed as µmol MDA g^{-1} FM. SOD activity was determined by monitoring the inhibition of photochemical reduction of nitroblue tetrazolium at 560 nm according to [17] with minor modifications. The activity of guaiacol-dependent peroxidase (POD) was determined according to [18]. The estimation of the specific activity of the enzymes was carried out on the basis of the wet weight of the plant material.

2.4 Statistics

Statistical data processing was carried out using the Statistics 10 software (StatSoft Inc., USA). The significance of differences between the mean values of the measured parameters was assessed using the Student's test. Calculations were performed at a given significance level $p \leq 0.05$. The table shows the arithmetic mean values with standard errors.

3. Results and discussion

3.1. Growth parameters

Wheat plants were grown on a sandy substrate without feeding in order to exclude the influence of other nutrients on the growth and other parameters of the seedlings. In experiments with Fe_2O_3, the effect of iron NPs was studied in the concentration range up to 500 mg L^{-1}. The pre-selected concentration of Fe_3O_4 (in the range of 0–200 mg L^{-1}), which gave a noticeable increase in growth and photosynthesis, was 200 mg L^{-1} (Table 1).

Table 1. Physiological parameters of the 2nd leaf of 20-day-old wheat seedlings grown without nutrients (Control, $-Fe_3O_4$) and from seeds treated with NPs ($+Fe_3O_4$ 200 mg L^{-1}). R is the respiration rate. FW – wet weight, DW – dry weight. Mean values from 5 biological replicates ± SD are shown. * – the values are significantly higher than the control ($p < 0.05$).

Parameters/Variant	$-Fe_3O_4$	$+Fe_3O_4$
FW, мг	43.0 ± 3.7	39.9 ± 5.8
DW, мг	8.4 ± 0.7	12.2 ± 1.6*
FW/DW	5.1 ± 0.5	3.3 ± 0.4*
Total leaf DW, mg	20.3 ±1.7	23.9 ±3.4
P_n, µM CO_2 m^{-2} s^{-1}	7.6 ± 0.3	8.8 ± 0.2*
R, µM CO_2 m^{-2} s^{-1}	3.8 ± 0.1	4.0 ± 0.2
SOD, rel. units g^{-1} FW	10.2 ± 0.2	5.1 ± 0.9*
POD, rel. units g^{-1} FW	2.25 ± 0.30	1.38 ± 0.23*
MDA, µM g^{-1} FW	4.4 ± 0.3	4.1 ± 0.2

On the 20th day from the beginning of seed germination, the plants formed the third leaf. The measurement of growth parameters was carried out for the 1st and 2nd leaves. Other physiological and biochemical parameters were determined only for the 2nd developed leaf (see Table 1 and Fig. 1).

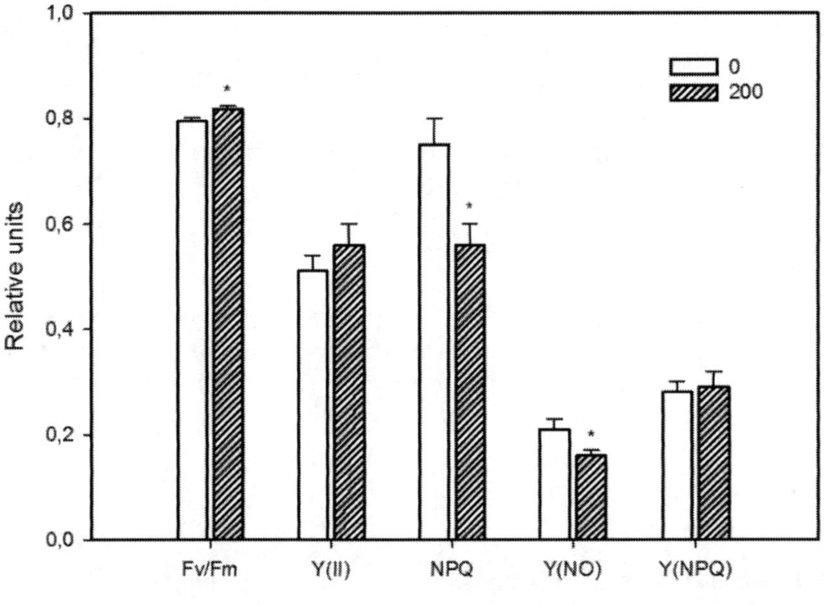

Fig. 1. Fluorescent parameters of the 2nd leaf of 20-day-old wheat seedlings grown without nutrients (0 – without Fe_3O_4) and from seeds treated with NPs (Fe_3O_4 200 mg L^{-1}). Mean values from 5 biological replicates ± SD are shown. * – the values are significantly higher than the control ($p <$ 0.05).

Treatment of wheat seeds with Fe_3O_4 nanoparticles improved the growth parameters of plant leaves, which primarily affected an increase in the accumulation of dry biomass (DW) of the 2nd leaf by 45% compared to untreated control plants. DW of the 1st leaf in the experiment with NPs was differed little from the control (weight 9.8 and 9.1 mg L^{-1}, respectively). At the same time, the total dry weight of all leaves was changed little, although a tendency to an increase was observed (20.3 mg in the control and 23.9 mg at 200 mg L^{-1}). There were no significant differences between the variants in terms of the accumulation of wet weight (FW). At the same time, the ratio of wet weight to dry weight was higher in the control (Table 1). In variants with 8 and 40 mg L^{-1} nanoparticles, no significant differences in the dry weight of the second leaf were found (data not shown). Also, no differences in growth rates were found between plants grown from treated and untreated seeds using a different form of Fe_2O_3 iron NPs (Fig. 2).

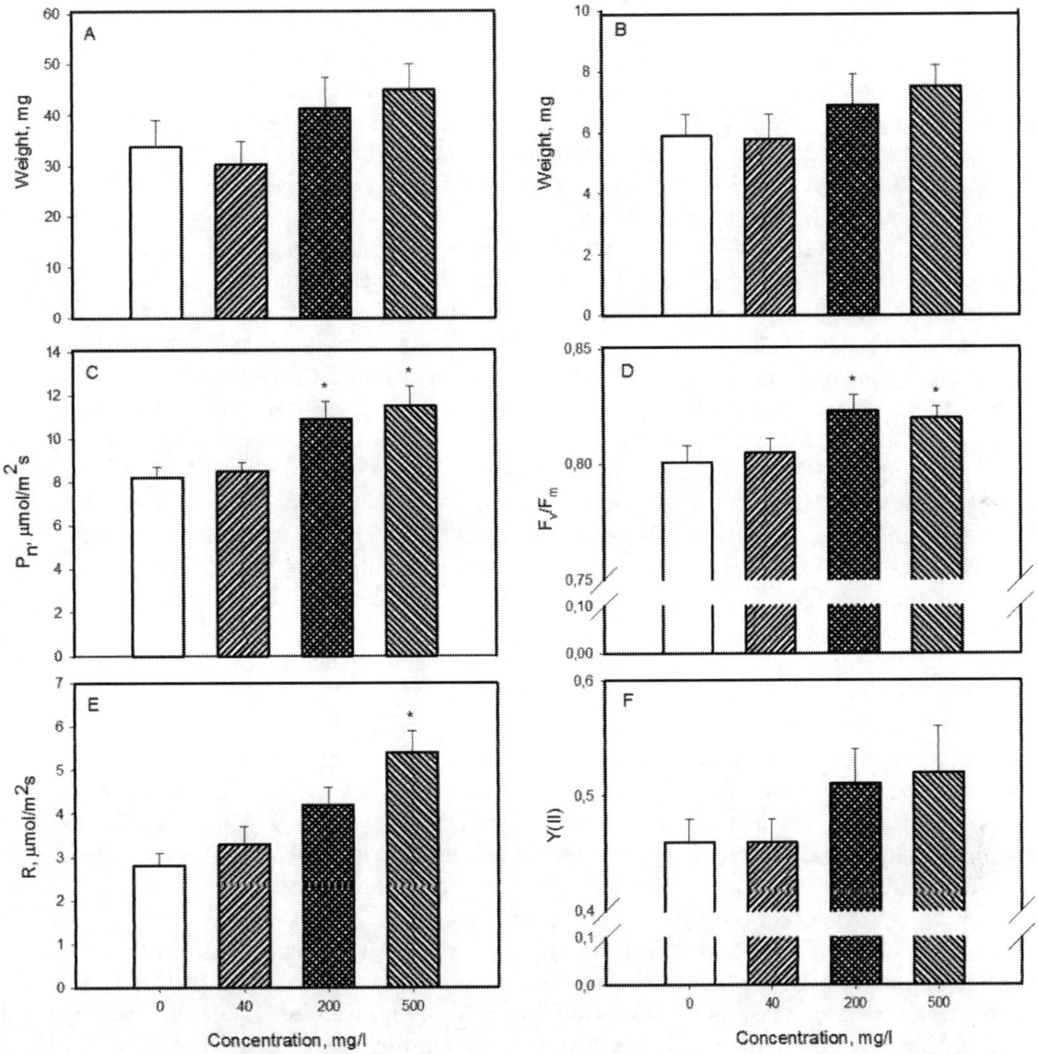

Fig. 2. Physiological parameters of the second leaf of 20-day-old wheat seedlings grown without nutrients (0, without Fe_2O_3) and from seeds treated with iron NPs (+Fe_2O_3 40, 200, 500 mg L^{-1}) A – average fresh weight of the leaf, B – average dry weight of the leaf. C is the rate of photosynthesis. D – maximal quantum yield of PSII (F_v/F_m). E - respiration rate (R). F – Y(II), the effective quantum yield of PSII. Average values from 3 biological replicates ± SD are shown. * – the values are reliably higher than the control ($p < 0.05$).

3.2. Photosynthesis and photochemical activity

Plants grown from seeds treated with Fe_3O_4 showed a 16% increase in photosynthesis (P_n), assessed by the rate of CO_2 fixation (Table 1). At the same time, the rate of dark respiration (R) remained approximately the same in both variants of the experiment. A change in some fluorescent parameters of the leaves of plants grown from seeds treated with Fe_3O_4 in comparison with the control was also revealed (Fig. 1). Thus, the values of regulated non-photochemical quenching (NPQ) and the quantum yield of non-regulated non-photochemical quenching of fluorescence (Y(NO)) in leaves treated with Fe_3O_4 decreased by 25 and 24%, respectively, as compared to untreated plants. On the other hand, the values of the maximum quantum yield F_v/F_m were higher than in the control. However, in both cases, the value of quantum yields (about 0.80) characterizes the plants as being within the physiological norm. The data obtained are consistent with the fact that under iron deficiency, the efficiency of the functioning of photosystem 2 is reduced [18]. The value of the parameter Y (NPQ), which reflects the efficiency of controlled dissipation of absorbed energy into heat, practically coincided in both variants of the experiment.

The treatment of seeds with Fe_2O_3 (200 and 500 mg L^{-1}) also led to an increase in the rate of photosynthesis of the second leaf as compared to the control (Fig. 2). Moreover, the maximum quantum yield in the control was also lower than in experiments with 200 and 500 mg L^{-1} of Fe_2O_3 NPs. The study of the effect of treatment with iron NPs on the pro/antioxidant balance in the 2nd leaf using the example of Fe_3O_4 showed that in the leaves of treated plants, the activity of such enzymes as superoxide dismutase (SOD) and guaiacol-dependent peroxidase (POD) significantly decreases relative to the control, by 50 and 68%, respectively (Table 1). At the same time, no significant differences were found between the control and experiment with Fe_3O_4 in terms of the content of malondialdehyde.

The effect of NPs on plant growth and development largely depends on the method of application of NPs, their size, and plant species [4, 7]. We hypothesized that under unfavorable conditions, iron NPs could support plant growth and photosynthesis. In our case, long-term cultivation of plants in sand under conditions of a lack of nutrients led to a lower activity of photosystem 2 and a reduced rate of photosynthesis compared to untreated plants. The higher photochemical activity of photosystem 2, estimated as the F_v/F_m ratio, and the decreased activity of antioxidant enzymes, all are consistent with the decreased heat dissipation in plants grown from treated seeds (Figs. 1 and 2).

It was shown in [2] that the use of Fe_2O_3 nanoparticles led to an increase in root length, height and biomass of plants, as well as chlorophyll content in peanut plants. The authors suggested that the use of Fe_2O_3 promotes the growth of peanuts by regulating the content of phytohormones and the activity of antioxidant enzymes (SOD, POD, CAT). In our study, the activity of the antioxidant enzyme SOD in control without Fe_3O_4 was increased, which, along with a lower value of the F_v/F_m value and a higher level of heat dissipation may indicate a weaker oxidative stress in the untreated plants.

Iron NPs, regardless of their shape (Fe_2O_3 or Fe_3O_4) and approximately equal particle size, stimulate wheat photosynthesis. At the same time, in the case of Fe_2O_3, there is a tendency to an increase in the accumulation of wet and dry biomass of the 2nd leaf, while in the case of Fe_3O_4, the wet to dry weight ratio significantly decreased, that is, an increase in the accumulation of dry matter occurred. An increase in biomass during treatment with nanoiron can be explained by an increase in the rate of photosynthesis and, in part, respiration, as well as a change in the pro/antioxidant balance due to the entry of iron ions into plants.

4 Conclusions

The positive effect of iron oxide NPs on growth processes can be associated with a more efficient use of nutrients from seeds in treated plants, as well as with an increase in the rate of photosynthesis and respiration in leaves, possibly due to a change in the redox balance in the presence of a sufficient amount of iron ions in the leaves. It can be assumed that the treatment of seeds with iron oxide NPs will be effective for maintaining the productivity of cereals (wheat, etc.) under conditions of a deficit of mineral nutrients.

Acknowledgements

The reported study was funded by RFBR and NSFC, project number 21-54-53015.

References

1. M.J. Abhilash, Int. J. Pharma Bio Sci., **1,** 1 (2010)
2. M. Rui, C. Ma, Y. Hao, J. Guo, Y. Rui, X. Tang, Q. Zhao, X. Fan, Z. Zhang, T. Hou, S. Zhu, Front. Plant Sci., **7**, 815 (2016)
3. K. Kornarzyński, A. Sujak, G. Czernel, D. Wiącek, Sci. Rep., **10**, 8068 (2020)
4. S.A. Elfeky, M.A Mohammed., M.S. Khater, Y.A.H. Osman, E. Elsherbini, Med. Plants - Int. J. Phytomed. Relat. Ind., **463**, 2051 (2013)
5. A. Al-Sherbini, H.G. Abd El-Gawad, M.A. Kamal and S.A El-feky. American-Eurasian J. Agric. & Environ. Sci., **15(7)**, 1435 (2015)
6. S. Ghasemi, A.H. Khoshgoftarmanesh, M. Afyuni, H. Hadadzadeh, Sci. Hortic., **165**, 91 (2014)
7. H. Wang, X. Kou, Z. Pei, J.Q. Xiao, X. Shan, B. Xing, Nanotoxicology, **5**, 30 (2011)
8. N. Siddiqi, J.K. Harrison, A. Clegg, E.A. Teale, J. Young, J. Taylor, S.A. Simpkins, Cochrane Database Syst. Rev., **3(3)**, CD005563 (2016)
9. M.F. Iannone, M.D. Groppa, M.L. de Sousa, M.B. Fernández van Raap, M.P. Benavides, Environ. Exp. Bot., **131**, 77 (2016)
10. S. Huang, L. Wang, L. Liu, Y. Hou, L. Li, Agron. Sustain. Dev., **35**, 369 (2015)
11. T.L Winder., J.N. Nishio, Plant Physiol., **108**, 1487 (1995)
12. V.D. Kreslavski, A.V. Lankin, S.I. Allakhverdiev, V.Y. Luybimov, G.K. Vasilyeva, G.N. Semenova, F.-J. Schmitt, T. Friedrich, J. Plant Physiol. Biochem., **81**, 135 (2014)
13. V.N. Goltsev, H.M. Kalaji, M. Paunov, W. Bąba, T. Horaczek, J. Mojski, H. Kociel, S.I. Allakhverdiev, Russ. J. Plant Physiol., **63(6)**, 869 (2016)
14. H. Kalaji, A. Jajoo, A. Oukarroum, M. Brestič, M. Zivcak, I. Samborska, M. D. Cetner, Izabela Łukasik, V. Goltsev, R. Ladle, Acta Physiol. Plant., **38**, 102 (2016)
15. M. Uchiyama, M. Mihara, Anal. Biochem., **86**, 287 (1978)
16. S.A. Gupta, R.P. Webb, A.S. Holaday, R.D. Allen, Plant Physiol., **103**, 1067 (1993)
17. B. Chance, A.C. Maehly, Methods Enzymol., **2**, 764 (1955)
18. F. Morales, A. Abadía, J. Abadía, Aust. J. Plant Physiol., **25**, 403 (1998)

Effect of light-converting coatings on the growth of Sarepta mustard (*Brassica* juncea L.) plants colonized by associative microorganisms

Natalia Zakharchenko[1,*], *Sergey* Anisimov[2], *Ivan* Dyadishchev[3], *Sergey* Ponomarenko[3], and *Robert* Khramov[2]

[1]Branch of Shemyakin-Ovchinnikov Institute of Bioorganic Chemistry, Russian Academy of Sciences, Science Avenue 6, 142290 Pushchino, Russia
[2]Institution of Theoretical and Experimental Biophysics, Russian Academy of Sciences, Science Avenue 3, 142290 Pushchino, Russia
[3]Enikolopov Institute of Synthetic Polymeric Materials, Russian Academy of Sciences, Profsoyuznaya str. 70, 117373 Moscow, Russia

Abstract. The effect of colonization by beneficial associative microorganisms Pseudomonas putida KT 2442 and Rhodococcus erythropolis X5 on the growth of Sarepta mustard (Brassica juncea L.) under a covering light-converting material containing organic photoluminophore, in vitro and in vivo, was investigated. The combined use of microbial colonization and photoluminophore coating led to stimulation of plant growth much stronger (2.4 times more) than separately only photoluminophoric coating (1.3 times) or colonization (2.1 times). These data indicate that when covering materials with photoluminophores are used in agrobiotechnologies, luminescent red light (610-730 nm) induces an increase in biochemical processes not only in plants, but also in microorganisms that supply plants with growth regulators and other useful metabolites. The data obtained are relevant for further study of the photobiological mechanisms of interactions between the plant-microorganism system in agrobiotechnologies.

1. Introduction

The use of light-correcting coatings is a promising method of increasing crop yields [1, 2]. The influence of coatings on the development of plants is associated with a change in the spectral composition of the electromagnetic radiation of the sun passing through them – the absorption of the ultraviolet component and its transformation into the red region of the spectrum [3]. At the same time, there is a change in the structure of membranes, their permeability increases, contributing to an increase in the effectiveness of cytokinins and gibberellins, leading to increased cell division. In natural conditions, plants exist in close association with a complex of soil microorganisms that have a stimulating effect on the growth and development of plants. This process is associated with the ability of microorganisms to nitrogen fixation, mobilize nutrients from the soil, produce

* Corresponding author: znata_2008@mail.ru

physiologically active substances, suppress the growth of phytopathogens and detoxify alien chemicals [4-6]. Previously, it was shown that sunlight that passed through the light-correcting film and transformed by photoluminophores stimulates growth processes, population dynamics and oxygenase activity of indigenous soil microflora during oil degradation on the example of Pseudomonas strains [7, 8]. It is possible that during the growth of agricultural plants under light-correcting films, not only the plants themselves, but also microorganisms associated with them – methylobacteria, pseudomonas, rhodococci and many others are exposed to the influence of transformed light. The role of microbiota in stimulating plant growth with light-converting coatings has not yet been investigated in details.

The purpose of our work was to perform a comprehensive investigation of the effect of light-correcting films with photoluminophoric coating with a maximum emission of 660 nm on the growth of plants associated with certain strains of microorganisms.

2 Materials and methods

2.1 Plant material

The objects of research were the plants of Sarepta mustard (*Brassica juncea* L.) of the Russian variety. The seeds were sterilized for 1 min in 70% ethanol, then 2 min in a 20% solution of sodium hypochlorite and washed 3 times (10 min each) in sterile distilled water. The seeds were germinated on a hormone-free Murashiga-Skoog (MS) medium containing 7 g/L of agar, 30 g/L of sucrose (pH 5.8) and a standard set of salts [9]. The plants were cultivated at a temperature of 22−24°C, a 16-hour day and illumination of 2000 lx. The source of illumination is fluorescent lamps LB-40 (Russia). In a series of experiments in the open ground (September-October, 2021), light-converting agrotextiles were used: polypropylene non-woven spunbond 30±3 g/m^2, coated with polylactide varnish with a content of 0.25% of the new organic photoluminophore (PL) with a concentration of 0.25% and absorption mainly in the band 460-560 nm, and re-emitting with a spectral maximum of 660 nm with a half-width spectrum of 610-730 nm. As PL, luminescent molecules of donor-acceptor type based on triphenylamine with effective radiation in the red range were used [10, 11]. Spunbond without PL was used as a control.

2.2 Colonization of plants by associative microorganisms

For colonization, bacterial strains *Pseudomonas putida* KT 2442 (pNF142::TnMod) [13] and *Rhodococcus erythropolis* X5 [14], obtained from the Laboratory of Plasmid Biology of the G.K. Scriabin Institute of Biochemistry and Physiology of Microorganisms of the Russian Academy of Sciences, were used. The bacteria were grown on LB medium [15]. Sterile monthly leaves of *Brassica juncea* plants were sprayed with cell suspension (10^3-10^5 CFU/mL) of the bacteria *P. putida* KT 2442 and *R. erythropolis* X5. The plants were placed in glass tubes (each variant of 40 pieces) with an agarized MS medium and cultured at 22°C - 24°C, illumination of 2000 lx in a 16-hour daytime photoperiod. Microbiological testing of various plant explants (leaves, root) was carried out after 7, 14 and 25 days after colonization by the Kandel method [16]. For this purpose, a plant extract obtained by homogenizing 100 mg of plant tissue was applied to the surface of a solid LB medium with selective antibiotics in Petri dishes and incubated at 22° - 24°C. After 2 days, the number of colony-forming units (CFU) per 1 g of wet weight was calculated.

2.3 Statistical analysis

Statistica v.6.0 and MS Excel 2007 were used for statistical data processing. Measurements were made in three biological and three analytical repetitions. Graphs and diagrams show the average values and their standard deviations. The reliability of the differences was assessed by the non-parametric Mann-Witney criterion.

3 Results and discussion

In the laboratory, sterile monthly mustard plants grown from seeds were colonized with bacterial strains of *P. putida* KT 2442 and *Rh. erythropolis* X5 under *in vitro* conditions. For this plant, a single spray with a suspension of one of the bacteria with a titer of cells of 10^3 - 10^5 was applied. After that, the test tubes with plants were covered with red (containing PL coating) and or white (without PL coating) spunbond. 20 tubes of colonized plants and 20 tubes of uncolonized ones were put under each of the spunbond. After 7, 14 and 21 days, testing was carried out to establish an associative relationship of the bacteria with plants *in vitro*. To do this, extracts obtained from colonized plants grown under different covering material – with PL and without PL, were applied to the surface of the nutrient medium with selective antibiotics in Petri dishes and incubated for 2 days at a temperature of 22-24°C. In 3 weeks after colonization of the mustard by *P. putida* bacteria, their population on the roots increased under the coating with PL from 5.0 x 10^2 (in the 1st week) to 4.2 x 10^3 CFU/g of raw mass, and under the coating without PL from 4.0 x 10^2 to 3.0 x 10^3. The photoluminophore coating on the spunbond stimulated the growth of bacteria by 1.5 times (Fig. 1A).

At the end of the experiment in the roots of plants, the population of bacteria became an order of magnitude greater than in the leaves of Fig. 1 A, B. When colonizing plants with the *Rh. erythropolis* strain, the bacteria were distributed almost equally in the leaves and in the roots, but with PL coating, the number of bacteria in the leaves (2.7 x 10^3) and roots (2.4 x 10^3) increased by more than 1.5 times compared to the coating without PL (1.6 x 10^3 and 1.8 x 10^3), which indicates the stimulating effect of red light 660 nm on the growth of bacteria during colonization of the plants.

Investigation of the rate of bacterial growth under the PL coating separately from the colonization (Fig. 2) revealed that irradiation of *P. putida* with light transformed by a light-correcting film increases their population by 1.5 times as compared to control variants (under coating without PL and growth without coating). The data obtained indicate on photostimulation of the growth of these microorganisms. These data also confirm the investigation results reported earlier on the example of strains of *Pseudomonas putida* and *Pseudomonas stutzeri* [8]. The use of *Rh. erythropolis* X5 strain for the colonization of plants we report here for the first time. Despite the fact that plant growth doubled with colonization by this strain together with the action of red light of the photoluminophore coating, its growth rate under the PL coating was lower than in the control experiment. This result could be related to its genetic modification, since this strain contains a plasmid with the green fluorescent protein (GFP) gene, the amplification of which is an energy-consuming process.

Plants colonized with strains of *P. putida* and *Rh. erythropolis* showed an increased growth rate as compared to control non-colonized plants, as can be seen in Figures 3 and 4. Moreover, the plants colonized with *P. putida* and growing under PL coating were more than 2 times taller than the non-colonized, cultivated without PL (4.3 ± 0.4 and 2.0 ± 0.2 cm, respectively). It was found an order of magnitude more bacteria in the roots than in the leaves. During the colonization of plants with the *Rh. erythropolis* strain, the bacteria were distributed almost equally in the leaves and in the roots. However, with the photoluminescent coating, the number of bacteria in the leaves and the roots increased almost 2 times as compared to those under the coating without PL, which indicates a stimulating effect of red light of 660 nm on the growth of bacteria during colonization of plants. Both strains studied stimulated the growth of plants under the PL coating and, accordingly, plants growing without PL coating lagged behind in growth. Red light is known to stimulate bacterial growth – enhancing nucleic acid synthesis and cell respiration intensity, which can cause certain changes in the metabolic regulation program [16]. We can assume that with the intensification of metabolic processes in bacteria irradiated with transformed PL light, greater accumulation of physiologically active substances occurs as well as mobilization of nutrients from the environment or soil is enhanced, which favorably affects the growth of plants. Thus, the average height of 1.5 monthly colonized *P. putida* (4.3 ± 0.4) and *Rh. erythropolis* (3.9 ± 0.2) plants under PL was 2.2 times greater than that of the noncolonized (2.0 ± 0.2 and 1.8 ± 0.2) (Fig. 3, 4).

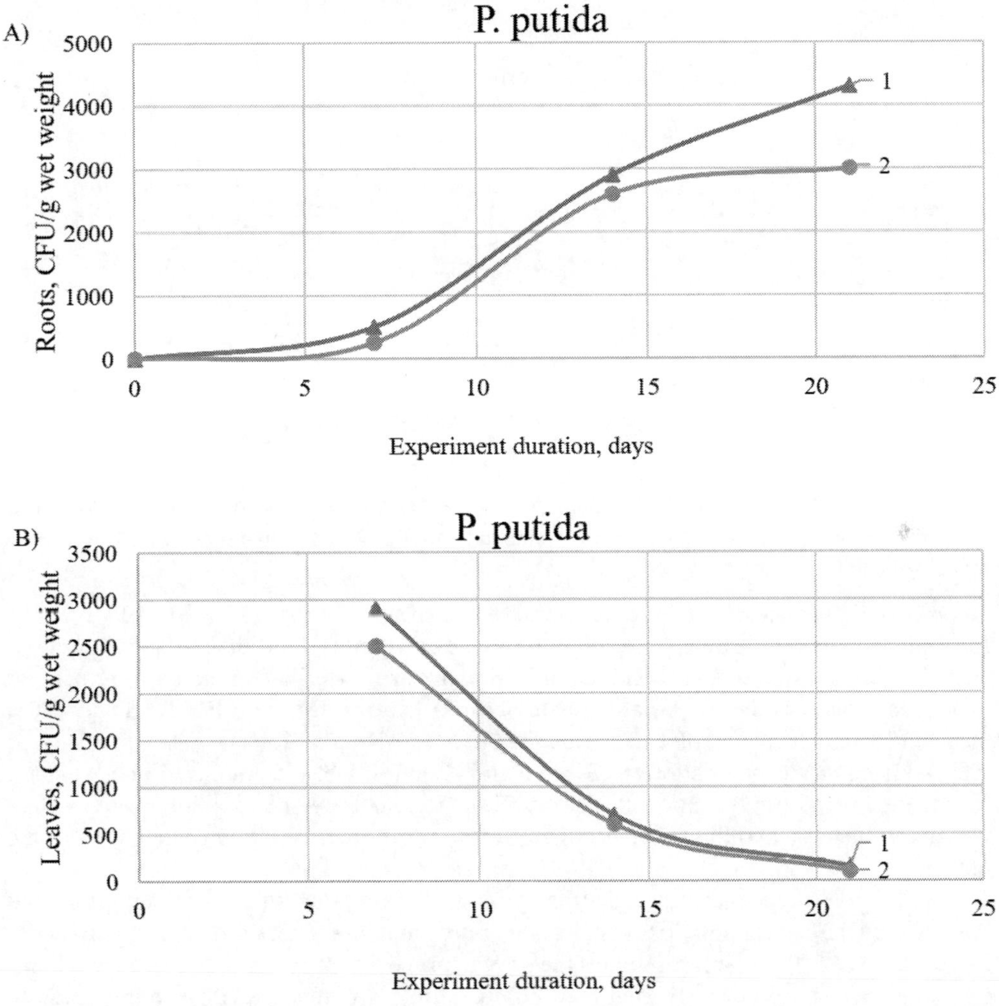

Fig. 1. The number of *P. putida* bacteria on mustard roots and leaves after colonization *in vitro* per 1 g of wet weight. A - the number of bacteria in the roots; B - the number of bacteria in the leaves. The numbers 1, 2 - respectively, indicate the options with PL containing and without PL containing spunbonds.

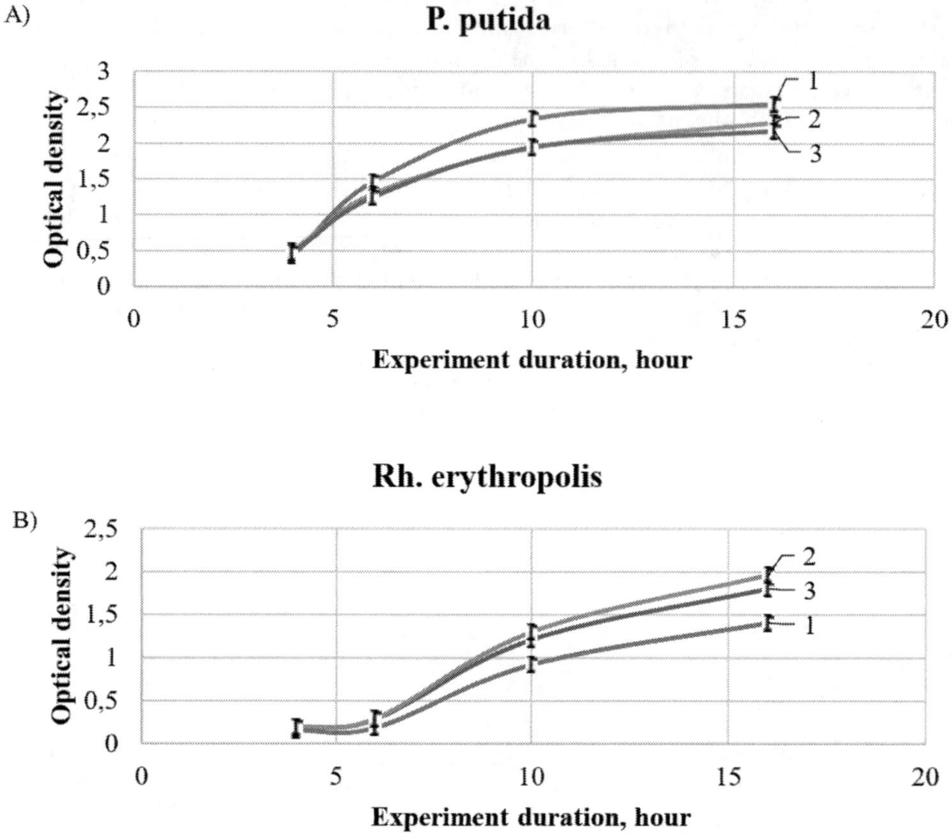

Fig. 2. Change in the optical density of *P. putida* (A) and *Rh. erythropolis* (B) during the growth of bacteria under different coatings. 1 - under the spunbond with PL; 2 - without a spunbond; 3- under the spunbond without PL.

In order to determine the effect of associative bacteria on the growth and development of mustard plants under the influence of photoluminescent coatings and their adaptation under conditions of closed and open ground, colonized plants from sterile conditions were planted in the closed ground of the Artificial Climate Station "Biotron" of the FIBH RAS and in the open ground on experimental plots in Pushchino, Moscow Region, Russia. When plants were transferred from *in vitro* conditions to Biotron and open ground, the colonized ones adapted better and the growth trend continued (Fig. 5 A, B). These data indicate a joint positive effect of associative bacteria and light-converting coatings on the growth and morphogenesis of the plants.

Thus, it was shown that the light that passed through the light-correcting film and transformed by the photoluminophores emitting with a maximum at 660 nm and a half-width of the spectrum at 610-730 nm stimulates the growth processes of plants and microorganisms, while combined application of microbial colonization of plants and PL-coating leads to stimulation of plant growth much stronger (2.4 times more) than separately only PL-coating (1.3 times) or colonization (2.1 times) according to the data shown in Fig. 4.

Fig. 3. Plants *Brassica juncea in vitro* (2 weeks), colonized by *Rh. erythropolis* X5 strain. A - 1-3 under spunbond with PL, 4-6 under spunbond without PL; B – not colonized. 1-3 – with PL; 4-6 without PL.

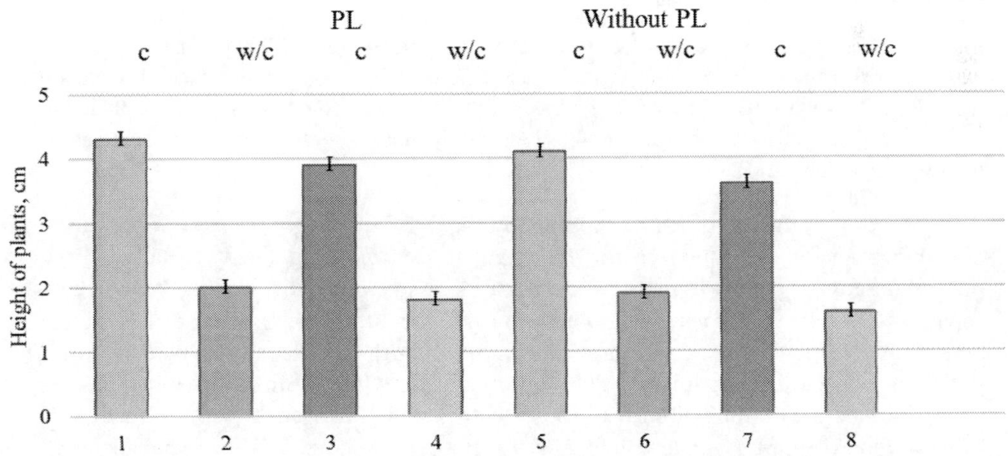

Fig. 4. Growth rate of Sarepta mustard plants (2 weeks) depending on colonization and photoluminophore coating (spunbond with PL or without PL). 1-4 PL.; 5-8 – without PL; 1,2 and 5,6 – *P. putida*; 3, 4 and 7, 8 – *Rh. erythropolis* X5; 1, 3, 5, 7 colonization; 2, 4, 6, 8 without colonization.

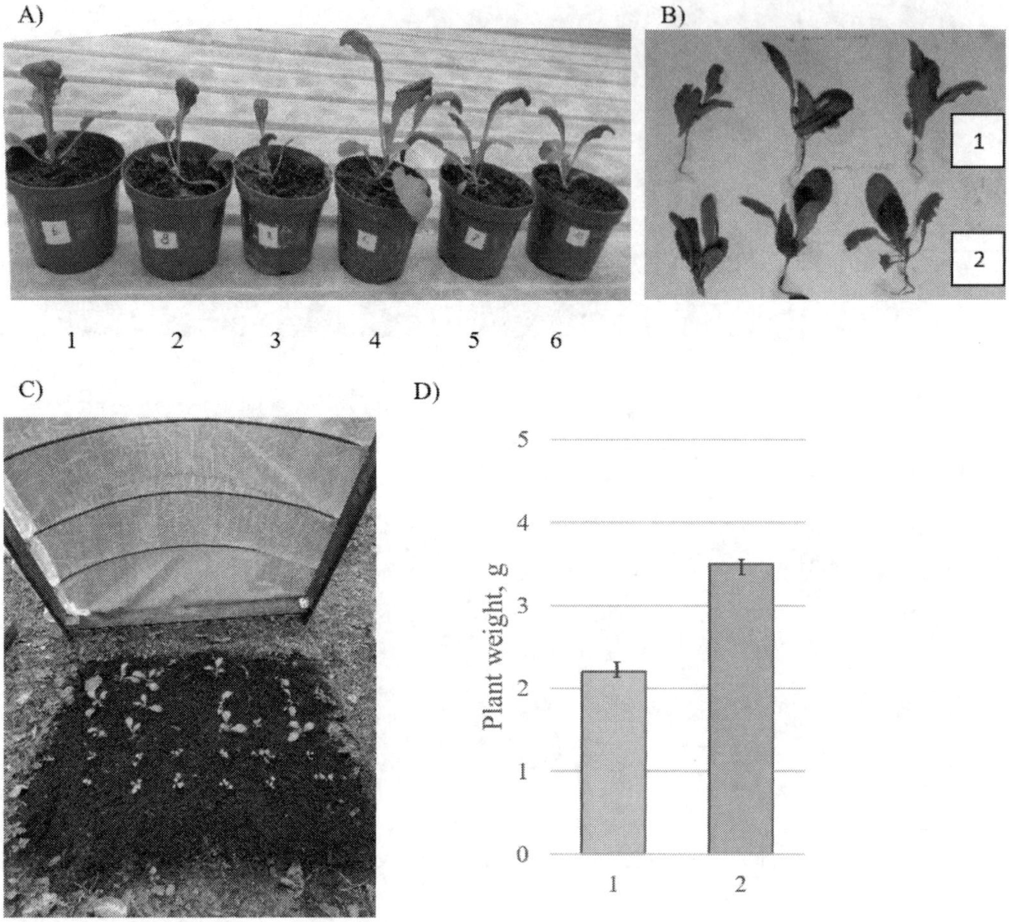

Fig. 5. Mustard plants *in vivo*. A – in the greenhouse of the Biotron artificial climate station; 1-3 – with PL, without colonization; 4-6 – with PL and colonization; B - the appearance of plants extracted from the soil: 1- with PL, without colonization; 2 – with PL and colonization; C – plants in the experimental site; D – weight of 2-month-old mustard plants. 1- with PL, without colonization; 2 – with PL and colonization.

Laboratory studies of the colonization and PL-coatings *in vitro* conditions are a convenient model for studying the mechanisms of photoregulatory systems. Previously, photoregulatory systems in plants and microorganisms under the action of PL were studied separately [8]. In this work we have shown for the first time the possibility of their joint action (synergy) – the stimulation of plant growth with the PL is improved in combination with individual microorganisms. The combined effect of colonization and light-correcting films exceeds the actions of each of these factors alone. Moreover, in the final study in the open ground (Fig. 5C), for the plants grown in the laboratory *in vitro* and transferred to the open ground to an experimental site with PL- coating, the weight was significantly (1.5 times) higher (3.5 ± 0.8 g) than in noncolonized plants (2.2 ± 0.6 g), respectively which confirms the stimulating effect of PL coatings in combination with colonization by beneficial microflora.

4 Conclusions

To sum up, from the results obtained and reported above the following conclusions can be drawn: 1) Red photoluminophore coating stimulates the growth of *Pseudomonas putida* bacteria. 2) Inoculation of Mustard plants by the bacteria *P. putida* and *Rh. erythropolis* X5 led to their stable association. 3) Colonized plants accelerate growth. 4) Red photoluminophore coating stimulates the growth of plants and especially colonized, both *in vitro* conditions and in open and closed ground conditions. 5) The development of studies of molecular and physiological processes of interaction of plants with associative microorganisms under the influence of red light is promising both for fundamental biological science and for agrotechnical practice.

Acknowledgements

The authors are grateful to D.O. Balakirev and Yu.N. Luponosov (Enikolopov Institute of Synthetic Polymeric, Russian Academy of Sciences) for the manufacture and provision of PL-containing agrotextiles for experiments. The work was financially supported by Ministry of Science and Higher Education of Russian Federation [the State Assignments Numbers 075-00381-21-00 and 0101-2019-0037] with the partial financial support from the Russian Foundation for Basic Research [Grants Numbers 18-29-17073, 19-08-00375].

References

1. I.F. Golovatskaya, R.A. Karnachuk, R.I. Leshchuk, A.S. Minich, S.A. Bol'shakova, G.A. Tolstikov, Sel'skokhoz, Biol., **5**, 45 (2002)
2. R. N. Khramov, V. D. Kreslavski, E. A. Svidchenko, N. M. Surin, A. A. Kosobryukhov, Opt. Express. **27(22),** 31967 (2019)
3. V.S. Raida, A.S. Ivanitskii, A.V. Bushkov, O.S. Andrienko, G.A. Tolstikov, Opt. Atmos. Okeana. **16(12),** 1126 (2003)
4. J.W. Kloepper, M.N. Schroth, T.D. Miller, Phytopathology. **70**, 1078 (1980)
5. G. Welbaum, A.V. Sturz., Z. Dong , J. Nowak, Crit. Rev. Plant Sci. **23,** 175 (2004)
6. C.L. Patten, B.R. Glick, Appl Environ Microbiol. **68(8),** 3795 (2002)
7. L.I. Svarovskaya, L.K. Altunina, D.A. Filatov, Prikl. Biokhim. Mikrobiol., Appl. Biochem. Microbiol. **44(6),** 585 (2008)
8. D.A. Filatov, L.I. Svarovskaya, V.S. Ovsyannikova, L.K. Altunina, Microbiology. **80(2),** 158 (2011)
9. T. Murashige, F. Skoog, Physiol. Plant. **15**, 473 (1962)
10. Yu.N. Luponosov, A.N. Solodukhin, D.O. Balakirev, N.M. Surin, E.A. Svidchenko, S. A. Pisarev, Y.V. Fedorov, S.A. Ponomarenko, Dyes and Pigments, **179**, 108397 (2020)
11. Yu.N. Luponosov, A.N. Solodukhin, A.L. Mannanov, P.S. Savchenko, B.A.L. Raul., N.M. Surin, S. M. Peregudova, A.V. Bakirov, M.A. Shcherbina, S.N. Chvalun, M.S. Pshenichnikov, D.Yu, Paraschuk, S.A. Ponomarenko, Materials Today Energy, **22**, 100863 (2021)
12. Chang C.Y., Solodukhin A.N., Liao S.Y., Mahesh K.P.O., Hsu C.L., Ponomarenko S.A., Luponosov Yu. N., Chao Yu-C., J. Mater. Chem. C **7,** 8634 (2019).
13. Smalla, H. Heuer, A Gotz, D. Niemeyer, E. Krogerreck-lenfort, E. Tietze, Appl. Environ. Microbiol. **66**, 4854 (2000)

14. Y. Delegan, L. Valentovich, K. Petrikov, A. Vetrova, A. Akhremchuk, V. Akimov, Microbiology. **8(48),** e01234-19 (2019)
15. T. Maniatis, E.F. Frisch, J. Sambrook. *Molecular cloning: a laboratory manual. New-York: Cold Spring Harbor Laboratory Press*, 2230 (1982)
16. S.L. Kandel, P.M. Joubert, S.L. Doty, Microorganisms. **5(4),** 77 (2017)
17. T.I. Karu, Health Phys. **56**, 1245 (1989)

Antigenic properties of serration bacteriophages

L.G. Molochaeva, *R.Kh.* Gayrabekov, and *T.I.* Gayrabekova

FSBEI HE "Chechen State University", 31, st. L. Yashina, Grozny, 364049, Russia

Abstract. The genus Serratia is one of the poorly studied among enterobacteria, which explains the interest in these microorganisms and bacteriophages active towards these microorganisms. The aim of this work was to study the antigenic properties of bacteriophages isolated by us from lysogenic cultures of serrations and from environmental objects. The work investigated 16 phages active against bacteria Serratia marcescens, isolated from various sources.

1 Introduction

In recent decades, special attention of health authorities has been drawn to the problem of nosocomial infections, which sharply reduce the effectiveness of therapeutic measures, increase the length of stay of patients in hospitals, and increase mortality. The problem of nosocomial infections is classified by WHO as one of the urgent and priority problems of modern medicine. Currently, the urgency of the problem of nosocomial infections has especially increased in the light of the new pandemic caused by the SARS-CoV-2 virus.

Coronavirus Disease 2019 (COVID-19) – a disease caused by a new coronavirus, now called *Severe acute respiratory syndrome-related coronavirus 2* (SARS-CoV 2); previously called 2019-nCoV), first identified during an outbreak of respiratory disease in Wuhan City, Hubei Province, China [12]. This was originally reported to the World Health Organization (WHO) on December 31, 2019, already on January 30, 2020, WHO declared the outbreak of COVID-19 as a global health emergency [13, 20], on March 11, 2020, WHO declared COVID-19 a global pandemic, first calling it such since the announcement of the H1N1 influenza pandemic in 2009 [22].

The risk of attaching a bacterial infection with COVID-19 disease increases if the patient is admitted to the hospital and at the same time has chronic pathology of the upper and lower respiratory tract or heart failure.

In conditions of reduced respiratory and motor activity, when mucus moves through the bronchi less, against the background of intoxication of the body and a decrease in immunity, bacteria from the nasopharynx descend into the lower parts of the respiratory tract and can begin to multiply there.

Therefore, first bacterial bronchitis can be caused, then bronchopneumonia with the transition to pneumonia. Development of bacterial pneumonia can also be facilitated by a long stay of the patient on the ALV. Despite the observance of all the strictest disinfection measures, ALV are risk factors for occurrence of ventilator-associated pneumonia.

Some species belonging to the genus *Serratia* are representatives of the group of opportunistic microorganisms that are most often isolated from patients during outbreaks of nosocomial infections; they are at the forefront of this process. Thus, for the first time, the role of *Serratia marcescens* in the pathological process was noted in 1913 [23], however, in recent decades, not only *Serratia marcescens* were isolated, but along with them, other representatives of this genus, such as *Serratia liquefaciens*, *S. plymuthica*, *S. ficaria, Serratia odorifer, Serratia plymuthica* were isolated [15, 16, 18, 22, 24, 25].

The genus *Serratia* is one of the poorly studied among enterobacteria, which explains the interest in these microorganisms and bacteriophages active towards these microorganisms.

Information on bacteriophages active against bacteria of the genus *Serratia* is very limited. In the available references, we came across only four papers in which such bacteriophages were mentioned. So in the paper of Seifert-Gertchman N. and Lippert W. [21], the moderate bacteriophage Kappa is described and some of its sizes are given. Matthews M. and Bradley D.E. [17] mention bacteriophage SM2. Bradley D.E. gave some sizes and morphology of bacteriophages SMB1, SM4, SMPA, SM2, SMP, SMP5, SMB3 [11].

Pitt T.L., Erdman I.J., Bucher C. described a system consisting of 10 bacteriophages active against bacteria of the genus *Serratia*, with the help of which they carried out typing of bacteria belonging to this genus in England [19].

Bacteriophages active against bacteria of the genus *Serratia* were studied by R.Kh. Gairabekov and coauthors. They examined the morphological, physiological, antigenic properties, the specificity of the bacteriophages isolated by them in relation to representatives of different species of the genus *Serratia*, as well as to representatives of different genera of the family *Enterobacteriaceae* [3-6], as a result of which they offered a typing scheme for bacteria belonging to the genus *Serratia*.

2 Materials and Methods

In this paper, 61 bacterial *strains belonging to the genus Serratia were used,* of which 57 cultures belonged to the species *Serratia marcescens* and 4 cultures — to the species *Serratia liquefaciens* [7]. These cultures were characterized by heterogeneity in a number of biological characteristics, including the presence of pathogenicity factors (sensitivity to antibiotics, lysozyme activity, antilysozyme activity, hemolytic activity, adhesive properties. The presence of plasmids, and etc.) [8, 9, 12].

The paper investigated 16 bacteriophages active against bacteria *Serratia marcescens*, isolated from various sources. Where, 15 bacteriophages were isolated from lysogenic cultures by cross-testing cultures on plates with 1.5 % nutrient agar. Wherein, all studied *Serratia* cultures were used both as indicator and as lysogenic.

One bacteriophage was isolated from the water of an open reservoir by seeding 1 ml of water with a two-layer method with 0.3-0.5 ml of a broth culture of the indicator strain and subsequent passaging of the bacteriophage from the resulting negative colonies.

When working with phages, we used the generally accepted methods of their isolation and study, given in the manuals of M. Adams [10] and I.M. Gabrilovich [1, 2].

The study of the antigenic properties of bacteriophages was carried out in the reaction of neutralization with rabbit sera according to the generally accepted method. To obtain antiphage sera, rabbits were immunized with bacteriophage suspensions containing1мл $10^8 - 10^9$ infectious particles.

Two courses of immunization were carried out, which consisted of three injections with intervals between injections of 2-3 days. The interval between courses of immunization was 7 days.

Neutralization of bacteriophages was carried out at the temperature of 37 °C in an ultrathermostat. The percentage of bacteriophage neutralization was noted and the neutralization rate constant was calculated using the formula:

$$K = \frac{2{,}3 \cdot lgP_0/P \cdot D}{t}$$

where P_0 and P — titers of bacteriophages, before and after incubation with antiphage serum, D — whey dilution, t — incubation time.

For the titer of antiphage serum, we took the highest dilution of it, which provided neutralization of at least 90 % of the homologous bacteriophage at the minimum exposure.

3 Study Results

To obtain antiphage sera, three rabbits were immunized with bacteriophages, which were selected for the difference in the morphology of negative colonies and the spectrum of action. The obtained sera had a relatively low activity, which indicates the weak antigenic properties of our bacteriophages active against bacteria of the genus *Serratia*, the rate constants for neutralization by the obtained sera of homologous bacteriophages were within 1.3-1.48 min^{-1}. The titers of the obtained sera were 1:10, the minimum exposure for all sera was 30 minutes (Table 1).

Table 1. Characteristics of antiphage sera.

Serum	Number of immunization courses	Phage injected	Serum titer	Exposure in minutes	% neutralization	Neutralization rate constant
32/23	2	17×10^9	1:10	30	98.1	1.3
P/5	2	25.8×10^9	1:10	30	99.5	1.33
29/10	2	74.1×10^9	1:10	30	99.4	1.48

Also in this paper, to test the specificity of the antigenic properties of bacteriophages serrations, we used serum obtained to the bacteriophage *Enterobacter* C_{14}.

The results obtained in the experiments on the neutralization of phages indicate the presence of common antigens in the studied phages active against bacteria of the genus *Serratia*. The data obtained make it possible to distinguish 4 serological types among the studied bacteriophages.

The first serotype includes 11 bacteriophages (32/23, 31/23, 30/23, 28/23, 23/23, 17/23, 27/21, 25/25, 24/24, 20/20, 12/12). Considering the differences in the antigenic properties of these bacteriophages, we considered it possible to distinguish subtypes within this type. Bacteriophages 32/23, 31/23, 30/23, 23/23, 17/23, 25/25, 24/24 and 12/12 are assigned to subtype Ia, since these bacteriophages were inactivated only by serum obtained to bacteriophage 32/23. Bacteriophage 28/23 was assigned to subtype Ib, on the basis that this bacteriophage was completely inactivated by sera obtained to bacteriophages 29/10 and P/5. Bacteriophage 27/21 was classified as subtype Ic, because it was completely inactivated by the serum obtained against bacteriophage 32/23, and partially inactivated by serum to bacteriophage 29/10. Bacteriophage 20/20 was isolated into a separate subtype Id, since it was completely inactivated by serum to bacteriophage 32/23 and partially by serum to bacteriophage P/5.

Table 2. Antigenic connections of serration bacteriophages.

Serotype	***Phage***	***Serums***			
		32/23	***29/10***	***P/5***	***C14***
Ia	32/23	■			
Ia	31/23	■			
Ia	30/23	■			
Ia	23/23	■			
Ia	17/23	■			
Ia	25/25	■			
Ia	24/24	■			
Ia	12/12	■			
Ib	28/23	■	▒	▒	
Ic	27/21	■	▒		
Id	20/20	■		▒	
II	29/10	■	■	■	
II	5/5	■	■	■	
II	P/5	■	■	■	
III	26/26		■	▒	
IV	3/3		▒		

■ - close antigenic relationship

▒ - distant antigenic relationship

☐ - lack of antigenic relationship

In serological type II, we included three bacteriophages: 29/10, 5/5, P/5. Bacteriophages of this type were inactivated by all three tested sera, but the sera obtained for bacteriophages 29/10 and P/5 did not inactivate bacteriophages classified as serological type I.

Bacteriophage 26/26, which was completely inactivated by serum to bacteriophage 29/10 and partially by serum to bacteriophage P/5, was classified as serological type III.

Finally, bacteriophage 3/3 was assigned to serotype IV, which was not completely inactivated by any of the tested sera, it was partially inactivated only by serum to bacteriophage 29/10 (by 34.85 %).

As indicated above, we used the serum obtained for the bacteriophage C_{14} active against bacteria of the genus *Enterobacter*. As can be seen from the data obtained (Table 2), bacteriophages of serrations were not inactivated by this serum, this suggests that bacteriophages of serrations and bacteriophages of Enterobacter do not have common antigens.

References

1. I. M. Gabrilovich, *Basics of bacteriophagy* (1973)
2. I. M. Gabrilovich, *A practical guide to bacteriophagy* (1968)
3. I. M. Gabrilovich, *Characteristics of phages active against bacteria of the genus Serratia,* 72 (1990)
4. I. M. Gabrilovich, Zh. Microbiology, Epidemiology and Microbiology, **2,** 10 (1992)
5. R. Kh. Gairabekov, Abstracts of the report. United scientific conf. of young sc., spec. and students of "Med. science - to practical health care". Makhachkala, 378 (1990)
6. R. Kh. Gairabekov, All-Union Conference. Microbiological and biotechnological foundations for the identification of crop and forage production. Alma-Ata, 15 (1990)

7. R. Kh. Gairabekov, Natural and technical sciences, **4,** 163 (2011)
8. R. Kh. Gairabekov, Topical issues of studying intestinal infections, Nalchik, 46 (1988)
9. R. Kh. Gairabekov, Natural and technical sciences, **4(54),** 163 (2011)
10. M. Adams, *Bacteriophages* (1961)
11. D. E. Bradley, J. Roy. Microscop. Soc., **84,** 257 (1965)
12. I. M. Gabrilovich, 2-nd International Meeting on Bacterial Epidemiological Markers, 268 (1990)
13. A. Gallegos, WHO Declares Public Health Emergency for Novel Coronavirus. Medscape Medical News (2020) https://www.medscape.com
14. CDC. 2019 Novel Coronavirus, Wuhan, China. Centers for Disease Control and Prevention. //www.cdc.gov/coronavirus
15. W. B. Harden, Arthritis and Reum, **23,** 946 (1980)
16. H. W. Horwits, J. Clin. Microbial, **25,** 1562 (1987)
17. M. Matthews, Proc. 3rd European red. conf. on electron microscopy, **2**, 543 (1969)
18. J. L. Pecarrere, Arch. Inst. Pasteur Madagascar, **48**, 37 (1979)
19. T. L. Pitt, I. J. Erdman, C. Bucher, J.Hyg., **84**, 269 (1980)
20. A. Ramzy, D. G. McNeil, W.H.O. Declares Global Emergency as Wuhan Coronavirus Spreads. The New York Times. https://nyti.ms
21. N. Seifert-Gertchman,Z. aalg. Microbiol., **1,** 87 (1960)
22. E. Serruses-Schontens, Eur. J. Clin. Microbial, **3,** 316 (1984)
23. The New York Times. Coronavirus Live Updates: W.H.O. Declares Pandemic as Number of Infected Countries Grows. The New York Times. https://www.nytimes.com
24. R. Verral, Infect. Contr., **4,** 469 (1983)
25. R. Zbinden, J. Clin. Microbial, **25**, 1409 (1938)

Influence of heat stress on the reproduction rates of sows of the Poltava meat breed, depending on the genotype

Natal`ya Garskaya[1], *Lidiya* Peretyatko[2], *Sergey* Pozyabin[3], *Sergey* Tresnitskiy[4], and *Anatoliy*Tresnitskiy[4]

[1]Lugansk State Pedagogical University, Oboronnaya Str., 2, 91002 Lugansk
[2]Institute of Pig Breeding and agroindustrial production of NAAS of Ukraine, Shvedskaya Mogila Str., 1, 36013 Poltava, Ukraine
[3]Moscow State Academy of Veterinary Medicine and Biotechnology – MVA by K.I. Scryabin, Akademika Scriabin str., 23, 109472 Moscow, Russia
[4]Don State Technical University, sq. Gagarin, 1, 3440003 Rostov-on-Don, Russia

Abstract. The article deals with the issues of the influence of heat stress that occurs in the summer on the indicators characterizing the reproductive qualities. Purebred sows of tested Poltava meat breed of different genotypes were investigated: Group I - individuals-offspring obtained by the method of introductory crossing using the blood of early maturing meat breed (blood content 12.5 %), Group II - individuals obtained by the method of introductory crossing using the blood of the Finnish Landrace (blood content of 12.5 %). It was found that in sows of the Poltava meat breed with the blood of early maturing meat breed, the number of live piglets at birth significantly exceeds the indices of sows with the blood of the Finnish Landrace by 8.27 % at $p \leq 0.05$. According to the analysis of Cv, these natural and technological conditions had the greatest influence on the duration of the service period in sows of Group I. Wherein, the least influence of these natural and technological conditions was found for the indicator of the duration of pregnancy. Animals with the blood of an early maturing meat breed were more adaptable to these natural and technological conditions.

1 Introduction

According to forecasts, by 2030-2052 the global climate change will lead to 1.5 °C rise in temperature. If it continues at the current pace, the number of hot days is projected to increase in most land regions [1]. It is also predicted that the frequency, intensity and duration of heat waves will increase in the future [2]. Currently, little is known about how this might affect the well-being of living organisms.

Extreme temperature events caused by climate change can increase the risk of exceeding the body's thermoregulatory capacity [3]. All animals have a thermal comfort zone to maintain physiological functions. However, under conditions when the temperature rise goes beyond the upper critical temperature range, animals begin to experience heat stress [4].

Heat stress caused by high ambient temperatures adversely affects the health and productivity of animals [5-7]. Pigs are especially vulnerable to heat stress [8]. In pigs, heat stress markedly increases respiration rate and body temperature, significantly reduces feed intake, endangers brain and intestinal health, and increases mortality and morbidity [5, 9, 10].

Depending on the severity, heat stress can have moderate or severe effects on many aspects of reproductive function in both males and females [11]. Short-term overheating of the testes in males can lead to temporary infertility [12]. Among the most noticeable effects of heat stress in females are low fertility [11], impaired fetal development, and reduced birth weight in humans, rats, cattle, and pigs [13].

The nature of the response of animals to meteorological parameters can differ depending on climatic conditions, production systems, or both [14]. In addition, it was found that the adaptive ability of animals to heat stress has a genetic component [15]. Highly productive modern pig breeds are usually bred in countries with continental or temperate continental climates, which means, that they are sensitive to heat. To minimize production and profitability losses in pig production associated with exposure to heat stress, it is necessary to search for preventive strategies that can mitigate the adverse effects of this natural factor.

The purpose of our paper is to study the effect of heat stress on the reproduction rates of sows of the Poltava meat breed, depending on the genotype.

2 Materials and Methods

The studies were carried out in the conditions of LLC «Plemzavod «Belovodskiy»» (Lugansk region, Ukraine) on a purebred livestock of tested sows of the Poltava meat breed. All animals belonged to the elite and first classes.

Poltava meat breed of pigs is a breed of meat productivity direction. Through introductory crossing in the breed, new genotypes were created with pigs of the Finnish Landrace breed (to improve meat qualities) and early maturing meat one (to improve adaptive qualities). For study, we formed two groups on the principle of analog pairs: Group I – sows-offspring of introductory crossing, with a share of early maturing meat breed blood of 12.5 % (n = 7), Group II – sows-offspring of introductory crossing, with a share Finnish Landrace blood of 12.5 % (n = 10). The data on the genotype and reproductive capacity of the sows were taken from the materials of the breeding and zootechnical records.

All sows received the same diet under the same conditions. The feeding conditions and maintenance of all groups corresponded to the feeding standards of the Institute of Pig Breeding and Agroindustrial Production of the National Academy of Sciences of Ukraine, taking into account age, live weight and physiological state. The type of feeding is concentrate, using feed of our own production. The keeping of animals is free-range.

According to climatic data in the Lugansk region, the warm period of the year is considered to be from May 1 to October 1. Based on long-term observations of the Lugansk meteorological station, it has been established that the summer months are the warmest in the region. The average temperature of the warmest month of the year, July, averages to + 21 °C. The number of days with a maximum air temperature above 25 ^{0}C is about 109 days a year. The number of days with a maximum air temperature above 30 ^{0}C is about 56 days a year (the effect of this temperature can be considered mild heat stress for animals [16]). For the study, sows tested were selected whose pregnancy took place in the warmest period of the year (from May to September).

Statistical data processing was performed using the STATISTICA (6.0) software package, with the assumption of a probability of $p \leq 0.05$.

3 Results

The data obtained for the indicators of reproductive qualities of the tested sows of the Poltava meat breed of various genotypes are shown in Table 1.

Table 1. Indicators of reproductive qualities of tested sows of Poltava meat breed of various genotypes, (M±m)

Indicator	Genotype	
	12.5 % early maturing meat breed	**12.5 % Finnish Landrace breed**
	Group I(n=7)	**Group II(n=10)**
Duration of pregnancy, days	119.57±3.08	114.7±0.56
Cv, %	6.83	1.54
Total piglets were born, heads	11.0 ± 0.53	12.5±0.76
Cv, %	12.82	19.36
Live piglets born, heads	11.0 ± 0.53	11.4 ± 0.72
Cv, %	12.82	19.91
Live piglets were born, %	100.0±0*	91.73 ± 3.04*
Cv, %	-	10.49
Big fetus, kg	1.0 ± 0	1.04 ± 0.04
Cv, %	-	12.11
Nest weight at birth, kg	11.0 ± 0.53	11.44 ± 0.72
Cv, %	12.82	19.84
Number of piglets weaned at 45 days, heads	10.14 ± 0.26	10.38 ± 0.31
Cv, %	6.8	9.44
Weight of one head weaned at 45 days, kg	9.61 ± 0.46	10.88 ± 0.72
Cv, %	12.69	20.86
Nest weight at weaning at 45 days, kg	82.43 ± 12.72	108.85 ± 8.08
Cv, %	40.83	7.42
Safety level, %	92.02 ± 4.32	89.74 ± 2.38
Cv, %	12.42	8.38
Service period, days	75.75 ± 21.36	52.56 ± 4.68
Cv, %	56.42	26.71

* - the probability of a difference between the groups $p \leq 0.05$

The results of the conducted studies show that the tested sows of the Poltava meat breed with a blood rate of 12.5 % for the early maturing meat breed (Group I) had the longest gestation period, surpassing animals with the blood of the Finnish Landrace by 4.87 days or 4.07 %. However, the established difference was not significant.

Animals of Group II, with the blood of the Finnish Landrace, exceeded the indicators of Group I in the number of born piglets by 1.5 heads or 8 %, in the number of live piglets born — by 0.4 heads or 3.51 %, respectively. However, the established difference was only physical, i.e. unreliable. Wherein, when calculating the relative number of live piglets born, we established a significant superiority of sows with the blood of early maturing meat breed by 8.27 % at $p \leq 0.05$.

Piglets of sows with Finnish Landrace blood were larger at birth and at weaning (by 0.04 kg or 3.85 % and 1.27 kg or 11.67 %, respectively). Piglet weight affected nest performance at birth and weaning, respectively. In the nests of animals of Group II, the mass was greater: at birth — by 0.4 kg or 3.63 %, at weaning — by 26.42 kg or 24.27 %. However, there were no significant intergroup differences in the indicators.

In the nests of sows with the blood of early maturing meat breed, the survival rate of piglets during the period of stay with the mother was slightly higher (by 2.28 %). However, after weaning, the sows of this group showed a slight increase in the service period (by 23.19 days or 30.61 %), but the established difference was only physical.

Based on the values of the obtained data of the coefficient of variation (Cv), it follows that these natural and technological conditions had the greatest effect on duration of the service period in sows of Group I. Wherein, the least influence of these natural and technological conditions was found for the indicator of the duration of pregnancy. The value of this indicator in sows with Finnish Landrace blood (Cv = 1.54 %), which may indicate the "pressing" effect of selection (artificial selection) is noteworthy.

Most of the studied indicators are homogeneous (Cv≤32 %) and average variability of indicators, i.e. depend both on the genotype of animals and on environmental factors. The indicator of the mass of the nest at weaning at 45 days and duration of the service period in animals with the blood of an early maturing meat breed is to a greater extent than other indicators due to environmental factors.

4 Discussion

The study is one of the attempts to find out the production potential of sows of the Poltava meat breed in terms of characteristics of their adaptability to the hot summer conditions of the breeding zone, by assessing the reproductive qualities of animals.

We found that the ability of sows of the Poltava meat breed to endure hot weather has a genetic dependence and significantly affects the number of live piglets at birth. Seasonal loss of piglets can become a serious problem in pig production in this region when using non-adapted genotypes or lead to economic costs to ensure microclimate indicators.

Sows with early maturing meat breed blood are more adapted to heat stress conditions. The established tendency to an increase in gestation in these animals is also probably associated with adaptive properties. It is known that among the factors that form the environmental conditions of prenatal and postnatal growth in the milk period, the maternal resource, directed to the growth and development of the fetus, probably plays the main role. Probably, sows of this group during pregnancy "spend" more body reserves, which further affects the postnatal growth of piglets, their weight and weight of the nest. In order to level the differences in the growth of piglets of sows of the Poltava meat breed with the blood of the early maturing meat breed, it is necessary to correct the conditions of feeding and keeping the animals.

The infusion of the blood of the Finnish Landrace, to improve the meat qualities of animals, created conditions for formation of better meat productivity in the offspring, but increased the "tension" of the body's defense systems in sows, which was manifested in a decrease in the number of viable piglets. Our findings are consistent with the opinion of Le HH et al. (2020) [17] that intensive genetic selection to increase productivity also makes animals more susceptible to heat stress due to the increased metabolic heat stress generated for rapid growth.

5 Conclusion

In these natural and technological conditions under the influence of summer heat stress, sows of the Poltava meat breed with a 12.5 % share of the blood of the early maturing meat breed will be optimal for breeding, subject to correction of the conditions of keeping and feeding.

References

1. IPCC, 2018. Summary for Policymakers. In: Global Warming of 1.5°C. An IPCC Special Report on the impacts of global warming of 1.5°C above pre-industrial levels

and related global greenhouse gas emission pathways, in the context of strengthening the global response to the threat of climate change, sustainable development, and efforts to eradicate poverty, Masson-Delmotte, V., P. Zhai, H.-O. Pörtner, D. Roberts, J. Skea, P.R. Shukla, A. Pirani, W. Moufouma-Okia, C. Péan, R. Pidcock, S. Connors, J.B.R. Matthews, Y. Chen, X. Zhou, M.I. Gomis, E. Lonnoy, T. Maycock, M. Tignor, and T. Waterfield (eds.), World Meteorological Organization, Geneva, Switzerland, 32 pp, (2018).

2. O.S. Iyasereab, M. Batesonc, A.P. Bearda, J.H. Guy,J. of Thermal Biology, **99** (2021)
3. F. Wang, J. Zhang, International Journal of Environmental Research and Public Health, **16**, 2202 (2019)
4. M.M. Rojas-Downing, A.P. Nejadhashemi, T. Harrigan, et al., Clim Risk Manag,**16**, 145-163(2017)
5. L.H. Baumgard, R.P. Rhoads, Annu. Rev. Anim. Biosci., **1**, 311-337 (2013)
6. I.BelhadjSlimen,T.Najar,A.Ghram,M.Abdrrabba, J. Anim. Physiol. Anim. Nutr.,**100**, 401-412(2016)
7. V. Sejian, R. Bhatta, J.B. Gaughan, F.R. Dunshea, N. Lacetera, Animal,**12**, 431-444(2018)
8. B. Xiaa,W. Wub, W. Fang, X. Wen, J. Xiea, H. Zhang, Heat stress-induced mucosal barrier dysfunction is potentially associated with gut microbiota dysbiosis in pigs,Animal Nutrition J. (2021)
9. R.P. Rhoads, L.H. Baumgard, J.K. Suagee, S.R. Sanders, Adv Nutr, **4**, 267-276 (2013)
10. Ch. Wen, S. Wei, X.Zong, Y. Wang, M.Jin, Animal Nutrition J., **7(4)**, 1329-1336 (2021)
11. P. J. Hansen, J. W. Fuquayb, Heat Stress: Effects on Reproduction, Encyclopedia of Dairy Sciences (Third edition), 285-296 (2022)
12. S.Pham, J. S. Schultz, J. of Thermal Biology, **99**,102954 (2021)
13. H. Guoab, Y. Yanga, Y. Qiao, J. Head, W. Yao, W. Zheng, Ecotoxicology and Environmental Safety, **227**, 112916 (2021)
14. M.J.Carabaño,B.Logar,J.Bormann,J.Minet,M.-L.Vanrobays,C.Díaz,B.Tychon,N.Gengler,H.Hammami, J. of Dairy Science,**99**, 3798-3814(2016)
15. G. Luna-Nevárez, A. L. Pendleton, R. I. Luna-Ramirez, S. W. Limesand, J. R. Reyna-Granados, P. Luna-Nevárez, J. of Thermal Biology, **101**, 103095 (2021)
16. A.M. Pereira, F.Jr. Baccari, E.A. Titto, J.A. Almeida, Int J Biometerol Jan. **52(3)**, 199-208 (2008)
17. H.H. Le, M. Shakeri, H.A.R. Suleria, W. Zhao, R.M. McQuade, D.J. Phillips, EVidacs., J.B. Furness, F.R. Dunshea, V. Artuso-Ponte, etal. Antioxidants, **9**, 1024 (2020)

Effectiveness of the Combined Use of Microelements and Tetravit in Calves of the Dairy Period

Alexey Trebukhov[1], *Elena* Shaganova[1], *Yulia* Chekunkova[2], *Nadezhda* Momot[3], and *Yulia* Kolina[3]

[1]Altai State Agricultural University, 656049, Barnaul, Russian Federation
[2]Federal Altai Scientific Centre of Agro-BioTechnologies, 656910, Barnaul, Russian Federation
[3]Primorsk State Agricultural Academy, 692510, Ussuriysk, Russian Federation

Abstract. Use of biologically active additives in feeding young animals is the most important element of resource-saving technologies. The aim of the study was to study the effectiveness of use of vitamin and mineral nutrition on the growth and development of young animals. 3 groups of calves were formed at the age of 10 days. Animals of the control group received the basic diet (OR), the II-experimental group received OR + salts of microelements and the calves of the III-experimental group received OR + salts of microelements and tetratwit in doses from 1 to 3 months – 1.5 mg, from 3 to 5 months – 2.0 mg and from 5 to 6 months – 3.0 mg. Tetravit was injected once every 2 weeks throughout the study. It was found that feeding the salts of microelements (copper, zinc, cobalt, manganese, iodine) to calves during the dairy period contributes to their more intensive development and growth of their meat productivity, both in combination with tetravite and without it. Use of salts of microelements separately, and in combination with a multivitamin preparation (tetravit) contributes to a significant increase in the studied hematological parameters in young animals, relative to control analogues.

1 Introduction

An integral part of the stable and effective development of livestock breeding is provision of full-value diets for farm animals, balanced in all essential nutrients. Departure from these conditions leads not only to a significant decrease in productivity, but also to development of various metabolic diseases (ketosis, osteodystrophy, hepatosis, and etc.) [1, 2] and, as a consequence, to a decrease in the productive longevity of livestock and its early culling [3].

Wherein, young animals are more sensitive to imbalance of mineral and vitamin components in the diet than adults [4, 5]. A growing organism depends on the qualitative composition of the diet, due to the high need for "building" elements to maintain proper growth and development, as a result, even a small excess or lack of necessary vitamins or minerals leads to development of clinically expressed pathologies [6-8].

At present, feed additives and premixes are widely used to optimize diets for deficient elements in formulations that include various biologically active substances (enzymes, vitamins, macro- and microelements, protein components of various nature, and etc.) [9-

11]. Use of feed additives containing vitamin and mineral components, allows to balance the diet of young animals according to these biologically active elements, ensuring the normal functioning of all metabolic processes in the body [12, 13].

The aim of the study was to study the effectiveness of use of vitamin and mineral nutrition on the growth and development of young animals.

2 Materials and Study Methods

The studies were carried out at Uchkhoz Prigorodnoye JSC in the city of Barnaul for 6 months in the winter-summer period. The object of the study was black-and-white calves at the age of 10 days, with a live weight of 26.6±0.5 kg. 3 groups of calves were selected. Young animals for the study were selected according to the principle of analogues, according to the methodology based on the selection of animals by weight: with an intergroup difference of no more than 5 % and an intragroup difference of no more than 10 %. The scheme of scientific and economic experience is presented in Table 1.

Table 1. Scheme of scientific and economic experience

Indicator	Number of heads	Feeding ration
I-control	5	Basic Diet (OR)
II-experimental	5	OR + microelement additives
III-experimental	5	OR + microelement additives + tetravit in doses, ml/head: from 1 to 3 months – 1.5, from 3 to 5 months – 2.0, and 5-6 months – 3.0; intramuscular injection interval is every 2-3 weeks

Throughout the study, the calves of the studied groups received the ration used on the farm (basic ration – OR). This diet included: colostrum, milk (whole and skimmed), corn silage, meadow hay, fodder beets, wheat turf.

Depending on the month of the study (age of young animals), the nutritional value of the diet varied in feed units within 2.1-4.5 g, metabolic energy – 9.4-16.4 g, digestible protein – 16.5-30.8 mJ, calcium and phosphorus – 9.4-16.5 g and 13.6-33.8 g, respectively. The diet was balanced according to the VASKhNIL standards for 17 most important indicators [14].

The juveniles of the second experimental group in the diet included sulfate salts of microelements (manganese, copper, cobalt, zinc), as well as iodide potassium salt.

In the third experimental group of calves, salts of microelements were added to the diet, similar to the animals of the second group, and intramuscular injections of tetravit were carried out once every 2 weeks throughout the study. Salts of microelements were given to young animals with concentrates once a day. This mineral additive was prepared before feeding by means of step mixing.

3 Study Results

In the course of our study, it was found that during the entire study period, young animals who received vitamin and mineral additives showed significant changes in live weight, both between the experimental groups and relative to the control group. The results of changes in the live weight of calves are provided in Table. 2.

Table 2. Live weight of black-and-white heifers in the milk period

Study month	Group		
	I-control	II-experimental	III-experimental
At birth	27.2±0.2	26.0±0.8	27.0±0.6
1 month	40.7±0.5	40.0±2.7	40.5±1.6
2 months	61.2±1.3	62.8±1.3	65.2±1.7
3 months	82.0±1.6	91.4±2.3	95.8±1.6
4 months	97.6±2.1	106.6±1.3	115.3±1.8
5 months	120.0±1.4	133.0±2.1	142.0±2.7
6 months	138.8±1.7	155.8±2.1	165.1±2.8

Analysis of Table 2 showed that the average live weight of calves of all studied groups significantly increased monthly. Wherein, by the second month of observation, the body weight of young animals of the third experimental group was higher than the control peers by 6.5 % ($P \leq 0.05$) and by 3.8 % relative to the analogs of the second group. By the third month of the study, the analyzed indicator of the calves of the second experimental group was significantly higher than the control peers during this period by 11.4 % ($P \leq 0.05$), and the young animals of the group receiving the vitamin and mineral supplement exceeded the weight of the control calves by 16.8 % ($P \leq 0.05$) and calves of the second experimental group — by 4.8 %. Wherein, there were no significant intergroup differences between the experimental groups.

The fourth month of observation was characterized by a significant excess of the live weight of calves in the third experimental group, relative to the control and the second experimental group, by 18.3 % and 8.2 %, respectively. The intergroup difference between the second experimental group and the control one was 9.2 % in favor of the experiment calves. A similar trend was observed in subsequent studies. So, five months after birth, the average live weight of young animals, in the diet of which the vitamin-mineral premix was added, was higher than the analyzed control parameter by 22.0 g (18.3 %), and a month later (6 study) — by 26.3 g (19 %). The average group values of the experiment ones during this period were higher in the third experiment group compared to the second one, by 6.7 % ($P \leq 0.05$) by the fifth month, and by 6 % ($P \leq 0.05$) by the final study.

In the group of calves that received the basic diet in combination with a mineral additive (experimental group 2), body weight indicators were higher than those of young calves in the diet, which did not have such an additive (control group), in the fifth study — by 11 % ($P \leq 0.05$).

Therefore, by the final study, the young animals of the 3rd experimental group exceeded their peers in the second experimental group by 9.3 kg ($P \leq 0.05$), and the control one — by 26.3 kg ($P \leq 0.05$).

The gain in live weight of calves is provided in Table 3.

Table 3. Average daily growth of heifers of black and motley breed in the milk period

Group	Live weight gain	
	average daily, g	absolute, kg
I-control	620±10	111.6
II-experimental	719±7	129.8
III-experimental	767±13	138.1

The average daily gain in live weight of calves of the second experimental group, which received only a mineral additive in addition to the main diet, was higher than the control calves by 99 g. Calves of the third experimental group, in accordance with the experimental scheme, in addition to the salts of microelements, who received an injection of Tetravit, showed the greatest daily weight gain, which exceeded the same indicator of the control group by 147 g and the second experimental group — by 48 g.

Blood plays an important role in the body of animals. It has a relatively constant composition, but to a certain extent it is capable of reflecting the changes occurring in the animal organism under the influence of various factors [15, 16]. For the purpose of multifactorial study of the effect on the growth and development of microelements and vitamins in young animals, a study of their blood was carried out.

Laboratory blood tests are provided in Table 4.

Table 4. Hematological blood tests in calves (M ± n; n = 15)

Age, months	Group		
	I-control	II-experimental	III-experimental
Erythrocytes, 10^{12}/l			
in 15 days	5.27±0.21		
2	7.42 ± 0.39	8.32±0.42	8.30±0.30
3	7.38 ± 0.2	7.52±0.3	7.71±0.4
4	7.36 ± 0.25	7.26 ± 0.17	7.46 ± 0.12
5	6.86±0.36	7.64±0.22	7.21 ± 0.21
6	6.84±0.09	7.24±0.22	7.28±0.08
Leukocytes, 10^9/l			
in 15 days	7.28±0.40		
2	6.8 ± 0.74	7.20±0.43	7.64 ± 0.58
3	6.4 ± 0.4	8.1±0.4	7.68 ± 0.5
4	5.26±0.27	9.52 ± 0.30	7.80±0.60
5	9.44±0.65	9.24±0.26	7.32 ± 0.58
6	6.7±0.3	7.36±0.63	7.74±0.32
Hemoglobin, g/l			
in 15 days	75.1 ± 4.2		
2	99.4±5.5	99.6±5.5	106.0±4.1
3	101.8±4.2	105.4±2.6	106.0±4.1
4	105.4±1.7	115.6±1.6	103.4±5.6
5	101.0±2.2	101.8±2.1	104.2±1.5
6	95.2±3.9	98.2±1.9	103.4 ± 1.0
Carotene, μmol/l			
2	7.5±0.4	10.6±4.1	11.3±0.8
3	8.3±0.2	10.3±5.6	10.7±0.6
4	10.7±0.1	10.1±0.4	10.3±0.1
5	8.0±0.9	8.8±0.6	10.4±0.5
6	11.1±0.2	8.6±0.3	10.2±0.3

Table 3 shows that the number of erythrocytes was significantly lower in the blood of calves from the control group during the entire experiment. So, in the second study, the level of erythrocytes was higher in the first and second experimental groups compared to the control one in the second study by 12.13 % and 11.9 %, in the fifth – by 11.4 % and 5 %, and in the sixth – by 5.8 % and 6.4 %, respectively. Wherein, there were no significant intergroup differences in the experimental groups, with the exception of the fifth study, in which the highest concentration of the analyzed indicator was in the first experimental group (by 6 %) relative to the second.

The number of leukocytes in the blood of calves from the experimental groups by the third study was higher than the control peers. By the final study, the content of leukocytes in the blood of young animals of the first experimental group was 9.8 % more leukocytes in the blood of the control, and in the second experimental group — by 15.5 %. There were no significant differences between the experimental groups, except for the fourth and fifth studies, during which the analyzed parameter was higher in the blood of calves from the first experimental group by 22 % and 26 %, respectively, compared to the second experimental group. Wherein, despite the higher values of this indicator during the entire experiment in the blood of calves of the first experimental group, relative to the second one, by the final study the number of leukocytes was less by 5 %.

The concentration of hemoglobin in the study groups did not have significant differences until the fourth month of observation, during which the content of this indicator was maximum in the first experimental group (115.6 ± 1.6 g/l) exceeding the control values by 9.7 % ($p \leq 0.05$) and the second experimental group — by 11.8 % ($p \leq 0.05$). Subsequently, the hemoglobin level in the first experimental group decreased almost to the level of the second study, but was still 3.2 % higher than the control peers. The hemoglobin content in the second experimental group, by the second month of the study, as in all the observed groups, increased by 1.4 times relative to the first study ($p \leq 0.05$), despite a slight decrease. By the final observation, the level of the described indicator was significantly higher than the control calves by 8.6 %, and the peers of the first experimental group — by 5.3 %.

It shall be noted that in the second study, the hemoglobin concentration increased in all groups relative to the baseline data, but there were no significant differences between the groups during this period.

The change in the blood carotene level of calves from the experimental groups as a whole had a similar dynamics with a decrease towards the end of the study. So, by the age of six months, the concentration of carotene was significantly higher in the blood of young animals of the second experimental group by 18.6 % relative to the first control one. It shall be noted that, despite significant fluctuations in the analyzed indicator in the control blood, 7.5 ± 0.4 - 11.1 ± 0.2 μmol/l, by the end of the study in this group, the carotene content in the blood was the highest among the peers of the studied groups, which allows to conclude that use of these drugs does not affect their concentration in the blood, providing, at the same time, a more stable level in the blood.

4 Conclusions

1. Feeding the salts of microelements (copper, zinc, cobalt, manganese, iodine) to calves during the milk period contributes to their more intensive development and growth of their meat productivity, both in combination with tetravit and without it.

2. Use of salts of microelements separately, and in combination with a multivitamin preparation (tetravit) contributes to a significant increase in the studied hematological parameters in young animals, relative to control analogues.

References

1. A. V. Trebukhov, Veterinary Medicine, **5,** 50 (2021)
2. A. V. Trebukhov, A. A. Elenshleger, IOP Conference Series: Earth and Environmental Science. The proceedings of the conference AgroCON-2019, 012152 (2019)
3. L. Redkozubova, Compound feed, **6**, 80 (2018)
4. E. I. Mashkina, Bulletin of the Altay State Agrarian University, **1(159),** 113 (2018)
5. O. B. Filippova, Chief zootechnician, **8**, 11 (2015)

6. A. Trebukhov, N. Momot, Yu. Kolina, A. Trebukhov, Igor Kamliya, E3S Web of Conferences, **254,** 09007 (2021)
7. A. V. Trebukhov, Bulletin of the Altai State Agrarian University, **6(200),** 44 (2021)
8. E. I. Mashkina, Innovations and food security, **3(17),** 85 (2017)
9. V. Fisinin, Animal Husbandry of Russia, **9,** 62 (2009)
10. L. Yu. Topuria, Bulletin of the Orenburg State Agrarian University, **4(16–1),** 82 (2007)
11. S. V. Moshkina, Chief zootechnician, **9,** 27 (2012)
12. A. V. Trebukhov, A. A. Elenshlager, S. P. Kovalev, V. N. Denisenko, G. G. Shcherbakov, A. V. Yashin, *Ketosis of cows and calves*, St. Petersburg, 132 (2019)
13. Yu. Luschai, O. Dutova, E. Shaganova, I. Pleshakova, E. Mashkina, International Research Conference on Challenges and Advances in Farming, Food Manufacturing, Agricultural Research and Education, 342 (2020)
14. A. P. Kalashnikov, *Norms and rations for feeding farm animals: a reference guide*, M.: Kolos, 456 (2003)
15. G. S. Azaubaeva, *Blood picture in animals and birds.* Kurgan, Trans-Urals, 168 (2004)
16. I. P. Kondrakhin, *Methods of veterinary clinical laboratory diagnostics.* M.: Kolos, 520 (2004)

Structure and properties of pectin substances of wild sea buckthorn (Hippophae rhamnoides L.) growing in Azerbaijan

E. B. Farzaliyev[1], *V.N.* Golubev[2], and *G.K.* Hafizov[3]

[1]Azerbaijan State University of Economics, UNEC, Baku
[2]University of Girona, Catalonia, Spain
[3]Research Institute of Horticulture and Tea Industry of the Ministry Agriculture, Zardabi, Azerbaijan

Abstract. The use of sea buckthorn fruits in medicine, cosmetology and as a source of ingredients that improve the quality of food has already been studied and continues to be studied. One of these ingredients can serve as pectin substances of sea buckthorn, isolated from it in its pure form. Considering the relevance of this direction, the main task of this work was the study and identification of pectin substances isolated from the fruits of wild sea buckthorn growing in Azerbaijan. Using cavitation-membrane technology, samples of pectin substances from sea buckthorn pomace were obtained under optimal technological conditions, after which the physicochemical characteristics were studied and their identification was carried out using physicochemical methods - elementary analysis, IR and NMR spectroscopy. Data have been obtained indicating that sea buckthorn pectin is a low-esterified pectin, which consists of a mixture of linear and highly branched polymers of α-D-galacturonane and other polysaccharides, whose macromolecules include galacturonic acid residues and neutral sugars. The high complexing ability of the obtained pectin in relation to lead ions has been established, which makes it possible to recommend it as an active ingredient for therapeutic and prophylactic products. The data obtained allow us to get a more complete picture of the biotechnological potential of local sea buckthorn as a source of pectin substances.

1 Introduction

Azerbaijan's vegetable raw materials are widely represented by food, vitamin and medicinal plants, from which hydrophilic extracts and beverages containing a wide range of nutrients and biologically active substances can be obtained [1-2].

In the list of wild plants growing on the territory of Azerbaijan, buckthorn buckthorn (Hippophae rhamnoides L.) stands out for abundant fruiting. Here, the potential volume of harvesting the fruits of this plant is 3,000 tons.

Sea buckthorn fruits are an excellent raw material for processing into lipophilic extracts and other products with high antioxidant, antibacterial, anti-carcinogenic and anti-radiation activity [3-4].

In recent years, interest in this plant has been growing everywhere as a source of biologically active substances capable of ensuring the normal functioning of the human body, increasing its resistance to oncological and viral diseases, stress, negative environmental influences and prolonging life [5-6].

Among the many natural components of sea buckthorn fruits, plant polysaccharides, in particular, pectin substances, are of particular interest.

Pectin substances (pectins) include protopectin, pectin polysaccharides and associated arabinans, galactans and arabinogalactans, which, as a rule, have a complex chemical structure and structure [7-8].

Pectin substances are a multifaceted family of complex plant polysaccharides that form a functionally important part of primary cell walls together with cellulose and hemicellulose. They ensure the strength of plant cells, plant resistance to drought and low temperatures, provide water-salt exchange, are characterized by a high gel-forming ability and play an important role in human nutrition as components of "dietary fibers" with complexing ability [9].

Changes in the composition and structure of pectin substances were revealed depending on their origin by types of raw materials, places of their cultivation and methods of manufacture [10]. This makes it necessary to conduct systematic studies aimed at clarifying the structure, properties and direction of use of pectin substances from various sources, and in particular from local wild sea buckthorn, which was the main task of this work.

2 Materials and Methods

2.1 Objects of research

The object of the study was the fruits of sea buckthorn harvested in the period September-November 2019 in the Babek administrative region of Azerbaijan, with a content of 10.32% (per absolutely dry mass) pectin substances in the form of a water-soluble form (hydropectin) and protopectin fraction.

2.2 Isolation of pectin substances

Modern trends in the development of technology of pectin substances provide not only economic aspects, but also environmental aspects related to both reducing the release of harmful chemicals into the environment and creating favorable working conditions. Another important aspect of technological improvement of the processes of obtaining pectin substances is the intensification of individual resource-determining stages.

One of the effective technological solutions is the use of the method of hydroacoustic treatment of pectin-containing raw materials in rotary cavitation extractors, since cavitation treatment of an aqueous extractant changes its physico-chemical properties, increases the pH of water, contributing to its activation, as a result of such treatment, water temporarily becomes an active solvent with acidic properties without the introduction of chemical reagents [11].

In a rotary-cavitation type extractor, under optimal conditions, the processes of grinding of pectin-containing raw materials (the area of the solid phase increases by 60-75 times), hydrolysis of the protopectin (water-insoluble) fraction of pectin substances and the actual extraction (diffusion) of pectin substances into the aqueous phase occur simultaneously.

Another important technological technique that allows preserving the nativity of pectin biomolecules is the use of membrane processes to purify them from ballast substances and concentrate them, since these processes take place at ambient temperature and without phase transitions.

To obtain pectin substances, freshly obtained extracts of wild sea buckthorn fruits were used, which were crushed to a size of 2-3 mm.

The extraction process of crushed pomace was carried out at the following parameters: hydromodule 5:1 – (deionized water: pomace); cavitation index 0.6; temperature - 65 ° C, process time - 20 min.

At the output, a white-beige (achromatic) powder was obtained, colorless both in dissolved form and in gels and emulsions, without absorption in the wavelength range with a length of 400-700 nm, with parameters L=90-92, a = (-3.7) – (-1), b = (+2)-(+15).

2.3 Identification and testing of pectin substances

To identify the isolated pectin substances, elemental microanalysis, IR spectra on the Impact 410 "Nicolet" brand infrared Fourier spectrometer in the range of wave numbers 400-4000 cm-1 in KBr tablets with a spectral resolution of 2 cm-1, NMR spectra of 1 H and 13 C on the Bruker Avantes device were used. Optical rotation was determined on a Perkin-Elmer 141 device in water at a temperature of 20 ° C.

The resulting pectin from wild sea buckthorn was tested for physico-chemical characteristics in accordance with the US Pharmacopoeia form "Pectin 9000 -69-5 FT-FSIJK254368".

The sorption capacity of pectin substances in relation to lead ions was determined by complexometric titration (reverse titration technique) [8].

The results of experimental studies were evaluated using Microsoft Office Excel 2013 and Statistica 8.0 for Windows application packages.

3 Results and Discussion

Microanalysis of isolated pectin substances from wild sea buckthorn pomace showed the following chemical parameters: C-28%, H-42%, O- 24%, which corresponds to the gross formula C14 H21 O12, i.e. it is really pure pectin.

Table 1 shows the physico-chemical characteristics of sea buckthorn pectin.

Table 1. Physical and chemical characteristics of sea buckthorn pectin.

Indicators	quantity %
Free carboxyl groups	21.80
Esterified carbonyl groups	7.15
Esterification degree	28.80
Uronid component	74.45
Acetyl groups	1.12
Methoxyl groups	8.14
Molecular mass, Da	$42.0 \cdot 10^3$
pH 1% solution	3,25
Jelly fortress, kPa	70.68

Table 1 shows that according to the degree of esterification, the resulting pectin belongs to the group of low-esterified pectins. Due to the sufficiently high content of free carboxyl groups (21.80%), it should have a high complexing ability.

This was confirmed during further study of the complexing ability of sea buckthorn pectin in relation to $Pb^{2}+$ ions. It amounted to 284.5 mg Pb/g, which can serve as a basis for creating functional products based on it with a high antidote potential in relation to heavy metals and radionucleides.

To identify the obtained pectin substances, IR-Fourier spectra and NMR spectra of ^{1}H and ^{13}C were taken (Figure 1). Figure 2 shows the results of their mathematical processing and interpretation.

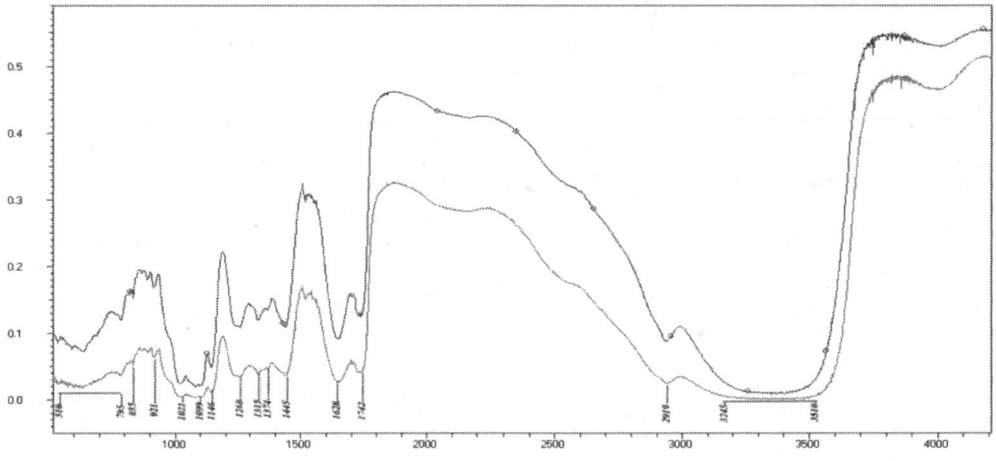

Fig. 1. IR spectra of sea buckthorn pectin samples.

Upon a detailed examination of the ir spectra, it can be concluded that pectin contains a large amount of galacturonic acid (intense absorption bands in the region of 1010-1150 cm^{-1}). The band in the region of 1374 cm^{-1} is caused by deformation fluctuations of the C-H groups of the pyranose ring. absorption bands are observed in the area of 1610-1740 cm-1, indicating the presence of free carboxyl groups. The available oscillation bands of the CH_3 groups indicate partial esterification of carboxyls.

The assignment of bands in the experimental IR absorption spectra of sea buckthorn pectin is presented in Table 2 based on a comparison of absorption bands with data from the NIST spectral database (ASTM).

Table 2. Assignment of bands in IR absorption spectra.

Band, cm^{-1}	Preferred types of oscillations
3245-3510	□$(OH)_C$, □(H_2O)
2919	□(CH)
1742	□$(C{=}O)_E$
1628	
1445	□$_{as}(CH_3)$□
1374	□$_s$(CH)□
1315	□(CH)□
1268	
1146	□(C–O–C)
1099	□,□ $(C–OH)_C$, □(C–C, C–O)□
1021	□$(C–C, C–O)_K$
921	□$(OH)_C$
835	□$(CH_3)_E$
510-795	Pulsating vibrations of pyranose rings

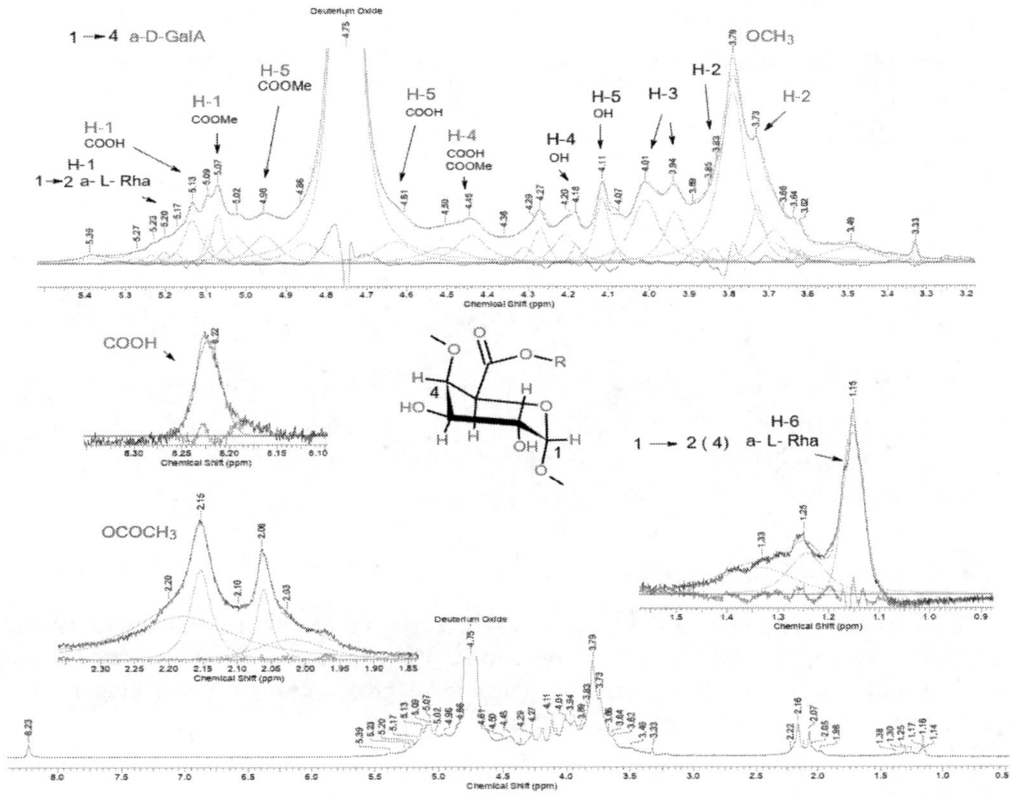

Fig. 2. The mathematical processing and interpretation.

It was found that in the proton spectrum of the sample in the region of a strong field (δ =0.97-1.47; δ = 1.47-1.73 m.d.), the manifestation of signals of CH groups located in position 4 is observed. Protons of the carbon atom of position 1 of the pyranose cycle resonate at 3.94-3.38 m.d., and chemical shifts 3.48-3.59 m.d. refer to carbon protons in positions 2 and 3 of the galactopyranosiluron fragment. The methoxyl and carboxyl groups are characterized by the manifestation of signals at 3.67 and 3.91 m.d, respectively. Analysis of the NMR spectrum on 13C nuclei showed the presence of carboxyl (δ =103.03 -103.64 m.d.), methoxyl (= 57.40 m.d.) and methine groups in the structure. Carbon atoms located at position 1 of the pyranose fragment give a signal in the strong field region at 63.78, 72.50 and 77.65 m.d. For 2 and 3 carbon atoms, the manifestation is characteristic at 77.33, 81.01 and 76.92 m.d. Carbon of position 4, participating in the connection of pyranose fragments by an oxygen bridge, resonates at 44.05, 40.44 m.d.

Figure 2 shows the NMR spectra of sea buckthorn pectin, as well as the data of their mathematical processing and interpretation.

It can be said that the NMR spectra clearly show a doubling of the signal group, which speaks in favor of the predominance of two different sections of the polymer molecule of pectin.

Based on this, the structure of oligosaccharide fractions is unambiguously α -1,4-D-glucans.

Mathematical processing of NMR spectroscopy data makes it possible to make assumptions about the structural features of pectin (Figure 3).

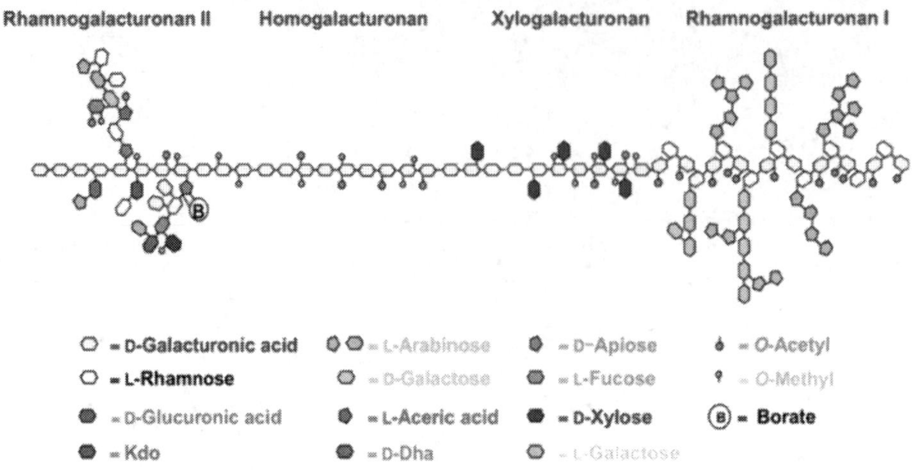

Fig. 3. Prospective structure of sea buckthorn pectin.

The abundance of carbonyl group signals indicates a very complex and heterogeneous structure of pectin, which has a rather branched structure. The presence of a large number of signals in the acetal carbon region also indicates the presence of a significant amount of neutral sugars, mainly rhamnose and galactose.

4 Conclusions

The data obtained indicate that pectin from wild sea buckthorn fruits mainly consists of a mixture of linear high-molecular polymers α-D-galacturonane and other polysaccharides, whose macromolecules include galacturonic acid residues and neutral sugars; this is a low-esterified pectin with a high antidote potential with respect to heavy metals and radionucleides.

References

1. G.K. Hafizov, S.G. Hafizov, Innovations in science, **41**, 27 (2015)
2. D.Sh. Mammadov, G.K. Hafizov, G.Sh. Abubekirov, Bulletin of the State Nikitsky Botanical Garden, **99**, 121 (2009)
3. A. Vilas - Franquesa, J. Saldo, B. Juan, Food Prod Process and Nutr, **2(17)** (2020)
4. E. Christaki, Food Pub. Health., **2,** 69 (2012)
5. B. Olas, B. Skalski, K. Ulanowska, Front Pharmacol. **9**, 232 (2018)
6. I.L. Novoselskaya, Chemistry of natural compounds, **1,** 3 (2000)
7. S. Perez, K. Mazeau, Plant Physiol. Biochem. **38**, 7.
8. G.A. Gubanenko, E.A. Pushkareva, E.A. Rechkina et al., IOP Conf. Series: Earth and Environmental Science, **640,** 062021 (2021)
9. Yu.S. Ovodov, Bioorganic chemistry, **35(3),** 293 (2009) http://rjbc.ru.
10. L. A. Mikheeva, A.V. Try, Bulletin of VSU, Series: Chemistry. Biology. Pharmacy. 2 (2013) http://www.vestnik.vsu.ru.
11. F. Meneguzzo, C. Brunetti, A. Fidalgo, et al., Processes. **7**, 581 (2019)

Genetic variation of CAST gene in Local Karnobat and Karnobat merino sheep breeds

Ivona Dimitrova[1,*], *Milena* Bozhilova-Sakova[2], *Margarit* Iliev[3], and *Atanaska* Teneva[1]

[1]Agronomy Faculty, University of Forestry, Sofia, Bulgaria
[2]Institute of Animal Science, Kostinbrod, Bulgaria
[3]Institute of Agriculture, Karnobat, Bulgaria

Abstract. Karnobat sheep plays an important role in the development of sheep breeding in Southeastern region of Bulgaria. They are valuable source of genetic material. The aim of present experiment was to determine the allele variation of CAST gene in Local Karnobat and Karnobat Merino sheep breeds. A total of 60 blood samples were collected – 30 per breed. DNA was extracted and genotypes of all animals were identified by means of PCR-RFLP technique. The restriction reactions were accomplished by specific enzyme *MspI*. As expected both breeds were characterized with low level of genetic diversity due to the fact that mostly maintaining selection has been implemented. In Local Karnobat sheep breed was identified only one heterozygous individual from all 30. In Karnobat merino were identified allele M with frequency 0,97 and allele N with frequency 0,03. Genotypes MM and MN were revealed with frequencies 0,93 and 0,07, respectively. According to the statistical analysis both breeds were in HWE equilibrium.

1 Introduction

Sheep farming is an economically important industry for agriculture. Sheep breeding efficiency is directly affected by the quality and quantity of produced meat. Providing of lamb meat with higher quality is a key priority for this industry, because the main part of the income in the sheep farms, in particular in Bulgaria, is formed from the sale of lambs [1]. The quantity and quality of the obtained meat are formed under the influence of a complex of genes and environmental conditions. Improving the meat quality is the main objective of livestock production whereby the meat tenderness is one of the most important factors for measuring the quality of meat [2]. However, the quality of meat can hardly be improved by traditional selection, as the inheritance of meat quality is low [3] and the measuring of quality characteristics is difficult, expensive and possible only after slaughter. The application of new methods based on DNA technologies allows to make sheep breeding a modern, competitive and profitable industry [4]. In this regard, DNA methods based on the latest advances in molecular genetics are increasingly used to assess the genotype of breeding animals [5].

* Corresponding author: ivonna_dimitrova@yahoo.co.uk

The genes CAST (calpastatin), MSTN (myostatin), FABP3 (fatty acid binding protein 3), CLPG (calypig), DGAT1 (diacylglycerol acyltransferase 1) and LEP (leptin) were studied as candidate genes for meat productivity and meat quality in sheep [7]. One of the promising markers for growth intensity and meat quality in sheep is the calpastatin gene (CAST) [6, 7]. The CAST gene is located at locus 5q15 on chromosome 5 of the sheep genome (*Ovis aries L.*), it is composed of 29 exons separated by introns and has a total size of 89553 bp [8]. Genetic polymorphism has been identified in the following regions: exon 1, exon 6, exon 12, intron 12, and domains II to IV of the CAST gene [9, 10]. CAST inhibits calpain activity and this effects regulation of meat post mortem tenderness, birth weight, and growth rate to weaning. The calpastatin gene shows large and significant effects on birth weight and fat thickness [11]. A significant association between CAST genotypes and growth rate and final body weight was reported, with lambs of the MN genotype having a higher mean daily gain and final body weight than lambs of the MM genotype. The genotypes of the CAST gene have a significant effect on some components of the carcass and the quality parameters of the meat - the MN genotype shows a lower total bone mass and a higher meat to bone ratio compared to the MM genotype [12]. The presence of polymorphism in the locus of the calpastatin gene indicates that the quality of meat in the studied breeds of animals can be improved by applying an appropriate breeding program. Genetic polymorphisms, which are significantly associated with certain interesting traits, are very useful [13].

Different breeds of thin-tailed sheep are bred in Bulgaria - including local breeds and others intended for specific purposes such as milk, meat and wool. Many of the commercial Bulgarian sheep breeds are created based on local breeds. Such is the case with the composite Karnobat merino breed, in the creation of which were also used local Karnobat sheep characterized by high quality meat [14]. The establishment of genetic polymorphism in existing breeds of sheep is a major task, on the one hand in local breeds in order to preserve genetic resources, and on the other in cultivated animal breeds in order to their effective management and exploitation

The present study was aimed to investigate the genetic polymorphism of the CAST gene in 60 ewes of one local and one merino sheep breed – Local Karnobat and Karnobat merino.

2 Materials and Methods

The present experiment was conducted with a total of 60 ewes belonging to two sheep breeds from the Institute of Agriculture, Karnobat – 30 animals from Local Karnobat breed and 30 animals from composite Karnobat Merino breed. Blood samples were collected from the jugular vein in vacuum tubes containing EDTA as anticoagulant.

Genomic DNA was extracted from whole blood using a commercial purification kit according to the manufacturer's instruction (Illustra Blood Genomic Prep DNA Purification Kit, GE Healthcare). DNA concentration and purity were determined using spectrophotometer Biodrop and agarose electrophoresis on 1% agarose gel (Bioline) and TBE buffer (Jena Bioscience).

PCR reactions were performed in total volume of 10 μl containing 40 ng genomic DNA, 0.2 μl dd H_2O, 20 pM of each primer and 5 μl of ready-to-use 2×(1.5 mM $MgCl_2$) MyTaqTM HS Red Mix (Bioline). For the amplification of CAST gene was used primer set suggested by Palmer et al., [15]:

forward primer: 5'-TGG GGC CCA ATG ACG CCA TCG ATG-3'

reversed primer: 5'-GGT GGA GCA GCA CTT CTG ATC ACC-3'.

Amplification process was performed on thermal cycler QB-96 r (Quanta Biotech). The specific PCR conditions were shown on table 1.

Table 1. Specific PCR conditions for amplification of ovine CAST gene

Locus	Primary Denaturation	Cycles	Denaturation	Annealing	Elongation	Final elongation
CAST	94°C for 5 min	30	94°C for 30 s	62°C for 45 s	72°C for 1 min	72°C for 10 min

The genotypes of tested individuals were determined using RFLP method. The restriction reactions were carried out in 10 µl final volume containing 6 µl PCR product, 10 U/µl speed enzyme *MspI* (Jena Bioscience), buffer and ddH_2O. The incubation was performed in heat-block at 37^{0}C for 15-20 min. The fragment sizes were validated by agarose gel electrophoresis using 50 bp DNA Ladder (Thermo) on 2,5% agarose gel (Bioline) stained by 10000x RedGelTM Nucleic Acid Stain (Biotuim) and 1x TBE buffer (Jena Bioscience). The results were visualized under UV light.

All steps in the present experiment were carried out in Laboratory of Genetics, part of Agronomy Faculty in University of Forestry, Sofia, Bulgaria.

3 Results and Discussion

A 622 bp fragments from exon 1 and intron 1 of the CAST gene of the investigated animals were amplified by the described technique. All PCR products were subsequently subjected to enzymatic digestion with restriction endonuclease *MspI.* According to authors, this enzyme specifically cleaves the particular site with sequence 5' C↓CGG 3' and as a result produces two fragments with sizes of 336 bp and 286 bp, which define the allele M. The nucleotide sequences of 622 bp where the specific site 5'C↓CGG3' is absent remains uncut and determines the N allele. As a result, three genotypes can be observed - MM (with fragment size of 336 and 286 bp), MN (with 622, 336 and 286 bp) and NN (with 622 bp) [15].

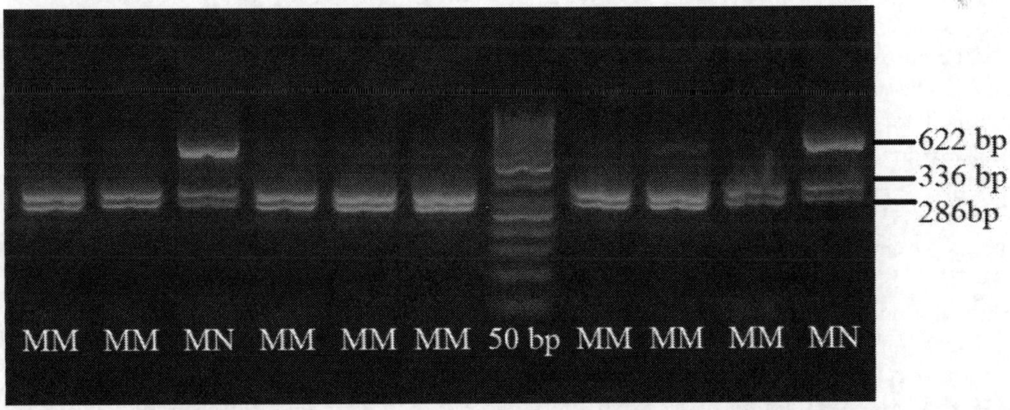

Fig. 1. DNA electrophoresis of CAST amplicons after digestion with *MspI* restriction enzyme: lane 7 - 50 bp DNA ladder, lanes 1, 2, 4 ,5, 6, 8, 9 and 10 - genotype MM, lanes 3 and 11 - MN genotype

In present experiment were determined two genotypes – homozygous genotype MM and heterozygous genotype MN (Figure 1). The most common allele in both breeds was the M allele. In the local Karnobat breed, the M allele was found with a frequency of 0.98 and the N allele with a frequency of 0.02. The Karnobat merino breed shows the following frequencies - of the allele M - 0.97 and of the N allele - 0.03 (Table 2). Both MM and MN genotypes were identified in both studied breeds, while the NN genotype was not established. In the local Karnobat breed, we observed that the MM genotype occurred with a frequency of 0.97, and the heterozygous MN genotype - with a frequency of 0.03. The observed value of the MN genotype was slightly higher in the Karnobat merino breed - 0.07, while the frequency of the MM genotype was 0.93. According to the statistical analysis both breeds were in HWE equilibrium.

Table 2. Animal number, values of allele and genotype frequencies, average heterozygosity (observed Ho and expected He) and degree of probability for CAST gene in investigated breeds.

Breeds	n	Allele frequency		Genotype frequency			Ho	He	P*
		M	N	MM	MN	NN			
Local Karnobat	30	0,98	0,02	0,97	0,03	0,00	0,033	0,039	>0,05
Karnobat Merino	30	0,97	0,03	0,93	0,07	0,00	0,066	0,058	

*non-significant difference

The results in this study showed that the heterozygous genotype MN is present in the local Karnobat breed with very low frequency (0.02), while a previous study on the same sheep breed showed the absence of the heterozygous genotype [16]. Studies in other native Bulgarian breeds showed differences in the frequency of alleles of the CAST gene [17]. In the Cooper-Red Shumen and Karakachan breeds, a lack of diversity and only the M allele and the MM genotype were found [17, 18]. Two genotypes were identified in the Stara Zagora and Breznik breeds - MM and MN. In the Stara Zagora sheep the frequency of the mutant allele N is 0.03, of the genotype MM - 0.97 and of the heterozygous genotype MN - 0.03 [16], while in the Breznik breed there was a higher diversity - the frequency of the allele N was 0.12, of the genotype MM - 0.77 and of the heterozygous MN - 0.23 [17]. In studies of thin-tailed Turkish sheep breeds Kivircik, Imroz, and Karayaka, a null frequency of the NN genotype was found similar to the results in the present study, and in the Imroz breed the frequency of the M allele was 0.96 and of the heterozygous MN genotype was also close to this established by us - 0.07 [8].

The current study of Karnobat merino sheep showed low genetic diversity of the studied locus. The obtained results slightly, but statistically insignificant, exceeded the results obtained for the local Karnobat breed. In a previous study of animals of this breed, two alleles were found with a frequency of 0.94 for the M allele and 0.06 for the N allele and two genotypes - with a frequency of 0.89 for the MM genotype and 0.11 for the MN genotype [19]. Results with only two genotypes were reported in a study of other merino breeds – Caucasian, Northeast Bulgarian Merino, Volgograd, Salskaya and Stavropol [17, 20, 21, 22]. Three genotype variants were identified in the Askanian, Soviet merino and Altai Mountain breeds [17, 22, 23].

Many publications are devoted to the possible association of calpastatin polymorphism with important traits of meat productivity in different sheep breeds. Only two genotypes, MM and MN, were found in a study of Salsk sheep, and slaughter analyzes showed better meat productivity results related to the MN genotype [5]. In other molecular studies in sheep of the Volgograd breed also found the presence of two genotypes MM and MN. The results in control slaughter, showed that sheep with genotype MN/CAST superiored their peers with genotype MM/CAST by 3.7 kg in weight before slaughter [21]. Animals from Pakistani breed Kajli of heterozygous MN genotype grown faster compare to animals with MM genotype from birth to four months of age [24].

Identifying genes associated with productive traits would help improving the quality and diversity of production in the livestock industry by optimizing breeding programs. The established polymorphism in the CAST gene shows that the study of this locus in Bulgarian sheep breeds must be studied in detail in order to be included in future breeding programs.

4 Conclusions

The results of this study showed low genetic diversity in the studied animals of both local Karnobat and Karnobat merino breeds. In both breeds, the presence of two allelic variants (M and N) and two genotypes (MM and MN) was found, with the M allele and the MM genotype predominance.

Acknowledgements

This research was part of the project КП-06-Н 26-9/18.12.2018 (2018–2022) "Investigation of DNA markers associated with production in sheep breeds reared in Bulgaria" financed by NSF - the Ministry of Education and Science, Republic of Bulgaria.

References

1. S. Slavova, N. Stancheva, S. Laleva, Y. Popova, P. Slavova, Bulgarian Journal of Animal Husbandry, **52(5)**, 90-97 (2015)
2. K. Jawasreh, R. Jadallah, A. Al-Amareen, A. Abdullah, A. Al-Qaisi et al., Indian J Anim Sci, **87(5)**, 635 (2017)
3. J. Knapik, K. Ropka-Molik, M. Pieszka, Ann Anim Sci, **17(1)**, 23 (2017)
4. M. Aali, M. Moradi-Shahrbabak, H. Moradi-Shahrbabak, M. Sadeghi, Gene, **537(1)**, 132-139 (2014)
5. I. Gorlov, N. Shirokova, A. Randelin, V. Voronkova, N. Mosolova et al., Turkish J. Vet. Anim. Sci., **40(4)**, 399 (2016)
6. S. Byun, H. Zhou, R. Forrest, C.Frampton, J.Hickford, Anim Genet **39**, 572 (2008)
7. G. Gebreselassie, H. Berihulay, L. Jiang, Y. Ma, Animals, **10(1)**, 33 (2020)
8. K. Avanus, Kafkas Univ Vet Fac Derg, **21(6)**, 789-794 (2015)
9. M. Greguła-Kania, Afr. J. Biotechnol, **10(61)**, 13082-13085 (2011)
10. H. Chung, M. Davis, AJAVA, **7**, 641 (2012)
11. E. Armstrong, G. Ciappesoni, W. Iriarte, C. Da Silva, F. Macedo et al., Meat Science, **145**, 202 (2018)
12. K. Jawasreh, R. Jadallah, A. Al-Amareen, A. Abdullah, A. Al-Qaisi et al., Indian J Anim Sci, **87(5)**, 635 (2017)
13. B. Ramadevi, B. Punya Kumari, K. Sudhakar, G. Gangaraju, U. Vinod, J. Anim. Res., **10(6)**, 881-887 (2020)
14. S. Boykovski, G. Stefanova, D. Dimitrov. Breeds and offspring of sheep bred in Bulgaria, Shumen (2006)
15. B. Palmer, N. Roberts, M. Kent, Proc NZ Soci Anim Prod, **57**, 294-296 (1997)
16. D. Hristova, S. Georgieva, S. Tanchev, AST, **7(1)**, 36 (2015)
17. I. Dimitrova, M. Bozhilova-Sakova, A. Teneva, T. Ivanova, I. Stoycheva. E3S Web of Conferences, **262**, 02026 (2021)
18. M. Bozhilova-Sakova, I. Dimitrova, *Athens: ATINER'S Conference Paper Series*, No: AGR2015-1824 (2016)

19. I. Dimitrova, M. Bozhilova-Sakova, M. Iliev, IOSR-JAVS, **10(8)**, 61 (2017)
20. M. Bozhilova-Sakova, I. Dimitrova, T. Tzonev, N. Petrov, Bulg. J. Agric. Sci., **26(2)**, 475–479 (2020)
21. Y. Kolosov, I. Gorlov, A. Kolosov, N. Shirokova, A. Kulikova et al., IOP Conf. Series: Earth and Environmental Science, **677**, 052112 (2021)
22. N. Shirokova, A. Kolosov, Y. Kolosov, L. Getmantseva, N. Bakoev et al., IOP Conf. Series: Earth and Environmental Science, **677**, 052113 (2021)
23. M. Selionova, L. Chizhova, E. Surzhikova, N. Podkorytov, A. Podkorytov, T. Voblikova, IOP Conf. Series: Earth and Environmental Science, **613,** 012130 (2020)
24. S. Khan, M. Riaz, A. Ghaffar, M.Ullah, Pak J Zool, **44(2)**, 377 (2012)

Development of sublimation drying modes of rose hip fruits Freezing drying of rose hip

Vladimir Alexandrovich Ermolaev*, and *Dmitry Evgenievich* Fedorov

FSBEI HE Kuzbass State Agricultural Academy, 650056, st. Markovtseva, 5, Kemerovo, Russia

Abstract. The aim of the study was the selection of effective modes of rosehip sublimation dehydration. An Iney-6M unit was used for freeze drying. The selection of the duration of the sublimation stage and the final drying temperature was carried out. At the same time, such indicators as the duration of drying, organoleptic assessment and the preservation of vitamin C were analyzed. The duration of freeze drying of rose hips with a period of sublimation of 4, 5, 6 and 7 hours was 6, 7, 8 and 8.5 hours, respectively. It was found that with a decrease in the duration of the sublimation stage, a decrease in the quality characteristics of dry rose hips is observed. It was found that when the drying temperature changes from 70 to 40 °C, the drying time increases from 6.5 to 9.5 hours. At the same time, the loss of vitamin C also decreases. Based on the research, the most favorable modes of sublimation dehydration of rose hips have been determined - the final drying temperature is 50 °C and the duration of the sublimation stage is 5 hours. Under these modes, the duration of lyophilization is 8 hours, the total organoleptic assessment is 18.25 points out of 20, and the preservation of vitamin C is 27%. Lyophilized rose hips can be used as an independent product, or ground into powder for adding to other products - confectionery, dairy, bakery products, as well as biologically active additives and functional food.

1 Introduction

Rosehips are a rich source of vitamins, minerals and essential amino acids [20, p. 108; 19, p. 1; 22, p. 1]. High content of vitamin C is especially noted in rose hips - its concentration ranges from 400 to 1500 mg/100g [9, p. 310; 16, p. 116]. This is several times more than in currants and tens of times more than in apples. The highest content of vitamin C is observed in mature, but firm rose hips. In addition to ascorbic acid, the rose hips also contain vitamins A, E and B1.

In addition, rose hips contain citric and malic acids, pectin substances, as well as micro - and macronutrients such as potassium, calcium, iron, manganese, etc. The chemical and mineral composition of rose hips is presented in table. 1 [2, p. 112].

* Corresponding author: ermolaevvla@rambler.ru

Table 1. Chemical and mineral composition of rose hips [2, p. 112].

Chemical composition		Mineral composition	
Water, %	64,89-67,89	Potassium, mg/g	19,60-20,60
Dry matter, %	32,11-35,11	Calcium, mg/g	4,36-4,96
Ascorbic acid, mg%	521,49-528,51	Magnesium, mg/g	4,76-5,36
Flavonoids, mg %	150,4-154,4	Copper, mg/g	0,10-0,20
Pectin substances, mg %	0,019-0,021	Iron, mg/g	0,03-0,15

The substances that make up the rose hips have a beneficial effect in inflammatory processes, accelerate tissue regeneration, better carbohydrate metabolism and improve vascular permeability [18, p. 3; 24, p. 564]. Rosehip oil helps to reduce gastric secretion and gastric acidity. In addition, the pectin substances that make up the rose hips remove heavy metals from the body [6, p. 120]. This is the reason for the use of rose hips not only in the production of confectionery products, juices, tea, etc., but also biologically active additives and functional food products [23, p. 45].

From rose hips, liquid concentrates and dry powders are produced. In the latter case, it is necessary to use drying plants.

2 Literature review

Drying of rose hips is carried out for technological reasons, or to extend its shelf life: fresh rose hips are stored for no more than 2-3 weeks. Drying in the open air is practically not used, since the temperature does not allow the rose hips to dry properly.

The most common method is convective drying [11, p. 7]. The disadvantage of this method is the relatively long duration of the process. An increase in the temperature of the drying agent is possible only up to a certain value - at too high temperatures, denaturation of thermolabile components of the product is observed. In the literature, there is evidence that the drying of rose hips should be carried out at a temperature of the order of 60-70 °C [9, p. 311; 8; 15, p. 672].

To intensify the drying of rose hips, ultrasonic radiation is used. So, in one of the works [1], the drying of rose hips was carried out in a combi steamer with an ultrasonic apparatus. Ultrasonic waves allow warm air to quickly destroy the boundary layer and penetrate to the center of the product, which causes uniform heating and uniform moisture removal. In the course of research it was found that the use of ultrasound can reduce the drying time by 2 times, as well as reduce the loss of vitamins by 17% [1, p. 83].

Microwave treatment is also used to destroy the cell structure of berries. The invention [4] proposes the processing of rose hips in a microwave field for 1.0-1.5 minutes, after which the rose hips are sent for drying at a temperature of 60-65 °C for 9-10 hours. This method allows you to speed up the process of removing moisture from the fruit and reduce the loss of vitamin C.

It is possible to intensify the process of drying fruits and berries by lowering the pressure - using the so-called vacuum drying [21, p. 3765; 10; 17, p. 42]. Vacuum drying is widely used not only for drying fruits and berries, but also for many other products [3, p. 121; 4, p. eighteen; 5, p. fourteen; 12; 13; 14]. Vacuum drying can be carried out in several stages. For example, in the invention [7] it is proposed to dry berries in 2 stages: in the first stage, the residual pressure is 8-12 kPa, and the drying temperature is 60-65 °C, in the second stage, the residual pressure is 3-4 kPa, and the temperature is 70-80 °C.

Lyophilization - freeze drying is a promising method of dehydration. It is carried out at a pressure below the triple point of water. The product self-freezes at this pressure and moisture sublimates from a solid state to a gaseous state. The advantage of freeze drying is the ability to carry out the process at a relatively low heating temperature, which allows you to preserve the valuable components of the product.

Thus, the purpose of this work was to select the modes of freeze drying of rose hips.

3 Materials and methods

The object of research was rose hips (harvest 2020).

For freeze drying, there was used an installation *Iney-6M*, the diagram of which is shown in Fig. 1.

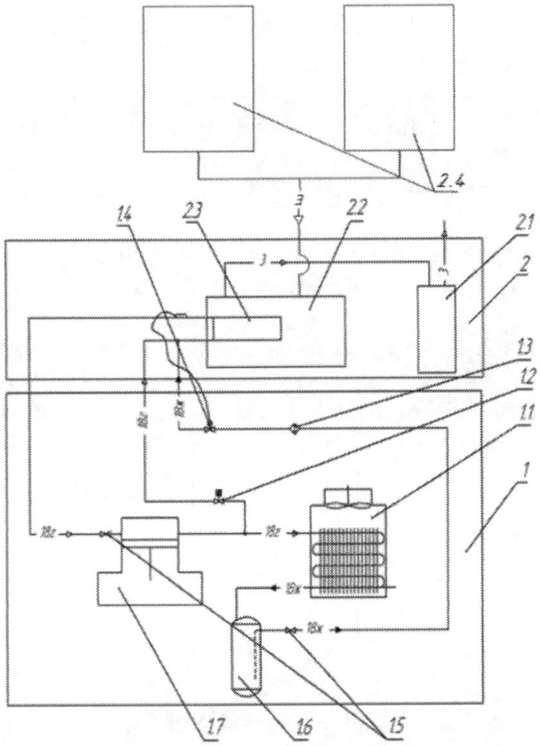

Fig. 1. Schematic diagram of a freeze drying plant:
1 - refrigerating machine; 1.1 - condenser; 1.2 - solenoid valve; 1.3 - filter drier; 1.4 - thermostatic valve; 1.5 - shut-off valve; 1.6 - receiver; 1.7 - compressor; 2 - vacuum unit; 2.1 - vacuum pump; 2.2 - desublimator; 2.3 - evaporator of desublimator; 2.4 - drying chambers

Operating this installation (Fig. 1), the product is put on pallets, which are placed in drying chambers. Then the chambers are closed and a vacuum pump is turned on, reducing the pressure to 50 Pa. To supply heat, heating lamps are used, placed vertically in the drying chambers. The desublimator freezes out the moisture contained in the removed air. Heat removal from the heat exchange surface of the desublimator is carried out due to the operation of the refrigerating machine, which consists of a compressor, condenser, filter-dryer, evaporator, receiver, thermostatic, solenoid and shut-off valves.

Drying was carried out in 2 stages: a sublimation stage and a final drying stage. During the sublimation stage, the heating lamps were turned off, and the moisture contained in the product was sublimated into the environment. At the second stage, the heating lamps were switched on and the product was dried up. Drying was carried out until a moisture content of about 5% was reached.

The moisture content was determined by an accelerated method using a device *Chizhov*, by drying a sample of the product in accordance with GOST 3626-73 and GOST R 51464-99.

The content of vitamin C was determined by the method of capillary zone electrophoresis using a capillary electrophoresis system "Kapel' 105".

Organoleptic evaluation of dry rose hips was carried out according to such indicators as taste, color, smell and texture. The maximum score for each of the indicators was 5 points. The evaluation was carried out by a tasting committee consisting of 4 people. The arithmetic mean of all 4 values was taken as the final value.

4 Results

According to the results of the analysis, the initial moisture content of rose hips was 67.13%, and the concentration of vitamin C was 4.83%.

As part of the research, the duration of the sublimation period and the drying temperature were selected.

Fig. 2 shows the graphs of changes in the relative mass of rose hips during freeze-drying at different durations of the sublimation period. The final drying temperature was 60 °C.

Table 2 shows the results of organoleptic evaluation of dried rose hips depending on the period of sublimation.

Table 3 shows the indicators of the effectiveness of freeze drying of rose hips depending on the drying temperature. In this case, the duration of the sublimation period was 5 h.

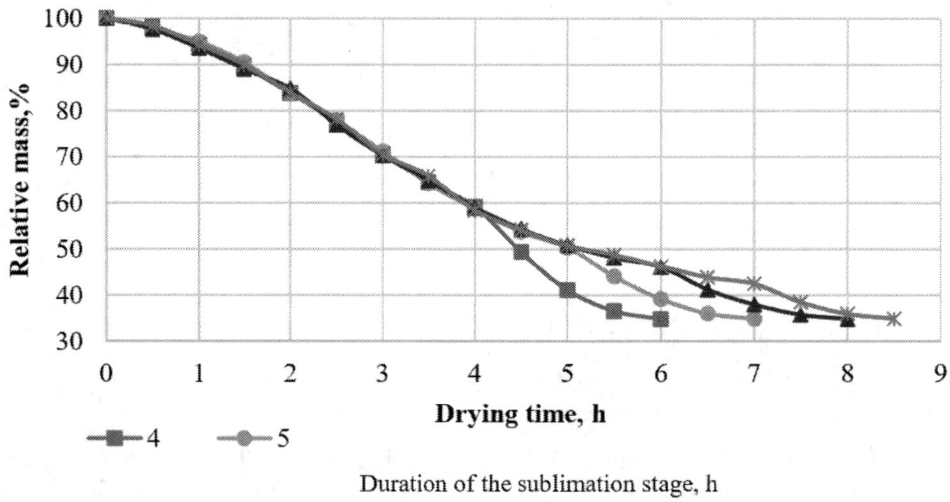

Fig. 2. Graphs of changes in the relative mass of rose hips during freeze drying with different duration of the sublimation period

Table 2. Organoleptic evaluation of dry rose hips depending on the period of sublimation.

Indicator	Duration of the sublimation period, h			
	4	5	6	7
Taste	3,75	4,25	4,5	5
Color	4,25	4,75	4,75	4,75
Smell	4	4,5	4,5	4,75
Consistency	4	4,25	5	5
Total	16	17,75	18,75	19,5

Table 3. Indicators of the rose hips drying effectiveness when selecting the final drying temperature.

Drying indicator	Final-drying temperature, °C			
	40	50	60	70
Drying time, h	9,5	8	7	6,5
Total organoleptic assessment, points	18,75	18,25	17,75	15,25
Moisture content, %	5,2	4,9	4,6	4,5
Vitamin C content, %	10,94	10,25	8,73	7,20

5 Discussion

The first stage of drying is the sublimation stage. During the first 15-30 minutes. the installation enters the operating mode: the vacuum pump then lowers the residual pressure in the drying chambers to 50 Pa. In this case, there is an intensive removal of heat from the product and it self-freezes. The relative mass at this stage changes insignificantly - by 1.5-2% (Fig. 2).

After the installation reaches the mode, the removal of most of the moisture contained in the product begins. The rate of change in the relative mass is about 10-14%/hour. 4 hours after the start of the drying process, the relative weight of the rose hips decreased to 60%.

At the third stage, the heating lamps turn on and the second stage of drying begins - final drying. After turning on the heating lamps, there is a noticeable jump in the rate of moisture removal. Moreover, the earlier the heating lamps turn on, the more dramatically the drying speed changes.

It is worth noting that the duration of the sublimation stage was determined as the total duration of freeze drying (including the setting of the installation to the operating mode) until the heating lamps were turned on.

The duration of freeze drying of rose hips with a period of sublimation of 4, 5, 6 and 7 hours was 6, 7, 8 and 8.5 hours, respectively.

The analysis of the organoleptic properties of dry rose hips made it possible to establish that with a decrease in the duration of the sublimation stage, a slight decrease in quality characteristics is observed. So, with the reduction of the sublimation stage from 7 to 6 hours, the overall organoleptic assessment decreases by 0.75 points. When the duration of the sublimation period is reduced from 6 to 5 hours, the organoleptic assessment is reduced by another 1 point. The greatest decrease in the quality of rose hips (by 1.75 points) occurs when the sublimation period is reduced from 5 to 4 hours. The change in the quality of the dry product is due to the fact that with a decrease in the sublimation period, the duration of the heating lamps increases.

Based on the foregoing information, it is recommended to lyophilize the rose hips with a sublimation period of at least 5 hours. In further experiments, the sublimation period was 5 hours.

To investigate the effect of the final drying temperature, corresponding experiments were carried out. According to the results of the experiments, it was found that a change in the final drying temperature entails a nonlinear change in the duration of moisture removal (Table 3). With a change in the final drying temperature from 70 to 40 °C, the drying time increases from 6.5 to 9.5 hours. The total organoleptic assessment also increases - from 15.25 to 18.75 points out of 20.

Within the framework of the research, the effect of the final drying temperature on the preservation of vitamin C in rose hips was also analyzed. Vitamin C is one of the most unstable vitamins during storage. Significant losses of this vitamin are observed when the pulp of the product comes into contact with oxygen. Therefore, in this case, the fruits were not crushed before freeze-drying.

The negative factors contributing to a decrease in the vitamin C content are freezing and exposure to heating lamps. According to known data, freezing has a relatively weak effect on the concentration of vitamin C. The storage of the product itself in a frozen or native state has a greater effect. With regard to the effects of heating lamps, high temperatures have a much more harmful effect than freezing. In this case, an important factor is not only the temperature, and the duration of its exposure. An increase in the final drying temperature, on the one hand, entails a reduction in the duration of dehydration, but on the other hand, an intensification of the temperature effect.

It was found that after drying, the concentration of dry matter increases, which also affects the content of vitamin C. Taking into account the increase in the concentration of dry matter, the loss of vitamin C at a final drying temperature of 40 °C was 21%. An increase in the final drying temperature to 50 °C entails an increase in vitamin C losses up to 27%. With a further increase in the drying temperature to 60 and 70 °C, the preservation of vitamin C decreases to 37 and 48%, respectively.

Based on the research results, it is possible to recommend a final drying temperature of 50°C.

6 Conclusion

In the course of the study carried out, the most favorable modes of sublimation dehydration of rose hips were determined - the temperature of additional drying is 50 °C and the duration of the sublimation stage is 5 hours. Under these modes, the duration of lyophilization is 8 hours, the total organoleptic assessment is 18.25 points out of 20, and the preservation of the vitamin C - 27%. Research results can be useful for food industry workers, technologists and researchers. A possible direction for further research may be the selection of freezing modes - self-freezing or preliminary freezing, as well as the selection of the drying layer thickness. Lyophilized rose hips can be used as an independent product, or ground into a powder for addition to other products - confectionery, dairy, bakery products, as well as biologically active additives and functional food products.

References

1. E.I. Verboloz, M.A. Ivanova, V.A. Demchenko, S. Fartukov, N.K. Evona, Technics and technology of food production, **1**, 79 (2020)
2. M.I. Gusalova, A.V. Khmelevskaya, S.K. Cherchesova, News of the Mountain State Agrarian University, **2**, 111(2019).
3. V.A. Ermolaev, O. N. Ivanenko, M.V. Onyushev, Bulletin of KrasGAU, **9**,121 (2016)
4. Ermolaev V.A. *Theoretical substantiation and practical implementation of the technology of dry cheese product* (2013)
5. V.A. Ermolaev, S.A. Zakharov. *Theoretical substantiation of the foundations of canning by drying and practical implementation of the technology of vacuum drying of cottage cheese: Monograph*, 176 (2009).
6. F.F. Kokaeva, D.N. Jatieva, News of the Mountain State Agrarian University, **1**, 120 (2018).
7. V.A. Ermolaev, D.E. Fedorov, G.A. Maslennikova, Patent RF No. 2462867 Method of vacuum drying of berries. IPC A23B 7/02 / Appl. 06.06.2011; Publ. 10.10.2012, Bul. No. 28.
8. T.A. Isrigova, M.M. Salmanov, Z.A. Abdurakhmanova, M.E. Akhmedov, A.B. Kurbanova, K.M. Salmanov, S.S. Isrigov, L.M. Bagavdinova, Patent RF No. 2648709 Method for the production of dry rosehip drink. IPC A23L 2/39, A23L 2/00, A23L 2/38, A23L 3/01 / Application. 04/12/2017; Publ. 28.03.2018.

9. E.A. Timanovsky, Product quality: control, management, enhancement, planning, 310 (2019)
10. S.M. Demarchi, R. Martín Torrez Irigoyen, S.A. Giner, A. Sergio, Journal of Food Engineering (2018)
11. Duygu Evin, International Journal of Engineering and Natural Sciences (IJENS), **1**, 7 (2019)
12. V.A. Ermolaev, I.A. Kechkin, E.V. Makhacheva, A.I. Romanenko, V.V. Tarakanova, BIO Web of Conferences, **27,** 00009 (2020)
13. V.A. Ermolaev, I.A. Kechkin, A.I. Romanenko, V.V. Tarakanova, Glebova I.A., Proceedings of the International Conference on Policies and Economics Measures for Agricultural Development (AgroDevEco 2020), Advances in Economics, Business and Management Research, 147 (2020)
14. V.A. Ermolaev, I.A. Kechkin, A.I. Romanenko, K.D. Buzetti, M.V. Ivanov, E.V. Makhacheva, IOP Conf. Series: Earth and Environmental Science, **640,** 032040 (2021)
15. C. Kadakal, T. Duman, R.Ekinci, Food Science and Technology, **38(4)**, 667(2018)
16. A. Leahu, C. Damian, M.Oroian, S.Ropciuc, R.Rotaru, Scientific Papers: Animal Science and Biotechnologies, **47(1)**, 116 (2014)
17. A. Mabellini, E. Ohaco, C.A. Marquez, Antonio DeMichelis, J.E. Lozano, International Journal of Food Studies, **1,** 42(2012)
18. I.Mármol, C.Sánchez-de-Diego, N. Jiménez-Moreno, C.Ancín-Azpilicueta, M. Jesús Rodríguez-Yoldi, International Journal of Molecular Sciences, **18**, 1137 (2017)
19. B. Medveckiene, J. Kulaitiene, E. Jariene, N.Vaitkevicien,E. Hallman, Applied sciences, **10**, 5337(2020)
20. D. Paunović, A. Kalušević, T. Petrović, T. Urošević, D. Djinović, V. Nedović, J.Popović-djordjević, Not Bot HortiAgrobo, **47(1)**,108(2019)
21. J.E. Safarov, Sh.A. Sultanova, G.T. Dadayev, D.I. Samandarov, International Journal of Innovative Technology and Exploring Engineering (IJITEE), **9 (1),** 3765(2019)
22. O.A. Starodub, V.M. Leontjev, G.R. Rybakova, I.V. Doiko, I.A. Zobnina, IOP Conf. Series: EarthandEnvironmentalScience, **421**, 022025 (2020)
23. Vasić, D. Paunović, B. ŠpirovićTrifunović, J. Miladinović, L. Vujošević, D. Đinović, J. Popović-Đorđević, Acta AgriculturaeSerbica, **25(49),** 45 (2020)
24. L.Zhiyong, C. Zhijian, Y. Zaiwang, A. Mingzhang, J. Wenwen, Y. Longjiang, Planta Med., **85,** 563(2019)

Influence of basic treatment of irrigated land on Fertility indicators and harvest formation of sweet pepper

Serazhutdin Aminovich Kurbanov*, and *Diana Sultanovna* Magomedova

FSBEI HE Dagestan State Agrarian University, 367032, st. M. Gadjiev, 180, Makhachkala, Russia

Abstract. The purpose of the research was to establish the most rational soil cultivation system for irrigated conditions, contributing to the preservation of ecological balance and increasing the profitability of pepper production. Experimental studies were carried out on meadow-chestnut soils of the plain zone of the Republic of Dagestan in 2017-2019. in a two-factor experiment to determine the most effective combination of soil cultivation and irrigation regime. The paper presents the results of the influence of plowing and disc cultivation on the water-physical properties of the soil and its biological activity, as well as on the yield of sweet pepper under drip irrigation. The results of three-year research have established that, under irrigation conditions, the most effective method of basic soil cultivation is moldboard soil cultivation to a depth of 0.23 ... 0.25 m, which contributed to the improvement of soil density, its permeability, structural state and increased biological activity in the treated layer. On plowing, the yield of sweet pepper fruits was obtained at the level of 6.23 t/ha. The use of disc processing by 0.10 ... 0.12 m leads to a deterioration of agrophysical indicators of fertility, phytosanitary state of irrigated fields by 1.7 ... 2.2 times, a decrease in yield by 5.1 ... 9.6% and profitability of pepper fruit production by 25.3 ... 31.0%.

1 Introduction

One of the main and labor-intensive elements in the technologies of cultivation of agricultural crops is the main tillage, which is designed to promote the rational use and preservation of its natural fertility, and the growth of crop yields. Among the yield-forming factors, it accounts for up to 20% [9] and at the same time, about 40% of the energy and 25% of the labor resources used to grow crops are spent on its implementation [21]. The problem of resource conservation with existing agro-ecological problems of agriculture requires the introduction of minimized and soil-protective methods and techniques of basic soil cultivation. According to a number of scientists, studies on minimizing soil cultivation carried out in many regions of Russia have shown that its efficiency, including in terms of resource conservation, depends on many factors [4, 10, 12, 14, 17, 20, 22, 23, 24]. The effectiveness of minimization in terms of preventing soil degradation and erosion, mineralization of or-

* Corresponding author: kurbanovsa@mail.ru

ganic matter, reducing soil moisture loss, reducing energy intensity is beyond doubt, but this is more related to rain-fed lands. The disadvantages of minimization include deterioration of the phytosanitary situation, soil compaction, a decrease in water permeability and moisture reserves, an increase in the consumption of pesticides, the problem of applying fertilizers, especially organic fertilizers [13], and on irrigated lands these disadvantages are increasing.

In the conditions of irrigated agriculture, the minimization of the main tillage is aggravated by the compacting effect of water, which worsens agrophysical indicators of fertility due to increased soil moisture, increased reproduction of weeds, pests and diseases, and increased costs for plant care [3, 5, 18]. Numerous studies have established the advantage of moldboard tillage for perennial grasses and row crops, and surface and flat-cut basic tillage are more promising for grain crops and forage crops for continuous sowing [1, 2, 7, 8, 11, 16].

2 Materials and methods

In 2017-2019, to achieve the research aim a field experiment was laid on the meadow-chestnut medium loamy soils of the educational and experimental farm of the Dagestan State Agrarian University, it was conducted to establish the effect of plowing at a depth of 0.23 ... 0.25 m (control), disc cultivation by 0.10 ... 0.12 m and different irrigation regimes (70, 80 and 90%) on the ontogeny of sweet pepper plants maintained using a drip irrigation system. When growing pepper, a zonal technology was used with planting seedlings in a wide-row way (0.7x0.3 m) in the first half of May at the age of 60...62 days. The method of observation, accounting and analysis is generally accepted in irrigated agriculture and vegetable production [15].

3 Discussions and results

One of the factors of soil fertility is its agrophysical properties, which largely determine the dynamics of soil processes, including biological activity. Under irrigation conditions, important indicators of the physical condition of the soil are the density and permeability of the soil. The bulk density is a complex indicator that reflects the entire set of physical properties of the soil, to a large extent, depends on the methods of soil cultivation. Studies carried out on medium loamy soils indicate that with a soil density of 1.0 ... 1.3 t/m^3, the best water-physical and air properties are created for the growth and development of plants.

Our research has established that the use of plowing by 0.23...0.25 m contributes to a decrease in the density of the arable layer to 1.19 t/m^3; while disc cultivation by 0.10...0.12 m compacts the arable layer up to 1.26 t/m^3. The use of different methods of soil cultivation determined the unequal composition of the arable and sub-arable soil layers. If, during plowing, the soil density during the growing season increases gradually with depth, then during disc processing the transition from the surface layer of 0.0...0.1 m to a layer of 0.1...0.2 m is sharp and amounts to 0.04 t/m^3. During the growing season, the arable layer gradually becomes denser and, during plowing, approaches the equilibrium density (1.27 t/m^3), while when working with discs, a significant overcompaction of the soil is observed, exceeding the equilibrium density by 0.06 t/m^3 (Table 1).

Table 1. Influence of the main tillage and moisture thresholds on soil water-physical properties (2017-2019).

Main tillage	Soil moisture threshold, %	Soil density, t/m³		Content of water-resistant units, %	Soil permeability, mm/h
		beginning of vegetation	end of vegetation		
Plowing at 0.23-0.25 m, control	70 K	1,19	1,27	46,6	153
	80	1,19	1,27	45,7	149
	90	1,20	1,25	45,4	143
Disc tillage at 0.10-0.12 m	70 K	1,26	1,35	43,5	123
	80	1,27	1,33	42,7	118
	90	1,27	1,32	41,8	106

Analysis of soil density with an increase in the threshold of soil moisture indicates its decrease by 0.02...0.03 t/m³ in comparison with the control (70%), regardless of the use of moldboard processing or disk tillage. Since porosity is a function of soil density, the lowest porosity (49.8%) was observed in the variant with disc cultivation, which, in our opinion, is associated with the compacting effect of increased irrigation rates.

One of the main characteristics of the water properties of the irrigated soil is its water permeability, the values of which depends on the aftereffect of the basic tillage techniques and determines the effectiveness of the studied pre-irrigation soil moisture thresholds. The use of disking reduced the permeability of the soil in comparison with plowing by an average of 32.7 mm/h, which, in our opinion, is associated with a higher layer density of 0.1...0.2 m ($y = -381.0x + 626.7$ $R^2 = 0.650$) and a 7.0% decrease in the number of water resistant units ($y = -37.62x + 93.13$ $R^2 = 0.639$). Agroecological assessment of soil permeability [19] showed that the transition to resource-saving disc cultivation transfers the irrigated soil from the group with high permeability (more than 150 mm/h) to the group with medium water permeability (50…150 mm/h).

An increase in the pre-irrigation moisture threshold up to 90% and the associated increase in vegetation irrigation had a greater effect on water permeability than the methods of basic soil cultivation. Determination of water permeability in the zone of closure of the moistening contours showed that against the background of disking at a pre-irrigation threshold of 90%, water permeability decreases by 13.8% or 2 times more than during plowing.

The content of water-resistant aggregates is of great importance in assessing the methods of soil cultivation in irrigated conditions. Carrying out plowing contributes to the preservation of 2.5...2.8% more water-resistant aggregates than disk processing, which increases the coefficient of water resistance during moldboard processing from 0.77 to 0.89. This is due to the fact that during moldboard processing as a result of technological operations with the soil, an even distribution of plant residues and organic matter is ensured in it. When processing with disks by structuring processes, only the upper 0.0...0.12 m layer is affected. A direct correlation was also noted between the content of water-resistant aggregates and water permeability ($y = 9.984x - 310.1$ $R^2 = 0.989$).

According to a number of researchers, dumping creates better conditions for the formation of the mass of the root system, an increase in the biological activity of the soil, which enhances the formation of humic acids and, as a consequence, an increase in the content of water-resistant aggregates [6, 17].

Our studies on the effect of plowing and disc cultivation indicated an increased biological activity of moldboard soil cultivation in comparison with disc cultivation, which was assessed by integral indicators: cellulose decomposition activity and "soil respiration" by the intensity of carbon dioxide release (Table 2).

It has been established that dumping to a depth of 0.23...0.25 m promotes an increase in the decomposition of cellulose of plant residues by 21.3%, and most intensively at 70%. At the same time, the rate of loss of the weight of linen fabric during moldboard processing is higher and if in the first month the difference between the methods of basic processing was 23.9%, then in 3 months it increased to 29.3%.

Table 2. Cellulose-destroying and CO2- capacity capacity of the soil, depending on the methods of-main tillage and irrigation (2017-2019).

Main tillage	Soil moisture threshold, %	Cellulose-destroying capacity, %			CO_2- capacity capacity, mg $CO_2/m^2{\cdot}h$
		1 month	2 months	3 months	
Plowing at 0.23-0.25 m, control	70 K	46,4	62,8	75,4	959
	80	44,7	59,9	72,3	928
	90	43,3	58,3	70,5	915
Disc tillage at 0.10-0.12 m	70	35,1	45,3	52,7	845
	80	34,0	43,7	51,4	829
	90	33,2	42,4	50,1	822

An increase in the pre-irrigation threshold to 90% reduces the activity of cellulose-decomposing organisms by 4.9% during moldboard tillage and by 2.6% during disc cultivation. nutrition of sweet pepper and provides, along with fertilizers, a significant increase in yield, and on the other hand, a decrease in the biological activity of the soil during disc cultivation slows down the decomposition of organic matter and it can be assumed that the processes of humification will intensify.

The most universal indicator of soil biogenicity is the production of CO_2 by soil organisms or "soil respiration". Our studies have shown that the highest level of carbon dioxide emission was observed during moldboard plowing - 934 mg $CO_2/m^2{\bullet}h$, which is 10.8% higher than with disc plowing, and we associate this with the deterioration of the water-physical properties of the soil during disc plowing. ... There was a direct correlation between the cellulose-destroying and CO_2-producing capacity of the soil ($y = 0.201x - 115.8$ $R^2 = 0.982$). As for the levels of pre-irrigation soil moisture, there is a tendency for the "soil respiration" to decrease with an increase in the moisture threshold from 70 to 90%, because the difference is in the range of 2.7...4.6%.

One of the disadvantages of the transition to resource-saving soil cultivation is an increase in the weediness of fields, and under irrigation conditions, not only the phytosanitary situation worsens, but also the accumulation of phytopathogens is noted [13, 18].

Our research found that minimization of the main tillage is a weak regulatory effect on the weed phytocenosis of sweet pepper plantings, since the weediness of crops increases both in quantitative terms, on average from 23.7 to 48.3 pcs/m^2, and in weight ratio - from 120.4 to 189.7 g/m^2, while the increase in weediness is associated with an increase in the number of juvenile weeds in the weed phytocenosis (white-backed weeds, garden purslane, chicken millet, green bristles, etc.). The use of disc treatment leads to a significant increase in the potential debris of the arable layer, which increases from 319 to 678 ml.pcs.seeds/ha. The transition from 70 to 90% practically did not affect the number of weeds and potential soil contamination, which is associated with the peculiarities of local soil moisture during drip irrigation.

Productivity is the most important indicator of the responsiveness of sweet pepper plants to the studied agricultural techniques. A field experiment showed that of the two factors affecting the yield of sweet pepper, the most effective methods of basic tillage, which provided an increase in yield of 14.2 t/ha (34.6%). An increase in the pre-irrigation moisture threshold of the active 0.5 m layer increased the yield of pepper by an average of only 26.7% (Table 3).

Table 3. The yield of sweet pepper and the structure of the yield, depending on the irrigation regime and methods of main tillage, t/ha.

Main tillage	Soil moisture threshold, %	Productivity, t/ha	Fruit weight from 1 plant, kg	Number of fruits per plant, pcs.	Weight of 1 fruit, g.	Marketability, %
Plowing at 0.23-0.25 m, control	70 K	48,4	1,02	17,3	53,9	93,4
	80	62,3	1,32	21,6	61,0	93,3
	90	54,8	1,16	20,2	57,6	91,7
Disc tillage at 0.10-0.12 m	70	35,4	0,75	14,1	53,0	92,5
	80	43,9	0,94	17,2	54,7	92,8
	90	43,7	0,93	16,4	56,9	91,4

HCP_{05} (t/ha): for processing methods - 2.9; for soil moisture thresholds - 2.3; for the interaction of factors - 3.4.

The decrease in yield during disk tillage is explained by the deterioration of the agrophysical indicators of the fertility of the meadow-chestnut soil, as well as the deterioration of the phyto-sanitary state of sweet pepper plantings. The highest yield of sweet pepper fruits was obtained with a combination of moldboard processing by 0.23...0.25 m and maintaining the pre-flood threshold not lower than 80% - 62.3 t/ha.

4 Conclusions

On irrigated meadow-chestnut soils for the cultivation of sweet pepper, the most effective is moldboard tillage to a depth of 0.23…0.25 m, which, when using drip irrigation and maintaining the moisture content of the active soil layer in the range of 80...100%, yields more than 62 t/ha of fruits.

References

1. S.I. Voronov, V.P. Zvolinsky, Yu.N. Pleskachev et al., Agriculture, **2,** 24(2020).
2. G.T. Balakay, *Soybeans on irrigated lands*, 138 (1999)
3. A.G. Bondarev, *Change of physical properties and water regime of soil during irrigation* (1982)
4. V.V. Borodychev, K.I. Pimonov, E.N. Mikhailichenko, Fertility, **1**, 34 (2020)
5. V.V. Borodychev, A.V. Shu-ravilin, V.T. Skorikov, Agriculture, **8**, 25 (2008)
6. A.F. Viter, V.I. Turusov, V.M. Garmashov, S.A. Gavrilova, *Soil cultivation as a factor in regulating soil fertility*, 173 (2014)
7. G.N. Gasanov, A.A. Aytemirov, *Resource-saving tillage for crops of field crop rotation in Dagestan: monograph*, 286 (2010)
8. R.A. Gish, G.S. Gikalo, *Vegetable growing of the south of Russia: textbook*, 632 (2012)
9. I. Gorbachev, Chief agronomist, **2**, 63 (2007)
10. N.I. Devterova, O.A. Well-being, Agriculture, **3,** 31(2019)
11. V.M. Zhidkov, O. G. Chamurliev, *The main tillage in irrigated crop rotations of the Lower Volga region: monograph* (Volgograd: Volgograd GAU), 152 (2013)
12. N.A. Kirillov, A.G. Lozhkin and A.I. Volkov, Agricultural science, 10, 8 (2015)
13. V.I. Kiryushin, Agriculture, **3**, 3 (2019)
14. N.L. Kurachenko, A.A. Kartavyh, Agriculture, **2**,17 (2017)

15. S.S. Litvinov, *Field experiment technique in vegetable growing. -M.: Publishing house of VNIIO*, 648 (2011)
16. N.R. Magomedov, G.Ya. Khalidova, Agricultural science, **1**, 9 (2017)
17. V.I. Turusov, V.M. Garmashov, N.A. necessary, etc., Agriculture, **6**, 25 (2020)
18. O.G. Chamurliev, G.O. Chamurliev, Fertility, **2**, 48(2017)
19. A.L. Chernogorov, P.A. Chekmarev, I.I. Vasenev et al., *Agroecological assessment of land and optimization of land use,* 268 (2012)
20. R. Sharushov. A. Dozorov, A. Naumov, M. Garanin, International Agricultural Journal, **4,** 45 (2017)
21. V.N. Sheptukhov, *Minimization of processing and direct sowing in crop cultivation technologies*, 208 (2008)
22. L.V. Yushkevich, D.A. Golovanov, Agriculture, **5**, 30 (2017)
23. P.W. Under, H. Blanco-Canqui, *Conservation tillage, Handbook of Soil Science: Resource Management and Environmental Impacts 2nd ed.* (2011)
24. G. Valiboa, Brandi G. Lagomarsino, et al., Soil and Tillage Research, **154,** 126 (2015)

Functional bread enreached with regional bioactive food additives

A. S. Khamitsaeva [1], *V. V.* Sadovoy [1], *Z. G.* Ramonova [1], *P. N.* Semenov [1], *A. O.* Nartikoeva[2], *E. S.* Dzodzieva [1], *V. A.* Gasieva [1], *Z. S.* Khamitsaeva [1], *V. Kh.* Sebetov [1,*], and *S. F.* Zokoeva[2]

[1]Gorsky State Agrarian University, Vladikavkaz, Russia
[2]North Ossetia State University named after K.L. Khetagurova, Vladikavkaz, Russia

Abstract. The scientific work presents materials of theoretical and experimental substantiation of the use of regional bioactive food additives, namely wild plants of the legume family (*Fabaceae*), permitted for use in the food industry: sweet clover (*Melilotus officinalis*), red meadow clover (*Trifolium pratense*), white acacia (*Robinia pseudoacacia*) in the production of functional foods [1]. Experimental studies of the safety indicators of food wild-growing plants (FWP) have been carried out. It was found that the studied FWP growing in ecologically clean areas of the North Ossetia-Alania are distinguished by a low mass fraction of toxic substances. Two fractions of cellulose were investigated: 1 - water-soluble (pectic acid, pectin, methylcellulose); 2 - water-insoluble (protopectin, cellulose, hemicellulose). The increased content of dietary fiber in WGP powders has been established, corresponding to the physiological norms of satisfying the daily requirement for FWP and, in this regard, indicating the expediency of using functional food in formulations for the prevention and treatment of diabetes mellitus, obesity, and atherosclerosis. The sorption capacity of WGP powders with respect to lead has been investigated. All studied powders of WGP modification products are characterized by high sorption capacity. The technologies of flour products with the addition of FWP powders have been developed. The results of clinical and preclinical studies of the developed functional bread show a corrective effect of glycemic blood parameters, as well as a decrease in the lead content in the blood of fed laboratory rat pups by almost two times, compared with the control.

1 Introduction

The purpose of this work was to expand the scope of using regional food wild-growing plants in the technology of functional flour products, which has not only scientific, practical and technological foundations, but also an important social argument in improving the human body. Fortified food products, including dietary fibers of medicinal wild plants, can prevent diseases caused by the anti-alimentation factor of nutrition, as well as harmful impurities in the environment.

Tasks to achieve this goal were set:

* Corresponding author: sebet2009@yandex.ru

- to perform theoretical substantiation of the FWP use in the production of functional flour products;
- to substantiate the choice of food wild-growing plants, to assess the technological and nutritional adequacy;
- to study the safety indicators of wild plants;
- to investigate the polysaccharide composition of FWP powders;
- to determine the sorption capacity of food wild plants;
- to investigate the sorption capacity of the developed flour products on laboratory animals;
- to investigate the glycemic profile of the developed functional bread with FWP in volunteer patients.

2 Methods and materials

Objects for development were selected according to their food, technological adequacy, resource availability, prevalence in the North Ossetia region.

Powders of dried wild-growing plants (red meadow clover flowers, red meadow clover leaves, white acacia flowers, sweet clover flowers, sweet clover leaves), as well as finished flour products were selected as objects of research.

The safety indicators of raw materials and finished products were investigated by atomic absorption spectrometry according to GOST 26932-86; GOST 32163-2013 - Food products. Method for determination of strontium content is Sr-90; mycotoxins - according to GOST R 50474-93 - Food products. Methods for the detection and determination of the number of bacteria of the group of Escherichia coli (coliform bacteria).

Soluble and insoluble polysaccharides in FWP powders were determined in accordance with: MUK 4.1.1596-03, patent 2434532 "Method for determining the mass fraction of pectin substances in plant raw materials" [7, 11].

Dietary fibers were investigated in a sulfuric acid medium with a colorimetric determination of the colored product by comparing the color intensity or light absorption of the test solution and the standard with a known content of the analyte.

The sorption efficiency of heavy metal salts by ballast substances was investigated using a modified technique.

Preclinical studies of laboratory rat pups biomaterials were carried out using methods on ARCHITECT 8000.

The change in the physicochemical indicators of the quality of FWP was assessed by the dynamics of the content of the polysaccharide composition of food wild-growing plants, with the aim of using them in the production of functional food products to enrich the composition of flour products with protective components. For the preparation of powders of edible wild plants, the aerial parts of the selected FWP objects (leaves, stems, inflorescences), collected in the Alania region in 2018-2020 during the flowering phase, dried according to the generally accepted technology, were used.

The importance of the research goal is within the framework of the "Concept of state policy in the field of healthy nutrition of the population of Russia", because one of the most important directions is indicated: "The development of new technologies for the production of food products with a directed change in the chemical composition should correspond to the needs of the human body, including to prevent various diseases, reducing the risk of exposure to harmful substances" [9, 11, 13].

A significant role in the technology of functional food products is assigned to flour products, which are the most affordable food products from the standpoint of the cost factor, consumed every day by all groups of the population of the Russian Federation.

To expand the range of functional flour products, it is important to introduce natural biologically active additives (BAA) into the formulations and not to use artificial fillers.

The results of such scientific research, associated with the use of biologically active additives, can guarantee the receipt of food products of valeological orientation.

Effective and expedient both from a practical-technological and from an ecological point of view to improve the quality of food products is the use of food wild plants rich in biologically active substances [6, 8, 11].

As a dietary supplement for enriching the composition of the most common food products, including bakery products, rich plant raw materials of the Alania region, modified in properties and composition, can be used, in contrast to synthetic food additives used all over the world.

The theory of rational and optimal nutrition - to ennoble the internal state of a person, at the level of the disease state of the population, it is possible by introducing such biological factors as enterosorbent, antioxidant, etc. into food products [2, 10, 12].

The study object of trophology is the general laws of assimilation of essential nutrients at all stages of any block of living systems.

The issue of impeccable food and optimal nutrition, coordination and measures of food production technologies based on trophological considerations refers to an applied problem, taking into account the peculiarities of the biosystem, the spare parts from it are only in the surrounding nature, in view of this, wild plant living systems were selected for use in development functional bread technology [2].

The products under development can take an extreme position between functional and traditional medicine, since health-improving products, providing homeostasis, regulating redox reactions, will become a means of general, prophylactic, therapeutic action [2, 4, 5, 14].

3 Results

Wild plant raw materials were modified using gentle convective drying, taking into account not only the intensity of moisture evaporation, but also the need to preserve the original composition of the biologically active substances.

In the course of drying, the mass fraction of moisture of the studied plant material was determined, taking into account which the moisture content ω (%), calculated according to formula 1:

$$\omega = \frac{W}{M_c} \cdot 100 \tag{1}$$

where W – sample moisture content, %;

M_c –absolutely dry matter content, %.

The amount of evaporated moisture $W_и$ in the process of drying the FWP was determined by the formula 2:

$$W_и = M_1 \cdot \frac{W_1 - W_2}{100 - W_2} \tag{2}$$

where M_1 – mass before drying;

W_1 – moisture before drying, %;

W_2 – moisture at time τ during drying, %.

Conversion of moisture content was carried out according to the formula 3:

$$w = \frac{100 \cdot \omega}{100 + \omega} \tag{3}$$

The moisture characteristics of fresh wild-growing raw materials are given in Table 1.

Table 1. Moisture characteristics of fresh FWP

Raw materials		W,%	M_c,%	ω,%
Red meadow clover	Inflorescences	89,90	10,20	900
	Stems, leaves	91,40	8,60	877
Sweet clover	Flowers	89,77	10,23	820
	Leaves			784
White Acacia	Flowers	90,7	11,3	880

Kinetic curves of the wild-growing raw material drying

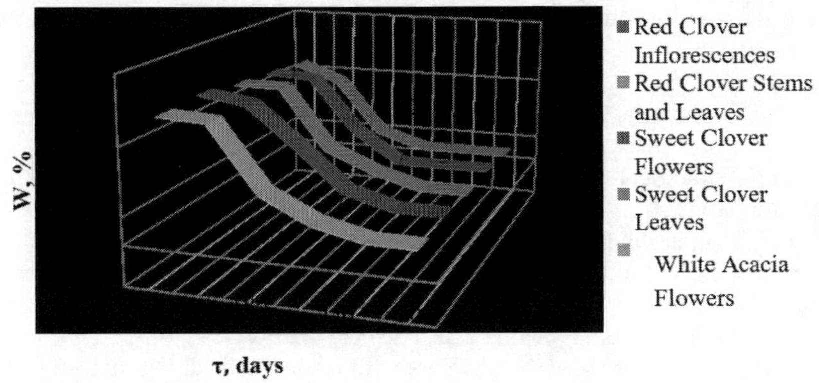

Fig. 1. Drying curves of the studied wild-growing raw material

The intensity of drying of wild plants is determined only by the parameters of the drying agent and does not depend on the moisture content of the dried material and its physicochemical properties.

The established parameters of the modes of heat treatment of food wild plants (FWP) characterize the preservation of the vitamin, mineral, polysaccharide composition and, as a consequence, increase the quality of the dried FWP.

Dried inflorescences, leaves, and stems of edible wild plants (FWP) were homogenized before turning them into flour on a roller mill. Dried wild plant flour is a finely ground powder. The degree of particle size reduction reached from 40 to 60 microns.

In order to obtain environmentally friendly products, the safety indicators of FWP were studied.

The obtained research data showed that the content of toxic substances in the FWP is below the permissible limit concentration, and this makes it possible to recommend the selected objects as environmentally friendly dietary supplements for inclusion in the formulations of functional flour products.

Indigestible carbohydrates of FWP powders are effective adsorbents, susceptible to bind and remove heavy metals from the body, and also have a great potential for use in the treatment of gastrointestinal and cardiovascular diseases.

In view of this, the content of dietary fiber in the obtained modified FWP powders was investigated (Table 2).

Table 2. Content of dietary fiber in FWP powders

Indicator	Mass fraction, %				
	Red Meadow Clover		Sweet Clover		White Acacia
	Inflorescences	Leaves , stems	Flowers	Leaves	Flowers
Alimentary fiber					
Methylcellulose	7,9	8,3	11,3	7,7	9,7
Pectin	7,5	6,0	5,7	4,9	6,6
Pectic acid	4,7	3,8	6,9	5,7	7,9
Protopectin	8,5	10,9	9,3	9,9	11,6
Cellulose	10,5	12,7	9,3	8,9	9,6
Hemicellulose	7,2	8,7	7,3	6,3	9,6

According to the data from the Table 2, wild plants (red meadow clover, sweet clover, white acacia) are characterized by a significant content of dietary fiber, both soluble and insoluble.

Increased content of dietary fiber in FWP powders: inflorescences, leaves, stems of red meadow clover; flowers, leaves of sweet clover; flowers of white acacia; in accordance with the physiological norms of satisfying the daily requirement for alimentary fiber indicate the advisability of using functional food in formulations for the prevention and treatment of diabetes mellitus, obesity, atherosclerosis.

As mechanical stimuli alimentary fiber stimulate secretion, motility, the formation of gastrointestinal hormones, slow down the rate of food digestion, and these properties can help people control appetite, blood sugar levels in the body.

The sorption capacity of FWP powders with respect to lead was investigated (Fig. 2).

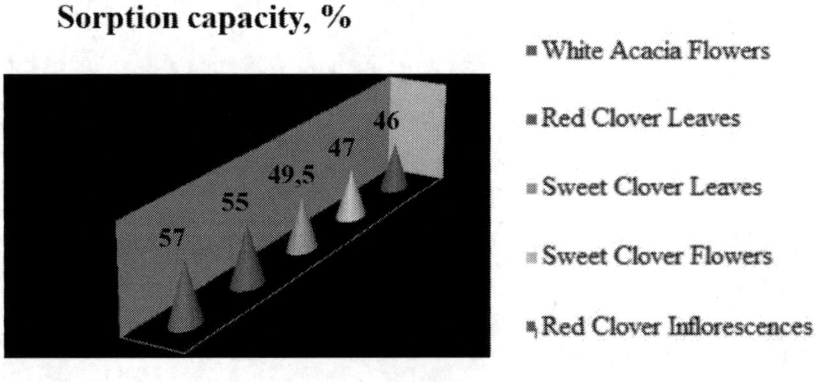

Fig. 2. Sorption capacity of powders of edible wild-growing plants in relation to lead, Pb

Finely ground powders of the FWP modification products showed a high enterosorbent capacity in relation to lead.

The line of efficiency of the studied FWP samples was revealed. A diagram of the sequence of decreasing Pb sorption capacity of FWP is presented: flowers of white acacia → leaves of red meadow clover → leaves of sweet clover → flowers of sweet clover → inflorescences of red clover.

The final results show broad possibilities of promising use for FWP modification products as enriching additives in the formulations of functional flour products.

Preclinical studies were carried out to study the sorption capacity of the developed functional bread. The results of studies of the physiological value are presented in table. 3.

Table 3. Study of the physiological value of functional bread with FWP

Flour products:	Lead content in blood, mg/l at the beginning of the experiment	Lead content in blood, mg/l at the end of the experiment
Control	2,79	2,30
Bread enriched with the FWP powders	2,90	1,47

Laboratory white rats were systematically feed for 28 days with a certain volume of flour products with FWP; compared with the control, they showed almost twice lower lead content in their blood.

Comparative analysis showed a pronounced therapeutic, functional sorbing effect of the developed bread with the addition of FWP.

The core device for the sorption of polysaccharides is their ability to waste and reduce the toxic effect of harmful substances.

The bioprotective and detoxifying properties of FWP, which include dietary fiber in physiologically standard doses, indicate an increase in the nutritional value of the developed functional bread and correspond to the preventive effect.

As powerful enterosorbents, they remove toxins and excess cholesterol from the gastrointestinal tract, regulate the glycemic profile, and are a substrate for the development of normal intestinal microflora.

Clinical and biochemical studies of the effectiveness of the action of functional bread on glycemic indicators (GI) were carried out.

A recruitment of patients was organized with some complaints of low increases in blood sugar levels in volunteers. All women were between the ages of 40 and 55.

In clinical trials, 28 respondents (volunteers), divided into 3 groups, took part; the first group (10 people) tested a new type of bread with the addition of FWP powders; the second group (10 people) - consummated traditional bread without functional additives; the third group (8 people) - according to research were treated with 50 g of pure glucose, measured blood sugar levels of these 8 patients are average.

The first experimental group of volunteers ate bread with a functional additive on an empty stomach in the amount of 145g, the second group ate traditional bread without functional additives in the amount of 140g, the third group or control consumed 50 g of pure glucose.

After consuming the studied bakery products every 15 minutes for the first hour and then every half hour, blood tests were taken to determine the level of sugar.

Based on the results of the analyzes, a graph of the blood sugar level after consumption of the test samples was compiled using a computer (Fig. 3). These graphs were compared with the analysis graph after consumption of 50 g.

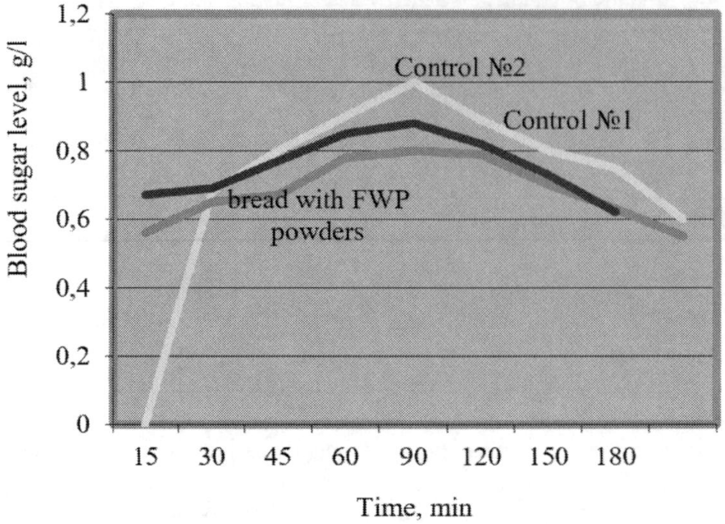

Fig. 3. Comparative assessment of blood sugar level of model objects of bread and glucose depending on time

The change in blood sugar over time after eating functional bread with FWP shows a slower digestion of carbohydrates in the studied foods and a moderate rise and fall in blood sugar compared to control samples.

The obtained research results indicate the functionality of the developed bakery product using FWP and can help patients with diabetes mellitus control sugar levels.

The results of studies of the developed functional bread with FWP allow us to propose it as correcting diabetic blood parameters, a promising enterosorbent in relation to lead compounds and to suggest their high sorbing effect in relation to other toxicants, including heavy metal ions.

Formulations and technologies of bakery products with the addition of FWP modification products in optimal ratios and proportions, characterized by increased nutritional value, have been developed.

Consumption of 100 g of the developed product with FWP powders will satisfy the daily need for dietary fiber by 40%, which indicates the advisability of using powders of edible wild plants in the production of functional bread.

4 Conclusion

Thus, the life quality against the background of a declining level of human health under the existing conditions of ecological relapse, economic crisis can be improved by enriching food with such biological factors as protective, macro- and micronutrient, etc.

The use of food wild plants: flowers of white acacia, red clover leaves, sweet clover leaves, sweet clover flowers, red clover inflorescences in the development of functional flour products, due to its biochemical properties, is promising for regulating the polysaccharide composition of the final product and will allow realizing the biological potential for the purpose of the nation's healthier indicator.

The indicated herbal ingredients can be considered as an essential resource in the food space of the Republic of North Ossetia - Alania for the technology of functional food products, including bread, while significant contributions are not required for their extensive and rational use.

References

1. Atlas of Medicinal Plants of Russia, 890 (All-Russian Scientific Research Institute of Medicinal and Aromatic Plants, 2006)
2. V.I. Vernadsky, *Biosphere*, 8 (1972)
3. M.B. Liga, I.A. Shchetkina, *Quality of life: organizational and managerial aspect*, 5-6 (2011)
4. V.V. Sadovoy, T.V. Shchedrina, V.V. Melentyeva, A.S. Khamitsaeva, Research Journal of Pharmaceutical, Biological and Chemical Sciences, **9(4)**, 1620 (2018)
5. V.V. Sadovoy, I.A. Trubina, T.V. Shchedrina, E.A. Skorbina, Research Journal of Pharmaceutical, Biological and Chemical Sciences, **8(1)**, 8925 (2017)
6. V.V. Sadovoy, M.A. Selimov, T.V. Slichedrina, A.A. Nagdalian, Research Journal of Pharmaceutical, Biological and Chemical Sciences, **7(5)**, 1861 (2016)
7. Sanpin 2.3.2.10 78-01, Hygienic requirements for the safety and nutritional value of food products, Ministry of Health of Russia, 165 (2002)
8. E.A. Strupan, V.S. Kolodyaznaya, O.A. Strupan, Bulletin of the Krasnoyarsk State Agrarian University, **7**, 161 (2020)
9. V.A. Tutelyan, V.A. Knyazhev, Nutrition issues, **3**, 4 (2003)
10. A.M. Ugolev, *Theory of adequate nutrition and trophology*, 8-12 (1991)
11. A.S. Khamitsaeva, *Development of theoretical foundations and development of technologies of flour and meat products using the products of modification of vegetable raw materials*, 71-77 (2011)
12. A.S. Khamitsaeva, Z.A. Khortiev, Prospects for the development of the agroindustrial complex in modern conditions, Materials of the 9th International Scientific and Practical Conference, 411 (2020)
13. A.S. Khamitsaeva, A.S. Tsarueva, Z.A. Khortiev, Kupat production method, Patent No. 266 3069 (2018)
14. N.A. Khloyan, I.A. Ephieva, Bulletin of scientific works of young scientists, graduate students and masters, Mountain State Agrarian University, 126 (2019)

Plutella xylostella (L.) population control in sowings of spring rapeseed using pheromones

Sergey Semerenko, and *Nadezhda* Bushneva*

V.S. Pustovoit All-Russian Research Institute of Oil Crops, 17 Filatova street, 350038, Krasnodar, Russia

Abstract. The diamondback moth *Plutella xylostella* (Linnaeus, 1758) is a common pest of rapeseed and other crops of *Brassicaceae* family. Annual yield losses and costs of pest control worldwide are estimated at $ 4-5 billion. The pest has an increased tendency to develop resistance to insecticides. The use of traps with synthetic sex pheromone is a modern instrumental method of monitoring *P. xylostella*. The use of the mating disruption method will effectively decrease pest numbers and reduce the application of insecticides. In 2017-2020, we researched the pheromone activity and evaluated the mating disruption method in the sowings of spring rapeseed at V.S. Pustovoit All-Russian Research Institute of Oil Crops (VNIIMK) (Krasnodar). We established that *P. xylostella* males were caught in traps with all tested dispenser types. The pheromone showed the greatest activity on the foil-film dispenser (F). The mating disruption method effectively decreased *P. xylostella* population in rapeseed sowing; the disruption effect by the end of crop vegetation was high and reached 82.5 %.

1 Introduction

Analysts predict a 30 % increase in global population (up to 9.2 billion people) by the middle of the 21st century, which will increase the demand for food production by 70 % and require further expansion of modern agriculture [1, 2]. At the same time, this will contribute t the accumulation of insect pests in sowings of agricultural crops, which are currently controlled by spraying insecticides or using genetically modified (GM) crops. However, the insects are rapidly developing resistance. In addition, insecticides and GM crops are prohibited in organic farming [3, 4].

There is therefore a real need for new, effective solutions for pest control. Within the concept of Integrated Pest Management (IPM), the use of pheromones is highly effective, environmentally friendly, and safe. It enables early detection of pests quickly and with a sufficiently high degree of reliability. By creating the same pheromones as the insects and releasing them over crop fields, it is possible to disrupt insect mating, thereby inhibiting insect reproduction and thus pest harmfulness. This approach is considered one of the promising ones [5-7].

* Corresponding author: naducha74@mail.ru

The cultivation of oil crops is considered to be the most profitable direction in the plant production industry. Rapeseed (*Brassica napus* L.) is the main oil crop in the *Brassicaceae* family and is grown all over the world. The seeds of rapeseed are used for the production of edible oil, a source of essential omega-3 and omega-6 fatty acids, as well as biological fuel. The meal is high in protein and is a source of fodder for farm animals. In addition, rapeseed is used as a green manure crop to improve the physical properties of the soil and its fertility [8-10].

The diamondback moth *Plutella xylostella* (Linnaeus, 1758) (Lepidoptera: *Plutellidae*) is a common polyvoltine insect pest of the crops of the *Brassicaceae* family. Global yield losses and control costs associated with *Plutella xylostella* are estimated at about $ 4-5 billion annually. Pest caterpillars can destroy rapeseed seedlings, damage leaves, flowers, the green outer layer of stems, and the developing seeds in pods, thereby significantly reducing crop yield. The pest has an increased tendency to develop resistance to insecticides because of its high reproductive capacity [11-14].

The strategy for effective control of *P. xylostella* on rapeseed is the timely detection of the pest imago. The use of synthetic sex pheromone traps is one of the least time-consuming and most modern instrumental methods of monitoring and predicting pest densities. The use of mating disruption, during which pheromones are released in significant quantities in the air of fields, resulting in a male disorientation effect – the inability to detect females and mate – will effectively decrease pest numbers and reduce insecticide use [15-16].

To improve the control of *P. xylostella* on the sowings of spring rapeseed, we studied the activity of a synthetic pheromone in several preparative forms for the timely detection of the pest and the possibility of decreasing its population numbers, using the mating disruption method.

2 Materials and methods

The material for writing this article was the results of evaluating the effectiveness under field conditions of a synthetic analog of sex pheromone for *P. xylostella* on different types of dispensers produced by AO "Schelkovo Agrohim" designed to catch insects in traps during monitoring, as well as to decrease the pest populations by mating disruption.

In 2017-2020, we carried out the research at V.S. Pustovoit All-Russian Research Institute of Oil Crops (Krasnodar) on the sowings of spring rapeseed of variety Ruyan. We tested the preparative forms in the experiment to study pheromone activity: rubber dispensers in the form of tubes of two colors (white rubber tube – WRT, red rubber tube – RRT) with a diameter of 5 mm and length of 7-8 mm, saturated with a solution of pheromone components in hexane by swelling method; foil-film dispenser (F) – hermetically sealed multi-layer bags of foil-film 2.5 × 2.7 cm in size, containing a solution of pheromone components between the foil-film and inner layers of the film, through which the substances are emitted.

We placed the dispensers in delta traps (1 dispenser per trap) and we used standard inserts with Pestifix glue to fix the attracted moths in the traps. Traps were set 10 days before the expected beginning of *P. xylostella* flight at the rate of 3 traps per 1 ha (one in the center of the plot and two at the edges of the plot). Rubber dispensers were replaced after 30 days, while foil-film dispensers were hung out for the whole season without replacement. We tested each type of dispenser in 3 replications, there were 27 traps in total. We conducted the counting daily until the first moths were caught in the pheromone traps, and then once a week.

We evaluated the mating disruption method on a 1 ha plot of spring rapeseed sowing with a manual distribution of 150 rubber tubes. Each dispenser contained 2 mg of pheromone, and the total dose per 1 has was 300 mg. A sowing area of spring rapeseed of 1 ha without pheromone application was the control. The pheromone traps under record were placed on the experimental and control plots at the rate of 10 pieces per plot. The counting of attracted males was conducted weekly from early May to late July. The disorientation effect was calculated from the number of males attracted by the pheromone traps at the experimental plots compared to the control.

3 Results and discussion

During the last ten years, a high infestation of spring rapeseed sowings by *P. xylostella* (about 5-9 specimens/plant) has been observed in the Krasnodar region. The conditions of the region are favorable for the development of two and sometimes three generations of the pest. Visual observations of the time of phytophage emergence in sowing and its quantity often lead to errors because the beginning of imago flight depends significantly on weather conditions. The optimal temperature for the development of the diamondback moth is 10 °C but it can also develop at 5 °C. The upper temperature limit for moth development is 35-37 °C.

We conducted experiments on studying the pheromone activity for the timely detection of *P. xylostella* and pheromonitoring in 2017-2018. In 2017, the first moths appeared in spring rapeseed sowings at the end of the last ten days of May, when air temperature reached 17.3 °C and relative humidity was 64 %. At the initial stage of the tests, the number of caught pest imago in traps with F and WRT dispensers was at the same level, and the number with RRT dispenser was the lowest (Table 1).

Table 1. Pheromone activity for *Plutella xylostella* (L.) on different dispensers, 2017.

Variant	Number of caught male moths, specimens/trap, by recording date						
	27.05	03.06	11.06	18.06	25.06	02.07	11.07
RRT dispenser	10	54	73	96	199	182	83
WRT dispenser	18	44	40	37	46	32	25
F dispenser	20	82	114	121	212	170	97

Subsequently, the ratio of caught males varied depending on the type of dispenser. The number of moths in traps with RRT dispensers increased from the first ten days of June and did not decrease by the end of 30 days, indicating pheromone activity on this dispenser. In traps with WRT dispensers the number of male diamondback moths gradually decreased, and by the end of the pheromone activity period, the average number of specimens was 37. After replacing the dispenser, there was a slight increase in the number of caught pests, but it was minimal in comparison with other variants. Throughout the season, the pheromone on foil-film dispenser (F) showed the highest activity, with the number of caught moths varying from 82 to 212 specimens/trap, depending on the date of recording (Fig. 1).

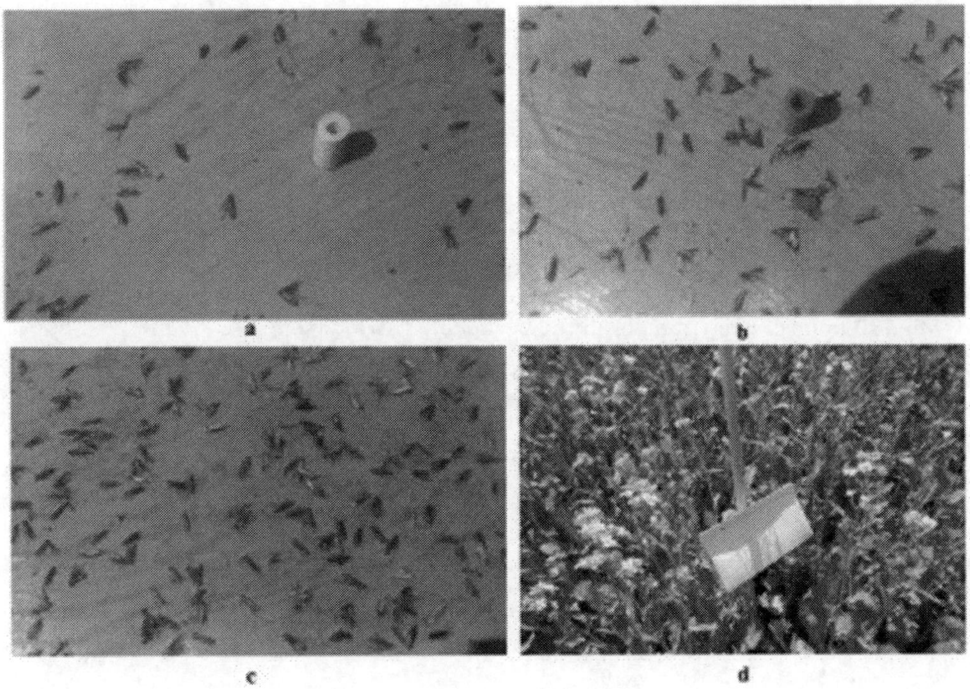

Fig. 1. Activity of pheromone of *Plutella xylostella* (L.) on different dispensers in traps: a) white rubber tube (WRT); b) red rubber tube (RRT); c) foil-film (F); d) traps appearance.

The recordings conducted in 2018 showed that *P. xylostella* moths began flying in spring rapeseed sowing from the second ten-day period of May. Nevertheless, the number of the diamondback moth imago was lower than in the previous year, which is explained by less favorable conditions for reproduction and development of the insect pest. Thus, in 2017, the sum of precipitation during the research period was 309.6 mm, while in 2018 it was half as much and was at the level of 196.8 mm. In 2017, the maximum air temperature by months was: May – 16-19 c; June – 27-32 °C; and July – 30-32 °C. In 2018, the indicators for the same period were higher: 27-33 °C; 40-41 °C; and 37-40 °C, respectively, which are higher than the upper temperature values of diamondback moth development.

The second testing of pheromone activity on different types of dispensers confirmed the preliminary data obtained the year before. The best catching dynamics of male diamondback moths were recorded in traps with F dispensers, the effectiveness of traps with WRT dispensers was lower (Table 2).

Table 2. Pheromone activity for *Plutella xylostella* (L.) on different dispensers, 2018.

Variant	Number of caught male moths, specimens/trap, by recording date						
	18.05	25.05	02.06	09.06	16.06	23.06	02.07
RRT dispenser	8	10	23	48	30	60	7
WRT dispenser	5	10	15	30	20	33	4
F dispenser	14	20	30	80	60	95	7

The results show that all tested dispensers can be used to detect *P. xylostella* in rapeseed sowings in time. Monitoring of the pest is better carried out with the pheromone on F dispenser as its activity remained high throughout the period of use or on the RRT dispenser.

Experiments on *P. xylostella* mating disruption were carried out in 2019-2020. The effectiveness of the mating disruption method with the application of pheromone to the station was evaluated by the almost complete absence of male moths attracted to the pheromone traps in the experiment.

The first moths were caught in traps located on the control plot from the first ten days of May and their number reached an average of 32.0 specimens/trap. Later, the number of pests increased and reached its maximum at the beginning of the first ten days of July with 112.6 specimens/trap. In the experiment, single specimens of the pest appeared at the end of the first ten days of June, their number gradually increased and reached its maximum in the middle of the second ten-day period of July. The resulting mating disruption effect was high (over 60 %), and in the period from the first to the third ten-day period of May it reached the maximum (100 %) (Table 3).

Table 3. Biological effectiveness of mating disruption of *Plutella xylostella* (L.) by pheromones, 2019-2020.

Date of recording	Number of insects, specimens/trap		Effect of mating disruption, %
	control	experiment	
04.05	32.0	0.0	100
10.05	37.0	0.0	100
25.05	49.0	0.0	100
08.06	93.1	0.4	99.6
15.06	61.5	0.9	98.5
20.06	100.1	1.3	98.7
01.07	112.6	2.2	98.0
07.07	101.0	8.0	92.0
14.07	83.0	11.2	86.5
20.07	54.4	9.5	82.5

4 Conclusions

Tests of the synthetic pheromone revealed that *P. xylostella* males were caught in traps with all tested types of dispensers. The pheromone showed the highest activity on the foil-film dispenser (F). The mating disruption method effectively decreased the number of *P. xylostella* males in the rapeseed sowing; the effect of disruption by the end of the crop vegetation was high and reached 82.5 %.

References

1. B. Lanz, S. Dietz, T. Swansone, Ecological Economics **144**, 260 (2018)
2. V. Kvakkestad, A. Sundbye, G.I. Klingena Environmental Science & Policy, **106**, 115 (2020)
3. K. Petkevicius, Ch. Löfstedt, I. Borodina Current Opinion in Biotechnology, **65**, 259 (2020)
4. M. Nasra, J.J. Sendi, S. Moharramipour, A. Zibaee Journal of the Saudi Society of Agricultural Sciences, **16**, 184 (2017)
5. C. Holkenbrink, B.-J. Ding, H.-L. Wang, M.I. Dam, K. Petkevicius, K.R. Kildegaard, L. Wenning, Ch. Sinkwitz, B. Lorántfy, E. Koutsoumpeli, L. França, M. Pires, C. Bernardi, W. Urrutia, A. Mafra-Neto, B. S. Ferreira, D. Raptopoulos, M. Konstantopoulou … I. Borodina Metabolic Engineering, **62**, 312 (2020)
6. J.K. Yuvara, M.N. Andersson, J.A. Corcoran, O. Anderbrant, Ch. Löfstedt Insect Biochemistry and Molecular Biology, **100**, 39 (2018)
7. M. Sumedrea, F.-C. Marin., M. Calinescu, D. Sumed, A. Iorgur Agriculture and Agricultural Science Procedia, **6**, 171 (2015)
8. E. Woźniak, E. Waszkowska, T. Zimny, S. Sowa and T. Twardowski Front. Plant Sci. **10**, 1423 (2019)
9. B. Zhao, R. Xu, Ma. Fengfeng, Y. Li, L. Wang J. of Environmental Management, **184**, 569 (2016)
10. Z. Yang, M. Xu, S. Zheng, J. Nie, J. Gao, Y. Liao and J. J. Xie of Integrative Agriculture, **11**, 655 (2012)
11. W. Jaleel, Sh. Saeed, M.N. Naqqash, U. Sial, M. Ali, S.M .Zaka, Z.M. Sarwar, M. Ishtiaq, M.A., Qayyum, Q.U. Aine, A. Anwar, M. Sarmad, R. Azad, M. Latif, A.Furqan., I. Waqar., K.A. Khan, H.A. Ghramh Saudi Journal of Biological Sciences, **27**, 1 (2020)
12. M. P. Zaluck, Sh. Asad, R. Silva, D. Adamson, Shu-L. Sheng, M. J. Furlong Economic Entomology, **105**, 115 (2012)
13. S.E. Squires, L. Hermanutz, P.L. Dixon Biological Conservation, **142**, 203 (2009)
14. S. Saeed, W. Jaleel, M.N. Naqqash, Q. Saeed., S.M. Zaka, Z.M. Sarwar, M. Ishtiaq, M.A Qayyum., M.U. Sial, Qurat-Ul-Aine, M. Batool, Kh. AliKhan, H.A. Ghramh, M. Hafeez, M.J. Ansari, G.K. Sharma. Journal of Biological Sciences, **26**, 1661 (2019)
15. Ch.E. Miluch, L.M. Dosdall, M. L. Evenden Crop Protection, **45**, 89 (2013)
16. S. Tewari, T.C. Leskey, A.L. Nielsen, J.C. Piñero, C.R. Rodriguez-Saona Integrated Pest Management, 141 (2014)

Carbon balance of sandy loam soils in arid landscapes of the Chechen Republic

Rustam Gakaev*

Chechen State University named after A.A.Kadyrov, Grozny, Russia

Abstract. The spatial variability of the stability of soil aggregates and its relationship with runoff and soil erosion were studied in a semi-arid environment in the field in order to assess the validity of the stability of structures as an indicator of soil erosion in soils of sandy loam ridges. The influence of soil and relief properties on the variability of aggregate stability was also investigated. Significant relationships were found in the number of water droplets required to break down the aggregate, as well as the rate of runoff and erosion. The most significant correlation was found between the number of droplet impacts and the soil organic matter content. The stability of aggregates in the upper soil layer is apparently a valuable indicator of field runoff and inter-season soil erosion of sandy loamy ridges in semi-arid conditions.

1 Introduction

The mountainous landscapes of the region represent the northern slope of the Main Caucasian ridge with its continuous orographic network. Various parent rocks of extremely rugged relief, a variety of mountain heights, the shape and direction of ridges, and various climatic conditions determine the presence of various soil varieties and vegetation formations in the mountains [1]. Mountains characterize a variety of parent rocks, which significantly influence the soil-forming process. On granites, for example, they form, poor in nutrients, acidic; nearby, on limestones with a slightly alkaline reaction - humus, dark-colored soils, rich in nutrients. On clay shales with different chemical composition, loamy multi-colored soils of a certain chemical composition are formed. The concentration of carbon dioxide in the atmosphere has increased from about 280 ppmv to about 405 ppmv today due to the burning of fossil carbon (coal and oil). The question of how to stop or reduce the increase in the concentration of CO2 in the atmosphere is currently being studied. One possibility is to use photosynthesis to release atmospheric CO2 as above-ground biomass carbon (ABC) and in roots. Thus, since 2003, almost 4 billion tons of ABC have been stored in the Chechen Republic from tree planting, which is 25% of its annual emissions from fossil fuels. Trees, which have a limited lifespan, are temporary stores of carbon for centuries only. In temperate climates, afforestation can compete with agriculture, resulting in reduced food production. It investigates a proposal to capture atmospheric carbon dioxide through afforestation in semi-arid regions as inorganic soil carbonate below

* Corresponding author: rustam.geofak@yandex.ru

the earth's surface in the unsaturated zone (USZ) [3]. In semi-arid regions, there is marginal agriculture, but the local flora (trees and shrubs with deep roots) can be supported by 250–600 mm of annual rainfall. In semi-arid regions, the unsaturation zone (USZ), which extends from the land surface down to the water table, is quite thick. Whereas in temperate regions, coniferous and deciduous forests take about 2.8 m deep rooting, in drier areas the plants sink their roots much deeper to reach the water. For example, the roots of the shepherd dog (Boscia albitrunca) and acacia tree can exceed 60 m, while the Israeli tamarisk (Tamarix aphylla) can sink 20 m [5]. Long-term storage of inorganic carbon below the earth's surface has generally been neglected. CO2 released through roots, combined with bacterial oxidation of soil organic matter, can create a partial pressure of CO2 in the soil several orders of magnitude higher than in the overlying ambient atmosphere. Some of this gas diffuses upward and leaves the soil. The rest can interact with soil moisture: high partial pressure of CO2 forms dissolved inorganic carbon (DIC), which later combines with divalent cations and precipitates as soil carbonate within the USZ [4]. With heavy precipitation, this precipitate dissolves. In semi-arid regions with sparse rainfall, precipitated calcite (soil carbonate) can remain stable for thousands of years. The pedogenic carbonate formed at the USZ and bicarbonate (DIC) together contain sequestered inorganic carbon.

2 Calculation of the carbon balance of soil ecosystems

Mountain meadows within the Chechen Republic, as noted above, are represented by subalpine and alpine meadows, which differ from each other in the species composition of vegetation, and, accordingly, in the reserves of phytomass and productivity. The harsh conditions of the alpine belt noticeably affect the vegetation. Low-growing alpine meadows prevail here. The height of the herbage does not exceed 10-15 cm. Large bright flowers of alpine species (bell, violet, gentian, etc.) significantly exceed the size of the leaves. The grass seems to be made up of only flowers. Thanks to this, low-growing high-mountain meadows, the basis of which is forbs, received a figurative expression - alpine carpets [6].

Soils are a significant sink of organic and inorganic carbon, which have not received adequate attention. Monger presents several compelling reasons for this claim. First of all, he believes that for soil carbonate to be a pure scavenger, calcium ions must come from silicate weathering or some other non-carbonate source. Moreover, he notes that it is technically difficult to easily determine the ratio of lithogenic to soil carbonates in soil. Soil inorganic carbon can be of two main sources, differing in their ability to extract and store

$$CO_2 + H_2O \rightarrow H_2CO_3 \rightarrow H^+ + HCO_3^- \quad (1)$$

atmospheric carbon. Lithogenic carbonates, primarily marine limestone, were formed in the distant past and have nothing to do with the modern atmosphere. Pedogenic carbonate is authigenic and forms in the soil profile when dissolved inorganic carbon (DIC) combines with calcium cations in soil moisture and eventually precipitates. Trees and shrubs breathe in atmospheric carbon dioxide for photosynthesis and breathe out some of it through the roots into the unsaturated soil zone. The partial pressure of CO2 in soil gas is much higher than in the surrounding atmosphere. A number of fractionation steps affect the carbon isotope ratios through different steps that start with the expiration of the C3 plants that are prevalent in the study areas and convert CO2 (g) to solid carbonate sediment. Here calcium will mean all other cations (such as iron and magnesium) that carbonate minerals can form. It should be noted that although the isotopic difference between precipitation and the atmosphere is insignificant for 13C (about 1%), for radiocarbon it is 100%. The enclosing sediments do not initially contain 14C, which forms in the upper atmosphere. Chemical equations of the soil dissolution process.

Gas into soil moisture, the formation of carbon dioxide and the precipitation of calcite can be represented in a simplified form as:

Soil CO2 combines with soil moisture to form a carbonic acid solution that rapidly dissociates into H + and HCO3. In this reaction, atmospheric CO2 becomes bicarbonate

$$CaCO_3 + H^+ \rightarrow Ca^{2+} + HCO_3^- \tag{2}$$

(DIC), usually the most abundant groundwater anion.

$$Ca^{2+} + 2HCO_3^- \rightarrow CaCO_{3\downarrow} + CO_{2\uparrow} + H_2O \tag{3}$$

The dissolution reaction is comparable to carbon sequestration since atmospheric CO2 is the source. Carbonic acid formed in the formula. (1), can dissolve existing calcite (mainly limestone) in the soil with the release of calcium ions. HCO3 is in the equation. (2) comes from relict carbonate not associated with modern atmospheric CO2. Therefore, this HCO3 does not capture modern atmospheric CO2.

$$2CO_2 + 3H_2O + CaSiO_3 \rightarrow H_4SiO_4 + 2HCO_3^- + Ca^{2+} \tag{4}$$

The cations are released in the equation. (2) combine with ICE. Through various processes, mainly CO2 degassing or H2O evaporation, secondary calcite is precipitated. Factors limiting the rate of secondary carbonate formation include Ca2 + content and CO2 partial pressure. However, the equation. (3) would mean that atmospheric CO2 is not captured if for every mole of calcite formed, 1 mole of CO2 is returned to the atmosphere. Thus, it was assumed that the formation of soil carbonate is neutral from the point of view of carbon capture and storage in the atmosphere. However, when this reaction occurs in the USZ soil column, which is usually thick in semi-arid regions, only the topmost fraction is in direct contact with the atmosphere. While this can be viewed as an "open system", the rate of gas evolution is diffusion limited. Most of the CO2 emitted by the formula. (3), especially due to precipitation at depth, ingress and mixing with CO2 with a relatively high partial pressure in the USZ soil gas should be expected. In the presence of non-carbonate sources of calcium ions, there will be a net binding of CO2 from the atmosphere. Wollastonite is used here as a typical calcium silicate mineral for purposes of illustration in equation (4).

With the weathering of silicates, all CO2, in contrast to carbonate weathering, is ultimately formed from atmospheric CO2. In this way, a net sequestration of atmospheric CO2 is unambiguously occurring. In nature, there are a variety of sources of calcium ions, in addition to silicate weathering or from relict limestone. Moreover, if the sediment contains other sources of calcium, which can adsorb as a cation on the surface of mixed clay minerals, weathering of feldspar or gypsum, sea spray or rain, then when Ca2 + combines, net binding of atmospheric CO2 can occur. with HCO3 - (equation (1)) for the precipitation of calcite. Therefore, it is important to evaluate sequestration in the field rather than in a theoretical or laboratory setting.

3 Results and Discussions

In the process of photosynthesis, plants absorb carbon and return some of it to the atmosphere through respiration. The carbon that remains in plant tissue is then consumed by animals or added to the soil as bedding when the plants die and decay. The main storage method for carbon in soil is soil organic matter (SOM). SOM is a complex mixture of carbon compounds, consisting of decaying plant and animal tissues, microbes (protozoa, nematodes, fungi and bacteria) and carbon bound to soil minerals. Carbon can be stored in the soil for thousands of years, or it can be quickly released back into the atmosphere. Climatic conditions, natural vegetation, soil structure and drainage all affect the amount and duration of carbon storage. The main sources of the mountain agricultural potential of the North-Eastern Caucasus are the presence in the mountains of 64.5 thousand hectares of arable land, 227.3 thousand hectares of pastures, 41.2 thousand hectares of hayfields, as well as 279.3 thousand hectares of mountain forests with an abundance of wild fruits and berries [11]. However, these huge reserves of agricultural and livestock production are underutilized and developed extremely slowly. For more than 40 post-war years, we have not reached the level of 1940 in terms of the population in the mountains, the availability of arable land and livestock there. Consider in more detail the potential of these reserves and the main ways of using them.

Potential mountain, arable land. As already noted, in the mountains of the region in 1940, 64.5 thousand hectares of arable land were cultivated (together with the household plots of the population), of which about 22 thousand hectares were collective farms. In 1987, the sown area here amounted to 16.4 thousand hectares, including 12.8 thousand hectares in collective farms and 3.6 thousand hectares in the individual sector. Thus, currently 48.1 thousand hectares of arable land are not used in the mountains. This is a considerable reserve for agricultural production [9]. Taking into account the high-rise location of mountain arable lands, according to preliminary calculations, in the future, with the full development of the mountains, the following amount of food products and fodder can be grown here annually. On the 64.5 thousand hectares of arable land available in the mountains, it is possible to produce annually: 43.6 thousand tons of grain and leguminous crops, 66 thousand tons of potatoes, 2.2 thousand tons of oil flax, 3.4 thousand tons of tobacco, 46.2 thousand tons of main and secondary feeds. In addition, on a mountain arable land without large costs, it is possible to lay in the order of the first stage orchards on an area of 10 thousand hectares and nut-bearing 1 thousand hectares. Forage resources in the mountains allow you to keep in the summer: 20 thousand cows, 100 thousand heads of young large stingray, 150 thousand sheep and 6837 horses: in winter: 19.8 thousand heads of young cattle, 100 thousand sheep and 6837 horses [10].

Land cover changes in arid regions are often associated with the displacement of herbaceous life forms by arboreal ones. This shift in vegetation is often a response to extensive grazing.

4 Conclusions

As the forest grows, carbon accumulates in woody tissues and soil organic matter. The net carbon sequestration rate is highest when forests are young, but declines over time. Old forests can capture carbon for a long time, but provide little or no net absorption. When a forest is cleared, the carbon it contains can quickly return to the atmosphere if the wood tissue is burned or converted into short-lived products such as paper. If wood is used for construction or furniture, then these products store carbon throughout their life and act as carbon sinks. A post-harvest approach that reduces waste and converts most wood into long-lived products is an effective strategy to help reduce global atmospheric carbon. However, the net carbon sink in long-lived wood products is still relatively small, so deforestation ultimately contributes to reducing the accumulation of carbon on land. In agricultural systems, the amount and duration of carbon storage is largely determined by how soil resources are managed. Many farming methods have been proposed that can improve carbon storage. Management decisions need to consider the benefits of these various practices, as well as their potential hidden costs. While not discussed here, there may also be direct or indirect monetary costs and benefits for farmers from adopting these methods. Benefits of soil carbon sequestration Removing CO2 from the atmosphere is only one of the significant benefits of improved soil carbon storage. Improved soil and water quality, reduced nutrient loss, reduced soil erosion, increased water savings, and increased yields can result from increased carbon stored in agricultural soils. Management practices that ensure net carbon sinks in soils include the following [12]:

• Conservation tillage minimizes or eliminates manipulation of the soil for growing crops. It includes mulch tillage, in which plant residues remain on the soil surface. These treatments usually reduce soil erosion, improve water-use efficiency, and increase the concentration of carbon in the topsoil. Conservative tillage can also reduce the amount of fossil fuels used by agricultural businesses. It has been calculated that it can capture significant amounts of CO2.

• Cover cropping is the use of crops such as clover and small grains to protect and improve the soil between regular cropping periods. Cover crops improve carbon sequestration by improving soil structure and adding organic matter.

• A crop rotation is a sequence of crops grown in a regularly repeated sequence on the same area of land. This mimics the diversity of natural ecosystems more accurately than intensive monoculture farming methods. Changing the type of crops grown can increase the level of organic matter in the soil. However, the efficiency of the crop rotation depends on the type of crops and the duration of the crop rotation.

On the calcareous soils of the Rocky and Pasture Ridges, which are also characterized by a lower climate humidity, short-legged forbs steppe meadows are also widespread. Tall-herb grass-forb meadows are usually found in small areas on the richest soils of subalpine meadows and the upper strip of the forest belt. In the composition of their herbage, a significant proportion belongs to tall, very valuable forage grains - Georgian bluegrass and long-leaved bluegrass, hedgehog, high ryegrass, etc. [13]. Among legumes, there are meadow rank, balanza vetch, fence peas and arcuate and similar clover. Forbs are represented by hogweed, Caucasian capitate, large-flowered letter, meadow geranium, bunch of anemone, etc. The height of the herbage reaches 150 cm, the yield of hay is 20–25 c / ha [14].

Acknowledgments

The work was carried out within the framework of the state assignment of the Ministry of Science and Higher Education of the Russian Federation (topic No. 075-03-2021-074 / 4).

References

1. A. I. Utkin, D. G. Zamolodchikov, O. V. Honest, Pools of phytomass carbon, biological carbon and soil nitrogen in the forest fund of Russia, 18-34 (2019)
2. A. M. Tarko, Mathematical modeling, 231(2019)
3. R. A. Gakaev, *Comprehensive assessment of the current state of the mountain-forest landscapes of the Chechen Republic and measures for their optimization*, 189-194 (2008)
4. R. A. Gakaev, I. A. Bayrakov, M. I. Bagasheva, *Environmental problems. Looking into the future*, 50-52 (2020)
5. A. M. Vladimirov, F. A. Imanov, *Principles for assessing the ecological flow of rivers*, 225-229 (2019)
6. Ya. P. Molchanova, *Hydrochemical indicators of the state of the environment*, 192 (2019)

7. N. I. Egorova, O. I. Mityakova, *Environmental Innovation and Sustainable Development*, 209-215 (2020)
8. P. P. Tans, D. W. R. Wallace, *Change and Pedogenic Carbonates*, 257-272 (2018)
9. R. T. Watson, I. R. Nobel, *Land Use, Land-use Change, and Forestry*, p. 236 (2018)
10. J. Meckling, L. Hughes, New Political Economy, 88-104 (2020)
11. K. A. Hibbard, S. Archer, D. S. Schimel, D. W. Valentine, Ecology, 82 (2019)
12. H. C. Monger, Soils as generators and sinks of inorganic carbon in geologic time, 27-36 (2020)
13. J. Meckling, L. Hughes, New Political Economy, 88-104 (2019)
14. I. Monasterolo, M. Raberto, Ecological Economics, 228-243 (2020)

Carbon capacity of gray forest soils in the Chechen Republic

Rustam Gakaev*

Chechen State University named after A.A.Kadyrov, 364024, st. Sheripova, 32, Grozny, Russia

Abstract. Common areas of research needed to measure soil carbon include: (1) the impact of forest management, (2) the impact of climate change, and (3) the impact of increased carbon dioxide. The organic matter of the soil has a great influence on the productivity of the site due to its effect on the physical (bulk density, water retention capacity), biological (microbial populations) and chemical (cation exchange capacity) properties of soils. Recently, there have also been claims of the benefits of dead wood for forest ecosystems. However, as with all nutrients, too much carbon in the soil can have negative consequences, especially with regard to nitrogen immobilization.

1 Introduction

Reclamation and reforestation can capture the most carbon per hectare. However, these activities come with high upfront costs, opportunity costs of land-use change (including implications for food security), onerous commitments to sustained use, and no longer generating carbon offset revenues once carbon equilibrium is reached. Carbon equilibrium is the point at which the rate of carbon accumulation equals the rate of carbon emission, so net bonding effectively stops. Marginal land projects that use logging systems to maintain income and employment on project land after carbon equilibrium has been achieved may pose less risk. Rangeland restoration usually has a low sequestration potential per hectare, but potentially has significant environmental benefits. Although the methodologies have not yet been approved, low sequestration levels mean that these actions should be targeted at areas with the highest sequestration potential and low verification costs, resource inputs and opportunity costs. Rangeland restoration activities are also associated with onerous rangeland conservation obligations. The main attention when increasing the SOC should be directed to improving agricultural productivity and the state of land resources [1].

When considering the role of natural forests in the global carbon cycle, it is useful to consider the "color" of carbon. Gray carbon (3) is the carbon stored in fossil fuels (coal, oil and gas deposits in the lithosphere). Green carbon is carbon stored in the biosphere. We call it "green" because carbon is absorbed by plants from the atmosphere through photosynthesis, which depends on the green pigment chlorophyll found in plant leaves (4). Here we use the term green carbon to refer to carbon captured through photosynthesis and stored in natural forests. Natural forests are here defined as forests that have not been

*Corresponding author: rustam.geofak@yandex.ru

disturbed by intensive human land use activities, including commercial logging. Brown carbon is carbon stored in industrial forests. These are forests that are harvested to obtain wood, which is used as a source of raw materials for industrial production processes. There are two types of industrial forests: 1) where the growth of trees occurs at the expense of the natural tree stock and seed fund; and 2) where the trees are planted by humans and usually consist of one tree species, which is very similar to a monoculture culture [4]. Industrial forests are a stock of organic carbon and are therefore part of the biosphere; however, we consider this carbon to be 'brown' rather than 'green' to highlight the fact that industrialized forests are a 'mixture' of green and gray carbon (5). Fossil fuels are consumed and hence gray carbon is emitted in the management of these forestry operations and associated manufacturing processes. Blue carbon refers to the inorganic carbon stored in the atmosphere and oceans. Despite the fact that the ocean contains significant reserves of marine green carbon (6), here we are interested in green carbon stored in terrestrial ecosystems and, in particular, in natural forests [5].

It was only after the United Nations Climate Change Conference in Bali in 2007 that international negotiations focused on the role of natural forests in storing carbon. The Intergovernmental Panel on Climate Change (IPCC) has identified the need for forest-based mitigation analyzes that take into account natural variability, use raw data and provide reliable baseline carbon accounts. In response, we are conducting a series of studies on carbon stocks in pristine natural forests over large geographic areas, including environmental factors at landscape and regional scales. We also look at the carbon impact of land use activities, including commercial logging. The key question we ask in our study is, "How much carbon can natural forests accumulate if they are not disturbed by intense human land-use activities?" This report summarizes the results of case studies in the forests of the Chechen Republic. At the same time, the highest technogenic and chemical pollution is typical for the central part of the republic, where the industrial centers Grozny, Argun, Gudermes are located, which are among the largest environmental pollutants [6].

2 What is the role of forests in the carbon cycle?

It is the biological, ecological and evolutionary aspect that distinguishes "green" carbon in natural forests from "brown" carbon in industrial forests, especially monoculture plantations. Relying on the same poetic license, we call inorganic carbon in the atmosphere (carbon dioxide) and oceans (carbonate) blue carbon. Natural forests are more resilient to climate change and disturbance than plantations because of their genetic, taxonomic and functional biodiversity [7]. This resilience includes fire recovery, pest and disease resistance and recovery, and adaptation to changes in radiation, temperature and water availability (including as a result of global climate change). Although the genetic and taxonomic makeup of forest ecosystems changes over time, natural forests will continue to absorb and store carbon as long as there is enough water and solar radiation for photosynthesis. Green carbon in natural forests is stored in more reliable stocks than in industrial forests, especially at ecological time scales. Carbon stored in industrial forests is more susceptible to losses than carbon stored in natural forests. Industrial forests, especially plantations, have reduced genetic diversity and structural complexity and therefore reduced resistance to pests, diseases and changing climatic conditions [8]. Carbon stocks in forests subject to commercial felling, and in particular in monoculture plantations, will, on average, always be significantly less (~ 40-60% depending on land use intensity and forest type) than carbon stocks in natural, intact forests. forests. The rate of carbon fixation by young regenerating plantations is high, but this does not compensate for the lower carbon stocks in the younger plantations of industrial forests compared to natural forests. Carbon accounts for industrial forests should include carbon emissions associated with land use and associated management, transport and processing activities.

The destruction of the Chechen forest by fires began more than 300 years ago, the command of the tsarist troops destroyed forest vegetation with arson. In place of oak geosystems, transformed herbaceous rows arose, as the fires resulted in significant changes in environmental conditions. "The change in the phytoclimate alone and the changes in the soil climate going along with it can be very noticeable and lead to such changes in the soil regime, which can be completely unfavorable for the restoration of the former vegetation, in our case, beech geosystems [9]. The total overlapping of time fluctuations of extreme values of the functioning of systems and the environment has created conditions conducive to structural changes in the geosystems of the region, which were activated in the process of anthropogenic impact. Under the influence of climate xerophytic factors in the eastern and southeastern parts of the territory, the areas of small-leaved and shrub geosystems are expanding and beech and oak geosystems are being pushed back to higher levels.

Access to relevant environmental field data is critical to accurately accounting for carbon in natural forests, otherwise misleading values will be generated. Models need to be developed and calibrated to reflect the fact that the carbon dynamics of natural forests differs significantly from those of industrial forests, especially monoculture plantations. Among other things, carbon in natural forests has a longer residence time. We demonstrated this by comparing our data with forest carbon accounts calculated from two commonly used sources. In terms of the global biome, Australian forests are classified as temperate forests. The IPCC default values for temperate forests are: carbon stock 217 t C ha-1 of total carbon, 96 t C ha-1 biomass carbon and NPP 7 t C ha-1 year-1. The IPCC default values for total carbon are about one third and for biomass carbon about a quarter of the averages for eucalyptus forests in southeastern Australia and one twentieth of the eucalyptus forests with the highest carbon density in biomass. We estimate that the total carbon stock that can be stored in the 14.5 million hectares of eucalyptus forests in our study region is 9.3 Gt (1) if not disturbed by intense human land use; applying the IPCC defaults will only give 3.1 Gt. The difference in carbon stocks between our estimates and the IPCC default values is a result of our using local data collected from natural forests not disturbed by logging. Thus, our estimates reflect the carbon capacity of natural forests. In heavily disturbed forests, current carbon stocks reflect the history of land use. The difference between the two is called the "carbon sequestration potential" - the maximum amount of carbon that can be sequestered as the forest regains its growth.

Scientifically, green carbon accounting and the protection of natural forests in all countries should be part of an integrated approach to tackling climate change. Ongoing international negotiations focus on reducing emissions from deforestation and forest degradation only in developing countries. Part of the current international negotiations on climate change includes discussions on a technical definition of key terms. "Forest degradation" should be defined to include the impacts of any human land use activity that reduces the carbon stock in the forest landscape relative to its natural carbon capacity. The definition of "forest" should also be revised to recognize the differences between the ecological characteristics of natural forests and industrial forests, especially plantations. These differences include higher biodiversity, ecosystem resilience, and carbon retention times in natural forests.

3 Results and Discussions

Terrestrial ecosystems, especially natural forests, play a critical role in regulating atmospheric greenhouse gases and therefore must be part of a comprehensive response to climate change. Understanding the importance of natural forests in the carbon cycle requires understanding how the Earth functions as a system. Since the Earth is a closed system in terms of chemical elements, the atomic components of the main greenhouse gases (water vapor, carbon dioxide and methane) are not created or destroyed. Rather, they inhabit reservoirs (also called "reserves" or "reservoirs") and move between them in the global carbon and hydrological cycles [10]. As they move between reservoirs, carbon and water change their state: water from liquid turns into gas or ice; and carbon from inorganic gases in the atmosphere to organic compounds in living and dead organisms on land and in the sea, to inorganic substances in the oceans and the earth's crust [14]. A simplified diagram of the global carbon cycle is shown in Figure 1. Estimates of global carbon stocks and fluxes are rough estimates due to lack of data. The annual uptake of carbon (in the form of carbon dioxide) by plants (through photosynthesis) from the atmosphere to plants and the soil reservoir (organic carbon) is about 120 Gt per year. Respiration of living organisms (including humans and their livestock) and oxidative combustion. Units of measurement - Gt of carbon, fluxes - per year. The colors of the arrows correspond to the color of the carbon. * Deforestation is ~ 2 Gt C year-1. 14 As a result of a fire, a little less than 120 Gt per year of reservoir carbon "plants and soil" is emitted into the atmosphere. The biosphere is considered a small carbon sink. Approximately 2 Gt of carbon emissions from plants and soil are associated with deforestation. This contributes to increased carbon loss and reduced carbon uptake by plants and soil. Therefore, over time, the size of the carbon pool in plants and soil decreases. Coal / oil / gas tanks (which provide most of the industry's energy needs) are also being reduced by about 6 Gt per year. If there is less carbon in the plant and soil reservoir, as well as in the coal / oil / gas reservoir, there should be more carbon in the atmosphere and ocean reservoirs. To date, humans have emitted about 300 Gt of gray carbon, but more than 5,000 Gt remain in the lithosphere that could potentially be available for human use. It is estimated that about 2,000 Gt of carbon is currently found in terrestrial ecosystems (plant and soil reservoirs), with about 75 percent of this carbon being stored in natural forest ecosystems. However, about 50 percent of the world's forests have been deforested because current land-based carbon stocks are well below their natural carbon capacity. Carbon cycles between the lithosphere, hydrosphere, atmosphere and biosphere, but its residence time in each of these reservoirs varies considerably. The concentration of carbon in the atmosphere due to the release of carbon from the lithospheric reservoir will remain at a high level for a long time, even if gray carbon emissions are stopped immediately. Two ways of carbon transfer from the atmosphere are: 1) dissolution in river and ocean water and, ultimately, incorporation into carbonate rocks; and 2) carbon uptake by plants and storage in the biosphere. The currents of the terrestrial biosphere and atmosphere occur on a faster time scale and are more under human control than the currents of the hydrosphere. Addressing climate change will require both reducing gray carbon emissions and maximizing carbon uptake by the biosphere. A healthy biosphere provides buffering capacity for changes in the carbon cycle [11].

Various processes allow forests to survive in the face of changing environmental conditions, including climate change. Natural forests are characterized by rich biodiversity at all levels: genetic, taxonomic and ecosystem. This is especially evident when, in addition to the diversity of plants and vertebrates, we consider invertebrates, bacteria and fungi, as well as the vast network of ecological and related interactions that together make up a functioning ecosystem. Genetic diversity within species provides, among other things, the possibility of microevolution, through which populations can quickly adapt to local conditions. High taxonomic diversity provides a pool of species with different life cycles and resistance to niches, from which natural selection can identify the plant or animal most suitable for new conditions. Natural selection affecting the rich biodiversity of natural forests can also lead to the optimization of plant physiological processes and the optimization of trophic interactions in response to environmental changes.

4 Conclusions

Natural forests play an important role in the global carbon cycle. Biomass and soil store about three times more carbon than the atmosphere currently contains, and the annual carbon exchange between the atmosphere and natural forests is 10 times the annual global carbon emissions from people burning fossil fuels. Although natural forests store such a significant amount of carbon, little attention has been paid to the role of forests in tackling climate change by policymakers to date. However, at the 2007 United Nations Climate Change Conference in Bali (UNFCCC 13), the international community recognized the need to reduce emissions from deforestation and forest degradation (REDD) as a vital component of a comprehensive solution to climate change. The importance of achieving REDD for climate change can be appreciated when one considers that about 35 percent of the greenhouse gases stored in the atmosphere are caused by deforestation in the past, and about 18 percent of annual global emissions are the result of ongoing deforestation (IPCC 2007) [13]. Moreover, even when the forest is not cleared to make way for other land uses, there are significant and ongoing carbon dioxide emissions from commercial logging and other land uses that reduce the carbon stored in the ecosystem. Consequently, there is now great interest and a truly urgent need for the development and application of methods to better determine the amount of carbon stored in natural forests and how these pools change as a result of human land use activities.

Acknowledgments

The work was carried out within the framework of the state assignment of the Ministry of Science and Higher Education of the Russian Federation (topic No. 075-03-2021-074 / 4).

References

1. A.I. Utkin, D.G. Zamolodchikov, O. V. Honest, *Pools of phytomass carbon, biological carbon and soil nitrogen in the forest fund of Russia*, 18 (2019)
2. A.M. Tarko, Anthropogenic changes in global biospheric processes. Mathematical modeling, **231** (2019)
3. R.A. Gakaev, *Comprehensive assessment of the current state of the mountain-forest landscapes of the Chechen Republic and measures for their optimization,* 189-194 (2008)
4. R.A. Gakaev, I.A. Bayrakov, M.I. Bagasheva, Environmental problems. Looking into the future, 50 (2020)
5. A.M. Vladimirov, F.A. Imanov, Principles for assessing the ecological flow of rivers, 225 (2019)
6. Ya.P. Molchanova, Hydrochemical indicators of the state of the environment, **192** (2019)
7. N.I. Egorova, O. I. Mityakova, Environmental Innovation and Sustainable Development, 209 (2020)
8. P.P. Tans, D.W.R. Wallace, Change and Pedogenic Carbonates, 257 (2018)
9. R.T. Watson, I.R. Nobel, Land Use, Land-use Change, and Forestry. Cambridge University Press, 236 (2018)
10. J. Meckling, L. Hughes, New Political Economy, 88 (2020)
11. K.A. Hibbard, S. Archer, D.S.Schimel, D.W. Valentine, Biogeochemical changes accompanying woody plant encroachment in a subtropical savanna. Ecology, **82** (2019)

12. H.C. Monger, Soils as generators and sinks of inorganic carbon in geologic time, 27 (2020)
13. J. Meckling, L. Hughes, Protecting Solar: Global Supply Chains and Business Power. New Political Economy, 88 (2019)
14. I. Monasterolo, M. Raberto, Ecological Economics, 228 (2020)

Oxygen regime of rivers in the area of sources of economic impact according to long-term monitoring of the Russian meteorological service

A.V. Goncharov[1,*], *V.A.* Isaev[2] and *N.M.* Yumina[1]

[1]Department of Land Hydrology, Lomonosov Moscow State University, Moscow, Russia
[2]OOO "New cloud technologies", Innopolis, Russia

Abstract. The results of long-term observations of the hydrochemical service of the Russian Meteorological Service on the content of dissolved oxygen in the rivers of Russia are considered. Data on the basins of the Amur, Lena, Ob, Volga for 1988-2007 is analyzed. We compared the average monthly oxygen concentrations above and below the sources of exposure – cities, large industrial enterprises – in total, about 100 paired sections were used. The study showed that in about half of these river sections, sometimes (in 14.8 % of cases) small changes in oxygen concentration (6.5-8.3 %) are observed. Wherein, both a decrease and an increase in oxygen concentration can occur below the sources of anthropogenic impact. The latter is typical for winter time, when the flow of heated wastewater into the river leads to formation of an air-hole and the flow of oxygen from the atmosphere into the water. Small rivers in cities experience the greatest stress; however, in general, for all the rivers under consideration, the anthropogenic change in the oxygen regime of the rivers is small and does not significantly affect the sustainability of the development of the territories.

1 Introduction

The sustainability of development of territories depends not only on the availability of the required volumes of water, but also on its quality, on the state of water bodies. The oxygen regime is an important indicator of the ecological state of rivers, which characterizes the quality of water, the habitat of aquatic organisms, and the trophic status of river ecosystems. The content of oxygen dissolved in water depends on the intensity of atmospheric reaeration of water masses and biological production and destruction processes. Spatio-temporal fluctuations in dissolved oxygen are associated with various geographic and hydrological factors: type of channel processes, geochemical characteristics of landscapes, runoff of allochthonous organic matter and biogenic elements, climate that determines the temperature regime of rivers and duration of the freeze-up period [3]. Many

* Corresponding author: mama15333@mail.ru

of these factors can be subjected to anthropogenic impacts, as a result of which the oxygen regime of rivers also changes.

The concentration of oxygen dissolved in water is an indicator that is mandatory in assessing water quality and monitoring water bodies and streams. The references contain information on dissolved oxygen for individual rivers [4, 8, 9, 11, 12]. The purpose of this paper is to clarify the nature of changes in the oxygen regime of Russian rivers in the area of sources of anthropogenic impact – according to the data of long-term monitoring of the Russian Meteorological Service. Such a study will help to highlight the environmental aspects of the sustainability of large areas development.

2 Materials and Methods of the Studies

The paper compared the oxygen concentrations measured above the sources of economic impact with those obtained below such objects (cities, towns, industrial enterprises). We used the data of the Russian Meteorological Service for 1988-2007. Only those posts at which observations were carried out regularly (more than 9 times per year) were considered; usually they were located in the area of large settlements.

Therefore, paired points were considered (see Figure 1). In the Volga basin there were 23 of them (on Vyatka, Zhizdra, Zusha, Oka, Protva, and etc.), in the Ob basin – 19 (on the rivers Berd, Biya, Kadat, Kazym, Kondoma, Ob, Tom, Chulym, and etc.); in the Lena basin – 15 (Aldan, Amga, Vilyui, Vitim, Kirenga, Lena, Tyry); in the Amur basin – 35 (on the rivers Amur, Argun, Bureya, Zeya, Tunguska, Shilka, and etc.).

The oxygen concentration in water was determined by the Winkler method [7]. The values obtained were converted to percent saturation in order to consider the effect of temperature. This makes it possible to reduce the number of factors affecting oxygen due to such a rather understandable phenomenon as the reaeration of water with a change in temperature and to draw attention to the biological processes of oxygen consumption (release), which are very difficult to consider.

Fig. 1. Location of observation points

The reliability of the difference in oxygen at paired sections was determined by the Mahalonobis distance [5, 10], which is used when there is a significant correlation between the compared parameters. The Mahalanobis distance ($d^2_{q,p}$) less than 3.84 indicates the unreliability of oxygen differences above and below the source of exposure).

3 Study Results and their Discussion

The results are collected in Table 1; the calculations used average long-term values for each month of the year. It can be seen that in most cases (53 out of 92) there was a significant change in oxygen below the sources of anthropogenic impact.

Table 1. The number of paired posts where there are significant changes in oxygen above and below the source of anthropogenic influence (average long-term values for each month are used).

River basins	Number of paired points											
	Number of paired points experiencing the influence	The number of months for which there are significant changes										Total paired points counted
		1	2	3	4	5	6	7	8	9	10	
Volga	15 (65 %)	9	3	2				1				23
Lena	5 (33 %)	2	2						1			15
Ob	16 (84 %)	4	3	3	2	1	2	1				19
Amur	17 (49 %)	4	6	4				1		1	1	35
Total	53 (100 %)	19	14	9	2	1	2	3	1	1	1	92

However, in each pair of points, changes are rarely observed. As can be seen in Table 1, there is not a single pair of points where changes were recorded for more than 10 months a year. On the contrary, in most cases such changes occur only 1 or 2 months a year. So, examining the Lena basin, it can be seen that there is 1 paired point, where the anthropogenic transformation of the oxygen regime takes place over 8 months; at two points – within 2 months and at one – within only one month.

Of all the compared pairs of monthly mean values, significant differences in oxygen were noted in 14.8 % only. Among them, 6.5 % refers to an increase in oxygen below the source of exposure and 8.3 % — to an increase (Table 2). That is, for the majority of monthly averages (85.2 %), there are no significant changes.

Table 2. The frequency of a significant decrease or increase in the concentration of dissolved oxygen in rivers below the sources of exposure

River basins	Total number of observations (months)	Increased oxygen concentration		Decreased oxygen concentration	
		Number of observations (months)	Number of observations (% of the total)	Number of observations (months)	Number of observations (% of the total)
Volga	262	10	3.8	18	6.9
Amur	379	21	5.5	33	8.7
Ob	222	30	13.5	21	9.5
Lena	122	4	3.3	10	8.2
On average			6.5		8.3

According to Table 2, a decrease in oxygen content below the source of exposure is noted only slightly more often than an increase. The latter is usually observed in winter – as a result of heat exposure. In this case, the river (or its section) does not freeze and oxygen enters the water from the atmosphere [2]. In summer, an increase in oxygen concentration below the source of exposure may occur as a result of photosynthesis enhanced by the discharge of nutrients into the river.

In addition to the characteristics discussed above, the magnitude of the decrease or increase in the oxygen concentration at the lower section as compared to the upper sections is of great importance. The analysis shows that such changes are relatively small: the average decrease is 9.3 %, and the increase is 10.4 %. Moreover, we are talking about a small array of all average monthly values (about 14.8 %). For the main part of the paired points (85.2), no significant change in oxygen was found.

Wherein, there are some of them, the oxygen regime of which is undergoing significant anthropogenic transformation. Therefore, further we will consider more detailed information on individual river basins.

In the densely populated Volga basin, changes in oxygen concentration in rivers below the sources of river impact occur in 65 % of the cases considered (Table 1). They are usually episodic and do not last more than 2-3 months a year.

One of the examples of a clear anthropogenic change in the oxygen regime here is the section of Sura river in the area of Penza (Fig. 2), where the differences in the monthly average values of the dissolved oxygen content are reliable for 7 months a year. The average annual decrease in the oxygen content in the lower section as compared to the upper section is 8.5 %. The source of pollution is the half-million city of Penza, which has a number of large industrial enterprises. In addition, the Surskoe reservoir is located above the city of Penza, which also affects the oxygen content in the downstream sections of the river.

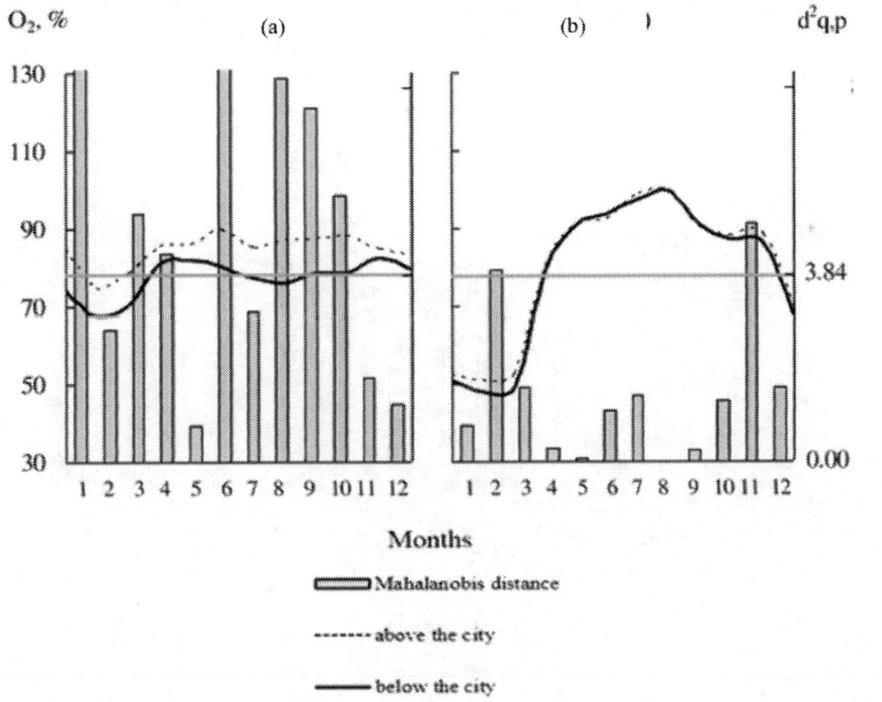

Fig. 2. Oxygen content (% of saturation) in the Sura River in the area of Penza (A) and in the Oka River above and below the city of Murom (B).

In the Oka River, anthropogenic changes in the oxygen regime are rarely recorded. In some months of the winter low-water period, when the water content and the self-cleaning capacity of the river decrease, significant differences are observed in the sections above and below the cities. A significant decrease in the oxygen content was found in the water of the

Oka River below the city of Murom in February and November (Fig. 2), in the city of Ryazan — in May. The opposite situation — an increase in the oxygen content in river waters, is observed in the sections below Orel — in January, Kaluga — in February, which is explained by atmospheric aeration of water in the river sections not covered with ice arising from the warming effect of urban wastewater.

In the Lena basin, a small population and industrial enterprises have almost no effect on the oxygen regime of rivers: it is absent in 67 % of the cases considered (Table 1). Exception is the city of Yakutsk, whose effluents reduce the content of dissolved oxygen in the Lena river for 8 months a year. However, this decline is not large and averages 5 % per year.

In the Ob basin, more often than in other regions, there are changes in the oxygen regime of rivers passing through cities (Tables 1, 2). A significant effect on the oxygen regime of watercourses at various points here can last from 1 to 7 months. Wherein, in the Ob basin, cases of an increase in the oxygen content below cities are more often detected than its decrease (13.5 % versus 9.5 % — see Table 2). An increase is observed, as a rule, in winter, when, due to the influx of warm waters, ice cover does not form on the river and active atmospheric aeration of water occurs (the Ob river in Novosibirsk, the Chulym river in Nazarovo, the Tom river in Kemerovo).

In the Ob river downstream the Novosibirsk hydroelectric power station in winter the air-hole forms, which in different years has a length of 10 to 30 km [6]. Therefore, during the freeze-up period, water saturation with oxygen, when the river passes through the city, significantly increases due to aeration in the open section of the channel (Fig. 3). In the warm season, the differences in the oxygen content downstream the hydroelectric power station are leveled.

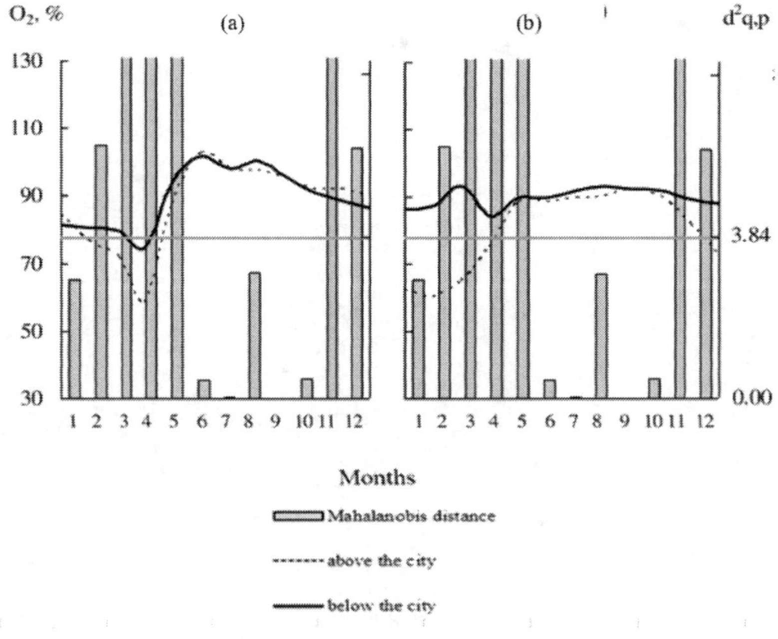

Fig. 3. The degree of water saturation with dissolved oxygen in the Ob river near Novosibirsk (A) and in the Chulym River near Nazarovo (B).

The state regional power plant located in Nazarovo has a powerful warming effect on the Chulym River. As a result, the river does not freeze in winter and atmospheric aeration leads to an increase in the oxygen content below the state regional power plant by 25 % compared to the upstream section of the river (Fig. 3).

In the Ob basin, there are small rivers, the oxygen regime of which has significantly changed: in the city of Tomsk it is the Ushayka River, in the city of Novosibirsk – the Tula River and the Kamenka River. In the latter, the usual saturation of water with oxygen at the level of 27-45 %; an increase (up to 67 %) occurs only during periods of increased water availability (floods).

Table 1 shows that the largest number of paired points (within the considered data set) is located in the Amur basin. At 51 % of the points, there are no reliable changes in oxygen in the rivers. Wherein, there are rivers in the Amur basin, the oxygen regime of which is most strongly transformed: on the Spasovka River near Spassk-Dalny, a significant decrease in oxygen is observed for 10 months a year, on the Arsenyevka River near the city of Arseniev – within 9 months (Fig. 4). Within floods (in April, November), the oxygen content increases as a result of dilution of pollution and increased aeration.

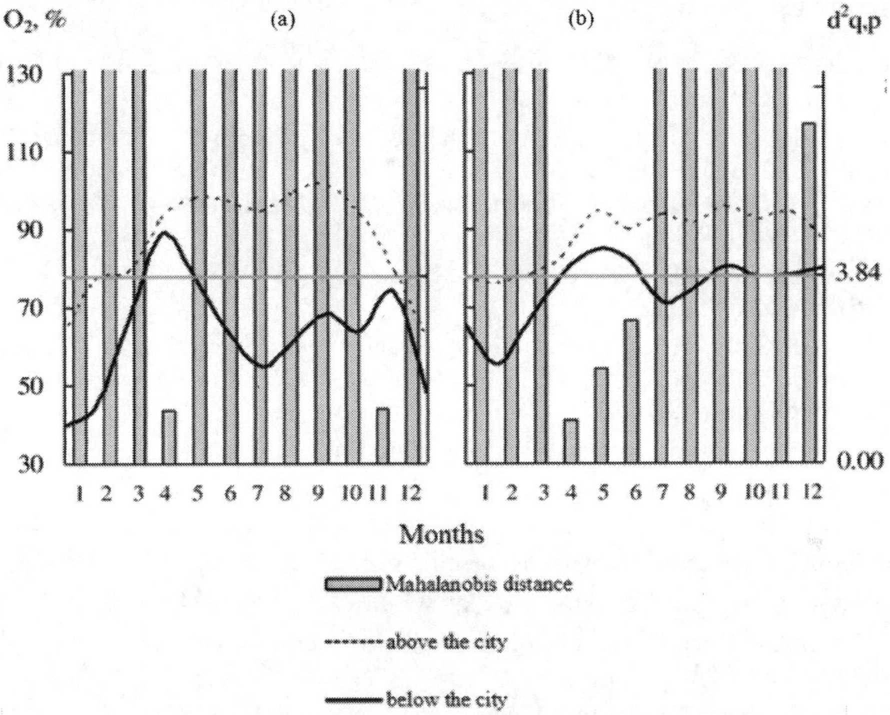

Fig. 4. The degree of saturation of water with dissolved oxygen in the Spasovka River near the city of Spassk-Dalny (A) and in the Arsenyevka River near the city of Arseniev (B).

4 Conclusions

1. Almost half of the considered observation points in the Ob, Volga, Lena and Amur basins are characterized by statistically significant changes in the content of dissolved oxygen below the sources of influence.

2. Wherein, such changes are rarely observed – mainly 1 - 2 months a year and never occur all year round. They are recorded on average for only 14.8 % of all monthly averages; wherein, in 6.5% of cases, there is an increase and in 8.3 % – a decrease in oxygen below the source of exposure.

3. Cases of an increase in dissolved oxygen are mainly related to the winter period. The reason is the influx of warm water, which prevents the establishment of ice, as a result of which there is an active supply of oxygen from the atmosphere.

4. Small rivers of large settlements are most susceptible to anthropogenic transformation of their oxygen regime.

5. The results obtained indicate that, in general, anthropogenic changes in the oxygen regime are not a factor that can have a significant impact on the sustainability of development of Russian territories.

Acknowledgments

The article was prepared as research work of the Department of Land Hydrology, Faculty of Geography, Moscow State University (Topic No. 121051400038-1).

References

1. A.V. Goncharov, M.B. Zaslavskaya, V.A. Isaev, E.E. Lobchenko, I.P. Nichiporova, Geography and natural resources, **3**, 69-76 (2013)
2. A.V. Goncharov, V.A. Isaev, E.E. Lobchenko, I.P. Nichiporova, Proceedings of the III All-Russian conference "Ice and Thermal Processes on Water Bodies of Russia", Onega, Arkhangelsk region, Russia, June 6-11, 2011, Institute of Water Problems, RAS, 286-290 (2011)
3. A.V. Goncharov, V.A. Isaev, E.E. Lobchenko, I.P. Nichiporova, Water Resources, **38(5)**, 564-570 (2011)
4. A.V. Goncharov, E.E. Lobchenko, I.P. Nichiporova, Proceedings of the scientific conference "Modern problems of hydrochemistry and the formation of water quality", Rostov-on-Don, 97-99 (2010)
5. A.A. Isaev, Statistics in meteorology and climatology, 248 (1988)
6. D.V. Kozlov, Fundamentals of Hydrophysics. Tutorial, 246 (2004)
7. RD 52.24.419-2005, Mass concentration of dissolved oxygen in waters. Methods for performing measurements by the iodometric method, 22 (2005)
8. M.A. Sokolova, V.E. Topnikov, Biol. int. water, **97**, 61-66 (1994)
9. L.M. Sorokovikova, Gas regime of the Yenisei in modern conditions, Water resources, **24(1)**, 86-89 (1997)
10. A.V. Khristoforov, Reliability of river flow calculations, 168 (1993)
11. M. Boeder, H. Chang, Journal of Environmental Management, **87(4)**, 567-581 (2008)
12. P. Chambers, J. Culp, N. Glozier, K. Cash, F. Wrona, L. Noton, Environmental Monitoring and Assessment, **113**, 117-141 (2006)

Thermal balance of greenhouse complexes of the V generation in the summer period

Vyacheslav Gulevsky, and *Vyacheslav* Makovichko

Voronezh State Agrarian University named after Emperor Peter the Great, 394087 Voronezh, Russia

Abstract. When growing various crops in greenhouses an important condition for obtaining high yields is compliance with the required parameters of the temperature regime of the air environment. The air conditioning systems currently used in greenhouses of the V generation "Ultra Clima" are equipped with adiabatic panels that cool the air entering the room by evaporation of moisture from their surface. However, in some cases, such systems are not able to support the required values. This is due to the large heat flows entering the greenhouse in the summer. The paper analyzes the temperature balance of the air environment of greenhouses of the V generation, evaluates the main heat flows, determines the operating modes of cooling systems that guarantee sufficient cooling capacity to achieve the required temperatures in the room.

1 Introduction

Modern semi-closed greenhouses of the V generation, equipped with microclimate adjustment systems, provide ample opportunities for their use in a number of regions of our country. The heating, cooling and humidification systems used in them, as well as CO_2 enrichment, are largely able to neutralize the negative impact of the environment. This becomes especially relevant in the development of such a direction as the biologization of agriculture. The relatively high isolation of the air environment of modern greenhouses, the increased pressure created in them, favorable air, solar and temperature-humidity regimes contribute to a significant reduction in the pesticide load when growing various crops.

At the same time, in some cases, during the hot season, the cooling systems that greenhouses are equipped with do not allow the indoor air to cool sufficiently, and therefore, its values may slightly exceed the temperatures set by the regulations, the values of which for some crops are shown in Table 1.

Table 1. Recommended air temperatures in greenhouses.

Culture	Air temperature before fruiting, °C	Air temperature during the fruiting period, °C
Cucumber	25 - 26	21 - 23
Tomato	24 - 26	20 - 22
The head lettuce	20 - 23	18 - 20
Radish	20 - 22	18
Peking cabbage	20	20
Dill, spinach	17 - 18	18 - 22

2 Materials and Methods

In order to determine the required cooling capacity of the air conditioning system, we will calculate the thermal balance of the greenhouse for the summer period. We will consider the most critical period of daylight in terms of the number of external heat accesses - from about 12 to 15 hours. The volume of air supplied to the greenhouse is limited by two factors: the air velocity in the plant placement area, which should be within 0.25 - 0.50 m/s and in rare cases may increase for a short time, but not more than 1m /s, as well as the level of CO_2 concentration, which with a large volume will tend to the street value of 440 ppm, whereas higher values are required for full photosynthesis.

Heating of the air environment inside the greenhouse is due to the receipt of energy from solar radiation, as well as from the lighting system. Cooling is carried out by a ventilation (air conditioning) system and heat transfer through the walls as long as the air temperature in the greenhouse is above street temperature. However, with a decrease in temperature inside the greenhouse below the outdoor heat flow unfolds and becomes positive - heats the greenhouse. Let's consider them in detail.

The value of heat gain with solar radiation can be determined by the formula:

$$Q_{pr} = K_{pr} Q_{sr} S_t \tag{1}$$

where Q_{pr} is the penetrating solar radiation, J; K_{pr} is the permeability coefficient of solar radiation. Its value depends on the time of year, geographical location, design features, surface cleanliness and is given in Table 2 for a two-layer glass clean surface; Q_{sr} - solar radiation on a horizontal surface, MJ/m2. Depends on the time of year and latitude. The values, in accordance with SN and R 23-01-99, are given in Table 2. S_t – greenhouse area, m2.

Table 2. Solar radiation per horizontal surface per month. The permeability coefficient of solar radiation.

Month	K_{pr}	Geographic latitude, degrees north latitude							
		Q_{sr}, MJ/м^2							
		40°	44°	48°	52°	56°	60°	64°	68°
May	0,8	893	872	862	850	840	825	824	809
June	0,8	897	889	881	880	873	877	864	865
July	0,8	891	886	877	882	875	856	855	889
August	0,75	803	768	736	719	695	660	641	639

The thermal load from the lighting system is defined as:

$$Q_{osv} = q_{osv} S_t \tag{2}$$

where Q_{osv} is the thermal load from the lighting system, J; q_{osv} is the value of specific heat flows per 1m^2 of the calculated room, W/m^2. In the case of incandescent lamps, its value is assumed to be 25 W/m^2, for fluorescent lamps – 10 W/m^2.

Negative constant (independent of temperature differences and other factors) heat flows include the heat flow supplied to the greenhouse room by the ventilation system - Q_{vent}. According to its energy component, it is really negative, because in the hot season it enters the greenhouse with a lower temperature, cooling it. We can find its value by the formula:

$$Q_{vent} = C\rho G_{vh} t_n \tag{3}$$

where C is the heat capacity of the air, J/kg· °C ; ρ is the air density, kg/m^3; G_{vh} is the air flow rate, m^3/s; t_n is the temperature of the outdoor (street) air, °C.

Part of the heat is removed from the room together with the air removed from it. Its quantity can be estimated by the formula:

$$Q_{out} = C\rho G_{out} t_v \tag{4}$$

G_{out} - the flow rate of air removed from the greenhouse through the ventilation roof vents, m^3/s; t_v - the temperature of the air removed from the greenhouse, °C.

Depending on the temperature difference inside the greenhouse and outside, there is an alternating heat flow that enters or exits the room due to heat transfer through the walls enclosing the greenhouse. In a stationary thermal regime , its value is determined by the formula:

$$Q_{ogr} = k_{ogr} F_{ogr} \left(t_n - t_v\right) \tag{5}$$

k_{ogr} is the heat transmission coefficient, W/(m2· °C). For double glazing, it is equal to 0.46 W/(m2· °C), and for triple glazing - 0.57 W/(m2· °C); F_{ogr} is the area of the fence, m^2.

Thus, in the general case, we obtain the equation of the energy balance in the greenhouse in the form:

$$Q_{pr} + Q_{osv} + Q_{vent} + Q_{out} + Q_{ogr} = 0 \tag{6}$$

3 Results and Discussion

Solving the last equation with respect to t_v, it is possible to determine the dependence of the required amount of G_{vh} air entering the room on its temperature. The calculation results are shown in Fig. 1.

It is known that adiabatic panels, which are equipped with cooling systems of modern Generation greenhouses, are able to cool the air entering the room by no more than 8-10 °C. As a result of calculations, it was found that with such a cooling depth, in order to achieve regulated temperatures in the greenhouse, in some cases it is necessary to provide an increased supply of cooled air in a volume of more than 250 thousand m^3/hour. In turn, this can lead to air velocity values in the plant growth zone above the maximum permissible.

Figure 2 shows the characteristics of a plate water vapor cooler in the form of a phase plane, which illustrates the nature of the dependence of air flow on temperature.

Each point of the phase plane carries information about the characteristics of the cooler. If the point lies on the graph of the function, then the temperature in the cooled object will be maintained at the specified level. If the point lies above the curve, then a lower temperature will be maintained in the room, which can be found from the balance equation.

In addition, air consumption is limited by several factors. Firstly, the minimum value of the air flow G_{min} is due to the minimum air exchange (this is the mass amount of air supplied to the room per unit of time to ensure the regulatory level of pollution of its air environment). Secondly, the maximum value of the G_{max} air flow is due to the maximum permissible air velocity in the plant placement area and the CO_2 concentration.

In the case of using a direct water vapor cooler, the temperature at the outlet of the cooler cannot be lower than the temperature of the wet t_{mt} thermometer.

Thus, the characteristics of the air conditioner at a given indoor temperature tv and a given outdoor temperature tn must satisfy the coordinates of the points located within the curved trapezoid ABCD

https://doi.org/10.1051/bioconf/20224202008

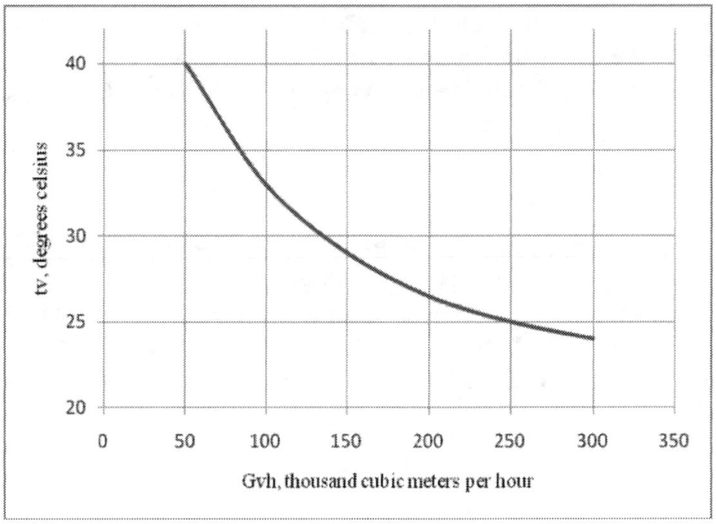

Fig. 1. The temperature regime of the greenhouse.

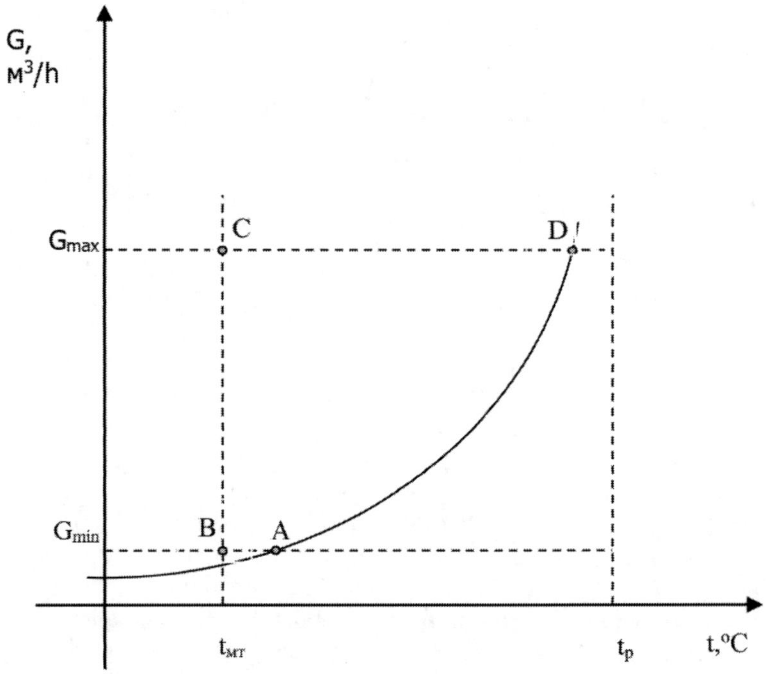

Fig. 2. The temperature regime of the greenhouse.

4 Conclusion

Thus, we come to the conclusion that in order to achieve regulated temperature values in the greenhouse in the hot season, it is necessary to use more efficient plate water-evaporating air coolers. Their use will allow to increase the depth of cooling of the supplied air flow, and therefore, it will be possible to reduce the volume of its supply, which will allow to withstand the regulated values of air velocity in the area of plant placement its CO_2 values.

References

1. V.V. Klimov *Greenhouse equipment for subsidiary and private farms*, 96 (1992)
2. Yu. Homutsky, Climate World, **98**, 78 (2016)
3. E.A. Aliyev, *Technology of cultivation of vegetable cults and mushrooms in protected ground* , 351 (1987)
4. V.A. Bryzgalov, *Vegetable growing of protected soil*, 351 (1995)
5. A.A. Autko, *Vegetable growing of protected soil*, 320 (2006)
6. R.A. Gish, KubGAU Scientific Journal, **123(09)** (2016)
7. *Methodological recommendations on the technological design of greenhouses and greenhouse plants for growing vegetables and seedlings RD-APK 1.10.09.01-14.* Moscow (2014)
8. V.P. Shishkin, Greenhouses of Russia, 2, 15-20 (2016)
9. V.A. Gulevsky, Russian Journal of Building Construction and Architecture, **2(42),** 6 (2019)
10. V.P. Shatsky, Russian Journal of Building Construction and Architecture, **4(36)**, 70 (2017)

Interactive methods of teaching mathematics, contributing to the formation of competencies in students of the field of training Agronomy format

Pavel Kiyko[1,*], and *Natalia* Shchukina[1]

Omsk SAU, 1 Institutskaya Square, 1, 644008 Omsk, Russia

Abstract. The article outlines aspects of the formation of general professional competencies (GPC-1) through the study of mathematical disciplines; reveals the features of the methodology for the application of tasks (tasks of profile content, task case, task-situation), which are the main means of assessing the formation of the corresponding indicators of achievement of competence: ID-1, ID-2, ID-3, etc. The work analyzed educational publications in mathematical disciplines for the presence of tasks for an objective assessment of the level of formation of general professional competence (GPC-1), the problem of developing the content of criteria and assessment tools for formation of general professional competencies. The authors provide examples of tasks in the disciplines of Higher Mathematics and Mathematics and Mathematical Statistics, designed to form skills in identifying and defining a problem, choosing research methods, analyzing, interpreting the result and transferring fundamental knowledge into the language of professional activity.

1 Introduction

In the existing FSES HE the mandatory requirements to results of development of sample principal educational program (SPEP) of a bachelor degree are declared, which is a combination of universal cultural and professional competences, which in turn should ensure the professional competence of the graduate, including future AIC expert.

1.1 Problem Statement

It is obvious that competencies are not formed by a single discipline or course, they are gradually formed and improved from discipline to discipline, from course to course. This statement applies to all competencies, including general professional ones, which are formed throughout education and further improved during professional activity. In that way considered under this article, GPC-1 (Capable of solving typical problems of professional activities on the basis of knowledge of the fundamental laws of mathematical, natural

* Corresponding author: pv.kiyko@omgau.org

scientific and general professional disciplines with the use of information and communication technologies) for major 35.03.04 Agronomy, according to the working curriculum for the bachelor begins its formation in the process of disciplines development – unit 1, taught by general education departments (of Mathematical Sciences), continues to evolve and develop with the help of disciplines taught by graduate departments, (in the cycle of general and professional disciplines), then in the process of educational practice – unit 2, monitoring and analysis of formation which is carried out in preparation for pass and the pass of state examination – unit 3.
Indicators of competence achievement of GPC-1, which are formed during the development of mathematical disciplines, are set in the SPEP:

AI-1 demonstrates knowledge of the basic laws of mathematical disciplines necessary for solving typical problems in the field of agronomy;

AI-2 uses knowledge of the basic laws of mathematical sciences to solve standard problems in agronomy;

AI-3 uses information and communication technologies to solve typical problems in the field of agronomy.

2 Research issues

2.1 Purpose of research

The purpose of this study is to develop the content of criteria and assessment tools for an objective assessment of the level of general professional competence formation in a student based on the diagnosis of relevant indicators at different stages of mathematical disciplines mastering on the example of the discipline "B1.O.09 Mathematics and Mathematical Statistics".

The main means of assessing the formation of the relevant indicators of GPC-1 are problems. It is important to note that students should be offered a set of problems in which the mental activity of students increases, and it increases when the thought comes across something new, for example, a problem condition with a profile context, an unusually formulated problem requirement, a non-standard representation of mathematical data, etc. Such conditions are met by profile content problems, case tasks, for which there is no ready-made template, considered in textbooks or in the classroom. The importance of problems in the practice of teaching with mathematical disciplines is quite difficult to overestimate. It is important for the student to understand not just how to solve certain mathematical problems, but also to master such techniques that will develop thinking and help to carry out the transfer of mathematical educational results for solving problems in professional activities. The result of the development of cognitive, mental activity is the mental independence of the individual, which is manifested in solving various kinds of problems and tasks.

3 Research methods

Consider the content of the discipline B1.O.09 Mathematics and Mathematical Statistics module "Statistical estimates of distribution parameters". The set of tasks for assessing the formedness of GPC is represented by the following types:

- problem with mathematical content.

Before analyzing and solving problems of profile content and case tasks, it is important to work out the student digestion of basic concepts and formulas, such as: variational series, numerical characteristics, etc., which are the foundation for mastering this topic. The content of such problems is formulated exclusively in mathematical language. It has one correct solution.

- problem with profile content.

This type of problems is aimed at the formation of techniques aimed at transferring mathematical knowledge and skills to the language of professional activity. The content of such problems is formulated exclusively in the language of future professional activity. It has at least one correct solution.

- case-task (problem-situation).

This type of problems is aimed not only at the formation of techniques aimed at transferring mathematical knowledge and skills to the language of professional activity, but also at assessing the formation of GPC. The content and requirements of such problems is formulated exclusively in the language of future professional activity. It has at least one correct solution, but only one of them is optimal.

Considering indicators of achievement of the competence of GPC-1, we present examples of these problem types on the above integrated unit and present levels of GPC development in accordance with the indicators.

It is obvious that the GPC formation is a level one. Problems with mathematical content are aimed at GPC forming at a critical and/or acceptable level. Problems with profile content are associated with the GPC formation at an acceptable and/or optimal level. Case tasks assume the GPC development at an optimal and/or high level.

Problem with mathematical content.

Problem 1.1. As a result of the tests, the value of X took several values:

8, 8, 7, 9, 9, 9, 10, 10, 10, 8, 9, 10, 12 12, 12, 10, 14, 9, 7, 7, 12, 14, 12, 12, 10.

The following is required:

1) to create a discrete variation series with the corresponding frequencies and relative frequencies. To build a relative frequencies polygon (AI-1);

2) write down the formulas for calculating the average variance, and mean square deviation of the sample population (AI-1);

3) to calculate the average variance, and mean square deviation of the sample population (AI-2);

4) to calculate the mode, median, coefficient of variation, evaluations of mathematical expectation, variance, and mean square deviation (AI-2);

5) to determine the confidence interval in which the mathematical expectation and the mean square deviation (AI-2) are located with a reliability of 0.99;

6) perform a verification of the implementation of cl. 2 and cl. 3. using ICT (AI-3).

Problem 1.2. As a result of the test, the random variable X took a number of values:

4.6 5.0 4.5 4.7 4.6 5.0 6.0 6.2 6.4 4.8 4.9 4.7 4.5 4.9
5.1 6.4 5.9 5.8 5.0 6.4 4.8 4.4 5.6 5.5 4.7 4.8 5.0 6.2 6.1
6.3 4.5 6.2 6.0 4.8 4.9 6.0 6.4 6.2 5.8 5.9 4.9 5.2 5.1 4.5
5.1 5.0 5.5 5.6 6.0 6.4

The following is required:

1) to create interval series, build a histogram of relative frequency density and empirical density curve (AI 1);

2) to calculate the average variance, and mean square deviation of the sample population (AI-1,2);

3) to calculate the mode, median, coefficient of variation, evaluations of mathematical expectation, variance, and mean square deviation (AI 1,2);

4) to calculate the average error of the sample average and the boundaries in which the average of the general population is located with a probability of 0.95 (AI 1,2).

5) perform a verification of the implementation of cl. 2 and cl. 3. using ICT (AI-3).

Profile content problem.

The influence of sowing methods on the development and productivity of carrots of the K variety is studied. Carrots were sown in two ways:

Method 1 (20 plots) This is the average weight of one carrot:

120, 135, 188, 176, 176, 135, 188, 188, 191, 190, 189, 189, 176, 176, 135, 184, 184, 188,190, 190

Method 2 (20 plots) This is the average weight of one carrot:

171, 176, 189, 178,178, 183, 125, 189, 190, 176, 190, 189, 176, 178, 176, 183, 183, 188, 190, 125

Problem 2.1.

1. Form the purposes and objectives of the study (AI-2),
2. Select the mathematical objects that correspond to the purpose of the study (AI 1),
3. Write down the mathematical formulas and laws necessary for the solution (AI 1)
4. Form a professional problem in mathematical language (AI 2).

Problem 2.2.

1. Build discrete distributions for each method of root crops planting (AI 1).
2. Calculate the sample characteristics for each distribution (AI 1,2),
3. Calculate the indicators that characterize the variability of the studied trait (AI 1,2).

Problem 2.3.

1. Build interval distributions for each method of root crops planting (AI 1).
2. Calculate the sample characteristics for each distribution (AI 1,2),
3. Calculate the indicators that characterize the variability of the studied trait (AI 1,2)
4. Perform a comparative analysis of the found indicators (AI 2) and perform a performance check using ICT (AI-3).
5. Reveal the influence of sowing methods on the development and productivity of carrots of the K variety and draw conclusions, give recommendations for sowing this variety of carrots (AI 2)

Case task. (AI 1) (AI 2) (AI 3)

The agronomist of the farm solves the problem of increasing the yield of carrots of the K variety. It is known that sowing carrots can be carried out in two ways: method 1 (wide-row single-line method), method 2 (four-line tape method). This farm could use any method of sowing carrots of this variety (financial costs are the same). The agronomist studies statistical data that characterize the influence of sowing methods on the productivity of carrots of the K variety on land plots that are similar in basic characteristics to the land fund of this farm. The statistical data of the methods are given above.

Analyze the situation. Come up with a technological solution that would increase the yield of carrots. Solution development should be based on knowledge of sample characteristics and statistical estimates of distribution parameters.

The correspondence of the levels of formedness of GPC-1 and indicators of competence achievement are presented in Table 1.

Table 1. The correspondence of the levels of formedness and indicators of achievement.

GPC-1 levels	Indicators of competence achievement	Number of points scored out of a maximum of 10
critical	AI-1 demonstrates knowledge of some sample characteristics necessary for solving typical problems in the field of agronomy, there is no independence AI-2 calculates some sample characteristics necessary for solving standard problems in agronomy, there is no independence AI-3 does not use information and communication technologies when searching for missing information and calculating some sample characteristics, there is no independence	0-4 (less than 50%)
acceptable	AI-1 demonstrates knowledge of the sample characteristics necessary to solve typical problems in the field of agronomy, but only based on recommendations AI-2 calculates sample characteristics for solving standard problems in agronomy, but only based on recommendations AI-3 uses information and communication technologies to search for missing information, based on recommendations	5-6 (more than 50 and less than 70%)
optimal	AI-1 demonstrates knowledge of sample characteristics and statistical estimates necessary for solving typical problems in the field of agronomy (pedagogical support of the teacher is possible) AI-2 calculates sample characteristics and statistical estimates for solving standard problems in agronomy (teacher support is possible) AI-3 uses information and communication technologies when searching for missing information (pedagogical support of the teacher is possible)	7-8 (more than 70 and less than 90%)
high	AI-1 demonstrates knowledge of sample characteristics and statistical estimates necessary for solving typical problems in the field of agronomy AI-2 independently calculates sample characteristics and statistical estimates for solving standard problems in agronomy AI-3 independently applies information and communication technologies when searching for missing information and calculating sample characteristics and statistical estimates	9-10 (more than 90%)

Reference to problems of this nature is due to the lack of such in the educational literature. Considering the textbooks and manuals recommended for the discipline B1.O.09 Mathematics and Mathematical Statistics for major 35.03.04 Agronomy, we noticed that in textbooks either missing the "Mathematical Statistics", or there are insufficient number of problems to solve, and relevant content problems and case tasks are missing completely, which does not allow enough time to work on formation of professional competences. Let's analyze the module "Statistical evaluation of distribution parameters".

The first edition of "Mathematics and Mathematical Statistics" by V.F. Komogortsev [3] contains chapter 6. Fundamentals of Mathematical Statistics §2. Point and interval evaluations of numerical characteristics of the general population [3, p. 151]. For an independent solution, students are offered 3 problems of mathematical content, there are no problems of profile content and case tasks.

Exercises

1. In order to study the size X of some similar products produced by the plant, 50 products were randomly selected. Their size distribution (statistical distribution of the sample) is as follows (Table 2.):

Table 2. Statistical distribution of products.

x_i (sm)	107.8- -108.0	108.0- -108.2	108.2- -108.4	108.4- -108.6	108.6- -108.8	108.8- -109.0
n_i	1	4	16	18	8	3

To find a confidence interval that estimates with reliability γ=0.95 the average size of products produced by the plant.

Answer: (108.39; 108.51).

2. When determining a certain value a experimentally, 5 repeated experiments were carried out, which gave the following results:

0.640; 0.652; 0.656; 0.664; 0.670

a) What is the probability that the true value a of the measured value differs from the average result of the measurements by no more than 0.01?

b) In what confidence interval $(x_6-\delta;\ x_6+\delta)$ with reliability γ=0.99 is the desired value a?

Answer: a) $\gamma\approx$0.87; b) (0.633; 0.680).

Quality control of a large batch of products is carried out selectively. What should be the minimum sample size so that with the reliability of γ=0.99 it can be stated that the deviation of the average service life of the product in the sample differs from the average service life in the entire test batch by no more than 3 hours (in one direction or the other)? According to the results of the preliminary (trial) sample, $s_в = 10$ hour was obtained.

Answer: 74. [3, pp. 159-160].

The second edition of the author Berezina, N.A. Mathematics [4] does not contain a section devoted to mathematical statistics.

The third edition of the author Nazarov A.I. The course of mathematics for non-mathematical specialties and bachelor's degrees [5, p. 533] contains section 3 Estimation of distribution parameters. The section is illustrated with examples, but the problems offered to students for solving are not included in the publication.

The fourth edition of the author Shipachev V.S. Higher Mathematics [6] as well as the edition of Berezina N.A. Mathematics does not contain the module "Statistical estimates of distribution parameters".

In such a way, we were convinced of the insufficient number of problems that contribute to the formation of general professional competencies through mathematical disciplines, which is the object of this study.

4 Conclusions

In this article, we have considered the possibility of applying profile content problems and case tasks in the study of the discipline B1.O.09 Mathematics and Mathematical Statistics by future farmers. It should be noted that in the educational publications available today, tasks of this nature are not reflected or contained in insufficient quantity. For the formation of GPC-1, it is not enough to learn only to determine the necessary formula, perform calculations, it is important to be able to formulate the problem, choose research methods, analyze the result, be able to interpret the data, justify the conclusions obtained, and make predictions. In our opinion, this approach will not only form the necessary mathematical apparatus for students but will also increase the strength and awareness of knowledge, which will contribute to the meaningful use of mathematical apparatus in professional activities.

References

1. FSES for Bachelor's Degree Programs: 35.03.04 Agronomy, http://fgosvo.ru/
2. O.B. Smirnova, N.V. Shchukina, Collection of tasks on probability theory and mathematical statistics (Omsk: Publishing house of FSEI HPE OmSAU) 64 (2006).
3. V.F. Komogortsev, Mathematics and mathematical statistics: textbook for bachelors in major 35.03.03, 35.03.04, 35.03.07 (Bryansk: Publishing House of the Bryansk SAU) 164 (2019)
4. N.A. Berezina, E.L. Maxina, Mathematics (Moscow: INFRA-M) 175 (2013)
5. A.I. Nazarov, I.A. Nazarov, Mathematics course for non-mathematical specialties and bachelor's degree courses (SPb.; Moscow; Krasnodar: Lan) 576 (2013)
6. V.S. Shipachev, Higher Mathematics (Moscow: SC INFRA-M) 479 (2021)
7. P.V. Kiyko, Economic and mathematical methods and models (Berlin: Direct-Media) 122 (2016)
8. P.V. Kiyko, N.V. Shchukina, Electronic scientific and methodological journal of Omsk SAU, **4(7)**, 22 (2016)
9. P.V. Kiyko, N.V. Shchukina, International Journal of Economic Research, **14(7)**, 59-75 (2017)
10. M. I. Grabar, K. A. Krasnyanskaya, Application of mathematical statistics in pedagogical research. Nonparametric methods (M.: Pedagogy) 71 (1977)
11. O.B. Smirnova, N.V. Shchukina, Tasks for typical calculations in mathematical disciplines (Berlin: Direct-Media) 146 (2016)
12. O. Korchinskaya, I. Ivanova, N. Shchukina, M. Mendziv, Proceedings of the International Scientific Conference The Fifth Technological Order: Prospects for the Development and Modernization of the Russian Agro-Industrial Sector (2019)
13. G.V. Redreev, P.V. Kiyko, N.V. Shchukina, 2019 International Multi-Conference on Industrial Engineering and Modern Technologies (FarEastCon) (Vladivostok, Russia) 01-04 (2019)

Influence of mineral and organic artificial substrates on productivity of covered ground tomato under additional lighting

P. M. Isaeva[1,*], *L. Sh.* Daulakova[1], *N. L.* Adaev[1,2], *A. G.* Amaeva[1,2], and *Zh. Kh.* Aiskhanova[1]

[1]Funded Institution of Higher Education Chechen State University named after A.A. Kadyrov, Grozny, Russia
[2]Research Institution Chechen Research Institute of Agriculture, Grozny, Russia

Abstract. The article presents the results of studying the reaction of tomato plants to an artificial substrate on which they are grown with all other things being equal — lighting, nutrition, temperature, and etc. Among the studied parameters, the plants reacted to the greatest extent with an average daily growth, filling of fruits, which mainly influenced the yield of tomatoes.

1 Introduction

Modern greenhouse vegetable growing is one of the most innovative and knowledge-intensive branches of the Russian agrarian and industrial complex. Industrial vegetable-growing complexes are created with the purpose of highly profitable production by saving fuel and energy resources, increasing yields, reducing production costs, and efficient use of investments [1].

The possibility to regulate a wide range of plant growth and development factors significantly reduces the climatic risks in soil production and makes it possible to maximize the genetic potential of varieties and hybrids of various vegetables, including tomatoes, planted by breeders. Controlled conditions include temperature, nutrient, gas, light and many other modes. Recently, special attention has been paid to the study of the effect of artificial substrates of various origins on the further ability to increase the productivity of vegetables in greenhouse and vegetable growing complexes. The most often opposed to each other are substrates of mineral and organic origin, the characteristics of which differ, namely, in density, structure, homogeneity, moisture capacity, thermal conductivity, and etc.

The high degree of vulnerability of greenhouse plants to disruptions in mineral nutrition is associated with the low stability of nutrient solutions (buffering). The buffering capacity of solutions is due to presence of buffer systems in them, represented by a mixture of weak acids, including carbonic acid [2].

* Corresponding author: fatima_isaeva74@mail.ru

Optimization of the conditions for the growth and development of plants in cultivation facilities is based on creation of a microclimate that covers the resulting effect of a system of technological equipment — heating, ventilation, irrigation, fertilizing nutrients and CO2, as well as artificial lighting [3, 4].

The microclimate inside the greenhouse is significantly influenced by environmental factors. These are solar radiation, the strength and direction of the wind, temperature and relative humidity, the amount of precipitation. The most important and determining factor in the external environment is solar radiation. It has a direct effect on the thermal regime of the greenhouse, serves as the main source of energy and determines such microclimate regimes as temperature, humidity, irrigation, food and carbon dioxide [5, 6, 7, 8].

In this regard, making evaluative comparisons in compliance with the scientific principle of a single difference is one of the most relevant areas of study in modern TOK.

2 Materials and Methods

The production experiment was carried out in the greenhouse complex of Yug-Agroholding LLC (Chechen Republic, Grozny) in 2020-2021, in unit 2 (1.8 hectares) when growing a tomato hybrid Merlis F1 with a density of 2.4 plants per square meter.

Substrates from GRODAN and BIOGROW were selected as objects for the assessment studies.

Mineral wool from the Dutch company GRODAN is a versatile substrate for growing vegetables in a greenhouse. Grodan substrates provide greater precision in fertilization and prevent root burns, are nutrient-free and have a neutral pH of 7. The product is characterized by increased porosity, which improves drainage and aeration of the roots, does not have buffering properties, therefore, it easily adopts the acidity and nutritional regime of the nutrient solution. Mineral wool has a temperature 2 degrees lower than other substrates. It has increased capillary properties, so it evaporates well and absorbs water. Most of the nutrient solution is retained in mineral wool, it needs one-fourth less water than peat. It is a sterile substrate, it does not contain pathogenic microorganisms, pests and toxic substances. Reuse of the substrate is allowed.

BIOGROW coconut mats are compressed coconut fiber substrate packaged in black and white polyethylene with UV protection and tightened with rubber bands for better shape retention during watering. Coconut fiber has excellent hygroscopicity and air permeability, allowing for an ideal air-water balance in the root zone.

Presence of coconut particles of different sizes in Biogrow coconut substract mats allows them to simultaneously function as a substrate and a drainage material. Fine – more moisture-absorbing – fiber fraction is located on top, allowing to reduce evaporation from the substrate surface; large - at the bottom, to improve removal of moisture from plant roots.

Coconut fiber substrates can easily replace peat. Wherein, if peat is depleted quickly enough, the coconut substrate has the ability to recover on its own, being saturated with fertilizer components. The moisture content of Bio-Grow coconut mats is higher than that of mineral counterparts. Fibers can absorb up to seven times their own moisture content.

The study program included the solution of the following tasks:

- to evaluate the effect of the substrate on the linear growth of tomato plants;
- the intensity of the fruit filling within 30 days;
- to study the beginning of flowering from the moment of sowing;
- to reveal the reaction of tomato plants to changes in Ec and pH of the solution;
- to determine the yield of tomato depending on the substrates used;

– to determine the economic efficiency of production depending on the substrates used.

3 Results and Discussion

The data obtained as a result of observations (Table 1) indicate a different reaction of tomato plants to the artificial substrate on which they are grown, all other things being equal - lighting, nutrition, temperature, and etc.

Table 1. Comparison result.

No	Name of substrate	Average daily linear growth, cm	Filling the fruit in 30 days	Days before flowering	Ec avg	PH avg	yield for vegetation
1	GRODAN	3.09	0.9-5.7	30	3.7	5.6	70.06
2	BIOGROW	3.19	0.9-6.2	30	3.6	5.6	73.40

Among the studied parameters, the plants reacted to the greatest extent with an average daily growth, filling of fruits, which mainly influenced the yield of tomatoes.

The growing season was 302 days. During this time, tomato plants on GRODAN mineral wool have reached 934 cm, which in terms of the average daily rate is 3.09 cm. The organic substrate based on coconut fiber influenced more accelerated growth of plants, the length of which was 966.7 cm (+3.5 %). The average daily gain was 3.19 cm.

Fruit size and filling rate are the leading parameters affecting the final yield in greenhouse complexes. Measurements and weighing the fruits during the growing season made it possible to calculate the average values. So, the highest productivity of tomato was revealed when grown on coconut fiber. For 30 days, the fruits gained an average diameter of 62 mm, while the diameter on mineral wool was 57 mm (-8.8 %). The size of the fruit also affected its weight. The effectiveness of the organic substrate was manifested in an increase in the mass of fruits by 6 % (140 g), in relation to the mass of fruits grown on a mineral substrate (132 g).

As a result, the final yield on coconut substrate was 4.8 % higher than on mineral wool, which in absolute terms is 3.34 kg/m ^{2}.

To a lesser extent, changes were noted in the concentration of EC salts of the nutrient solution in the mats. Although the difference in EC is only 0.1 units, nevertheless, this may indicate a higher utilization rate of minerals from the nutrient solution by plants growing on organic substrates.

The origin of the substrate did not in any way affect the rate of passage of phenological phases by plants. So in both cases, the flowering phase began at the same time — 30 days after sowing.

Despite the fact that the average indicators of the acidity of the solution are relatively equal, nevertheless, during the season, trends were noted indicating the presence of a greater buffering capacity of the coconut substrate, expressed in smaller fluctuations in the pH of the solution during the day.

Despite the fact that the initial cost of an organic substrate is higher than mineral wool, the economic efficiency of production with coconut fiber is significantly higher. So, with an average cost of a tomato 95 rubles/kga, n additional increase in monetary terms per hectare is 3,173,000 rubles.

4 Conclusions

R&D activities arranged on a permanent basis in production conditions at Yug-Agroholding LLC allow to develop cost-effective solutions and improve production technologies, leading to a decrease in production costs from season to season. So, according to the results of testing various artificial substrates, it was decided to increase use of coconut substrate in the greenhouse complex and to continue the search for scientific ways to implement the genetic potential of tomato hybrids, which reaches 95-105 kg/m^2.

References

1. P.N. Vinogradov, S.S. Shevchenko, O.L. Sedov, S.G. Tishkovets, A.V. Ipatov, V.V. Grishechko, V.G. Pritula, D.Yu. Mozgovoy, A.V. Izhutkin, *Guidelines for the process design of greenhouses and greenhouse complexes for growing vegetables and seedlings*, 103 (2014)
2. L.S. Gil, A.I. Pashkovsky, L.T. Sulima, Modern indoor and outdoor vegetable growing. Practical guide, 468 (2012)
3. M.T. Dzhabrailov, Proceedings of Kazan University, **5**, 24-28 (1983)
4. F.S. Dzhalilov, D. Amini, Gavrish, **2**, 24-25 (2003)
5. M. Andreev, Kh. Ganev, Solskostop. Tech., **36(1)**, 14-18 (1999)
6. E.V. Mamonov, *Vegetable crops*, 244 (2001)
7. *State and problems of scientific support of vegetable growing in greenhouses*, 152 (2003)
8. E. Czuper, Halso ogrodn., **58(ll)**, 17 (1991)

Realization of the bioresource potential of the middle late hybrid of corn Beshtau in the steppe zone of the Chechen republic

A.G. Amaeva[1,2,*], *N. L.* Adaev[1,2], *V. N.* Bagrintseva[3], *J. H.* Aiskhanova[1], and *K.M.* Salamova[1]

[1]Federal State Budgetary Educational Institution of Higher Education Chechen State University named after A.A. Kadyrov, Grozny, Russia
[2]Federal State Budgetary Scientific Institution" Chechen Research Institute of Agriculture", Grozny, Russia
[3]FGBNU All-Russian Research Institute of Corn, 357528 Pyatigorsk, st. Ermolova, 14b, Russia

Abstract. The article presents the results of studying the bioresource potential of the medium-late hybrid of corn Beshtau at the All-Russian Research Institute of corn on dry land and at the Chechen Research Institute of Agriculture on irrigation. In the steppe zone of the republic, against the background of the use of irrigation, the application of mineral fertilizers at a dose of N90P120K60 in combination with pre-sowing seed treatment with the agrochemical Bioplant Flora (1.0 l / t) and foliar top dressing with a tank fertilizer mixture Kristalon (3 kg / ha) + Brexil Zn (0, 15 kg / ha) + urea (7 kg / ha) showed high efficiency in realizing the potential of the Beshtau hybrid. Due to irrigation and the use of these fertilizers, it turned out to be possible to increase the grain yield of the Beshtau corn hybrid to 12 t / ha. The application of mineral fertilizers to the soil at a dose of N90P120K60 in combination with foliar feeding of plants with the agrochemical Bioplant Flora at a dose of 1.0 l / ha in 5 leaves and 2.0 l / ha in 8 leaves increased the starch content in the grain from 62.63 to 72.08%, protein - from 10.65 to 12.28%.

1 Introduction

Corn is a high-yielding grain and forage crop. Due to accumulation of a large amount of organic matter within the growing season, plants show a high need for water and mineral nutrients. Lack of water leads to significant losses in grain yield [1].

The most common and effective way to meet the needs of corn plants for nutrients and to increase productivity is application of mineral fertilizers to the soil [2-6]. In addition to traditional fertilizers, plant nutrition with various agrochemicals is now of great importance in the nutrition system. The modern agrochemicals market is represented by a variety of organic, organomineral and mineral fertilizers containing not only basic nutrients (nitrogen, phosphorus, potassium), but also a complex of microelements, organic acids,

* Corresponding author: aset-6666@mail.ru

phytohormones. According to many researchers, use of such fertilizers can significantly increase the yield of green mass and corn grain [7-9]. Mineral fertilizers and agrochemicals for foliar dressing also significantly improve the quality of corn grain [10-12].

2 Materials and Methods

Scientific research was carried out at the All-Russian Research Institute of Corn in 2012-2014 and the Chechen Research Institute of Agriculture in 2015-2017. We studied the potential yield of a mid-late hybrid of Beshtau corn bred by All-Russian Research Institute of Corn in different soil and climatic conditions.

The experimental field of the All-Russian Research Institute of Corn is located in the zone of sufficient moisture in the Stavropol Territory. Soil – ordinary calcareous heavy loamy chernozem. The average long-term rainfall for the growing season of corn is 343.6 mm, including: 79.4 in May; June — 87.1; July — 70.4; August — 58.7; September — 48 mm.

The experimental field of the Chechen Research Institute of Agriculture is located in the steppe zone of the Chechen Republic. The soil is meadow-chernozem, calcareous, weakly humified, weakly washed away. The average long-term rainfall for the growing season of corn here is 355 mm, including: 82 in May; June — 95; July — 74; August — 62; September — 42 mm.

The climatic conditions of the steppe zone of the Chechen Republic, in comparison with the zone of sufficient moisture in the Stavropol Territory, are characterized by uneven moisture in the middle of summer, long periods of dryness without precipitation.

Beshtau is a medium-late simple hybrid of corn for universal use, both for grain and for silage. The plants are tall, well leafy, with a strong stem, and are distinguished by their permanent flowering capacity. The hybrid is included in the State Register of Breeding Achievements approved for use in the Russian Federation since 2012.

The commercial grain of the hybrid is tooth-like. The grain contains 69.85 % starch, 8.17 % protein, 3.95 % fat, 2.28 % sugar, 1.12 % fiber, 0.77 % ash. The value of the grain of the Beshtau corn hybrid lies in the increased starch content and its suitability for production of starch and alcohol.

The objective of the study was to identify the reaction of the medium-late hybrid Beshtau under irrigation to mineral fertilizers, the effect on seed material and crops with agrochemicals in order to implement its genetic potential in the Chechen Republic.

3 Results and Discussion

In an experiment conducted in the zone of sufficient moisture in the Stavropol Territory to test corn hybrids of the All-Russian Research Institute of Corn, the Beshtau hybrid showed the highest yield of green mass and grain [13]. In addition, this hybrid was highly responsive to nitrogen fertilization (Table 1).

On average, over 3 years of testing from nitrogen fertilization at a dose of N60, the plant height increased by 11 cm, the yield of green mass – by 6.0 t/ha (12.8 %), grain – by 0.75 t/ha (7.0 %).

In the steppe zone of the Chechen Republic, the factor limiting the yield of corn is moisture availability [14]. As it is known, corn consumes a large amount of water within the growing season. In the experiment carried out at the Chechen Research Institute of Agriculture, the total water consumption of the mid-season hybrid Krasnodar 385 MB on average for 2010-2012 on dry land it was equal to 3,207 m^3/ha, and on irrigation it reached 5,790 m^3/ha [14].

Table 1. Plant height and yield of the Beshtau corn hybrid without fertilizers in the experiments of the All-Russian Research Institute of Corn

Indicator	2012	2013	2014	On average
No fertilizers				
Plant height, cm	266	250	245	254
Productivity of green mass, t/ha	51.6	40.7	48.7	47.0
Grain yield, t/ha	12.02	11.35	8.76	10.71
N60				
Plant height, cm	275	256	263	265
Productivity of green mass, t/ha	56,.8	48.7	53.5	53.0
Grain yield, t/ha	12.45	12.09	9.83	11.46

Without irrigation in this area, corn grain yields are very low. So, on average for 2010-2012, without irrigation and use of fertilizers, the grain yield of domestic corn hybrids Krasnodar 382 MB and Krasnodar 385 MB varied within 2.3-2.8 t/ha (Table 2).

Irrigation of 500-800 m^3/ha with an irrigation rate of 2,050-2,800 m^3/ha to maintain soil moisture at the level of 70-80 % HB, even without use of fertilizers increased the yield of hybrids on average for 3 years by 1.9-2.0 times.

Importance of irrigation was especially great in the extremely dry 2011, when 1.7-2.8 t/ha of grain were obtained without irrigation on a fertilized background. By artificially improving the moisture supply of corn, the grain yield of the Krasnodar 382 MB hybrid increased 3.8 times, and the Krasnodar 385 MB – 6 times.

In the steppe zone of the Chechen Republic, irrigation combined with an efficient fertilization system is the most important reserve for increasing corn yields.

In conditions of irrigation, the importance of the nutrition system increases, which shall be considered as a priority element of the corn cultivation technology. The use of macro- and micronutrient fertilizers, as well as other agrochemicals, in turn, is a condition for the effective use of irrigation.

In order to use the bioresource potential of the high-yielding corn hybrid Beshtau, the Chechen Research Institute of Agriculture studied the effect on the yield and grain quality of different options for using mineral fertilizers and agrochemicals for pre-sowing seed treatment and foliar top dressing of plants during the growing season.

The decisive importance provided use of a complete mineral fertilizer in a dose of N90P120K60, due to which the yield of the Beshtau hybrid in 2015 increased by 1.2 t/ha (14.1 %), in 2016 — by 1.6 t/ha (18.8 %), in 2017 — by 2.8 t/ha (36.8 %). On average, in addition to the control, 1.9 t/ha (24.1 %) were obtained.

Two foliar top dressing with Bioplant flora fertilizer against the background of mineral root nutrition gave an additional 1.6 t/ha over 3 years, which was 15.8 % compared to the N90P120K60 option.

Table 2. Productivity of corn hybrids depending on irrigation and fertilization.

Irrigation	2010		2011		2012		On average	
	1	2	1	2	1	2	1	2
Hybrid Krasnodar 382 MB								
No irrigation	4.6	5.7	1.9	2.8	2.1	2.9	2.8	3.8
With irrigation	8.7	13.0	6.9	10.5	9.0	12.8	8.2	12.1
Hybrid Krasnodar 385 MB								
No irrigation	4.6	5.8	1.3	1.7	1.0	2.4	2.3	3.3
With irrigation	9.1	13.4	6.7	10.2	9.4	13.6	8.4	12.4
$HCP_{0.05}$	0.14		0.19		0.21		-	

Note: in column 1 – no fertilizers; 2 – N90P120K60.

Also, foliar top dressing of plants with a tank mixture of Kristalon (3 kg/ha) + Brexil Zn (0.15 kg/ha) + Carbamide (7 kg/ha) fertilizers was effective. On average, over the years of studies in relation to the background, an increase in yield was obtained equal to 1.3 t/ha (12.9 %).

Table 3. The effect of fertilizers on the grain yield of the Beshtau hybrid corn under irrigation (t/ha)

Experience options	2015	2016	2017	Average	Increase	
					t/ha	%
Fertilizer-free (control)	8.5	8.5	7.6	8.2	-	-
N90P120K60 (Background)	9.7	10.1	10.4	10.1	1.9	2.1
Background + top dressing Bioplant flora (1.0 l/ha in 5 leaves + 2.0 l/ha in 8 leaves)	10.8	12.6	11.7	11.7	3.3	40.1
Background + top dressing (Kristalon 3.0 kg/ha + Brexil Zn 0.15 kg/ha + urea 7 kg/ha)	11.2	11.2	11.7	11.4	2.5	28.0
Background + top dressing (Kristalon 3.0 kg/ha + Brexil Zn 0.15 kg/ha + urea 7 kg/ha) + seed treatment Bioplant Flora (1.0 l/ha)	11.7	12.3	12.4	12.1	3.9	47.4
HCP $_{0.05}$, (t/ha)	0.13	0.18	0.18	0.85	-	-

The same option of use of fertilizers in combination with pre-sowing seed treatment with Bioplant Flora fertilizer gave the maximum increase in the grain yield of the Beshtau hybrid. Compared to the control without fertilizers, the yield increase was 3.9 t/ha (47.4 %). Compared to application of mineral fertilizers to the soil, due to seed treatment and plant top dressing grain yield increased on average by 2.0 t/ha (by 19.8 %).

The effect on the yield of the hybrid was significant both for introduction of mineral fertilizers into the soil before sowing and use of fertilizers for seed treatment and plant top dressing during the growing season.

The offered options for the fertilization system influenced not only the grain yield, but also its quality. The lowest starch content in grain (62.63 %) was found in the control option of the experiment without fertilizers. Application of mineral fertilizers to the soil at a dose of N90P120K60 increased the starch content by 10.39 %. However, at the same time, compared with the control, the protein content in the grain decreased by 2.35 %. The most optimal option of the experiment, providing a simultaneous increase in both starch and protein, turned out to be a combination of mineral fertilizer with treatment of plants with the agrochemical Bioplant Flora at a dose of 1.0 l/ha in 5 leaves and 2.0l/ha in 8 leaves.

Table 4. Chemical composition of corn hybrids (on average for 2015-2016)

Index, %	Experiment option				
	1	2	3	4	5
Starch	62.63	73.02	72.08	65.91	67.6
Protein	10.65	8.30	12.28	5.06	8.95
Fat	5.59	5.45	5.46	5.80	5.67
Ash	1.27	1.23	1.32	1.3	1.31
N	1.70	1.33	1.96	1.63	1,.43
P_2O_5	0.31	0.31	0.32	0.33	0.36
K_2O	1.96	1.45	1.41	1.99	3.34

Note: 1 – Without fertilizers (control); 2 – N_{90} $P_{120}K_{60}$ – Background; 3 – Background + top dressing Bioplant Flora; 4 – Background + top dressing (Crystalon + Brexil + urea); 5 – Background + top dressing (Crystalon + Brexil + carbamide) + seed treatment Bioplant Flora.

4 Conclusions

1. Against the background of irrigation, application of mineral fertilizers at a dose of N90P120K60 in combination with pre-sowing seed treatment with the agrochemical Bioplant Flora (1.0 l/t) and foliar top dressing with a tank fertilizer mixture Kristalon (3 kg/ha) + Brexil Zn (0.15 kg/ha) + urea (7 kg/ha) showed the high efficiency in implementing the potential of the medium-late hybrid Beshtau. Due to irrigation and use of these fertilizers, the yield of the hybrid turned out to be possible to increase the yield of corn grain up to 12 t/ha.

2. Application of mineral fertilizers to the soil at a dose of N90P120K60 in combination with foliar top dressing of plants with the agrochemical Bioplant Flora at a dose of 1.0 l/ha in 5 leaves and 2.0 l/ha in 8 leaves increased the starch content in the grain from 62.63 to 72.08 %, protein – from 10.65 to 12.28 %.

References

1. V. V. Melikhov, I. P. Kruzhilin, N. V. Kuznetsova, *Guidelines for the cultivation of corn for grain*, Volgograd: State Institution Publisher, 88 (2003)
2. A. F. Stulin, Corn and sorghum, **1,** 19 (2012)
3. G. N. Sukhoyarskaya, *Productivity of hybrids of corn of different ripeness groups when using fertilizers on ordinary chernozem of the Central Ciscaucasia*, 23 (2009)
4. V. N. Bagrintseva, I. N. Ivashenenko, Agrochemistry, **1,** 72 (2018)
5. V. N. Bagrintseva, S. V. Nikitin, M. A. Cherkasova, Agrochemistry, **10**, 46 (2019)
6. R. S. Sharifi, R. Taghizadeh, J. Food Agricult. Environ, **7(3-4),** 518 (2009)
7. D. V. Lomovsky, *Corn productivity depending on seed treatment with dressing agents, microfertilizers on leached chernozem of the Western Ciscaucasia,* 24 (2007)
8. Z. P. Okazova, M. M. Tokbaev, Agricultural Science, **10,** 22 (2008)
9. D. A. Taran, *Productivity of corn hybrids depending on the pre-sowing application and top dressing with nitrogen and potassium humate on leached chernozem of the Western Ciscaucasia,* 24 (2013)
10. Zh. M. Yakhtanigova, V. M. Kokova, New technologies, **5,** 58 (2008)
11. Z. I. Usanova, I. V. Shalnov, Feed production, **2,** 21 (2013)
12. A. I. Volkov, N. A. Kirillov, L. N. Prokhorova, Feed production, **7,** 16 (2013)
13. V. N. Bagrintseva, Agrochemistry, **11,** 47 (2015)
14. A. G. Amaeva, Implementation of the bioresource potential of highly productive corn hybrids of domestic and foreign selection, depending on the conditions of soil moisture in the steppe zone of the Chechen Republic, 18 (2013)

Effectiveness of biotechnology on the basis of associative microorganisms at the early stages of forestry process of the Pinus sylvestris L.

Marina Yu. Sautkina[1*], *Nina F.* Kuznetsova[1], *Michael A.* Semenov[1] and *Andrew S.* Khoroschev[2]

[1]All-Russian Research Institute of Forest Genetics, Breeding and Biotechnology, Department of Forest Genetics and Biotechnology, 394087 Voronezh, Russia
[2] Tambov Regional State Government Institution "Vernadsky forestry", 393976 Tambov Region, Russia

Abstract. The transition to biologically based technologies is one of the priority scientific areas of agriculture and forestry in Russia. The aim of the work is to develop and modify the technology of using biological products on forest tree species and to test it on Scots pine. The effect of pre-sowing inoculation of pine seeds with biological products on soil germination, safety of one- and two-year-old seedlings and their biometric characteristics was studied. The results of the analysis of soil germination of seeds and the safety of 1-year-old seedlings showed that these biological products can be introduced into the technology of growing planting material of Scots pine. It was revealed that the seeds of sensitive trees of the Stupinskaya population turned out to be the most responsive to inoculation with biological products. A stimulating effect has been established on the survival rate of seedlings, preservation, height, growth and diameter of 2-year-old plants in a forestry area (Vernadsky forestry, Tambov region). The height of 2-year-old seedlings of pine 'Ostrogozhskaya' exceeds the control by an average of 18.1%. The use of this biotechnology in forestry production will create favorable conditions for the germination of pine seeds, the growth of seedlings and their higher survival rate.

1 Introduction

The high rates of climate warming in the Central Black Earth Region impose additional requirements on the species composition, seed and planting material when organizing forestry works in this region. In Russia, the problem of obtaining high-quality planting material in a volume sufficient for forestry production still remains [1]. The high rates of climate warming in the Central Black Earth Region impose additional requirements on the species composition, seed and planting material when organizing forestry works in this region. Measures to improve the sustainability of newly created forests include a set of environmental, technological and biological measures to increase the longevity and

* Corresponding author: sautmar@mail.ru

sustainability of forest plantations in more arid conditions (biological soil reclamation, the use of varietal seeds, forms and hybrids of certain forest species) [2].

In recent decades, there has been a deficiency of basic nutrients in the soil. The difficult economic situation with the rather high cost of mineral fertilizers intensifies the need to search for alternative sources of plant nutrition. One of the ways to solve this problem is the use of biological products based on associative microorganisms in the technology of growing planting material [3, 4]. Plant-microbial interactions are activated under the influence of these biological products. The main mechanisms of the beneficial effect of microorganisms on plants include fixation of molecular nitrogen, optimization of phosphorus nutrition, stimulation of plant growth and development, increase in plant resistance to environmental stress factors, suppression of phytopathogens, etc. In general, the action of associative biological products can be divided into three types: 1) direct stimulation of plant growth and development due to metabolites synthesized by bacteria; 2) an indirect effect due to the displacement of pathogenic microorganisms from the soil microflora; 3) fixation of molecular nitrogen available to plants after bacteria die off [5]. In recent years, a large number of microbial preparations have been created, the effectiveness of which has been proven in many agricultural crops [6-9]. This direction of nature-compatible biotechnology is practically absent in the Russian forest industry. The effectiveness of the influence of associative biological products on forest tree species has been little studied. For reforestation and afforestation of the main forest-forming species, this biotechnology is promising due to its low cost and environmental safety. This was the basis for its development, optimization and testing on Scots pine.

The purpose of research was to study the effect of pre-sowing inoculation of Scots pine seeds with biological products based on associative microorganisms on seed germination, safety of one- and two-year-old seedlings and their biometric characteristics; development and modification of the technology for the use of biological products on forest tree species and testing of its stages on pine seeds of different varietal categories.

2 Materials and Methods

In May 2020, in the forest park area of All-Russian Research Institute of Forest Genetics, Breeding and Biotechnology, experience was laid in testing biotechnology based on associative biological products on the seeds of Scots pine of two populations: Stupinsky and pine 'Ostrogozhskaya'. The control population is a typical plantation for the forest-steppe zone, grows in ecologically favourable conditions. The droughty-tolerant variety pine 'Ostrogozhskaya' is located on the southern border of the forest-steppe zone. Earlier we the resistant and drought-sensitive trees were identified in these populations [10, 11].

The day before sowing, the seeds went through the process of inoculation with biological products based on associative microorganisms. These preparations were obtained from the All-Russian Institute of Agricultural Microbiology (St. Petersburg). They are registered as fertilizing biological products of complex action of the "ULTRAFIT" group. Each of them is an original biological product with specific mechanisms of interaction with plants. We used: mizorin (strain 7), azorizin (strain 8), strain 17-1, strain 18-5, flavobacterin (strain 30), rhizoagrin (strain 204), strain PG-5, strain 2P-7. Then the inoculated seeds are sown in a prepared soil consisting of ordinary chernozem, sand and peat in a ratio of 1: 1: 1. The experimental scheme for 2020 is described in work [12].

3 Results and Discussion

Table 1 shows the data of the analysis of the soil germination of the seeds of the harvest of the optimal year of the two studied pine populations.

Table 1. The effect of biological preparations on the germination of Scots pine seeds of different breeding categories

Experience variant	Germination, % / + to control			
	Sr	Sd	Or	Od
Control	65.0	29.0	40.0	49.0
Strain 7 mizorin	70.0/7.7	35.0/20.7	44.0/10.0	66.0/34.7
Strain 8 azorizin	62.0/0	56.0/93.1	37.0/0	55.0/12.2
Strain 17-1	65.0/0	36.0/24.1	25.0/0	48.0/0
Strain 18-5	40.0/0	34.0/17.2	40.0/0	64.0/30.6
Strain 30 flavobacterin	55.0/0	30.0/10.3	44.0/10.0	83.0/69.4
Strain 204 rhizoagrin	54.0/0	35.0/20.7	62.0/55.0	49.0/0
Strain PG-5	83.0/27.7	55.0/89.7	48.0/20.0	58.0/18.4
Strain 2P-7	73.0/12.3	37.0/27.6	59.0/47.5	59.0/20.4

Note: Seeds of resistant (Sr) and drought-sensitive (Sd) trees of the Stypinskaya population; resistant (Or) and drought-sensitive (Od) trees of pine 'Ostrogozhskaya'.

It was shown that the most responsive to inoculation with biological products were the seeds of sensitive trees of the Stypinskaya population (Table 1). In all variants of the experiment, an increase in indicators of soil germination relative to control was noted. The maximum stimulating effect is shown in variants with azorizin (strain 8) and strain PG-5, where the number of seedlings was almost twice as high as the control: +93.1 and 89.7%, respectively (Table 1). Also, a pronounced effect from the use of biological products was noted on the seeds of sensitive trees of the varietal population. It has been shown that the maximum stimulating effect on increasing soil germination was revealed when seeds were inoculated with flavobacterin (strain 30) – 69.4%. Inoculation of seeds of resistant pine trees with pine 'Ostrogozhskaya' with rhizoagrin (strain 204) made it possible to increase soil germination by 55% relative to control.

In the third decade of August, the safety 1-year-old pine seedlings of control and experimental variants was determined (Figures 1, 2).

In almost all variants of the experiment, high rates of seedling safety were noted. In general, it was revealed that the highest safety is in the offspring of sensitive trees of the Stupa population. The maximum result was found in the variant with azorizine (pcs. 8): the preservation was 89.3%. In other variants of inoculation of seeds of this population, the safety ranged from 66.7 (strain 17-1) to 86.7% (strain 30 flavobacterin).

The high preservation of the seedlings of the 'Ostrogozhskaya' pine was noted. Thus, the maximum preservation rates were observed in seedlings of sensitive trees in variants with strain 204 (rhizoagrin) and 17-1 – 83.7 and 81.3%, respectively (Figure 2). For resistant trees of this population, a high preservation was also revealed in variants with strains 17-1, 30 (flavobacterin) and PG-5 - 80.0, 79.5 and 79.5%, respectively (Figure 2).

In autumn, seedlings of resistant trees of the experimental and control populations were planted on a forest-cultural area in the Vernadsky Forestry of the Tambov region. In May 2021, measurements of the biometric characteristics of pine seedlings were made and their safety was determined.

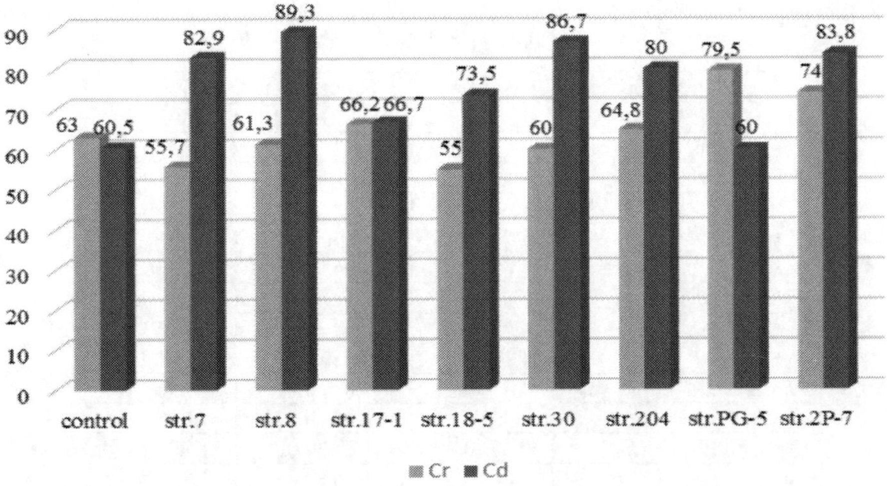

Fig. 1. Influence of biological products based on associative microorganisms on the safety of Scots pine seedlings, Stupinskaya population, %

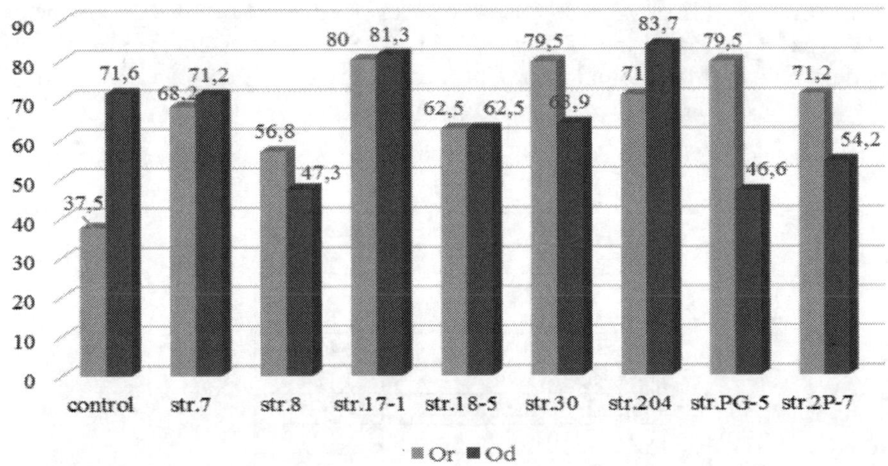

Fig.2. Influence of biological products based on associative microorganisms on the safety of Scots pine seedlings of the varietal pine 'Ostrogozhskaya', %

The results of the analysis of biometric characteristics of control (untreated with biological products) variants of the field experiment showed that the height of seedlings of the pine 'Ostrogozhskaya' is 18.1% higher than that of seedlings of the Stupinsky population (Table 2).

It was found that inoculation with associative biological products had a stimulating effect on the growth of seedlings of the Stupinsky population in all treatment options, except for str. 17-1. The maximum plant heights are noted in the variant with str. 204 (rhizoagrin) – 6.47 ± 0.42 cm, which is 37.1% higher than the control (Table 2). Also, a significant increase in height was noted in variants with str. 18-5 and 2P-7 — by 30.3 and 20.8%, respectively. In other variants, the increase in the height of seedlings relative to the control ranged from 15.2% (str. PG-5) to 24.1% (str. 30 flavobacterin). For seedlings of pine 'Ostrogozhskaya', a positive effect of biological products on plant height was also noted. So, the maximum result was shown by azorizin (str. 8) – 6.65 ± 0.40 cm, which is 33.8% higher than the control (table 2). The stimulating effect after inoculation with strains 7 (mizorin) and 18-5 was 7.2 and 13.5%, respectively.

Table 2. Biometric indicators of 2-year-old seedlings of scots pine on the forest area (measurement date 21.05.2021)

Experience variant	Height, cm	Diameter, mm	Growth, cm
Sr control	4.07±0.32	0.93±0.02	0.80±0.01
Sr+str.7(mizorin)	4.91±0.28	1.01±0.02	1.18±0.08
Sr+str.8(azorizin)	4.79±0.37	0.99±0.04	1.56±0.14
Sr+ str.17-1	3.84±0.35	1.05±0.03	1.24±0.20
Sr+str.18-5	5.84±0.43	0.99±0.01	1.33±0.16
Sr+str.30(flavobacterin)	5.05±0.37	0.95±0.02	1.35±0.23
Sr+str.204(rhizoagrin)	6.47±0.42	1.07±0.01	1.95±0.23
Sr+str.PG-5	4.69±0.22	0.98±0.03	1.51±0.13
Sr+str.2P-7	5.14±0.27	0.98±0.02	1.70±0.15
Or control	4.97±0.22	1.01±0.02	1.39±0.13
Or+str.7(mizorin)	5.33±0.38	1.00±0.03	1.61±0.17
Or+str.8(azorizin)	6.65±0.40	1.21±0.08	1.99±0.18
Or+str.18-5	5.64±0.27	1.24±0.08	1.98±0.19

The average diameter of plants at the root collar in the experimental variants of the Stupinsky population exceeds the control from 5.4% (str. PG-5, 2P-7) to 15.1% (str. 204 rhizoagrin). In the seedlings of pine 'Ostrogozhskaya', the excess of this parameter is in variants with str. 8 (azorizin) and 18-5 by 19.8 and 22.8%, respectively (Table 2).

The effectiveness of the biological products used can be judged by the size of the annual growth of seedlings in the first year after their transplantation to forestry area (Table 2, 3).

Table 3. The influence of biological products based on associative microorganisms on the safety of 2-year-old Scots pine seedlings

Experience variant	Safety, %
Sr control	66.7
Sr+str.7(mizorin)	88.9
Sr+str.8(azorizin)	73.3
Sr+ str.17-1	44.4
Sr+str.18-5	100
Sr+str.30(flavobacterin)	76.5
Sr+str.204(rhizoagrin)	100
Sr+str.PG-5	80.6
Sr+str.2P-7	67.6
Or control	85.7
Or+str.7(mizorin)	70.6
Or+str.8(azorizin)	66.7
Or+str.18-5	90.9

The maximum stimulating effect is shown when processing str. 204 (rhizoagrin). The growth of seedlings is 1.95 ± 0.23 cm, which is 2.4 times higher than the control (0.80 ± 0.01 cm). A significant increase in growth was noted in variants with str.2P-7, 8 (azorizin) and PG-5 - by 112.5; 95 and 88.8%, respectively (Table 2). The greatest result of varietal seedlings was obtained in variants with str. 8 (azorizin) and 18-5 – 1.99 ± 0.18 and 1.98± 0.19 cm, respectively, which is higher than the control by 23.6 and 23.0%. In general, it should be noted that the growth of seedlings in the experimental version of the pine 'Ostrogozhskaya' is 73.8% higher than in the Stupinsky population. At the end of May, the safety of seedlings was taken into account. Compared with the control, the safety of varietal seedlings is 19% higher. 100% survival of plants of the Stupinsky population was noted in variants with str. 18-5 and 204 (rhizoagrin) (Table 3). In other variants of seed treatment of the Stupinsky population, the safety of seedlings was 67.6% (str. 2P-7) – 88.9% (str.7

mizorin). Of the seedlings of the pine 'Ostrogozhskaya' planted on the forestry area, the plants in the variant with str. 18-5 had the greatest preservation – 90.9 % (Table 3).

4 Conclusion

Thus, the two-year results obtained on the approbation and study of the effect of biological products based on associative microorganisms on the quality of 1- and 2-year-old Scots pine seedlings confirm the expediency of their use in the forestry of the Russian Federation. The use of these biological products is cost-effective, environmentally friendly and contributes to the production of high-quality planting material. This biotechnology is environmentally compatible, as it imitates real natural processes. It can be introduced into the technology of forestry production at the stage of obtaining planting material, will create favorable conditions for seed germination and seedling growth, as well as increase the yield of standard planting material per unit area.

References

1. T.S. Ustinova, Act. Prob. of the For. Compl. **23** (2009)
2. S.N. Kryuchkov, A.S. Stolnov, Sc.-agr. J. **103,** 2 (2018)
3. M.Yu. Sautkina, Yu. I. Cheverdin, 6th International Conference on Agriproducts processing and Farming, Voronezh (2020)
4. A. Yu. Cheverdin, M.Yu. Sautkina, Yu. I. Cheverdin, 9th International Symposium «Steppes of Northern Eurasia», Orenburg (2021)
5. I.A. Tikhonovich, A.A. Zavalin, G. G. Blagoveshchenskaya, A.P. Kozhemyakov, Plod.J. **3** (2011)
6. E.A. Blinkov, Actual aspects of modern microbiology: collection of articles scientific. tr., Moscow (2010)
7. A.A. Belimov, N.Yu. Zinovkina, A.I. Shaposhnikov, et al, Methods for assessing agricultural risks and technologies for mitigating the effects of climate change in agriculture: collection of articles scientific. tr., St.Petersburg (2011)
8. A.A. Zavalin, N. S. Almetov, The use of biological products and biological nitrogen in agriculture in the Non-Black Earth Region (2009)
9. A.P. Kozhemyakov, S.V. Timofeeva, Agr. Exp.J. **2** (2007)
10. N. F. Kuznetsova, Izv. Saratov Univ. (N. S.), Ser. Earth Sc. **19**, 3 (2019)
11. E.S. Klushevskaya, N.F. Kuznetsova, Rast. Res. **55**, 3 (2019)
12. M.Yu. Sautkina, N.F. Kuznetsova, Bull. of MichGAU **65**, 2 (2021)

Features of cultivation technology for organic pumpkin certified by JAS (Japan) in the conditions of Primorsky region

T.N. Kirtaeva[1,*], *A.E.* Komin[1], *A.I.* Grishko[2], *E.A.* Evseeva[3], and *P.A.* Komin[4]

[1]Federal State Budgetary Educational Institution of Higher Education «Primorskaya State Agricultural Academy», Usuriysk, Russia
[2]Peasant farm of A.I. Grishko, Sergeevka village, Partizansky district, Russia
[3]Primorsky EM-Center LLC, Vladivostok, Russia
[4]Non-Profit Organisation Association Far Eastern Organic Union, Ussuriysk, Russia

Abstract. In the modern world, special attention is paid to rational use of natural resources and provision of the population with high-quality food. In this connection, the principles of organic farming are relevant both in our country and abroad, as they allow reducing the negative impact of agriculture on the environment. Primorsky Territory has significant potential for development of organic farming, as well as the possibility of exporting organic vegetables to the countries of the Asia-Pacific region, including Japan. Based on the results of the studies carried out, the authors found that the soil and climatic conditions of the Partizansky District of the Primorsky Territory are suitable for cultivation of pumpkins considering the requirements of the Japanese JAS standards. Compliance with the recommended agricultural practices made it possible to obtain a yield of marketable pumpkin fruits of 11.1-12.2 c/ha with high quality indicators.

1 Introduction

According to the FAO forecast, by 2050 the world's population will increase by 34 % and will amount to 9.1 billion people. In this connection, in order to provide food for the growing population, the production of agricultural products will need to be increased by 60 % [1].

In the modern world, there has been a change in paradigms and ideas about healthy food. The focus is not on volumes, gross harvests, harvest, but healthy nutrition and quality of life. Movement for the so-called "ecological vegetables" is growing, while the share of organic vegetable growing is constantly growing [2]. According to the National Organic Union of the Russian Federation, the market for organic products is one of the most dynamically developing in the world and over the past twenty years has grown more than five times (from 18 to 97 billion euros). According to forecasts by Grand View Research, the market will continue to grow at a rate of 10-12 % per year and will reach about $212-230 billion in 2025 [3].

*Corresponding author:kirtaevat@mail.ru

A great deal of practical experience has been accumulated in the cultivation of organic crops. So, in Russia organic grain crops, soybeans, peas, potatoes and even artichoke are grown [4-6]. However, only since 2020, after the entry into force of Federal Law No. 280 "On Organic Products", this direction is regulated at the legislative level [7].

Despite the differences in the certification procedure and in the requirements for organic production in different countries, the technology of organic farming is the same, since it excludes use of synthetic agrochemicals, pesticides, hormonal preparations, and genetically modified (genetically engineered, transgenic) organisms.

According to the FIBL in 2019, 187 out of 239 countries in the world cultivate organic agriculture (78 %), 72 of them have their own laws in the field of production and circulation of organic products. The most developed systems of regulation of organic agriculture exist in Europe, the USA, Japan, India, Canada and China [8]. However, despite the fact that the *organic JAS* brand in Japan began operating in 1999, the organic market in the country is only only 185 million yen and it is produced on an area of 9,956 hectares, i.e. 0.22 % of all cultivated land. Mainly they grow fodder crops, fruits and tea [9].

Features of the geographical location of Primorsky Territory – the proximity of the markets of the Asia-Pacific region with a high demand for organic products, including Japan, presence of priority development areas and large investment projects contribute to the focus of production on such export crops as soybeans, corn, feed, vegetables, and etc., including pumpkin.

Pumpkin (lat. *Cucurbita*) is one of the valuable melons and gourds, the fruits and seeds of which are of great nutritional value, as products that provide dietary (due to the high content of carotene, dietary fiber, macro- and micronutrients) and therapeutic and prophylactic nutrition (reduces the risk of cardiovascular, oncological and gastrointestinal diseases, is a raw material for the processing and pharmaceutical industries [10, 11]. In addition, pumpkin is an indispensable vegetable in baby food and an ideal complementary food product at an early age, which is noted not only by Russian, but also by Japanese experts.

The purpose of the study is to improve the technology of pumpkin cultivation, taking into account the requirements of organic certification JAS (Japan) in the conditions of the Primorsky Territory.

2 Materials and Methods

Experimental studies to improve the technology of cultivation of organic pumpkin were carried out on the production fields of the farm of A.I. Grishko in Sergeevka village, Partizansky district, Primorsky Territory in 2021. The area of production experience is 4 hectares, the area of the experimental plot is 1 hectare (Figure 1). The object of the study is table pumpkin plants.

Fig. 1. Sowings of pumpkin in the farm of A.I. Grishko (August of 2021).

In the course of the study, field surveys, observations and measurements were carried out using B.A. Dospekhova [12] and V.F. Belik techniques [13].

Cultivation of pumpkin was carried out taking into account the requirements of organic production and in accordance with the JAS standards (Japan), which included a set of such measures as the selection of pumpkin varieties resistant to diseases; use for sowing organic seeds; placement of the crop according to the best predecessor in the crop rotation; mechanical inter-row cultivation of pumpkin crops, as an alternative to herbicides and protection against a complex of weeds at the early stages of its development; using biological fertilizers to improve soil quality and provide nutrition for pumpkin plants; optimal terms of harvesting and storage conditions for pumpkin fruits.

3 Results and Discussion

Biological features of pumpkin. Pumpkin is a melon crop demanding for heat, moisture and soil fertility. Pumpkin seeds begin to germinate at a temperature of 13-14 °C, but the optimal temperature for the growth and development of plants and pumpkin fruits is 22-25 °C. Frosts (-1 ^{0}C), even short-term ones, lead to the loss of plants. All pumpkin varieties are drought tolerant due to their strong root system. The highest water consumption by plants occurs during the period of mass flowering and fruit formation. Pumpkin grows and bears fruit with a short daylight hours. It tolerates a little shading, but tastier fruits are formed on plants well-lit by the sun. Highly fertile loamy soils are suitable for growing crops, with a reaction of 6.5-7.5. On heavy wet soils, plant growth is inhibited. Pumpkin responds positively to fertilization, especially organic fertilization [14].

Soil and climatic features of the pumpkin growing zone. Partizansky district is located in the southern part of the Primorsky Territory. The proximity of the Sea of Japan has imparted a peculiar monsoon character to the climate. Despite the rather high percentage of humidity (average annual rate of 71 %), it can be considered soft. The meteorological conditions for the growing season in 2021 significantly differed from the average long-term values. The amount of precipitation during the period of active growth and development of pumpkin plants was more than three times lower than the norm, and the temperature regime was slightly higher than the average long-term values.

The soil cover of the experimental site is represented by brown forest podzolized soils with a high degree of stonyness [15]. According to agrochemical analysis, the content of humus in the plow horizon was 1.8 %, mobile phosphorus – 125 mg/kg, exchangeable potassium – 140 mg/kg, pH – 5.6.

Selection of varieties. In organic farming, use of chemical means of protection is excluded, and the emphasis is on biologization and organic fertilizers, and, therefore, the efficiency of the entire production largely depends on the correct choice of varieties (hybrids) of agricultural crops. According to the study program in 2021 in the conditions of the farm of A.I. Grishko one pumpkin variety Vnuchka (Far Eastern variety) and three pumpkin hybrids of Japanese breeding: Kurishogu nneo F1, TSX-891 F1, Sweet mama F1 were cultivated.

Pumpkin variety Vnuchka is a mid-early variety of pumpkin of local selection (author A.S. Kornilov), biological ripeness occurs 90-100 days after germination. The fruit is flat-round, green, with a cream-colored pattern in the form of stripes and spots, smooth one (Figure 2). The average fruit weight is 1.7 kg. The pulp is orange, thin, starchy, dense, slightly juicy. The taste is excellent. Transportable. The fruits retain their commercial qualities for 80-90 days after picking. It is drought-resistant and resistant to waterlogging.

Fig. 2. Pumpkin variety Vnuchka.

Pumpkin hybrid Kurishogun neo F1 – a hybrid of Japanese pumpkin selection. Bush type table pumpkin. Forms 2-3 fruits per plant. Ripening period is 90-95 days. The fruit is flat-round, dark green, weighing up to 2.0 kg (Figure 3). The pulp is bright yellow with a high sugar content.

Fig. 3. Pumpkin hybrid Kurishogun neo F1.

Pumpkin hybrid TSX-891 F1 – mid-early hybrid of Japanese selection for table use. The plant is climbing, on average forms 2 fruits per plant. Ripening period is 90-95 days. The fruit is flat-round, dark green, weighing 1.7-2.2 kg (Figure 4). The pulp is bright yellow with a high sugar content.

Fig. 4. Pumpkin hybrid TSX-891 F1.

Pumpkin hybrid SweetmamaF1– **a** hybrid of Japanese table pumpkin selection. High-yielding bush pumpkin. Forms 2-3 large fruits per plant. Ripening period is 85-90 days. The fruit is flat-round, dark green, weighing up to 1.8-2.0 kg (Figure 5). The pulp is bright yellow with a pronounced nutty taste. Possesses high keeping quality.

Fig. 5. Pumpkin hybrid Sweet mama F1

Place in the crop rotation. The basis of any crop rotation is the alternation of crops in time and territory. This achieves the prevention of appearance and active manifestation of diseases and pests, excludes one-sided soil depletion, creates the prerequisites for the rational use of fertilizers and the fight against weeds, which is of paramount importance in organic farming due to the impossibility of using chemical plant protection products. In the conditions of the farm of A.I. Grishko in 2021, pumpkin was cultivated according to one of the best predecessors – Solidor oat variety.

Main soil cultivation. The tasks of cultivation of the soil for pumpkin are the same as in cultivation of other agricultural crops: maintaining and increasing soil fertility, improving the physical properties of the soil, incorporating fertilizers, preventing the massive spread of pests, diseases and weeds, creating optimal conditions for seed germination, plant growth and activity of beneficial microorganisms. The main cultivation of the pumpkin soil included fall plowing, which was carried out in October 2020.

Spring soil preparation. The task of spring cultivation is to level the surface of the field as much as possible, destroy weeds and preserve soil moisture. In the conditions of the farm of A.I. Grishko after the onset of physical ripeness of the soil (March), harrowing was carried out with medium harrows in 2 tracks across the autumn cultivation. We used MTZ-82 tractor, ZBS-1 tooth harrows and BZS harrow coupler.

After harrowing, 2-7 days before sowing, "EM • 1 microbiological fertilizer" Vostok EM-1 "(hereinafter referred to as a biological product) is introduced into the soil at the rate of 5 l/ha in order to reduce the phytopathogenic background; increasing soil fertility; providing plants with the necessary nutrients that microorganisms produce by fermenting organic residues in the soil; normalization of air and moisture exchange in the soil.

Treatment with a solution of a biological product was carried out with MTZ 82 tractor + OM-600 sprayer at the rate of 200-300 l/ha. Immediately after introduction of the biological product with the smallest time gap, pre-sowing cultivation was carried out to a depth of 4-5 cm (MTZ-82 + KRG-3.6) (Figure 6).

Fig. 6. Presowing cultivation, April 2021

Sowing terms and methods. The timing of sowing pumpkin depends on the biological characteristics of the crop and varieties, the planned timing of the sale of products, climatic and weather conditions.

In the Primorsky Territory, pumpkin is sown in the third decade of May, when the soil warms up to 8-12 ^{0}C and the danger of frost damage to seedlings passes.

In a production experiment, the pumpkin was sown on May 25, 2021 according to the scheme 270 cm x 30-35 cm with a seeding rate of 1.5 kg/ha and a plant density of 12,000 pcs./ha (Figure 7).

Fig. 7. Sowing pumpkin, 2021

Caring for pumpkin crops. In the system of organic farming, the use of chemical herbicides is excluded and the main way to control weeds is inter-row cultivation and manual weeding. In the early stages of development, pumpkin plants are susceptible to weeds, therefore, pre-emergence and inter-row cultivation for emergence give a good effect. In 2021, in the farm of A.I. Grishko four inter-row cultivations were carried out (MTZ 82+ KRN 2.8) in the following terms: 1st – June 16, 2nd – June 21, 3rd – July 01, 4th – July 8th. In addition, after closing the rows, manual weeding of pumpkin crops was carried out as needed.

To increase plant immunity, crop yield and quality of cultivated products, pumpkin was treated with biological product "EM-1 microbiological fertilizer "Vostok EM-1" on 6-7 real leaves at a dose of 1 ml/l.

Harvesting and accounting of the harvest. The pumpkin was harvested in the third decade of October 2021. The fruits were cut off with a pruner and left in the field for 2-3 hours until the heel dries up. Collection of fruits was carried out manually in wooden boxes with a capacity of 600 kg, followed by transportation and placement in warehouses for ripening (2 weeks).

An indicator of the effectiveness of cultivation techniques in traditional agriculture is the size of the yield obtained; in organic production, the main emphasis is on product quality. Analyzing the yield of pumpkin, it shall be noted that this indicator varied from 15.5 to 17.1 t/ha (total) and from 11.1 to 12.2 t/ha (marketable fruits). At the same time, the average weight of one pumpkin fruit was 1.63-1.98 kg (Table 1). Under the conditions of 2021, the maximum yield values were obtained for the pumpkin cultivar of local breeding Vnuchka.

Table 1. Pumpkin yield in the farm of A.I. Grishko, 2021

Variety/hybrid	Productivity, t/ha		Average weight of one fruit, kg
	general	marketable fruits	
Vnuchka	**17.1**	**12.2**	**1.98**
Kurishogun neo F1	15.5	11.1	1.63
TSX-891 F1	16.4	11.7	1.83
Sweet mama F1	15.8	11.3	1.67

Note: according to the organic standard JAS (Japan), the mass of one pumpkin fruit shall be in the range of 1.1-2.7 kg.

One of the indicators of the quality of organic products is the content of nitrates in it. Studies have shown that in pumpkin fruits grown according to the Japanese JAS standard, the nitrate content varied from 72 to 88 mg/kg, which did not exceed the permissible MPC value [16].

Economic calculation. The economic calculation of the production of organic pumpkin was carried out within the framework of the project of the Union of Organic Agriculture "Organic agriculture – new opportunities. The system and practices of responsible land use, sustainable development of rural areas" with use of the grant of the President of the Russian Federation for development of civil society provided by the Presidential Grants Fund [17].

The data in Table 2 show that with an average pumpkin yield of 12 t/ha, the cost of organic production is 212,160 rubles/ha, and the cost of pumpkin is 17 rubles 68 kopecks

Table 2. Economic indicators of organic pumpkin production in the farm of A.I. Grishko

Name of indicator	Measuring unit	2021
Yield	t/ha	12
Seed costs (1.5 kg/ha)	rub/ha	29,640
Fuel costs	rub/ha	10,920
Costs of EM drugs ("Vostok EM-1" and "EM-5")	rub/ha	2,000
Container - wooden boxes of 500 kg	rub.	65,000
Electricity in warehouses	rub.	2,200
General running costs	rub/ha	11,800
Salary,	rub/ha	90,600
Total costs per hectare	rub.	212,160

Preparing and conducting a certification audit at the farm of A.I. Grishko according to the Japanese standard JAS were carried out with the sponsorship of the Federal State Budgetary Institution Rosselkhoztsentr in the Primorsky Territory, and, therefore the costs were not reflected in the calculation of economic indicators of production.

4 Conclusions

Based on the results of the studies, it was established that the soil and climatic conditions of the Partizansky District of the Primorsky Territory are suitable for the cultivation of pumpkins, taking into account the requirements of the Japanese standards JAS. Compliance with the recommended agricultural practices in the cultivation of melons and gourds made it possible to obtain a yield of marketable pumpkin fruits of 11.1-12.2 c/ha with high quality indicators.

Acknowledgments

The authors express their gratitude to the Union of Organic Agriculture and personally to Sergey Aleksandrovich Korshunov for his help in arranging the scientific study.

References

1. Food and Agriculture Organization of the United Nations (FAO), https://www.fao.org
2. S.S. Litvinov, Plant protection and quarantine, **4**, 3 (2015)
3. National Organic Union of the Russian Federation, https://rosorganic.ru
4. A.N. Panasyuk, AgroEkoEngineering, **2(99)**,194 (2019)

5. Yu. P. Boyko,Russia: trends and development prospects, **12-1,**712 (2017)
6. O.A. Starovoitova, Agroengineering, **6(76),**42 (2016)
7. Federal Law "On Organic Products and on Amendments to Certain Legislative Acts of the Russian Federation" dated August 03, 2018 No. 280-FZ (latest edition).
8. THE DAIRYNEWS. https://www.dairynews.ru
9. S.B. Markaryan, Eastern Analytics,**3,** 71 (2020)
10. N.P. Sidorova, Potatoes and vegetables,**8,**28 (2012)
11. T.I. Zavyalova, IzvestiaSPbGAU, **39,** 45 (2015)
12. B.A. Dospekhov. *Field experiment methodology (with the basics of statistical processing of research results): Textbook for higher agricultural educational institutions*, M.: Alliance, 351 (2014)
13. V.F. Belik, *Experimental methodology in vegetable and melon growing,*M.: Agropromizdat, 312 (1992)
14. V.P. Kotov, *Vegetable growing with the open ground: tutorial,* SPb.: Prospect Nauki, 360 (2012)
15. E.P. Sinelnikov. Soils of the agricultural zone in the south of the Far East: Text of lectures, Ussuriisk, 60 (1987)
16. Technical Regulations of the Customs Union TR CU 021/2011 "On Food Safety".https://www.rst.gov.ru
17. Organic Farming Union, https: //soz.bio

Sunflower diseases control by biological fungicides

Nadezhda Bushneva*, and *Alexander* Bushnev

V.S. Pustovoit All-Russian Research Institute of Oil Crops, 17 Filatova street, 350038 Krasnodar, Russia

Abstract. The demand for organic food has grown rapidly over the last 25 years in many countries of the world. Organic farming standards require the application of biological products of disease control. The article presents the material on studying the effectiveness of fungicides based on microbiological antagonists against sunflower diseases. We carried out the study in 2020-2021 at V.S. Pustovoit All-Russian Research Institute of Oil Crops (VNIIMK) (Krasnodar). We established that the fungicides Geostim Fit type E, WS based on microbiological antagonists did not reduce the germination of sunflower seeds and actively suppressed seed infection. Their effectiveness against *Alternaria* spp. reached 50-74 %, against *Rhizopus* spp. – 74–100 %, against bacteria – 76–80 %. Double treatment of sunflower plants with BFTIM, WS (different strains) reduced the frequency of bacterial blight to 37 %, of dry rot – to 60 %, of Alternaria blight – to 39 %, of Phoma rot – to 41 %, and of Fusarium blight – to 75 %. The application scheme of Geostim Fit E (G), WS; BFTIM (E), WS; BFTIM (Zh), WS showed the best economic effectiveness with a seed yield of 3.56 t/ha, net income of 122 910 rubles, and profitability of 311 %.

1 Introduction

The demand for organic food has grown rapidly over the last 25 years in Europe, North America, and Asia. This is due to the fact that organic agriculture is good for the environment, soil, biodiversity, food quality, and food safety compared to traditional agriculture. Organic agriculture is also recognized as one of the alternative farming systems for overcoming climate change problems [1-4].

Organic farming standards, as defined by government laws and regulations in most countries, prohibit or limit the use of many external factors that are commonly used in traditional agriculture [5]. In particular, the standards of organic crop production prohibit the use of synthetic plant protection chemicals (fungicides, herbicides, insecticides, plant growth regulators) and basic mineral fertilizers (all sources of N, KCl, and superphosphate), which are commonly used in traditional farming systems [6-8]. Instead, standards require the regular application of organic fertilizers (e.g., manure and compost), the use of legumes in crop rotation (to increase nitrogen levels and balance nitrogen to phosphorus ratios in the soil), and the use of preventive and non-chemical plant protection methods, such as the use

* Corresponding author: naducha74@mail.ru

of diverse crop rotations, resistant (tolerant) varieties, mechanical weeding as well as biological products for control of diseases and pests [5].

The harmfulness decrease of disease pathogens by fungal and bacterial antagonists is becoming increasingly common in crop protection [9-11]. For instance, antagonistic bacteria can secrete chitinase and protease enzymes, which are responsible for their effectiveness against many fungal pathogens [12]. Mixtures of microorganisms and their metabolites in the form of wettable powders and liquids help to reduce disease affection, increase seed germination and productivity indicators, obtain a preserved yield and additional income [13]. Polyfunctional preparations have a wide range of antifungal action, they include fungicides based on micromycetes of the *Trichoderma* spp. genus, which are effective in the control of fungal diseases of plants. In addition, it is established that *Trichoderma* spp. fungi increase plant growth, disease resistance, and abiotic stress [14].

Sunflower is a widespread oil crop, including the Krasnodar region, where the natural and climatic conditions are favorable for their cultivation. However, high sunflower yields are hindered by several diseases, which disrupt normal growth and development, and sometimes kill the plants. Annual phytosanitary inspections of the sunflower sowings in the Krasnodar region show that infestations by dry (Rhizopus) rot and bacterial blight are increasing. The aim of producers to reduce the number of pesticides used in crops cultivation, including sunflower, is stimulating the search for effective preparations against diseases based on microbiological antagonists.

Thereby, the purpose of this research was to evaluate the effectiveness of organic fungicides in the protection system of sunflower sowings against diseases in the conditions of the Krasnodar region.

2 Materials and methods

The article is based on the results of field trials of biological fungicides of the company OOO "Biotechagro" (Timashevsk, Russia) on the basis of microbiological antagonists against diseases of confectionery sunflower varieties of VNIIMK breeding (variety SPK), conducted in 2020-2021 on the leached chernozem of the Western Ciscaucasia in the central agroclimatic zone of the Krasnodar region at V.S. Pustovoit All-Russian Research Institute of Oil Crops (VNIIMK) (Krasnodar).

Sunflower is affected by diseases throughout the growing season, so we included a seed treatment and two plant treatments in the research scheme to control the diseases more effectively. We tested 4 variants of sunflower crop protection schemes against diseases. Each scheme included: treatment of sunflower seeds with fungicides under the trademark Geostim Fit type E based on *Trichoderma viride, Pseudomonas koreensis, Bacillus subtilis* (different strains of antagonists), and treatment of plants at the stage of 2-3 pairs of leaves and at the budding stage with biofungicides under the BFTIM trademark based on *Bacillus amyloliquefaciens* (titer 1×10^9 CFU/ml) (different strains of antagonists). The variant with untreated sunflower seeds and plants was the control. The standard was the variant with the recommended chemical fungicides (for seed treatment – Fludioxonil, SC (25 g/l) + Mefenoxam, AQ (350 g/l) – 5.0 + 3.0 l/t; for the first plant treatment – Azoxystrobin + Cyproconazole, SC (200 g/l + 80 g/l) – 0.8 l/ha; for the second plant treatment – Dimoxystrobin + Boscalid, SC (200 g/l + 200 g/l) – 0.5 l/ha) (Table 1).

Table 1. Scheme of field trials of sunflower crop protection systems against diseases in the Krasnodar region, 2020–2021.

№	Variant	Application rate, l/t, l/ha	Method of application
1	Control	–	without treatment
2	Fludioxonil, SC + Mefenoxam, AQ	5.0 + 3.0	seed treatment
	Azoxystrobin + Cyproconazole, SC	0.8	treatment at the stage of 2-3 pairs leaves
	Dimoxystrobin + Boscalid, SC	0.5	treatment at the budding stage
3	Geostim Fit E (A), WS	2.0	seed treatment
	BFTIM (A), WS	2.0	treatment at the stage of 2-3 pairs leaves
	BFTIM (B), WS	3.0	treatment at the budding stage
4	Geostim Fit E (B), WS	3.0	seed treatment
	BFTIM (V), WS	2.0	treatment at the stage of 2-3 pairs leaves
	BFTIM (G), WS	3.0	treatment at the budding stage
5	Geostim Fit E (V), WS	3.0	seed treatment
	BFTIM (D), WS	3.0	treatment at the stage of 2-3 pairs leaves
	BFTIM (D), WS	3.0	treatment at the budding stage
6	Geostim Fit E (G), WS	5.0	seed treatment
	BFTIM (E), WS	2.0	treatment at the stage of 2-3 pairs leaves
	BFTIM (Zh), WS	3.0	treatment at the budding stage

We carried out the presowing treatment of sunflower seeds with fungicides on a laboratory encrustator Hege. We sowed the seeds using a GASPARDO-MT 8 seeder according to the scheme with a seeding rate recommended for this variety: 400.000 pcs/ha, the number of experiment replications was 3, the area of one plot was 112 m^2. We recorded all manifestations of diseases and made calculation of the biological effectiveness of fungicides according to the guidelines for registration testing of fungicides in agriculture [15]. Sunflower was harvested using an Acros-530 combine, the yield was corrected to 100 % of purity and 10 % of moisture content. We processed the research data by analysis of variance using the SPSS software.

3 Results and discussion

We determined the sowing qualities of sunflower seeds and their infection with pathogens before sowing. We found that the laboratory germination of seeds untreated with fungicides was lower and at the level of 87 % due to the presence of a seed infection complex, consisting of the fungi *Alternaria* spp., *Rhizopus* spp., and bacteria (*Xanthomonas* spp., *Pseudomonas* spp.) with 34, 15, and 10 % of affected seeds and seedlings, respectively. The seed treatment with fungicides increased germination by reducing pathogen infection. In variants with the treatment of seeds with Geostim Fit E, WS (different strains), germination exceeded the control and reached 90-92 %, germination was highest with seed treatment with Fludioxonil, SC + Mefenoxam, AQ – 96 %. Compared with the control, the variants showed the decrease of the infection of seeds and seedlings: with the fungi of *Alternaria* spp. – by 50-88 %, with *Rhizopus* spp. – by 44–100 %. The chemical standard and the variant with seeds treatment with Geostim Fit E (A), WS showed the best results against fungal infection. We noted the decrease of seed infection with bacterial blight by 76-80 %

when seeds were treated with biological fungicides; chemical preparations had the lowest effectiveness – 30 %.

During the period from seedling emergence to sunflower maturity, we recorded the occurrence of the diseases: the crop plants were infected with bacterial blights (*Xanthomonas* spp., *Pseudomonas* spp.), dry rot (*Rhizopus* spp.), Alternaria blight (*Alternaria* spp.), Fusarium blight (*Fusarium* spp.), and Phoma rot (*Plenodomus lindquistii* (Frezzi)).

The prevalence of dry rot in the experiment was 33.3-32.9 %. It was lower in variants № 6 Geostim Fit E (G), WS; BFTIM (E), WS; BFTIM (Zh), WS) and № 4 (Geostim Fit E (B), WS; BFTIM (V), WS; BFTIM (G), WS) – 13.0 and 15.7 %, respectively (Table 2).

Table 2. Effect of fungicides on disease prevalence (%) in sunflower sowing, V.S. Pustovoit All-Russian Research Institute of Oil Crops, 2020–2021.

№	Variant	*Xanthomonas* spp., *Pseudomonas* spp	*Rhizopus* spp.	*Alternaria* spp.	*Plenodomus lindquistii* (Frezzi)	*Fusarium* spp.,
1	Control (without treatment)	54.0	32.9	24.0	17.0	20.0
2	Fludioxonil, SC + Mefenoxam, AQ; Azoxystrobin + Cyproconazole, SC; Dimoxystrobin + Boscalid, SC	42.0	20.6	14.0	9.1	5.0
3	Geostim Fit E (A), WS; BFTIM (A), WS; BFTIM (B), WS	37.2	24.0	16.0	12.0	9.0
4	Geostim Fit E (B), WS; BFTIM (V), WS; BFTIM (G), WS	32.0	15.7	18.2	10.0	7.2
5	Geostim Fit E (V), WS; BFTIM (D), WS; BFTIM (D), WS	34.7	19.0	20.0	13.6	9.0
6	Geostim Fit E (G), WS; BFTIM (E), WS; BFTIM (Zh), WS	34.0	13.0	14.6	13.0	5.0

We observed the maximum number of sunflower plants infected with bacterial blight in the control and the standard – 54.0 and 42.0 %, respectively. The bacterial blight in other variants was at the level of 32.0-37.2 %. The prevalence of Fusarium blight in the control reached 20.0 %, while in the variants with fungicides it was significantly lower, averaging 5.0-9.0 %. Sunflower infection with Alternaria blight in the control was 24.0 %, while in the variants with treatment it was slightly lower – 14.0-20.0 %. By the end of the sunflower growing season, the highest Phoma rot prevalence was in the control (17.0 %). The sunflower treatment with fungicides kept the disease occurrence at the level of 9.1-13.6 %.

Based on the results of records of the prevalence of the diseases, we calculated the biological effectiveness of fungicides. The application of chemical preparations reduced the bacterial blight prevalence by only 20 %, while the biological fungicides showed effectiveness at the level of 31-40 %. The variant № 6 (Geostim Fit E (G), WS; BFTIM (E), WS; BFTIM (Zh), WS) was the most effective against dry rot – 60 %. The variant № 3

(Geostim Fit E (A), WS; BFTIM (A), WS; BFTIM (B), WS) had the minimum biological effectiveness – 27 % (Table 3).

Table 3. Biological effectiveness (%) of fungicides against diseases on sunflower, V.S. Pustovoit All-Russian Research Institute of Oil Crops, 2020–2021.

№	Variant	*Xanthomonas* spp., *Pseudomonas* spp	*Rhizopus* spp.	*Alternaria* spp.	*Plenodomus lindquistii* (Frezzi)	*Fusarium* spp.,
1	Control (without treatment)	54.0*	32.9*	24.0*	17.0*	20.0*
2	Fludioxonil, SC + Mefenoxam, AQ; Azoxystrobin + Cyproconazole, SC; Dimoxystrobin + Boscalid, SC	22	37	41	46	75
3	Geostim Fit E (A), WS; BFTIM (A), WS; BFTIM (B), WS	31	27	33	29	55
4	Geostim Fit E (B), WS; BFTIM (V), WS; BFTIM (G), WS	35	32	24	41	64
5	Geostim Fit E (V), WS; BFTIM (D), WS; BFTIM (D), WS	35	42	16	20	55
6	Geostim Fit E (G), WS; BFTIM (E), WS; BFTIM (Zh), WS	37	60	39	23	75

* – the diseases prevalence in the control, %

The effectiveness of fungicides against Alternaria blight was 16-41 %, the best variants being № 2 (Fludioxonil, SC + Mefenoxam, AQ; Azoxystrobin + Cyproconazole, SC; Dimoxystrobin + Boscalid, SC) and № 6 (Geostim Fit E (G), WS; BFTIM (E), WS; BFTIM (Zh), WS) – 41 and 39 %, respectively. The standard and the variant № 6 (Geostim Fit E (G), WS; BFTIM (E), WS; BFTIM (Zh), WS) had the maximum decrease of Fusarium blight prevalence – 75 %; in the remaining variants, the decrease was within 55-64 %. The variants № 2 (Fludioxonil, SC + Mefenoxam, AQ; Azoxystrobin + Cyproconazole, SC; Dimoxystrobin + Boscalid, SC) and № 4 (Geostim Fit E (B), WS; BFTIM (V), WS; BFTIM (G), WS) were the best against Phoma rot with biological effectiveness of 46 and 41 %, respectively.

We determined the economic effectiveness of the studied protection schemes after sunflower maturing. Sunflower seed yield in the variants Geostim Fir E (A), WS; BFTIM (A), WS; BFTIM (B), WS, and Geostim Fit E (B), WS; BFTIM (V), WS; BFTIM (G), WS was lower compared to the control by 0.20 and 0.33 t/ha, respectively, but we observed no significant differences between them. The variant № 6 (Geostim Fit E (G), WS; BFTIM (E), WS; BFTIM (Zh), WS) gave the highest sunflower yield, with a preserved yield of 0.50 t/ha. We calculated the economic effectiveness based on the results of trials of plant protection schemes against diseases on sunflower (Table 4).

Table 4. Economic effectiveness of plant protection systems against diseases on sunflower, V.S. Pustovoit All-Russian Research Institute of Oil Crops 2020–2021.

№	Variant	Yield, t/ha	Production costs per 1 ha	Net income per 1 ha, rub	Profitability, %
1	Control (without treatment)	3.06	38 396	102 364	267
2	Fludioxonil, SC + Mefenoxam, AQ; Azoxystrobin + Cyproconazole, SC; Dimoxystrobin + Boscalid, SC	3.32	47 441	105 279	222
3	Geostim Fit E (A), WS; BFTIM (A), WS; BFTIM (B), WS	2.86	40 034	91 526	229
4	Geostim Fit E (B), WS; BFTIM (V), WS; BFTIM (G), WS	2.73	39 226	86 354	220
5	Geostim Fit E (V), WS; BFTIM (D), WS; BFTIM (D), WS	3.31	40 457	111 803	276
6	Geostim Fit E (G), WS; BFTIM (E), WS; BFTIM (Zh), WS	3.56	39 470	122 910	311

The results show that the cost of cultivating 1 ha of sunflower was in the range of 38396 - 47441 rubles/ha, including the additional cost of preparations – 839.14-8915.0 rubles/ha. Without fungicide treatments, costs were lowest and reached 38 396 rubles/ha, while the standard had the highest costs – 47 441 rubles/ha. The highest net income per 1 ha in the experiment was received in the application of Geostim Fit E (G), WS; BFTIM (E), WS; BFTIM (Zh), WS – 122 910 rubles/ha, which exceeded the standard by 17 631 rubles/ha. Profitability in this variant was 311 %.

4 Conclusions

The research showed that the fungicides Geostim Fit of type E, WS based on the microbiological antagonists did not reduce the germination of sunflower seeds and actively suppressed seed infection. Their effectiveness against *Alternaria* spp. reached 50-74 %, against *Rhizopus* spp. - 74–100 %, against bacteria - 76–80 %. Double treatment of sunflower plants with BFTIM, WS (different strains) reduced the frequency of bacterial blight to 37 %, of dry rot - to 60 %, of Alternaria blight - to 39 %, of Phoma rot - to 41 %, and of Fusarium blight - to 75 %. The application of the scheme of Geostim Fit E (G), WS; BFTIM (E), WS; BFTIM (Zh), WS in the cultivation technology of confectionery sunflower showed the best economic effectiveness with a seed yield of 3.56 t/ha, net income of 122 910 rubles, and profitability of 311 %.

References

1. H. Willer, L. Kilcher, the World of Organic Agriculture. Statistics and Emerging Trends 2011. FiBL-IFOAM Report IFOAM, Bonn and FiBL, Frick (2011)
2. Y. Joshi, Z. Rahman, International Strategic Management Review, **3(1–2)**, 128–143 (2015)
3. A. Naorem, J. Maverick, P. Singh, Sh. K. Udayana, Advances in Organic Farming Agronomic Soil Management Practices, 47-58 (2021)

4. T. Gomiero, M.G. Paoletti, D. Pimentel Critical Reviews in Plant Sciences, **27**, 239–254 (2008)
5. P. Bilsborrow, J. Cooper, C. Tétard-Jones, D. Średnicka-Tober, M. Barański, M. Eyre, Ch. Schmidt, P. Shotton, N. Volakakis, I. lCakmak, L. Ozturk, C. Leifert, S. Wilcockson, European Journal of Agronomy, 51, 71-80 (2013)
6. B. P. Baker, C. M. Benbrook, E. Groth III K. Benbrook Food Additives & Contaminants, 19, 427–446 (2002)
7. H. Willer, L. Kilcher, The World of Organic Agriculture. Statistics and Emerging Trends 2016 Bonn and FiBL, Frick (2016)
8. H. Willer, L. Kilcher, The World of Organic Agriculture. Statistics and Emerging Trends 2017 Bonn and FiBL, Frick (2017)
9. A. Majeed, M.K. Abbasi, S. Hameed, S. Yasmin, M.K. Hanif, T. Naqqash, A. Imran, Microbiological Research, 216, 56–69 (2018)
10. Sh. Moin, S.A. Ali, K.A. Hasan, A. Tariq, V. Sultana, J. Ara, S. Ehteshamul-Haque, Crop Protection, 130 (2020)
11. M. Kgatle, B. Fletty, M. Truter, T. Aveling, Crop Protection, 132, 105–146 (2020)
12. W.T. Chang, M.L. Chen, S.L. Wang Microbiology and Biotechnology, 26, 945–950 (2010)
13. N.A. Nafady , M. Hashem, E.A. Hassan, H. Ahmed, S.A. Alamri, Biological Control 138 (2019)
14. R.N. Silva, V.N. Monteiro, A.S. Steindorff, E.V. Gomes, E.F. Noronha, C.J. Ulhoa, Fungal Biology, 123, 565–583 (2019)
15. K.V. Novozhilova, A.I. Anisimov, *Guidelines for registration testing of fungicides in agriculture* (2009)

Organic agriculture as a factor in the development of rural tourism in the Sverdlovsk region

A. A. Oshkordina, *E. I.* Okhrimenko, and *N. A.* Goncharova
Ural State University of Economics, 620144, Yekaterinburg, Russia

Abstract. The authors in the article consider sectoral changes and structural transformations in the field of agriculture associated with the influence of global trends and growth factors in agricultural productivity. One of these trends, the authors identify rural tourism (ecotourism), the development of which increases the socio-cultural, financial and economic level of agricultural areas. The factors of development and implementation of agricultural technologies of organic agriculture, which have a positive effect on the development of rural tourism and form the demand for tourism products, have been identified. The ways of improving the organization and implementation of event and tourist activities as effective tools for attracting attention and popularizing rural areas have been identified, which in the context of the prevention of coronavirus infection around the world is an urgent factor for many categories of the population. In the context of the prevention of coronavirus, infection in many countries, domestic tourism is becoming a priority area for the development of the tourism industry, focusing, among other things, on the development of new products for rural and agro-industrial tourism.

1 Introduction

As the world practice shows, the agro-industrial complex is the most important, rapidly developing sector of the economy, which determines the economic security of any country. As the head of Rosstat Pavel Malkov noted: "Agriculture is one of the key sectors of the economy. This is economic and food security and this is the independence and confidence in the future of the country. «The development of the Russian agro-industrial complex will be determined by market-significant technologies of the agricultural production process, focused on informatization, automation, robotization and integration of the activities of organizations of various industries and subsectors of the national economy. In modern socio-economic conditions, the multiplier effect of tourism can make a significant contribution to the development of the economy of rural areas as a factor in additional sources of financial resources, the formation of the social infrastructure of the area, as well as the popularization and enhancement of the cultural, cognitive and patriotic consciousness of the population.

Rural tourism (ecotourism) is not only a certain tribute to the widespread fashion, but also a certain necessity for the urban population living in an unfavorable ecological and psycho-emotional environment of the modern and dynamically changing rhythm of life of every city dweller. Rest in the countryside allows you to get closer and enjoy nature, combined with an introduction to the historical, cultural and ethnographic aspects of the heritage. Get acquainted with customs, crafts, national cuisine, clothes, etc. the population living in a given territory. In addition, rural tourism can also carry a research orientation in the form of collecting folklore, studying the local language and dialect, collecting plant and mineral raw materials and materials, etc. According to Smirnova I.V. the population, tired of the bustle of a big city, seeks to relax physically, mentally, enjoy the beauty of nature, touch the cultural heritage of historically significant, but not as popular places as in large tourist centers [1].

The purpose of the study is to determine the main directions of the development of ecotourism as a factor in increasing the economic efficiency of the development of agricultural technologies in organic agriculture.

Thus, the territory of our country, including the Sverdlovsk region, has a powerful potential of natural, landscape, climatic, cultural, historical, ethnic and economic resources, which involves the implementation of various kinds of tourism activities in rural areas. At the same time, it should be noted that the implementation of ideas and activities in rural areas is possible only in ecologically clean areas that are not covered by industrial production, polluting the environment in rural areas.

2 Materials and Methods

The materials and research base were statistical, accounting and reporting and information sources of the activities of agricultural organizations and enterprises of the tourism industry of the Sverdlovsk region. The materials of scientific research of domestic and foreign scientists in the field of development of rural and agro-industrial tourism were used. When processing and systematizing information and analytical material, the methods of grouping, detailing, synthesis, the historical and logical research method, as well as methods of economic analysis were used.

3 Results and Discussion

Based on the study of the information and statistical material of the UN comtrade dadabase databases and industry experts, prepared by the International Independent Institute of Agrarian Policy, 4 groups are distinguished in the trends in the development of high technologies for the next 10 years:

1) demography - an increase in the level of urbanization and population growth, as well as a high level of medical technologies in health care, which increases the life expectancy of the population, causes an increase in agricultural production throughout the world, which will entail an increase in the intensity of production processes while minimizing labor costs;

2) climate and resources - a decrease in land and water resources, environmental degradation, changes in climatic factors cause a reorientation from traditional agricultural technologies to irrigation technologies;

3) market - the globalization of world food trade, the formation of vertically integrated transnational agricultural holdings, the development of legislation governing world food trade;

4) technologies to improve the economic production, social efficiency of agricultural production [2].

In the 70s of the last century, technologies of organic agriculture are being introduced into the agro-industrial complexes of economically developed countries, involving the organization of a social-production system with a focus on maintaining the health of the population, ecology, soil and food. Since this time, the volume of sales of organic agriculture products has tended to increase steadily and by 20012 reached $ 63 billion worldwide. This demand led to a similar increase in organic farmland, which increased by 8.9% from 2001 to 2011. Annually [3].

By 2019, more than 70 million hectares on Earth are used in accordance with the principles of organic agriculture, which is approximately 1.4 percent of the total agricultural land in the world [4]. Moreover, more than half of this volume is accounted for by Australia [5]. Since 2015, the world market for organic products has been steadily increasing on average by about 15% [6].

According to the Union of Organic Agriculture, the main incentives for the development of organic production are: public health (24%), high added value (27%), environmental improvement (22%), personal health (15%), humanitarian attitude towards animals (7%) and other (7%) and is currently the most promising direction in the development of agriculture [7]. At the same time, the transition to environmentally friendly agriculture is currently among the priorities of the scientific and technological development of Russia, but it is not happening at a fast enough pace. The Strategy [8] notes the problem of "increasing anthropogenic pressures on the environment to a scale that threatens the reproduction of natural resources, and the increase in risks to the life and health of citizens associated with their ineffective use".

Table 1. Quantitative indicators of the development of the agro-industrial complex of Russia [9, 10]

Indicators	2002 year	2012 year	2020 year
Number of agricultural organizations, thousand	253,1	140,6	41,9
Number of peasant farms and individual entrepreneurs, thousand	32,8	39,0	144,9
Number of non-profit associations and societies, thousand	79,8	76,3	80,4
Number of subsidiary farms, million	22,6	18,2	16,5
The share of products produced on personal subsidiary plots of the total agricultural production,%	53,8	43,2	28,2
Total area of land plots of personal subsidiary plots, million hectares	6,2	6,9	8,0

Within the last two decades, the indicators of table 1 indicate a dramatic reduction in the number of agricultural organizations by more than 5 times against the background of an increase in the number of peasant farms and individual entrepreneurs in almost equal proportion - 4 times, which may be due to factors of the regulatory and tax nature of maintaining accounting and reporting documentation various organizational and legal forms of activity of enterprises of the agro-industrial complex of Russia. Attention is also drawn to a significant change in the structure of agricultural production. So, in 2002, more than half of the total agricultural production fell on personal subsidiary plots, while in 2020 the share of agricultural production was less than 30% in its total volume. At the same time, it is necessary to note an increase in the area of land plots owned by Russians from 6.2 million hectares to 8 million in 2020.

The ratio of the indicators under consideration may indicate a decrease in the population's interest in the production of agricultural products with their own hands due to the saturation of the market, as well as the reorientation of the needs of citizens to use homesteads and dacha estates as objects of aesthetic and contemplative rest and entertainment. In addition, the pandemic of coronavirus infection has made its own adjustments to increase the demand for the acquisition of land plots in order to grow crops and food with our own hands due to economic, physiological and environmental factors that can strengthen the body's immune system and increase its resistance in the process of carrying out preventive measures for fight against coronavirus infection. This thesis is confirmed by a member of the State Duma Committee on Economic Policy and Entrepreneurship, Chairman of the Union of Gardeners of Russia Oleg Valenchuk, who in his interview reflected the high interest of the population in acquiring abandoned plots in order to grow agricultural products with their own hands. At the same time, Andrey Tumanov, editor-in-chief of the newspaper Vashi 6 sotok, a member of the public council of Rosreestr, urged not to rush into forecasts due to the lack of reliable statistics on this issue, the database of which has not yet been formed over the past 2 years. In addition, Andrei Tumanov noted the high demand for the population of only those land plots that can be capitalized in the future, are distinguished by a high level of transport accessibility, are gasified, that is, absolutely suitable for permanent or temporary residence and recreation, which does not at all imply the goal of collecting a high yield [11]. This assumption is indirectly confirmed by the statistical indicators of Table 1, reflecting a decrease in the number of subsidiary farms by more than 30%, as well as a decrease in the share of agricultural products produced on personal subsidiary farms from the total agricultural production by almost 2 times.

Thus, over the past 2 decades, significant structural transformations have taken place in the structure of the agro-industrial complex of Russia, associated with an increase in the number of peasant farms and individual entrepreneurs by almost 5 times, while the number of agricultural organizations has decreased by more. It should be noted a very important factor, which in turn affects the rapid growth of changes and transformations in the field of agriculture - this is a reorientation of consumer preferences associated with an increase in the population's demand for the consumption of environmentally friendly, healthy products, as well as cultural, entertainment and contemplative pastime in the bosom nature in rural areas with the aim of improving the effect of the psychoemotional state of a person. In addition, today, uniform approaches to taste preferences and global brands and brands are gradually becoming outdated, yielding to the formation of demand for regional, local products that reflect the cultural and historical heritage of a given territory. At the same time, an urgent approach to the formation of a healthy lifestyle and the prevention of many diseases with the use of occupational therapy technologies in ecologically safe agricultural areas is also increasingly important. Therefore, ecotourism today is becoming an increasingly popular trend not only in Russia, but also in other economically developed countries.

The territory of the Sverdlovsk Region was formed on January 17, 1934 and is located on the border of Europe and Asia at the intersection of transcontinental flows of raw materials, goods, financial, labor and information resources. The Sverdlovsk Region today is a large economically developed territory of Russia with a high level of business, cultural and social activity, one of the most promising subjects of the Russian Federation. According to official statistics, the resident population of the region at the beginning of 2021 is 4 million 290 thousand people, of which 85% is urban population, and only 15% of the population of the total population of the region is rural residents, which once again confirms the high level of urbanization and the uneven ratio of urban and rural populations. On the territory of the Sverdlovsk region there are 47 cities, 26 workers' settlements and urban-type settlements, 1804 rural settlements [12]. Despite the harsh climatic features, agriculture is rapidly developing in the Sverdlovsk region. Therefore, at the beginning of 2021, 350 agricultural organizations are carrying out production and economic activities, about 800 peasants and farms and entrepreneurs, about 419 thousand citizens have personal subsidiary plots. Thus, on the territory of the Sverdlovsk region. According to the latest data in the Sverdlovsk region, more than 818.0 thousand hectares are allotted for sown areas for agricultural crops. By industry, the average annual number of employees as of 01.01.2020 is: in agriculture - 51,524 people, in fish farming and fishing - 217 people, in food production - 32,566 people, in the production of drinks - 2837 people. Thus, according to the main socio-economic development criteria, the Sverdlovsk Region is included in the top ten regions of the Russian Federation. In addition, the socio-cultural infrastructure formed on the territory of the region allows many rural areas to implement event, sports and tourist events.

Analyzing the information and analytical material of the Tourism Development Center of the Sverdlovsk Region on the implementation of event activities in rural areas and settlements, one can identify a rather low level of activity among the population and rural administrations. So, in 2021, more than 370 events of various orientations were planned, from cultural, ethnic and religious to sports, entertainment and educational. Only 31 events of an eventful nature were implemented in the rural areas of the Sverdlovsk region, which amounted to about 8% of their total number. At the same time, it is necessary to note as a positive trend the annual increase in the implementation of projects in the field of event events, as well as the formation of new products of rural, eco and gastronomic tourism, which indicates an increase in consumer demand and the involvement of an increasing number of rural settlements in the tourism industry. At the same time, it is necessary to note the fact of expanding the range of tourist products. So, for example, in the Artinsky district, the Festival of Mowers in July is already righteous 12 times, the festival "Valenok 21" was held for the first time, and also in 2021, in July, the participants of the motocross celebrated the sixth anniversary, and this year the motocross was held for the first time and in the winter period ... The following events can be noted as positive examples: Motocross in memory of B.F. Zefirova in the village of Melnikova; the massive descent "Avalanche" in the village of Uralets; "Maslenitsa" and "Trinity Festivities" in the villages of Pokrovskoye, Chernoskutovo, Kadnikovo; competition "Dog Race" in the village of Visim; "Georgievskaya Fair" in the village of Mironovo; Festival "Broad Cossack Prowess" and "Sabantuy" - a bright and colorful holiday that contributes to the development of national cultures, strengthening the friendship of peoples in the settlements of Lesnoy and Novaya Lyalya; Festival Lyalinskoe Porechye "; Festival "Konzhakovsky chord" in the village of Kytlym; in the picturesque area of the village of Kashino, a holiday dedicated to the Day of Family, Love and Fidelity is taking place. At the central site of the Kashinsky Round Dance, newlyweds are honored and an old wedding ceremony is held. Creative teams perform here and various competitions are held; Festival "Park of National Cultures" The village of Bulanash is a multinational settlement where representatives of 30 national cultures live; "Demidov - fest" in the village of Chernoistochinsk; the Visimsky picnic gastronomic festival, restoring forgotten recipes of the Ural cuisine and rediscovering the Ural culinary brands; Semeonovskaya fair in the village of Raskat; Stone festival "Samotsvetnaya side" in the village of Murzinka; Festival "At the outskirts" in the village of Kordyukovo; Fair "Krasnogorsk torzhok" in the village of Krasnogorskoe; Birthday of Ural Moroz in the village of Aramil and others.

If we talk about the degree of active participation in event activities of all rural areas of the region, then we have to state a sad fact: only about 30 rural settlements of the region out of 1804 take any part in the organization and implementation of tourism activities, which, first of all, indicates a low level of interest, both the rural population itself and the leading personnel of rural administrations and agricultural enterprises. At the same time, in recent years, the government of the Sverdlovsk Region has provided organizers of events in sparsely populated cities and rural areas with a unique opportunity to apply for financial, organizational and legal support free of charge.

One cannot ignore the young unique project in the village of Mramorskoye "Day of Marble Music" within the framework of the "Marble Mile" project, which was successfully launched in 2019 with the support of the Idea Social Activity Fund. Another unique project, known for over 40 years and attracting the attention of tourists not only of the Sverdlovsk region, but also guests from other regions "Nizhnyaya Sinyachikha and the Museum-Reserve of Wooden Architecture". Moreover, attention is drawn to the fact that the implementation of this project was initiated by one and only person, Ivan Danilovich Samoilov. To date, a stable tourist flow per month has been formed in Nizhnyaya Sinyachikha, averaging 400-500 people. Nizhnyaya Sinyachikha is proof of the possible organization and implementation of projects in the presence of personal interest and activity of the local population. Nizhnie Tavolgi is an old Ural village with a population of about 350 people, since ancient times it was famous for its products of pottery. Together with the Upper Tavolga, a center of folk crafts was organized. A modern tourist infrastructure has been created, providing a camphor stay on the territory of these settlements. There have been developed tourist routes of various duration and thematic component. The presence of a tourist social infrastructure makes it possible to hold regional and international festivals, such as the tea eco-festival “When Tavolga Blossoms”, the Festival of arts and crafts “Secrets of the Gem Ring”, etc. Thus, the territory is actively developing and creates additional jobs through the sale of tourism products and agricultural business.

At the same time, it is necessary to note the fact of the irregular, one-time nature of event events during the entire calendar year, which does not imply an even distribution of tourist flows, which in turn affects the abrupt formation of financial flows (both from the side of income and from the side of expenses) ... In addition to the financial and economic component, one should not forget about the formation of cultural, ethnic and cognitive functions in the implementation of event and tourist activities. So, in most of the above-listed settlements, creative and amateur groups, circles and creative communities are being created that are interested in observing local culture, traditions, customs, religious beliefs, as well as studying the historical aspects of the development of the territory of their place of residence. This undoubtedly increases the level of involvement in the creative process of children and adolescents, forming from an early age love for their small homeland, patriotism, craving for knowledge in all spheres of human life, as well as acquiring skills in the production of folk art items, souvenirs and agricultural products.

In the formation of regular flows of eco-tourists to the countryside, the production of organic agricultural products plays a vital role, which is becoming increasingly popular among the population of all countries of the world. A person wants to consume environmentally friendly, natural products, as well as study the process and recipe for cooking food and folk art. For these purposes, many farms organize short-term and long-term stays in rural areas and in many constituent entities of the Russian Federation, financial resources are allocated in the form of grant support. The Sverdlovsk region cannot boast of the rapid introduction of such tourist products and today only 7 organizations have been organized in the region, these include: Reindeer Nursery "Northern Lights", Verkhne-Pyshminsky District; "GreenPoll" in the Gornouralsky urban district, the village of Nikolo-Pavlovskoye; sled dog kennel "Elbrus" in the village of Medny; the first eco-farm "Harmony" in the region in the village of Bolshebrusyanskoye, which produces and sells products of organic agriculture; the eco-farm "Valley of the goats" in the village of Parshino, which implements weekend gastronomic tours, where you can actively relax, caring for animals, get acquainted with the intricacies of cooking dishes from animal and plant products, contemplate the beauty of the Ural nature and take part in contests and quizzes; A sturgeon farm in the village of Kosulino, which successfully implements gastronomic tours with the participation of tourists in the preparation of national dishes of the Ural cuisine; The Slavyanskoe Gnezdo eco-farm, which opened in July 2021, in the village of Beklenishcheva, where you can taste traditional dishes of the Ural cuisine and get acquainted with the oriental recipe of camel milk dishes.

Studying the reviews of tourists who used the services of eco-farms and nurseries of the Sverdlovsk region, it is necessary to note the high marks of the organization and implementation of these tours. In addition, eco and gastronomic tours are focused on all age groups of the population and, above all, on family vacations, where both children and adults are involved in the production of agricultural products, which undoubtedly also affects the formation of family values in society and makes the vacation unforgettable.

4 Conclusion

In the modern socio-economic conditions of the development of rural areas, a certain role can be played by the organization and sale of products of rural and gastronomic tourism. At the same time, the share of agritourism in the Russian tourist market does not exceed 2%. The presence of a huge variety of climatogeographic, natural, cultural-ethnic, historical resources of rural areas of Russia, including the Sverdlovsk region, has the ability to organize and introduce tourism products with a high degree of competitiveness, which in turn will allow solving various problems and problems of social and economic and cultural nature in rural areas. At the same time, rural tourism today has not received a mass character on the territory of Russia, but is so-called point distribution and is directly proportional to the activity and initiative of the local population and rural administrations. However, the implementation of ideas and projects of rural tourism requires support from the government.

Despite the dynamically developing rural tourism on the territory of the Russian Federation, in recent years, there are serious problems in the industry that require immediate solutions. So, according to I.L. Polyakova and M.P. Grigorieva. The state of rural tourism in Russia and its development trends allow us to single out a number of factors that hinder its further development:

- insufficient support for rural tourism as a separate branch of the economy;
- undeveloped legislative and regulatory framework;
- lack of a comprehensive system of incentives for rural tourism activities;
- poorly developed tourist infrastructure, including the transport system;
- the low standard of living of the rural population of the Russian regions;
- insufficiently effective mechanism for managing the rural tourism sector;
- unfavorable climatic conditions;
- terrorism, criminal situation;
- insufficiently developed infrastructure of rural areas;
- weak system of promoting and informing agricultural products in the domestic and international markets;
- lack of a recognizable brand;
- low level of investment in the development of rural tourism;
- lack of support programs for rural areas [13].

To solve these problems, in our opinion, it is necessary:

- development of regulatory documents regulating tourism activities at all stages of its implementation, taking into account the specifics of the functioning of economic entities in rural areas;
- development of regional target programs focused on financial support of current ideas and projects for the organization of rural tourism;
- attraction of financial resources of private organizations and investors on a preferential or gratuitous basis;
- raising the level of awareness about the possibility of participation of the population in various events in rural tourism through the mass media;
- update the development of short-term weekend tourist products using elements of event tourism (hiking, making organic products and folk art items with your own hands, engaging in agricultural work, collecting medicinal herbs, making agricultural products, etc.);
- popularization of themed corporate events, weddings, banquets, children's parties, etc. on the territory of agricultural enterprises. In our opinion, the set of measures developed will increase the scale of the spread of ecotourism in rural areas, as well as increase the speed of

their popularization among the population. In modern conditions of development of rural areas, agritourism should be perceived as an effective economic tool for their development from various aspects of activity.

References

1. I.V. Smirnova, Bulletin of the Mari State University, **1(13),** 87 (2014)
2. Global Seed. Organic farming (2021) http://www.globalseed.info
3. J. Paull, Journal of Social and Development Sciences, **2(3),** 111 (2011)
4. The World of Organic Agriculture 2019. FiBL and IFOAM (2019)
5. J. Paull, Journal of Environment Protection and Sustainable Development, **5(2),** 70 (2019)
6. A. N. Anischenko, Food policy and security, **2,** 97 (2019)
7. Organic agriculture: innovative technologies, experience, prospects. / scientific. analyte. Overview, 92 (2019)
8. Strategy of scientific and technological development of the Russian Federation (approved by the Decree of the President of the Russian Federation of December 1, No. 642) (2016)
9. Dyatlovskaya E. Rosstat published the first operational results of the agricultural census 2016. Agroinvestor (2016) https://www.agroinvestor.ru
10. E. Manukyan, In Rosstat told about the tasks of the agricultural census in 2021. Russian newspaper (2021) https://rg.ru
11. Yu. Ignatieva, Pitchfork on the ground: what the last microcensus will show. Izvestia. (2021) https://iz.ru
12. Official site of the government of the Sverdlovsk region (2021) http://midural.ru
13. I.L. Polyakova, M.P. Grigorieva, Service in Russia and abroad, **11(5),** 31 (2017)

The Influence of the Degree of Saturation of Crop Rotations with Alfalfa (*Medicago Sativa* L.) and Winter Wheat (*Triticum Aestivum* L.) with Ncp on the Accumulation of Plant Mass and Nutrients in the Energy Storage System of Soil Content in the Western Precaspian

A. A. Guseynov[1], *G. N.* Gasanov[1,2], *M. A.* Arslanov[1], *K. M.* Gadzhiev[2], and *T. A.* Asvarova[2]

[1]Dagestan State Agrarian University named after M.M. Dzhambulatov, Makhachkala, Russia
[2]Dagestan State Federal Research Center of the Russian Academy of Sciences, Makhachkala, Russia

Abstract. The energy storage system of soil maintenance in the irrigation landscapes of the Western Precaspian is based on the use of the second half of summer for the formation of a natural crop phytocenosis (NCP) and plowing it for green fertilizer. The article deals with the formation of plant mass in grain-grass crop rotations, the concentration and accumulation of nutrients in it. The degree of saturation of alfalfa (*Medicago sativa* L.) and winter wheat (*Triticum aestivum* L.) with natural crop phytocenosis (NCP) crop rotations ranges from 0 to 100%. The most productive of the studied crop rotations is a four-field with one output field of alfalfa and three fields of winter wheat, after harvesting which in the second half of summer a NCP for green fertilizer is formed. In this crop rotation, 97.4 t of agricultural products were produced on 1 ha of the crop rotation area, including 30.2 t/ha of its non-alienable part from the soil. Accordingly, the removal of N, P_2O_5 and K_2O from the soil increases. The proportion of the returned amount of nutrients from the phytomass inalienable from the soil was (%): N-17.7; P_2O_5 and 45.3.

1 Introduction

In the southern regions of the country (the North Caucasus, the Lower Volga region), after harvesting winter grain crops, there are 100-120 days with a warm period of the year. The sum of positive air temperatures for this period is 2500-2600^{0}C, active temperatures above 10^{0}C - 2200-2300^{0}C, photosynthetically active radiation (FAR) - 20-22 kcal/cm^2, or, respectively, 60-62%; 55-60% and 33-37% of their annual amount. This period is currently used for tillage for the next crop according to various systems, the main purpose of which is the destruction of germinating weeds. At the same time, the above-mentioned climatic resources are irrevocably and without benefit for agricultural production. Researchers recommend using this period to provoke the germination of weeds (natural phytocenosis), and use the resulting mass for green fertilizer [1, 2]. According to [2, 4], this makes it possible to obtain more than 20 t/ha of aboveground and underground phytomass, and using it as a siderate increases the yield of winter wheat (*Triticum aestivum*) with repeated sowing by 15-17%. It is even proposed to have a purely grain crop rotation, where every year the grain crop is sown according to the crop natural phytocenosis (NCP). [5, 6, 7] it is believed that NCP in the irrigation landscapes of the south of our country is a kind of busy steam, but in a different time interval (in the second half of summer) and on a different technological basis (not sown siderates, but NCP). In our opinion, the energy storage system of soil maintenance proposed by these authors should become the basis of agrotechnologies of organic agriculture in our country.

2 Materials and Methods

This article discusses the issues of accumulation of plant mass and nutrients in it in grain-grass crop rotations with varying degrees of saturation with alfalfa (*Medicago sativa*) and winter wheat (*Triticum aestivum*) with NCP.

Table 1. Scheme of experience.

Crop rotation	Degree of saturation of crop rotation, %	
	Alfalfa (*Medicago sativa*)	winter wheat (*Triticum aestivum*) with NCP
1	100	0
2	75	25
3	50	50
4	25-output field	75
5	0	100

The plot area is 100m^2, the repetition is 4 times, crop rotations are 4 complete, the years of research 2013-2019.

The soil of the experimental site in "LLC Vimpel2002" of the Khasavyurt district of the Republic of Dagestan is light-chestnut, contains in the arable layer: humus 2.77%, P_2O_5-2.21, K_2O- 32.8 mg /100g of soil, its density is 1.24 g/cm^3. The lowest moisture capacity of the 0-0.6 m layer is 29.2%.

Agrochemical and agrophysical properties of the soil were determined according to existing methods [8, 9], observations, records, analysis of the crop structure were carried out according to [10, 11], statistical data processing was carried out according to [12]. Watering of the soil after harvesting winter wheat for the formation of NCP was carried out in the first decade of July, NCP harvesting - in the phase of onset of milk ripeness of seeds of dominant phytocenoses (late August - early September), sowing of winter wheat - in the first decade of October (Grom variety) with a norm of 5 million. germinating seeds per 1 ha. Superphosphate was introduced under alfalfa: for plowing P_{40} for one year of using grass and P_{10} for sowing, nitrogen and potash fertilizers were not given. For winter wheat, P_{40} was applied for plowing, when sowing with seeds P_{10}, for pre-sowing cultivation (harrowing) - N_{30}, the same amount was added to the top dressing in the tillering phase in early spring. Potash fertilizers were not applied, given the high content of exchangeable potassium in the soil. Otherwise, the technology of cultivation of alfalfa and winter wheat corresponded to the zonal recommendations. Statistical processing of the results was carried out according to B.A. Dospekhov [12], the Microsoft Exsel computer program was also used.

3 Results and Discussion

The degree of saturation of crop rotations with alfalfa (*Medicago sativa*), winter wheat (*Triticum aestivum*) with NCP and, especially, the accumulation of phytomass in the soil, inalienable from the soil, plays a crucial role in the accumulation of nutrients in the arable soil layer. In our studies, an increase in the share of the most productive crop – alfalfa - in the structure of sown areas to 100% did not lead to an increase in the phytomass accumulated by plants compared to other crop rotations, where winter wheat occupied part of the area with the subsequent formation of NCP (Table 2).

Table 2. Accumulation of plant mass in crop rotations with varying degrees of saturation with field crops, t/ha, 2014-2019.

Crop rotation and crop ratio, %*	Culture**	Main	Secondary	Crop fields	Root	All phytomass	Including not alienated from the soil
		Products		remains			
1. 100-0-0	1	38,76	0	4,75	11,90	55,41	16,65
	2	0	0	0	0		
	3	0	0	0	0		
2.75-25-25	1	36,63	0	3,25	9,35	67,76	18,23
	2	5,57	4,32	1,66	3,63		
	3	0,51	0	0,12	0,20		
3.50-50-50	1	29,72	0	4,75	11,90	69,00	27,81
	2	10,98	7,66	2,98	6,44		
	3	1,10	0	0,24	0,40		
4.25-75-75	1	38,7	0	0,87	11,90	97,37	30,17
	2	16,50	12,00	4,65	10,08		
	3	1,71	0	0,36	0,60		
5.0-100-100	1	0	0	0	0	75,84	21,80
	2	22,28	15,48	6,08	12,96		
	3	1,76	0	036	0,64		

*here and further in the tables: The ratio of alfalfa - winter wheat - NCP; **1- alfalfa (*Medicago sativa* L.), 2-winter wheat (*Triticum aestivum* L.), 3- NCP.

With the annual alternation of winter wheat with NCP, the production of phytomass on 1 ha of the crop area amounted to 75.84 t/ha, or more than with alfalfa monoculture by 36.9%. In the case when 25% of the area was allocated for grain crops with NCP (crop rotation 2), and alfalfa was used for three years, the amount of accumulated phytomass increased in relation to alfalfa monoculture by 22.3%. There were two reasons for this: firstly, an increase in the yield of alfalfa from three years of use compared to four years by 0.51 t/ha (11.74 t/ha versus 11.23), and secondly, additional production of 15.18 t/ha of winter wheat products with NCP.

The largest amount of phytomass per hectare of crop rotation area - 97.37 t/ha - was obtained in the case of four-year use of alfalfa in the output field (25% of the area) and the use of the rest of the area for growing winter wheat with its annual alternation with NCP for green fertilizer. The increase in the phytomass yield in relation to the alfalfa monoculture was 75.7%. Respectively, crop rotations are 29.1% and 30.4% inferior to the best option, where both these groups of crops occupy 50% of the area and with 75% of alfalfa and 25% of winter wheat for grain, followed by the formation of NCP for green fertilizer (crop rotations 3 and 2). Alfalfa monoculture (No. 5) reduces the production of phytomass in relation to the best crop rotation with the alfalfa output field by 22.1%.

Important in this aspect is the question of the amount of plant mass remaining in the soil after harvesting, since only it is a reserve for increasing nutrients in the soil. In our studies, the maximum amount of products not alienated from the soil was obtained in the same crop rotations in which more than all plant products accumulated. First of all, this applies to the crop rotation with an output field of alfalfa and three fields of winter wheat with NCP for green fertilizer, where such a mass was obtained by 30.17 t /ha, or more than with a monoculture of alfalfa by 81.2%. The second position on this issue is occupied by crop rotation, where alfalfa and winter wheat with NCP occupy 50% each. The excess in relation to the monoculture of alfalfa in it is 37.1%. Only 9.3% more plant mass, inalienable from the soil, accumulates compared to four-year cultivation of alfalfa in the same field in crop rotation with 75% of alfalfa and 25% of winter wheat.

The nitrogen concentration in the dry mass of alfalfa, winter wheat and NCP in our studies did not change significantly depending on the placement in crop rotations (Table 3). Most nitrogen was contained in hay and plant residues of alfalfa, NCP and winter wheat grain, phosphorus - in winter wheat grain and alfalfa roots, potassium - in the vegetative mass of alfalfa, NCP, in winter wheat grain and straw.

Table 3. Concentration of chemical elements in the plant mass of field crops, % of dry weight, 2014-2018

Nutritional elements	Culture	Main products	Straw	Crop residues	Roots
N	Alfalfa	2.18±0,09	0	2.11±0,09	2.24±0, 11
	winter wheat	1.51±0,06	0.66±0,03	0.58±0,04	0.41±0,03
	NCP	1.70±0,08	0	0.95±0,04	0.51±0,05
P_2O_5	Alfalfa	0.59±0,04	0	0.44±0,03	0.62±0,07
	winter wheat	0.77±0,07	0.25±0,02	0.22±0,02	0.31±0,04
	NCP	0.62±0,5	0	0.38±0,02	0.40±0,03
K_2O	Alfalfa	1.41±0,12	0	0.93±0,04	1.43±0,06
	winter wheat	1.16±0,10	1.00±0,07	0.61±0,04	0.70±0,05
	NCP	1.22±0,10	0	0.85±0,05	0.61±0,05

Our studies on the accumulation of nitrogen in the harvest of various crop rotations have shown that the maximum amount of it is contained in the products obtained in crop rotations 2 (with 75% alfalfa and 25% winter wheat with NCP) and 4 (with the output field of perennial grass): 122.53 and 122.19 t/ha, respectively (Table 4).

Table 4. N content in total and non-alienated plant mass in crop rotations with varying degrees of saturation with field crops, t/ha, 2014-2019.

Crop rotation and crop ratio,%*	Culture**	Main	Secondary	Crop fields	Root	All phytomass	Including not alienated from the soil
		products		remains			
1. 100-0-0	1	84,50	0	4,75	6,34	95.59	11,09
	2	0	0	0	0		
	3	0	0	0	0		
2.75-25-25	1	79,85	0	6,86	20,94	122,53	31,42
	2	8,41	2,85	0,96	1,49		
	3	0,87	0	0,11	0,19		
3.50-50-50	1	64,80	0	4,60	21,24	119,15	32.69

Table 4. Continued

	2	16,60	5,06	1,73	2,64		
	3	1,87	0	0,23	0,38		
4.25-75-75	1	17,79	0	4,75	6,34	122,19	21,66
	2	24,92	7,92	2,70	4,13		
	3	2,91	0	0,34	0,57		
5.0-100-100	1	0	0	0	0	56,80	12,76
	2	33,64	10,22	3,51	5,31		
	3	2,99	0	0.34	0,61		

Crop rotation 3 is not much (by 2.8%) inferior to them, with the same share of both groups of crops. It is characteristic that with alfalfa monoculture, the nitrogen content in the phytomass decreases by 28.2% compared to the best crop rotation in this respect. A feature of crop rotation 5 without alfalfa is the minimal accumulation of nitrogen in the phytomass obtained in it, which is inferior to one of the best crop rotations (fourth) by 40.6%.

If we evaluate the studied crop rotations by the nitrogen content in phytomass that is not alienated from the soil, then crop rotation 3 has an indisputable advantage with the same proportion of the studied crop groups - 32.69 t/ha. In comparison with it, in crop rotation 2 with 75% of alfalfa and 25% of winter wheat with NCP, the content of this element decreases by only 3.9%, more significantly - by 33.7 and 61.0%, respectively - it decreases in crop rotations 4 and 5 with the output field of alfalfa and the annual alternation of winter wheat with NCP. The minimum amount of nitrogen accumulates during alfalfa monoculture, it is possible that this is due to the removal of a large amount of this food element with the harvest of phytomass.

In the accumulation of P_2O_5 in the phytomass harvest in the studied crop rotations, a slightly different picture is observed: most phosphates accumulate in the phytomass of the third crop rotation with the same number of alfalfa and winter wheat fields with NCP - 38.20 t/ha (Table 5), the second and third positions are occupied by crop rotation 2 with 75% alfalfa and 25% winter wheat with NCP and crop rotation 4 with an output field of alfalfa and three fields of wheat with NCP. Similar values were obtained in contrasting crop rotations: with a monoculture of alfalfa and annual alternation of winter wheat with NCP.

Table 5. The content of P_2O_5 in the total and non-alienated from the soil plant mass in crop rotations with varying degrees of saturation with field crops, t/ha, 2014-2019.

Crop rotation and crop ratio,%*	Culture**	Main	Secondary	Crop fields	Root	All phytomass	Including not alienated from the soil
		products		remains			
1. 100-0-0	1	22,86	0	2,09	7,38	32,33	9,47
	2	0	0	0	0		
	3	0	0	0	0		
2.75-25-25	1	21,61	0	1,43	5,88	35,41	7 9,43
	2	4,29	1,08	0,36	1,12		
	3	0,52	0	0,04	0,08		
3.50-50-50	1	17,53	0	0,96	5,80	38,20	
	2	8,40	1,92	0,66	2,00		
	3	0,68	0	0,09	0,16		
4.25-75-75	1	4,81	0	0,38	7,38	33,85	13,34
	2	12,70	3,00	1,02	3,12		
	3	1,06	0	0,14	0,24		
5.0-100-100	1	0	0	0	0	33,00	21,80
	2	22,28	3,87	1,34	4,02		
	3	1,09	0	0,14	0,26		

Alfalfa monoculture is not the best option for the accumulation of exchangeable potassium in the phytomass harvest (Table 6). In this regard, the most preferable is crop rotation 3 with an equal number of fields of perennial grass and winter wheat with NCP, the phytomass of which contained 85.57 t /ha of K_2O, its increase in relation to alfalfa monoculture in it was 13.2%.

Table 6. K_2O content in total and non-alienated plant mass in crop rotations with varying degrees of saturation with field crops, t/ha, 2014-2019.

Crop rotation and crop ratio,%*	Culture**	Main	Secondary	Crop fields	Root	All phytomass	Including not alienated from the soil
		products		remains			
1. 100-0-0	1	54,18	0	4,38	17,02	75,58	21,40
	2	0	0	0	0		
	3	0	0	0	0		
2.75-25-25	1	51,24	0	3,02	13,37	82,80	20,78
	2	6,46	4,32	1,01	2,54		
	3	0,62	0	0,10	0,12		
3.50-50-50	1	41,61	0	1,89	13,56	85,57	23,56
	2	12,74	7,66	1,82	4,51		
	3	1,34	0	0,20	0,24		
4.25-75-75	1	11,42	0	2,03	17,02	67,96	31,72
	2	19,14	12,00	2,84	7,06		
	3	2,09	0	0,31	0,37		
5.0-100-100	1	0	0	0	0	58,13	15,63
	2	25,75	15,48	3,71	9,07		
	3	2,15	0	0,31	0,39		

A reduction in the share of winter wheat with NCP in this crop rotation to 25% (crop rotation 2) leads to a decrease in K_2O content by 4.7%, but remains higher than the alfalfa monoculture by 8.5%. In crop rotations with alfalfa output field (No. 2) and in its complete absence (No. 5), the accumulation of exchangeable potassium decreased by 10.1 and 23.1%, respectively.

Crop rotation with the alfalfa output field is also the best in this case for the accumulation of exchangeable potassium in products not alienated from the soil. On 1 ha of the crop rotation area, 31.72 t/ha of exchangeable potassium accumulates in the phytomass of this crop rotation, or 2.9 times more than with alfalfa monoculture. Similar indicators were obtained in the second crop rotation with 75% alfalfa and 25% winter wheat with NCP, exceeding the monoculture of perennial grass by 2.8 times In the fifth crop rotation, where alfalfa was completely excluded from the structure of sown areas, the inalienable part of the product contained 15% more of this plant nutrition element than with the monoculture of alfalfa.

4 Conclusions

1. The development of an energy storage system for soil maintenance in irrigation landscapes of the Western Precaspian, based on the use of the second half of summer for the formation of NCP and plowing it for green fertilizer, is an effective way to increase plant mass per unit area. The most productive of the studied crop rotations is a four-field with one output field of alfalfa (*Medicago sativa* L.) and three fields of winter wheat (*Triticum aestivum* L.), after harvesting which in the second half of summer a NCP for

green fertilizer is formed. In this crop rotation, 97.4 tons of agricultural products were produced on 1 ha of the crop rotation area, including 30.2 tons/ha of its non-alienable part from the soil.

2. The ratio of the products produced in crop rotations and the part of it that is not alienated from the soil is the widest (3.48:1) was in the crop rotation without alfalfa, the narrowest (2.47:1) - in the crop rotation with the same area of alfalfa and winter wheat with NCP. In other crop rotations, it ranges from 3.23:1 (with an output field of alfalfa) to 3.72:1 (with three fields of alfalfa and one field of winter wheat with NCP).

3. Crop rotations with three and two fields of alfalfa, respectively, with one and two fields of winter wheat with NCP, accumulate the largest amount of nitrogen in the plant mass not alienated from the soil - 31.42 - 32.69 t/ha. The minimum indicator was obtained with a monoculture of alfalfa - 11.09 t /ha, with constant alternation of winter wheat and NCP, it increases by 15.1%, in crop rotation with the output field of perennial grass - by 95.3%.

4. In phytomass that is not alienated from the soil, the maximum amount of P_2O_5 accumulates in a crop rotation without alfalfa - 80t/ha, and the minimum amount - with a monoculture of alfalfa and three fields of its four crop rotation fields – 9.47-9.43 t/ha. Phosphates accumulate 41-61% more in crop rotations with an output field of alfalfa and an equal amount (50% of the area) of it and grain crops with NCP.

5. Crop rotation with an output field of alfalfa and three fields of winter wheat with NCP provides the output of the maximum amount of K_2O from 1 ha of the crop rotation area - 31.72 t. With the annual alternation of winter wheat with NCP, its production decreases by 50.7%, and with alfalfa monoculture and in crop rotation with 75% of this crop - by 32.5 and 34.5%, in crop rotation with an equal number of studied crops - by 25.7%.

References

1. Zh.N. Abdullaev, G.N. Gasanov, A.A. Beksultanov, Agrarian science, **3**, 9-12 (2012)
2. I.T. Tamazaev, M.R. Musaev, G. N. Gasanov, Bulletin of the Russian Agricultural Science, **4**, 44-47 (2018)
3. E.N. Pakina, G.N. Gasanov, T.A. Asvarova, K.M. Gadzhiev, R.R. Bashirov, Problems of agroindustrial complex development in the region, **4(44)**, 30-36 (2021)
4. E.N. Pakina, G.N. Gasanov, Res. On Crops, **21(2)**, 152-158 (2020)
5. E. N. Pakina, G. N. Gasanov, T. A. Asvarova, Agrochemistry, **6**, 73-783 (2021)
6. E.N. Pakina, G.N. Gasanov, T.A. Asvarova, Fertility, **2(119)**, 42-45 (2021)
7. E.N. Pakina, Fertility, **5(116)**, 21-24 (2020)
8. Workshop on agrochemistry, 2nd, 689 (2001)
9. I.P. Vasiliev, A.M. Tulikov, G.I. Bazdyrev, *Practicum on agriculture*, 424 (2005)
10. M.A. Fedin, *Methodology of the State Commission on variety testing of agricultural crops*, 239 (1985)
11. *Methodological guidelines for conducting field experiments with forage crops*, 198 (1987)
12. B.A. Dospekhov, *Methodology of field experience*, 416 (1985)

Agroparks as a network form of business organization in agriculture

Andrey Plakhin*, *Tatiana* Kochergina, and *Maria* Selezneva

Ural State University of Economics, 8 Marta Str, 62, 620144 Ekaterinburg, Russia

Abstract. In this study, the authors substantiate the relevance of the creation of agro-industrial parks as a network form of organizing the business of agricultural producers. Agro-industrial parks in modern conditions are becoming a kind of growth points for the innovative development of agricultural producers, providing accelerated growth in production and import substitution in order to achieve food security in the Russian Federation and its regions. The results of the study allow us to conclude about the growing interest in this form of business organization.

1 Introduction

The tasks associated with the intensification of the development of both individual constituent entities and the Russian Federation as a whole require more and more innovative conjugation of participants in the process of production and processing of agricultural products in the interests of import substitution and ensuring food security. Based on the theoretical provisions of the development of organizational forms of business integration, the concept of uniting the participants in the process of production and processing of agricultural products through the formation of agro-industrial park structures was formed [1, 2, 3].

The arguments in favor of this thesis are the following main opportunities provided by agro-industrial parks:

- coordinating the actions of participants in the agro-industrial sector, creating conditions for cooperation, inter-firm networking, clustering and building an effective chain of consumer value formation, obtaining business advice and data on the conjuncture of sales markets;
- provision of financial incentives and tax preferences - reduction of tax rates, exemption (temporary) from certain types of taxes;
- provision of the necessary engineering, transport, production, sales and logistics infrastructure;
- development of new technologies and new products with improved consumer properties, etc.

Agricultural parks perform an important function of uniting entrepreneurs working in different sectors of the agri-food market into production and distribution structures. Such structures allow uniting agricultural producers and processing enterprises, implementing the provisions of the network theory and forming a number of relational effects for the

* Corresponding author: apla@usue.ru

participants [4]. At the same time, agro-industrial parks expand the innovative component of entrepreneurial activity by stimulating the exchange of innovations both between residents of the agro-park and among other participants in production and distribution networks [5].

2 Materials and Methods

Based on the provisions of the network theory, it can be assumed that the use of resources, including land within the framework of agro-industrial structures is more efficient in comparison with the competitive form of interaction between agricultural producers. The resource component of the agricultural park, as a rule, includes significant land, road and engineering infrastructure for the comprehensive provision of industrial enterprises that process agricultural products. Also, a number of authors emphasize the need to provide services to residents in research, marketing, legal and other areas [6,7].

Analysis of the experience of the formation of agro-industrial parks allows us to identify three main models used in Russia. Figure 1 shows a model of an agricultural park operating on the basis of public-private partnerships.

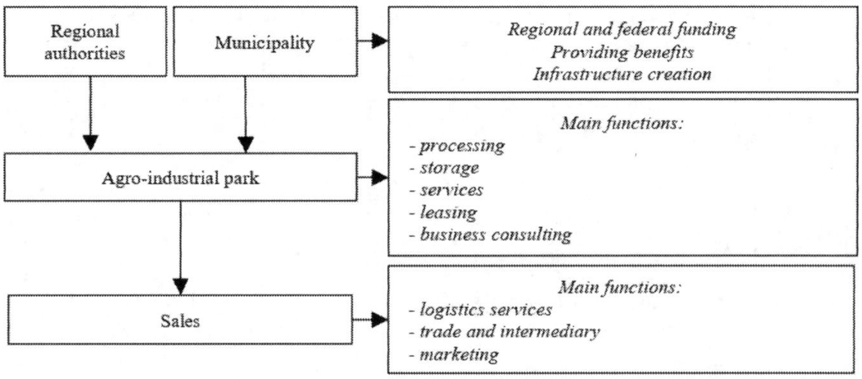

Fig. 1. Model of an Agropark, created on the basis of public-private partnership

In modern practice, the most often used model is the formation of an agro-industrial park, created on the basis of a public-private partnership, by the initiators of projects of this kind is based on the initiative of state and municipal authorities.

The second model of the emergence of agro-industrial parks is based on a complex of processing industries that form an agricultural holding with a different ownership structure, Figure 2.

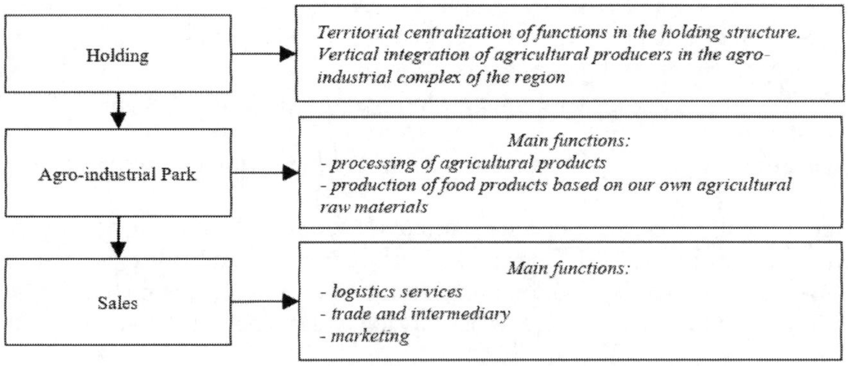

Fig. 2. Model for creating an agricultural park on the basis of an agricultural holding

As a rule, agro-industrial parks of this type are formed in traditional agricultural regions of Russia. The most rare is the model of the formation of agro-industrial parks based on the creation of organizations providing complexes of service functions to agricultural producers, Figure 3.

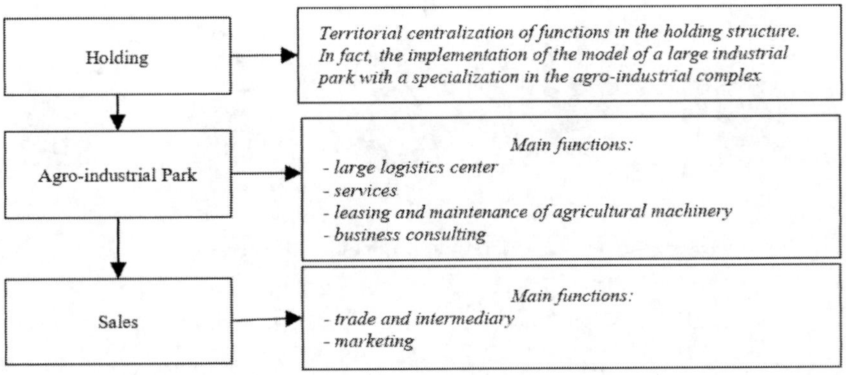

Fig. 3. Model for creating an agricultural park based on a service holding

All the presented models of aroparks make it possible to create regional centers that combine functions within the production and distribution network, which makes it possible to form and use a wide range of additional effects based both on the scale of production and on the effect of specialization.

3 Results and Discussion

Let's present the data on the formed base of agricultural parks in the Russian Federation. Figure 4 shows the data on the distribution of agricultural parks by form of ownership

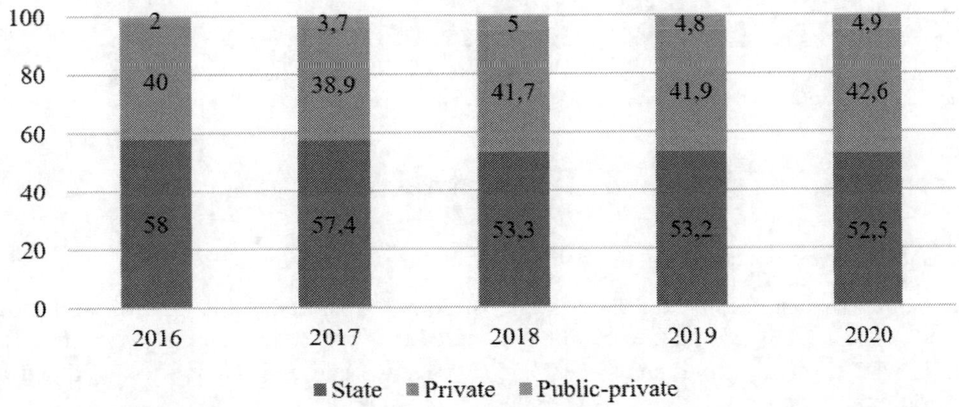

Fig. 4. Data on the distribution of agro-industrial parks by ownership, 2016-2020, %

As one can see in the figure, the share of state agricultural parks is gradually decreasing, giving way to projects initiated by commercial structures and within the framework of public-private partnership projects. Next, we will characterize the structure of agricultural parks by the stage of project implementation, Figure 5.

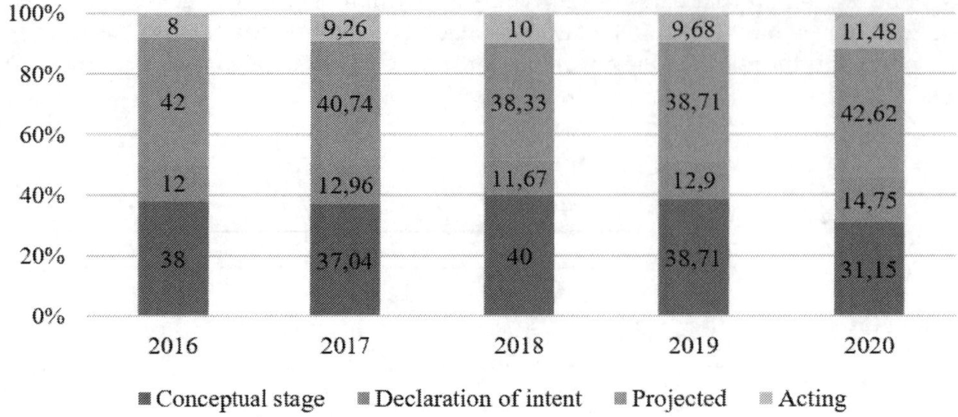

Fig. 5. Data on the distribution of agro-industrial parks by the stage of project implementation, 2016-2020, %

As one can see in the figure, most agricultural parks are at the design and construction stage, this circumstance indicates the demand for this format of doing business. As a rule, the agro-park format is most actively used in resource-abundant regions specializing in the production of agricultural raw materials. Interesting is the information on the distribution of agricultural parks by area of land resources included in the turnover, Figure 6.

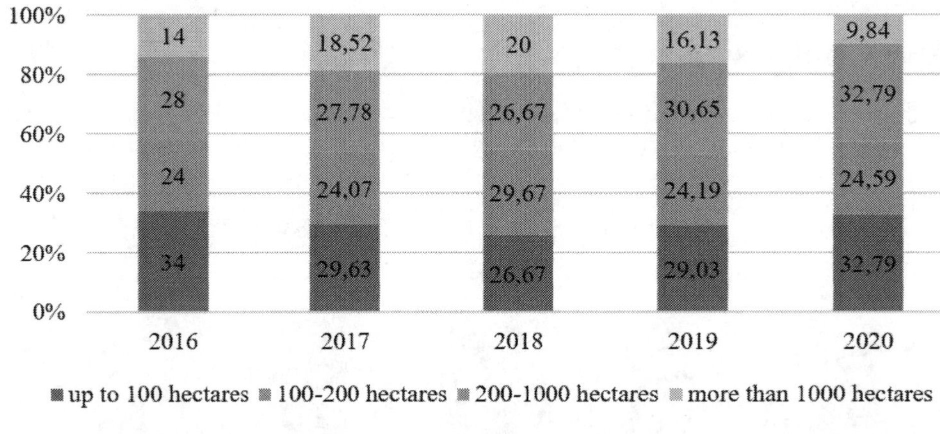

Fig. 6. Data on the distribution of agro-industrial parks by area of land plots, 2016-2020, %

As can be seen in the figure, there is almost complete parity in the area of land plots included in the agricultural park. It can be concluded that projects are equally attractive for investors with relatively small areas of up to 100 hectares and with large areas included in the agricultural park.

4 Conclusion

At present, agro-industrial parks built on the principles of clustering are showing increasing efficiency. The point is that in the process of creating an agro-industrial park, the possibilities of concentrating technologically related enterprises on its territory are being investigated.

In a narrow sense, a cluster is understood as "the geographical concentration of enterprises that benefit from their joint location on the same territory" [8]. In a broader sense, a cluster is "a collection of geographically concentrated interconnected enterprises and organizations in a specific area or industry." Clusters can be a collection of interconnected manufacturing enterprises and specific organizations that result in a higher level of competitiveness. Such organizations include, for example, suppliers of specialized materials such as components, mechanisms and services, as well as companies that form specific infrastructure.

Clusters are also characterized by vertical expansion "down" in relation to distribution channels and consumers of products and horizontal expansion towards both producers of additional components and companies in industries related to education, science, technology or other general needs. It should be noted that in the leading industrialized countries, clusters actively interact with government agencies, universities, standards setting and development agencies, think tanks, vocational education organizations, trade associations, which together provide specialized training, education, information, research, and technical support, which is very important for the agricultural sector and the agricultural parks created here.

References

1. I. Doronina, V. Borobov, E. Ivanova, E. Gorynya, B. Zhukov, International journal of economics and financial issues, **6**, 1 (2016)
2. A. Sharnin, O. Frolova, G. Klychova, N. Nigmatullina, A. Iskhakov, Formation and development of clusters in the Russian regional agro-industrial complex In E3S Web of Conferences, 91 (2019)
3. D. V. Zavyalov, O. V. Saginova, N. B. Zavyalova, Journal of Environmental Management & Tourism, **8** (2017)
4. A. Plakhin, I. Semenets, E. Ogorodnikova, The asymmetry of income distribution between stakeholders within the framework of industrial park structures In MATEC Web of Conferences, **184** (2018)
5. A. Plakhin, E. Ogorodnikova, M. Khokholush, T. Guseva, M.Selezneva, Revista Inclusiones, **7** (2020)
6. A. Semin, A. Truba, The economics of agriculture in Russia, **7** (2019)
7. R. Akhmadeev, A. Redkin, N. Glubokova, Entrepreneurship and sustainability issues, **7** (2019)
8. M.Graubner, I. Ostapchuk, T. Gagalyuk, European Review of Agricultural Economics, **48**, 1 (2021)

Experience of using sewage sludge in soil for green building

Irina Elshaeva[2], *Vera* Titova[1], *Alexander* Vetchinnikov[1], *Anna* Pinaeva[2], and *Oksana* Vetchinnikova[1]

[1]Nizhny Novgorod State Agricultural Academy, Gagarina, 97, 603107 Nizhny Novgorod, Russian Federation
[2]Saint-Petersburg State Agrarian University, Petersburgskoe shosse, 2, 196601 Pushkin, Saint-Petersburg, Russian Federation

Abstract. Wastewater sludge from municipal wastewater treatment plants in Nizhny Novgorod and soils based on them, prepared with different ratios of sludge, sawdust and peat, have been investigated. Sewage sludge-based soils contain plant nutrients and are suitable for use as fertilizing materials in green building. Under the conditions of a three-year lysimetric experiment on cereal lawn grasses, a significant efficiency of organic soils was noted in comparison with traditional soils. The lawn maintenance regime revealed differences in the mixtures of fertilizing materials in terms of the effect on the productivity of the phytocenosis. Use of sewage sludge containing large amounts of heavy metals as one of the soil components undoubtedly leads to increase in the amount of these toxic elements in the soil as a whole.

1 Introduction

Analysis of the environmental situation in the production area of the Nonchernozem belt shows that the main reason for the negative impact on the environment of industrial and municipal enterprises is the low efficiency of waste management. Wherein, being a rich source of organic material, energy and macronutrients, they can be used to solve a wide variety of environmental and economic problems [1,2]. The studies carried out to assess some industrial wastes containing calcium indicate the possibility of their use as ameliorants on acidic soddy-podzolic and podzolic soils [3, 4, 5].

Along with traditional organic fertilizers in modern agriculture, alternative fertilizing materials are also involved in operation with strict adherence to scientific recommendations for use. The degree of influence of organic waste on the agrochemical properties of the soil depends on the application dose, culture and duration of application. The batteries in them are in organic form, which reliably prevents them from being washed out and promotes prolonged action. In general, they provide a positive balance of nitrogen, phosphorus and potassium in the soil, increase enzymatic activity, the total number of microorganisms and soil respiration [4, 5, 6].

Study in the field of environmental assessment and wastewater sludge disposal technology (WWS) has been going on since the 80s of the twentieth century [7]. Recommendations have been developed and limits have been determined for introduction of waste silts into the soil for agricultural industrial crops. It is noted that sewage sludge has a noticeable direct and long-term aftereffect [7, 8]. The effect of direct action is all the more pronounced on impoverished soils. In some cases, the aftereffect of sediment is long-term and ultimately contributes to the long-term preservation of soil fertility.

The issues of using organic waste as a secondary material resource are becoming increasingly important as the modern housing complex and green building facilities develop. Waste silt finds its way into the regeneration of urban landscapes and land reclamation, making its utilization of environmentally and economically sound [9-15].

The purpose of the paper is to study the possibility of using sewage sludge from sewage treatment plants of public utilities in the city of Nizhny Novgorod and composts based on them in green construction and for the creation of cereal grass.

Study objectives: to determine the agrochemical and toxicological indicators of sewage sludge and soils based on them. To assess the possibility of using sediments and composts as unconventional organic fertilizers in urban soils. To study the effectiveness of WWS and composts and their effect on development of lawn grasses under the conditions of vegetation lysimetric experiments.

2 Materials and Methods

The objects of study were the waste silts of the Nizhny Novgorod aeration station and soils prepared on their basis: with soil, peat and sawdust in various ratios. As a control option, a standard soil was used, consisting of 60 % peat, 20 % soil, and 10 % manure and sand. The experiment scheme consisted of five options:

1. Standard soil
2. WWS: sawdust: peat - 0.5: 1: 1
3. WWS: sawdust: peat - 0.5: 2: 1
4. WWS: sawdust: peat - 0.5: 1: 2
5. WWS: sawdust: peat - 1: 1: 1

The studies were carried out under the conditions of a vegetation-lysimetric experiment in vessels 70 cm high and 80 cm in diameter. The area of each vessel is 0.5 m^2 . The vessels were located at a distance of 50 cm from each other, so that their edges were 5 cm above the soil level and filled with prepared soil. The experiment was repeated three times. The effectiveness of fertilizing materials was studied in many years of experience on a culture traditionally used to create lawns for various purposes.

Meadow fescue (Festuca pratensis) is a perennial cereal plant, which is characterized by early and abundant bushing out, long-term preservation of the ability to shoot. Under certain conditions of agricultural technology, grasses form a dense turf cover and a uniform density of herbage, which exhibits high resistance with systematic mowing. The projective cover is 50-60 %. They tolerate unfavorable meteorological conditions and soil salinization.

The experiment provided for the study of two possible methods of using the lawn – with harvesting the mowed mass from the territory of the lawn and leaving it (mowed chopped grass) on the soil surface. Therefore, the lysimeters were divided into two separate parts.

Therefore, in one part of the vessel (series "a"), finely chopped cut grass was left on the surface, that is, the biomass returned to the "soil-plant" system, and in the other part (series "b"), it was alienated from the system and transported for its limits.

The characteristics of the components used for preparation of soils are shown in Table 1.

Table 1. Initial characteristics of soil components

Components	Organic matter, %	Content of nutrients, % on dry matter			The gross content of heavy metals in the original substrates taken for preparation of soils, mg/kg			
		N	P_2O_5	K_2O	Cd	Pb	Cu	Zn
WWS	228	2,65	7.00	0.32	7.0[1] / 15.0	31.0 / 250.0	205.0 / 705.0	553.0 / 1750.0
Sawdust	31	2.24	0.09	0.16	5.0	85.0	276.0	720.0
Peat	32	1.84	0.16	0.24	1.6[2] / 2.0	18.0 / 32.0	13.0 / 132.0	71.0 / 220.0
Standard Soil	24	108[3]	42[3]	180[3]	0.9[2] / 2.0	11.0 / 32.0	29.0 / 132.0	61.0 / 220.0

[1] Above the line there is the total content of HM in the substrate, below the line - MPC according to GOST R 54651-2011 [16]
[2] Above the line there is the total content of HM in the substrate, below the line - MPC (APC)
[3] according to SanPiN 2.1.7.573-96 [17]

3 Results and discussion

Lawn grasses in multi-cut cultivation consume a large amount of nutrients, especially nitrogen and potassium. During the growing season, sod grasses carry 4.5 % nitrogen, 3 % potassium and 0.7 % phosphorus. The total nitrogen content in WWS is at the level of 2.65 %, while most of it is in a form inaccessible to plants. Presence of a significant amount of phosphorus (7 %) in the WWS determines its active effect on the phosphate regime of soils. In the process of mineralization, the degree of mobility of phosphates in the form of organic compounds increases, which is especially manifested in the aftereffect. The potassium content in sediments is much less, only 0.32 %, therefore, use of WWS practically does not affect the change in the availability of potassium in soils. The valuable properties of waste sludge are associated with the content of a large amount of organic matter in them.

Use of sewage sludge containing large amounts of heavy metals as one of the soil components undoubtedly leads to increase in the amount of these toxic elements in the soil as a whole. So, the content of heavy metals in soils is presented in following Table 2.

Table 2. Gross content of heavy metals in WWS-soils, mg/kg

No of option	Option		Heavy metals					
			Cd	Co	Cu	Ni	Pb	Zn
1	Control	a	0.6	3.1	19.7	14.3	9.0	82.6
		b	0.7	3.0	14.2	13.5	7.0	49.7
WWS: sawdust: peat								
2	0.5: 1: 1	a	74.5	3.2	455.3	260.1	174.6	2693.0
		b	60.2	2.7	411.0	237.2	147.5	1885.2
3	0.5: 2: 1	a	66.1	3.3	448.4	246.7	154.7	1747.7
		b	88.7	3.7	547.3	300.0	215.8	2220.6
4	0.5: 1: 2	a	10.1	2.6	220.9	155.1	54.6	781.3
		b	12.2	2.4	216.1	148.5	64.5	798.4
5	1: 1: 1	a	128.9	3.9	798.7	433.5	310.6	2448.3
		b	130.8	3.1	785.7	412.9	297.3	2358.4

It can be seen from several years of study that use of composts in any proportions increased the content of heavy metals in soils. The exception was the indicator of the content of cobalt, its concentration did not change relative to the control. Sewage sludge is characterized by a high zinc content, therefore, in options with use of WWS-soils, its gross content ranges from 71 to 2693 mg/kg. The content of heavy metals of the first hazard class in composts from sewage sludge significantly exceeds their content in standard soil. The return of biomass to the soil-plant system has an ambiguous effect on the accumulation of toxicants in the soil.

One of the main indicators determining the suitability of precipitation for their use as fertilizer is a positive effect on the productivity of culture phytocenosis. Long-term observations indicate that at normal plant density, the yield of herbage according to the options of the experiment during the growing season varied from 6.9 to 13 kg/m^2 (Table 3). Introduction of organic waste in all options contributed to an increase in plant productivity in comparison with the control option. The maximum increase in biomass (156 % to the control) showed that for formation of a lawn culture, the best option is to apply compost in an equal ratio of WWS: sawdust: peat. At the same time, when the biomass was returned to the system, an increase in the proportion of silt in the compost from 0.5 to 1.0 had little effect on plant productivity.

Table 3. Total productivity of meadow fescue in 2019

Option No.	**Option**	**Green mass yield**		
		In total for two mows, kg/m^2	**Difference with control ($^+_-$)**	
			kg/m^2	**%**
Series "a" - return of biomass to the system				
1	Control	6.90	-	-
WWS: sawdust: peat				
2	0.5: 1: 1	11.67	+4.77	+69.13
3	0.5: 2: 1	13.02	+6.12	+88.70
4	0.5: 1: 2	7.92	+1.02	+14.78
5	1: 1: 1	6.93	+0.03	+0.43
Series "b" - alienation of biomass from the system				
1	Control	5.82	-	-
WWS: sawdust: peat				
2	0.5: 1: 1	10.89	+5.07	+87.11
3	0.5: 2: 1	8.91	+3.09	+53.09
4	0.5: 1: 2	11.40	+5.58	+95.88
5	1: 1: 1	14.91	+9.09	+156.19

Sewage sludge, like any organic fertilizers, has a significant and long-term aftereffect [1, 4, 9]. In a long-term lysimetric experiment, the largest yield in the second cut was provided by options with the ratios of fertilizing materials: 0.5: 2: 1 and 1: 1: 1 (Fig. 1.2). The biomass yield in these options in the first cut is significantly higher than the control and exceeds the results of the second cut of the control option by more than two times. It shall be noted that the aftereffect was more noticeable in series b, with the removal of the cut grass. In the option with the full dose of WWS (1: 1: 1), the yield of green mass of the second cut was 2.5 times higher than the control, and exceeded the biomass of the first cut.

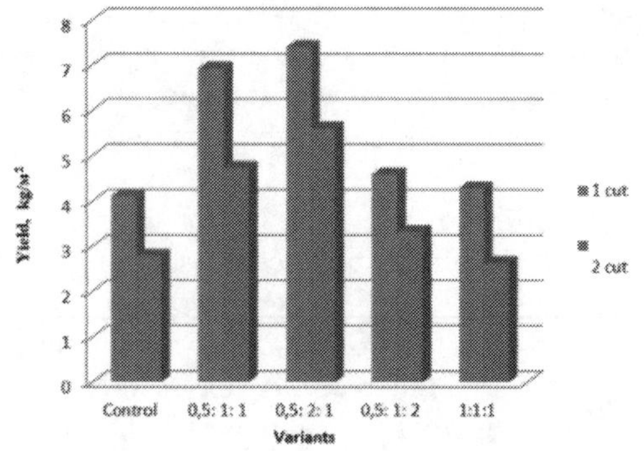

Fig. 1. The yield of fescue when returned to the lysimeter

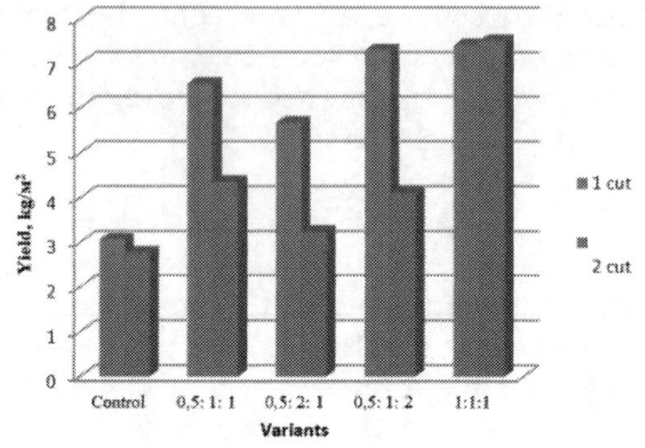

Fig. 2. Fescue yield when alienated from a lysimeter

The lawn maintenance regime largely determines the water-physical and acid-base properties of soils and grounds, as well as the balance of nutrients. All of the above is inevitably reflected in changes in the productivity of the phytocenosis. The higher efficiency of the mixture of fertilizing materials with the ratio of WWS: sawdust: peat - 0.5: 2: 1 in the series of options "a" is probably associated with the optimal pH value, as well as the physiological characteristics of meadow fescue [18] (Table 4).

Table 4. Comparison of the maintenance regimes for the lawn sown with meadow fescue in 2019

No of option	Option	Series "a" - return of biomass to the "soil-plant" system	Series "b" - alienation of over-terrestrial biomass	Mode "a" in comparison with mode "b"	
				kg/m²	%
Productivity of green mass, kg/m²					
1	Control	6.90	5.82	+1.08	+15.65
WWS: sawdust: peat					
2	0.5: 1: 1	11.67	10.89	+0.78	+6.68
3	0.5: 2: 1	13.02	8.91	+4.11	+31.57
4	0.5: 1: 2	7.92	11.40	-3.48	-43.94
5	1: 1: 1	6.93	14.91	-7.98	-115.15

4 Conclusions

As a result of the study carried out, the following conclusions can be drawn up:

1. Soils based on sewage sludge contain plant nutrients and are suitable for use as fertilizing materials in green building and for creating grass stands.

2. Use of sewage sludge as one of the soil components leads to an increase in the amount of heavy metals in the soil as a whole. The zinc content is especially high (up to 2693 mg/kg). Excessive levels of cadmium and lead in prepared soils can be considered environmentally hazardous.

3. The high efficiency of WWS and composts in the cultivation of cereal grasses under the conditions of vegetative lysimetric experiments was noted. The maximum yield increase, at the level of 156 %, was obtained in the variant with the introduction of compost in an equal ratio of WWS: sawdust: peat. At the same time, the aftereffect was more noticeable in series "b", with the alienation of the aboveground biomass.

References

1. V. I. Titova, I.A. Pitina, M. V. Dabakhov, International Agricultural Journal, **62**, 24–27 (2019)
2. A. Burges, I. Alkorta, L. Epelde, C. Garbisu, International Journal of Phytoremediation, **20**, 384–397 (2018)
3. S. Gintaras, K. Danute, A. Jurate, Zemdirbyste-Agriculture, **106**, 213–218 (2019)
4. L. D. Varlamova, I. D. Korolenko, Agrochemical Bulletin, **2**, 17–20 (2017)
5. A.Lavrishchev, A.Litvinovich, V.Bure, O.Pavlova, E. Saljnikov, Biological communications, **3**, 163-173 (2018)
6. V.N.Efimov, V.P.Tsarenko, Eurasian soil science, **2**, 57-67 (1993)
7. N. A. Zaitseva and A. N. Pyrsikova, International Research Journal, **3-1**, 104–105 (2015)
8. N. T. Chebotarev, N. D. Naydenov, A. A. Yudin, Science. Thought: electronic periodical, **6**, 31–36 (2016)
9. I. V. Elshaeva, E. V. Voropaeva, A. S. Pinaeva, The International Scientific and Practical Conference BAICSEM 2020 (2020)
10. V. I.Titova, I. A.Pitina, E. U.Sudakov, International agricultural journal, **62(4)**, 111-120 (2019)
11. V. I. Smetanin, V. N. Zemskov, Scientific Life, **1**, 38 (2015)
12. A. Kwarciak-Kozłowska, Industrial and Municipal Sludge, 337-360 (2019)
13. R. F. Baybekov, G. E. Merzlaya, O. A. Vlasova, Agriculture, 34–36 (2015)
14. A. Grobelak, M. Jaskulak, *Sludge multifunctions in a phytobiome - Forest and plantation application including microbial aspects,* Industrial and Municipal Sludge (United Kingdom , Oxford : Butterworth-Heinemann), 323–336 (2019)
15. L.V. Bryndina O. V. Baklanova, Forestry Engineering Journal, **9**, 23–30 (2019)
16. GOST R 54651-2011, Organic fertilizers on the basis of sewage sludge
17. SanPiN 2.1.7.573-96, Soil, cleaning of inhabited areas, domestic and industrial waste, sanitary protection of soil
18. V. F. Drichko, M. A.Efremova, A. A.Izosimova, Radiation biology. Radioecol-ogy, **2**, 166-171 (2009)

Development of functional caramel using system packages MathLab

Nikolay Nikolaevich Berezhnov[1,*], *Marina Gennadievna* Kurbanova[2], and *Anastasia Yurievna* Kolbina[1]

[1]FSBEI HE Kuzbass State Agricultural Academy, 650056, st. Markovtseva, 5, Kemerovo, Russia
[2]FSBEI HE "Kemerovo State University" (KemSU), 650000, st. Krasnaya, 6, Kemerovo, Russia

Abstract. The article presents substantiation for expanding the market for functional food products, in particular, confectionery - caramel. Proper nutrition is an integral part of a modern person diet; due to their properties, they increase the immunity and immune status of a person. The aim of the work is to create composite mixtures for sugar-free caramel using modern software. The studies were carried out using the MathLab software package, in particular the Statistica 10.0 software. A methodology for creating recipes using system packages has been developed. To enrich the caramel, the following components were selected: apple juice, coltsfoot extract, orange juice, mint and burdock extract, cranberry and lingonberry juices. An optimal recipe has been designed and equations are presented for calculating the dosages of the components of the enriched sugar-free caramel based on a natural sweetener isomalt with the addition of plant-based additives containing macro and micronutrients necessary for the normal functioning of the human body with a diagnosis of type II diabetes mellitus.

1 Introduction

In recent decades, consumer demand for healthy foods such as re-formulated foods (such as low-fat or fiber-added foods) and functional foods has grown rapidly. Consumer demand has risen in part due to socio-economic changes such as increased life expectancy, rising health care costs, the social costs associated with non-communicable diseases, and widespread desire for a better quality of life. According to experts, the world market for food products with functional action in 2020 is $ 168 billion with an average annual growth rate of 8.5% [1, p. 142]. Food companies, attracted by this market growth and high profitability, are investing in the development of new functional products.

In recent decades, consumer demand for healthy food products such as re-formulated food (such as low-fat or fiber-added) and functional food has grown rapidly. Consumer demand has risen in part due to some socio-economic changes as increased life expectancy, rising health care costs, the social costs associated with non-communicable diseases, and widespread desire for a better quality of life. According to experts, in 2020 the world market of food products with functional properties is $ 168 billion with an average annual

[*] Corresponding author: n.berezhnov@mail.ru

growth rate of 8.5% [1, p.142]. Food producing companies, attracted by this market growth and high profitability, are investing in the development of new functional products.

Functional food play an important role in the development and expansion of the food market. This is mentioned, in particular, in the presidential decree of January 30, 2010 "On the adoption of the Food Security Doctrine of the Russian Federation" and the Order of the Government of the Russian Federation of June 29, 2016 No. 1364-r "On the approval of the strategy for improving the quality of food products of the Russian Federation to 2030 "[2, p.113].

The main principle of developing functional products is to expand the range of traditional products by adding enriching components on natural basis to their composition.

Eating habits acquired in childhood affect a person's lifestyle and behavioral changes [7, p. 110]. Insufficient intake of macro- and micronutrients caused by improper diet is the cause of the development of diseases associated with the immune system. In addition, lack of knowledge about healthy food raion among young people and children also leads to unhealthy eating habits that negatively affect their health [8, p.169].

Type II diabetes mellitus is one of the most common diseases of the 21st century. According to research by scientists, in 2019 there were about 285 million patients all over the world (6.4% of the world's adult population), and by 2020 this number had increased to 438 million (7.8% of adults) [9, p. 105]. Over the past decades, the prevalence of diabetes has increased especially in countries with low and middle living standards. According to experts, diabetes mellitus is the 9th and 21st cause of death in the third world countries among men and women, respectively [9, p.105]. Lifestyle modification, including weight management, physical activity and diet, plays an important role in reducing the prevalence of type II diabetes mellitus [10, p. 736].

In Russia, to control the number of cases, a special website was created for monitoring and helping patients - www.diaregistry.ru, it contains all the relevant information. So, according to the site, as of June 1, 2020, the number of registered cases is 4,833,923 people, of which the increase in the number of cases for the year was 169,451 people. At the same time, it is worth noting that patients with type II diabetes mellitus make up 92.5% of the total number, and 90.5% of the number of patients with diabetes since the beginning of 2020.

When suffering from type II diabetes mellitus, a person should exclude sucrose containing food from the diet. To restrain from eating products with a large amount of sugar, e.g. confectionery, is also preferable for people who monitor their health and go in for sports [3, p.10, 4, p.6]. Nevertheless, an analysis of the Russian market for specialized food products revealed a deficit in the assortment of confectionery products for these groups of population.

Also, in the diet of diabetics, it is necessary to include foods containing inulin - a natural analogue of insulin, which is found in large quantities in burdock (57%), foalfoot, garlic, Jerusalem artichoke, chicory and other plant species. According to scientists, the inulin contained in burdock is better absorbed and has a greater positive effect on the human body than other types of plant materials [6, p.355].

When developing any recipes, it is also relevant to take into account the consumers' taste preferences. Thus, according to the results of research of the Research Institute of Nutrition of the Russian Academy of Medical Sciences, the clinic of the Federal State Budgetary Institution "Federal Research Center of Nutrition and Biotechnology", the taste preferences of patients with diabetes mellitus were determined, which are recommended for consideration when developing a line of specialized confectionery products, these are orange, green apple, mint, vanilla, peach, lingonberry, cranberries, blackberries [5, p.140].

Purpose of the study is creation of composite mixtures for sugar-free caramel using modern software.

2 Materials and Methods

The studies were carried out using the MathLab software package, in particular the software Statistica 10.0.

3 Results and Discussions

In the course of the research, based on the scores of organoleptic quality indicators, there were determined the maximum allowable dosages of fortifying additives. On their basis, regression dependences have been obtained, which make it possible to determine with a high degree of reliability the optimal dosages for obtaining a product with high organoleptic properties. Orange juice, mint extract and burdock extract were chosen as the main enrichment additives. The choice of the components was made due to their chemical composition, as well as the recommendations of dietitians. In addition, the bulk of the components used is got from the raw materials growing on the territory of Kuzbass and as a result it is readily available.

The task of the study was to determine the dosage of the component mixture by the method of mathematical modeling. For this, we used the data obtained from the control group of subjects based on the results of evaluating the product organoleptic properties. The analysis of consumer preferences of the target audience was carried out by the method of a sociological survey using a questionnaire based on quota samples, taking into account gender, age and educational level of the composition of the control group of consumers. According to the results of the survey, the respondents were given a point score, which serves as a comprehensive assessment criterion for the organoleptic properties of the product.

The factors influencing the receipt of high score organoleptic assessment of quality indicators are presented in table 1.

Table 1. Factors and levels of their variation.

Designations	Apple juice, cm^3/kg	Extract of foalfoot, cm^3/kg	Cranberry juice, cm^3/kg	Orange juice, cm^3/kg	Lingonberry juice, cm^3/kg	Burdock extract, cm^3/kg	Mint extract, cm^3/kg
	X_1	X_2	X_3	X_4	X_5	X_6	X_7
Upper level (+)	7,0	4,0	7,0	6,0	7,0	5,0	1,0
Main level (0)	5,0	3,0	5,0	5,0	5,0	4,0	0,5
Lower level (-)	4,0	2,0	4,0	4,0	4,0	2,0	0,3

Implementation of the experiment on the design matrix, presented in Table 1, made it possible to obtain the values of the optimization criteria and mathematical models in the form of a regression equation.

The results are shown in Table 2.

Table 2. The results of the organoleptic score assessment of the product.

Experiment No.	Mixture components (X factors)							Organoleptic score (response Y)
	X_1	X_2	X_3	X_4	X_5	X_6	X_7	
Mixture №1								
1	4	2	-	-	-	5	-	27,5
2	4	4	-	-	-	2	-	26,8
3	7	2	-	-	-	2	-	28,3
4	5	3	-	-	-	4	-	27,6
Mixture №2								
5	-	-	4	-	7	2	-	28,4
6	-	-	4	-	4	5	-	26,9
7	-	-	7	-	4	2	-	29,1
8	-	-	5	-	5	4	-	29,1
Mixture №3								
9	-	-	-	6	-	2	0,3	29,8
10	-	-	-	4	-	2	1	28,6
11	-	-	-	4	-	5	0,3	27,4
12	-	-	-	5	-	4	0,5	28,9

The processing of experimental data in the study was carried out by statistical methods of analysis using the applied programs "*Statistica*" and "*Microsoft Excel*". The adequacy of the obtained regression equations was assessed using *Fisher's criterion* according to the condition $F_R > F_{Ti}$. To obtain reliable results, the required number of experiments was calculated. In the experiments, three parallel determinations were reproduced in four replicates. The numerical values that are presented in tables and graphs reflect the arithmetic mean, with a confidence level of P=0.95 and a confidence interval of ±10%.

Based on the experiments carried out, screening out statistically insignificant coefficients, using the stepwise regression method, after appropriate mathematical processing, the following equations were obtained:

- mixture №1

$$Y = 25{,}667 + 0{,}389x_1 - 0{,}167x_2 + 1{,}236x_1x_2 + 0{,}815x_1x_6 + 0{,}122x_2x_6 - 1{,}318x_1^2 + 0{,}935x_2^2 + 1{,}241x_6^2 \quad (1)$$

$$R = 0{,}832,\ F = 1{,}054$$

- mixture №2

$$Y = 18{,}143 + 0{,}729x_3 - 1{,}467x_5 + 1{,}231x_3x_5 + 0{,}709x_5x_6 + 0{,}0048x_3x_6 + 0{,}802x_3^2 - 1{,}012x_5^2 + 0{,}357x_6^2 \quad (2)$$

$$R = 0{,}826,\ F = 2{,}676$$

- mixture №3

$$Y = 26{,}702 + 0{,}636x_4 - 0{,}376x_6 + 0{,}104x_4x_6 + 1{,}111x_4x_7 + 1{,}41x_4^2 - 0{,}415x_6^2 - 1{,}115x_6^2 \quad (3)$$

$$R = 0{,}802,\ F = 1{,}425$$

A graphical interpretation of the dependence of the organoleptic parameters of candy caramel enriched with extracts on their quantitative ratio is presented in Figures 1-3.

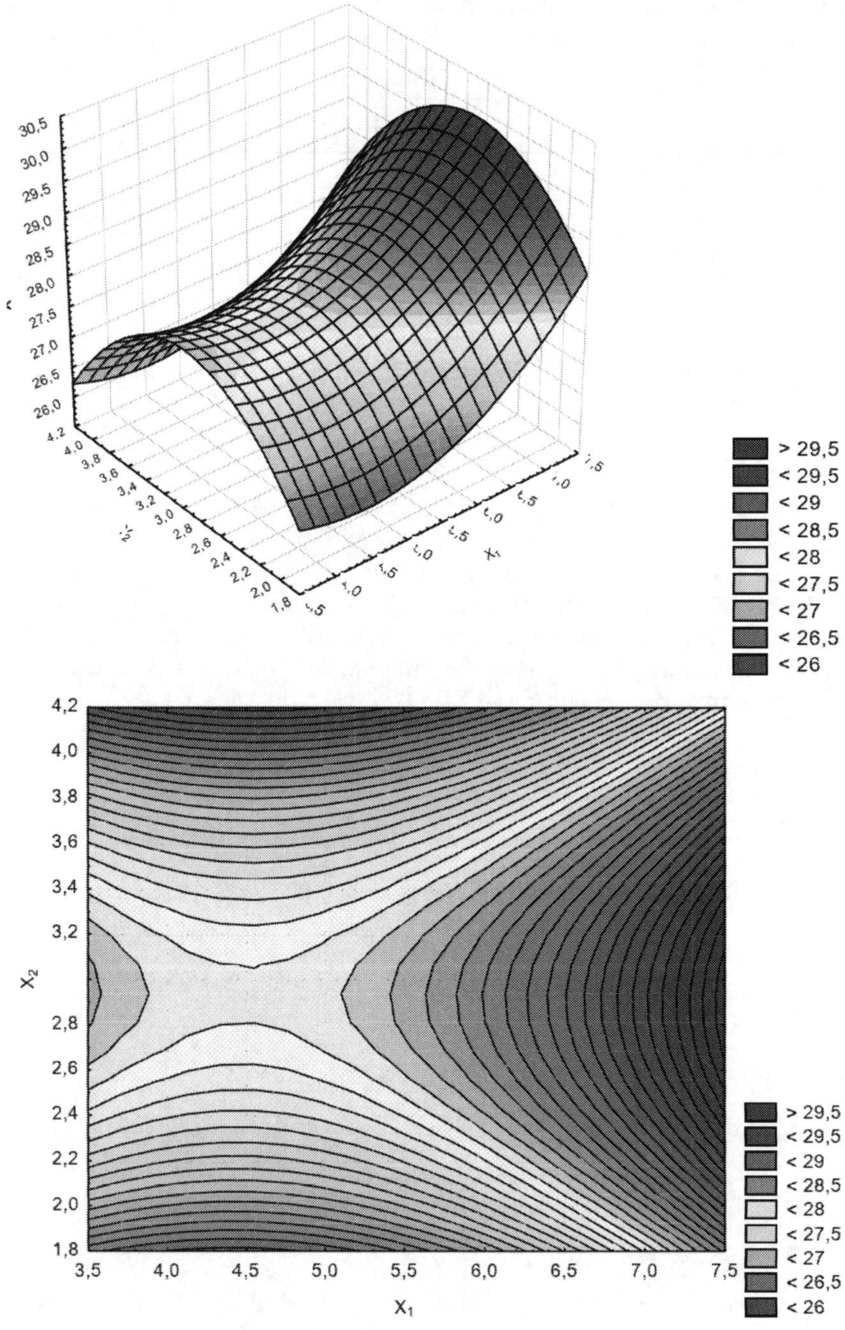

Fig. 1. Surface response (mixture 1) Y=f (X1, X2, X6 = 4,0).

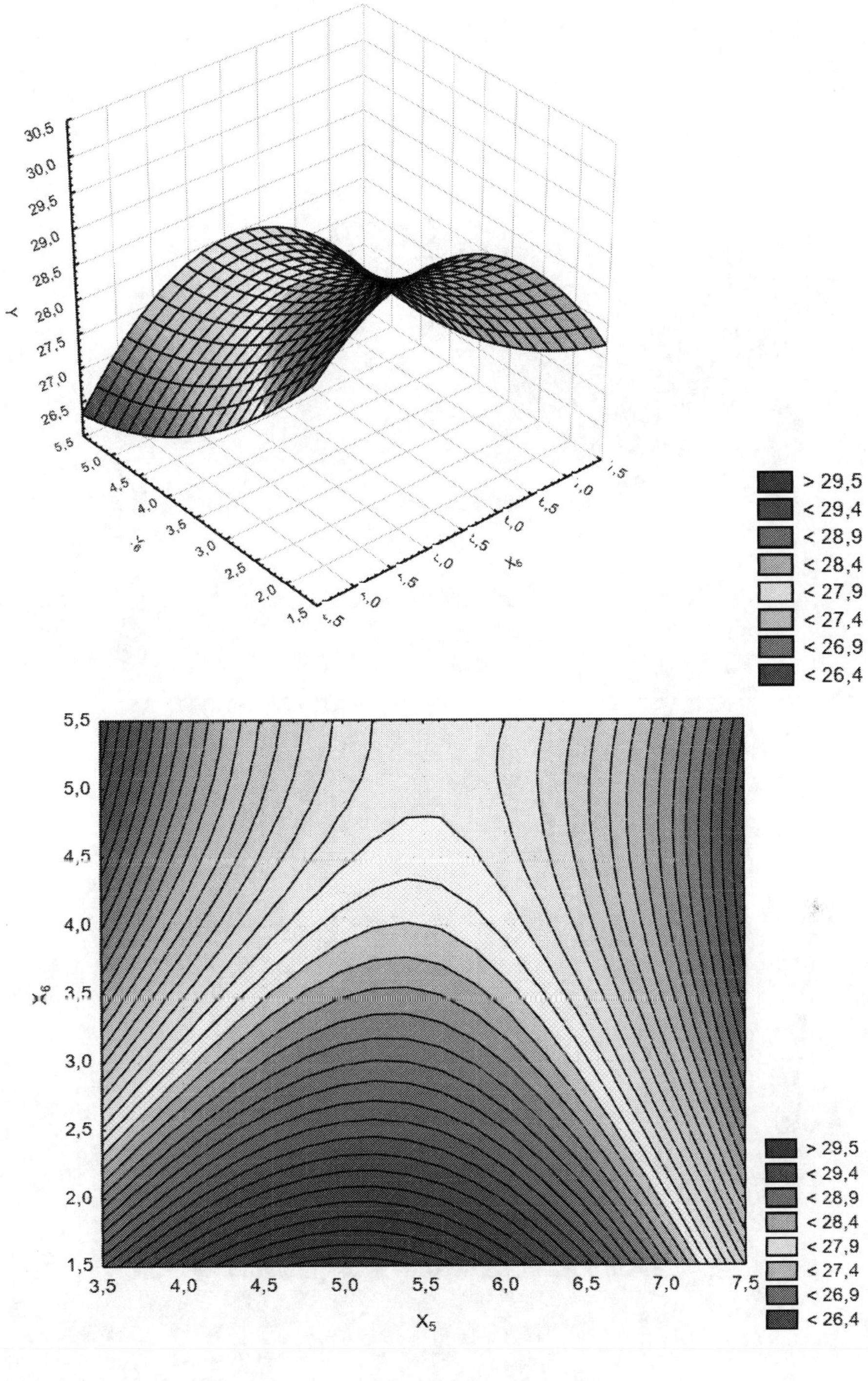

Fig. 2. Surface response (mixture 2) Y=f (X5, X6, X3=5,0).

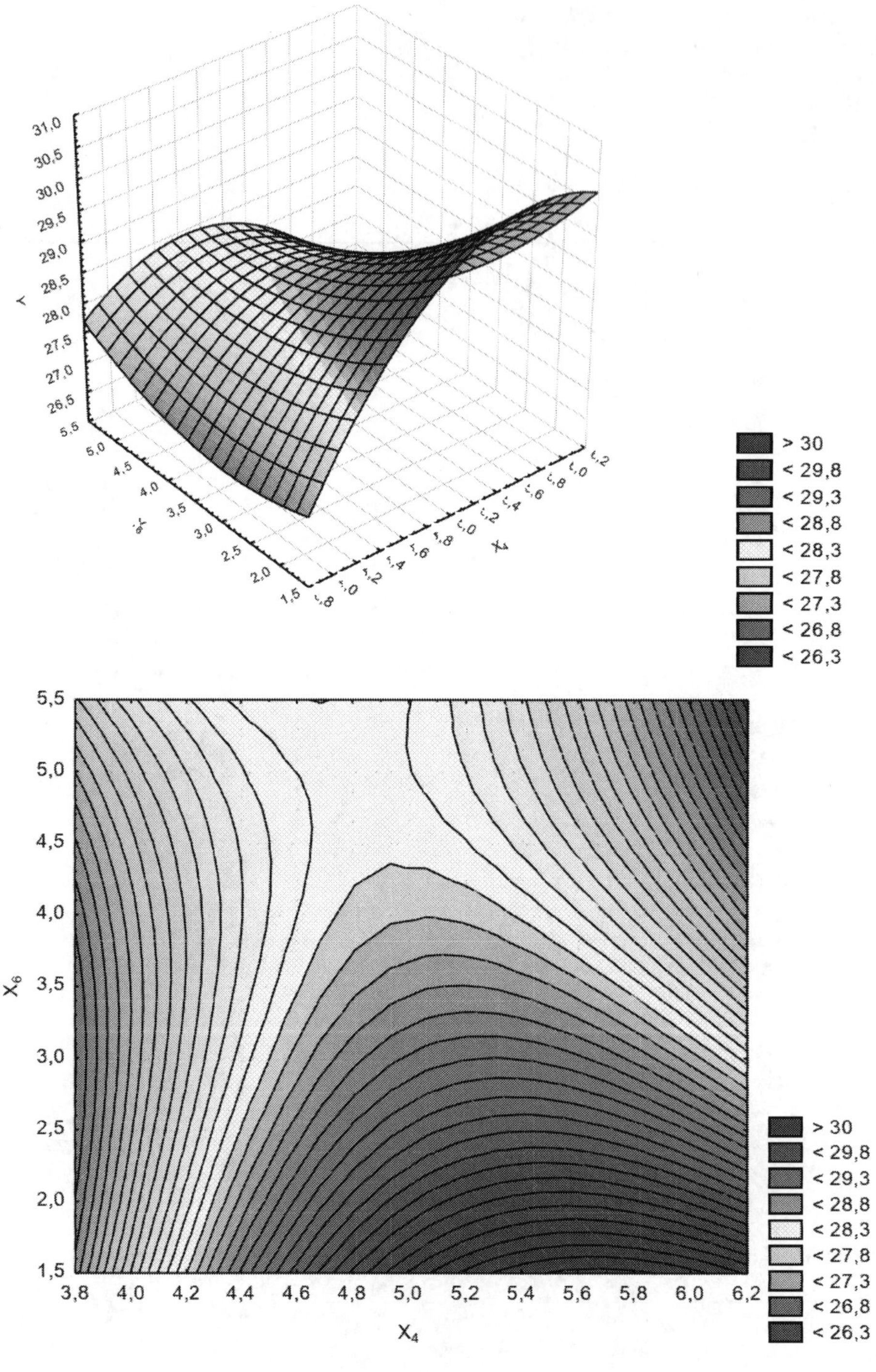

Fig. 3. Surface response (mixture 3) Y=f (X4, X6, X7=0,5).

Thus, the conducted studies allow concluding that the following ratio of extracts is a more accurate ratio of extracts to obtain enriched caramel with high organoleptic characteristics (table 3).

 https://doi.org/10.1051/bioconf/20224203002

Table 3. Ratios of enriching extracts (in shares)

Designations	Apple juice, cm^3/kg	Extract of foalfoot, cm^3/kg	Cranberry juice, cm^3/kg	Orange juice, cm^3/kg	Lingonberry juice, cm^3/kg	Burdock extract, cm^3/kg	Mint extract, cm^3/kg
	X_1	X_2	X_3	X_4	X_5	X_6	X_7
mixture №1	0,39	0,94	-	-	-	0,12	-
mixture №2	-	-	0,73	-	0,71	0,35	-
mixture №3	-	-	-	0,64	-	0,42	0,05

Therefore, the recipe for the enriched caramel will be as shown in table 4.

Table 4. Enriched caramel recipe

Designations	Apple juice, cm^3/kg	Extract of foalfoot, cm^3/kg	Cranberry juice, cm^3/kg	Orange juice, cm^3/kg	Lingonberry juice, cm^3/kg	Burdock extract, cm^3/kg	Mint extract, cm^3/kg
mixture №1	42	53	-	-	-	10	-
mixture №2	-	-	64	-	48	7	-
mixture №3	-	-	-	82	-	9	3

4 Conclusions

Based on the results of calculating the recipes and using programs for analysis, the optimal dosages of enriching additives for creating candy caramel for diabetic nutrition have been determined. According to the revealed regression equations, the calculation method can be used to determine the amount of added additives, depending on the mass of the required finished products.

References

1. N. Aktaş, Y. Özdoğan, Reserachers World, **7,** 46 (2016)
2. F. Bimbo, A. Bonanno, G. Nocella, R. Viscecchia, G. Nardone, B. De Devitiis, D. Carlucci, Appetite, **113,** 141 (2017)
3. S. Emami et al., International Journal of Preventive Medicine, **10** (2019)
4. S. Firouzi, M. Y. Barakatun-Nisak, K. N. Azmi, J Res Med Sci., **20,** 40 (2015)
5. L. Guariguata, D. R. Whiting, I. Hambleton, J. Beagley, U. L. Linnenkamp, J. E. Shaw, Diabetes research and clinical practice, **103,** 137 (2014)
6. I. Kalkan, Nutrition research and practice, **13(4),** 352 (2019)
7. I. I. Dedov, M. V. Shestakova, G. R. Galstyan, Diabetes Mellitus, **19,** 104 (2016)
8. A. Yu. Kolbina, M. G. Kurbanova, Agroindustrial complex - new ideas and solutions, 168 (2020)
9. G. G. Oganesyan, A. V. Topchyan, A. A. Barsegyan, L. V. Revazova, N. G. Grigoryan, R. M. Bagumyan, Bulletin of the Medical Institute, **2,** 103 (2006)
10. E. V. Smirnov, Food flavorings. Directory, 736 (2008)

Influence of natural production conditions on efficient operation of wheel tractors

Aleksandr A. Shuravin*, *Vladimir* Leonov, *Elena* Polikutina, *Sergey* V. Shchitov, and *Evgeny* E. Kuznetsov

Far Eastern State Agrarian University, Polytechnicheskaya street, 86, Blagoveshchensk, Russia

Abstract. For the zones of "risky farming" characteristic of the Far East of the Russian Federation, the natural production conditions of the region are an important problem in preparing the soil for further basic agricultural work. So, when carrying out early spring agricultural work, due to presence of a solid underlying layer in the form of permafrost, they shall be completed in operational terms no more than 10 days, until the permafrost base thaws and the soil has not lost its bearing capacity. In addition, due to the peculiarities of the relief, the soil does not thaw equally in depth everywhere, which reduces the quality of field work and harrowing, as the most common operation, namely. This article provides theoretical and experimental studies on the adaptation of a wheeled tractor as part of a machine-tractor unit (MTU) used in harrowing to natural production conditions by installing a device that automatically regulates the load on the working body of the disc harrow or on the propellers of the energy device, depending on the conditions of use or the state of the motion surface.

1 Introduction

The primary cultivation is a set of measures, the main purpose of which is to provide favorable conditions for the growth of agricultural crops. Among the main functions of this operation are cutting weeds, accumulating moisture, combating pests and diseases of agricultural crops, creating favorable conditions for biological processes, increasing fertility due to conservation and productivity. It shall also be noted that the planning and conduct of operations related to the primary cultivation shall be carried out taking into account the natural production conditions of the region, as well as the general physical and physical and mechanical properties of the cultivated soil (aggregation, porosity, density, hardness, weediness, and etc.) [1,2].

In the Amur Region, soil preparation for sowing begins as the soil thaws to the depth of cultivation, usually in late April or early May of the current year. This operation, including sowing, must be completed as soon as possible, since as daytime temperatures increase, intensive thawing of the permafrost subsoil occurs, and precipitation at the same time sharply reduces the bearing capacity of the soil.

* Corresponding author: sh.aleksandr.2019@mail.ru

It shall be noted that the moisture content of the upper soil layer allows agricultural work, and the lower soil layer, due to the increased moisture caused by the above conditions, has a low bearing capacity. In these conditions, in order to implement the possibility of MTU movement, it is necessary to adapt the used machine and tractor units to the above specific features of the region, by reducing the normal pressure of the propellers on the soil in particular.

In the region under consideration, as previously conducted studies have shown, the most acceptable ways to reduce the normal pressure of the wheel propellers on the soil is to install double wheels or redistribute the adhesion weight [3].

Due to the fact that the soils of the region are heavy in texture and have a high degree of stickiness, such an agricultural operation as harrowing with heavy disc harrows is most widespread in early spring agricultural works. The choice of this tool is justified by the fact that in the process of snow melting on clay soils, the upper layer is "pulled in" and a crust is formed.

Use of light harrows for these purposes is impractical, since they do not always cope with the task due to the low weight of the harrow, which does not allow the working bodies to be deepened to a sufficient depth. In this connection, depending on the condition of the soil, ballasting of the harrow with additional weights is used to improve the depth of cultivation [4, 5, 6]. Considering that during the harrowing period, due to the different relief of even one agricultural landscape, the soil thaws unevenly, it becomes difficult to choose the mass of ballast and the positions of its distribution on the harrow. Therefore, in these conditions, it is necessary to use other methods of correcting the weight on the working body of the harrow (disc).

In previous studies, it was proved that the most acceptable way to increase the load on the working bodies of an agricultural machine is to install devices that make it possible to correct the redistribution of weight in the system of the machine-tractor unit itself. In other words, such devices make it possible to use part of the weight of an agricultural machine to increase the traction-coupling properties of an energy device, and, if necessary, to reload its working bodies, the weight of an energy device [7, 8].

In this connection, the purpose of the offered work is creation, manufacture, testing and implementation of a device that allows to automatically increase the traction and coupling properties of the energy device and adjust the weight on the working body of the disc harrow.

2 Material and Methods

As an example of the aforementioned device capable of automatically adjusting the traction and coupling properties of an energy device and the load on the working body of a disc harrow, the use of a device of a new promising design, which, when installed on a tractor, allows to adjust and redistribute the weight of the MTU between the tractor and the harrow as necessary to increase the vertical load on a tractor or harrow with an inventive level, high industrial applicability and novelty, which will allow solving the task of increasing the efficiency of the MTU (Figure 1-2).

Fig. 1. MTU diagram with installed device.

Fig. 2. General view of the MTU with the installed device

When carrying out studies on optimization and redistribution of weight in the scheme of a machine-tractor unit, an energy tool was taken as the object of research - a wheeled tractor of class 1.4 and a disc harrow BDT-3, as the most often used in peasant farms of the region with a small volume of sowing capacities, the offered device was installed at the MTU.

Within the study, the well-known and separate methods were used with application of specialized programs for mathematical processing, experimental modeling and methods of statistical and regression analysis [9, 10]. Within the experiment, the following parameters were measured: adhesion weight, distance traveled, rotational speed of the driving wheels of the power plant, tractive effort, weight of the harrow's working body, and the quality of the treatment performed.

The above parameters were measured using specialized instruments and equipment. Processing the obtained values was carried out using well-known methods of mathematical statistics and information technology.

3 Results

Theoretical and experimental studies on use of a machine-tractor harrow unit, carried out on the basis of the Far Eastern GAU, Amur Region, the city of Blagoveshchensk, the Russian Federation and the peasant farms of the region, have shown the validity and correctness of the chosen method for increasing the redistribution of weight within the machine-tractor unit in order to increase the traction-coupling properties of the energy means and additional loading of weight on the working body of the harrow.

It was found that use of the offered device made it possible to redistribute the mass of the MTU (Figures 3 and 4).

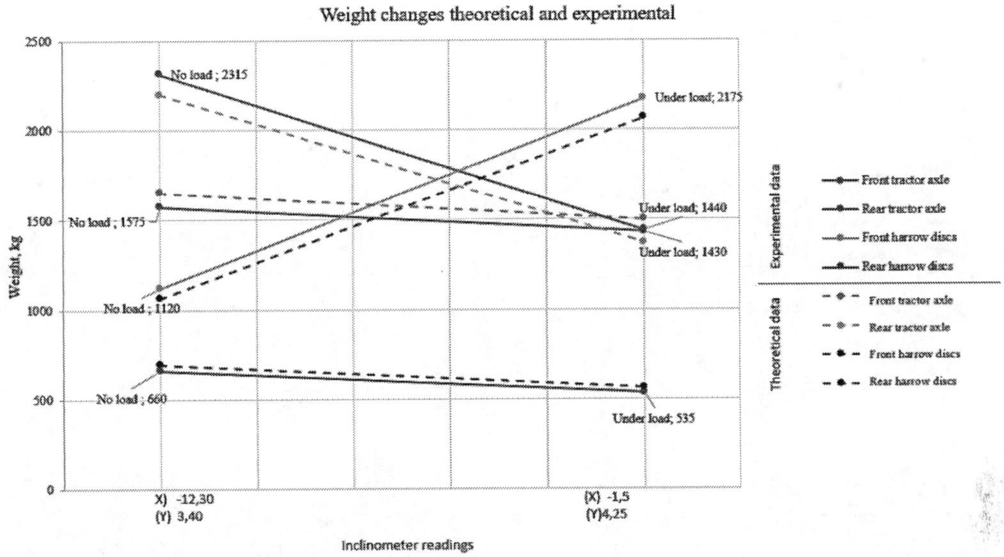

Fig. 3. Results of experimental studies on the redistribution of mass when the device is operating inside the MTU

Analyzing the data obtained (Figure 3), the following conclusions can be drawn that when the device was operating in order to reload the harrow sections, the following redistribution of mass within the MTU scheme occurred: the load on the rear axle of the tractor decreased from 2,315 kg to 1,440 kg; on the front axle of the tractor, 1,575 kg decreased to 1,440 kg; on the front sections of the harrow increased from 1,120 kg to 2,175 kg; on the rear sections of the harrow decreased from 660 kg to 535 kg. The above data and the results of graphical displays allow to conclude that use of the offered device allows to redistribute the mass in the MTU system and thereby adjust its technological properties and energy indicators during soil cultivation.

In order to check the parameters of the additional loading of the tractor, studies of the offered design were also carried out in the mode of weight redistribution from the harrow to the tractor running system (Figure 4).

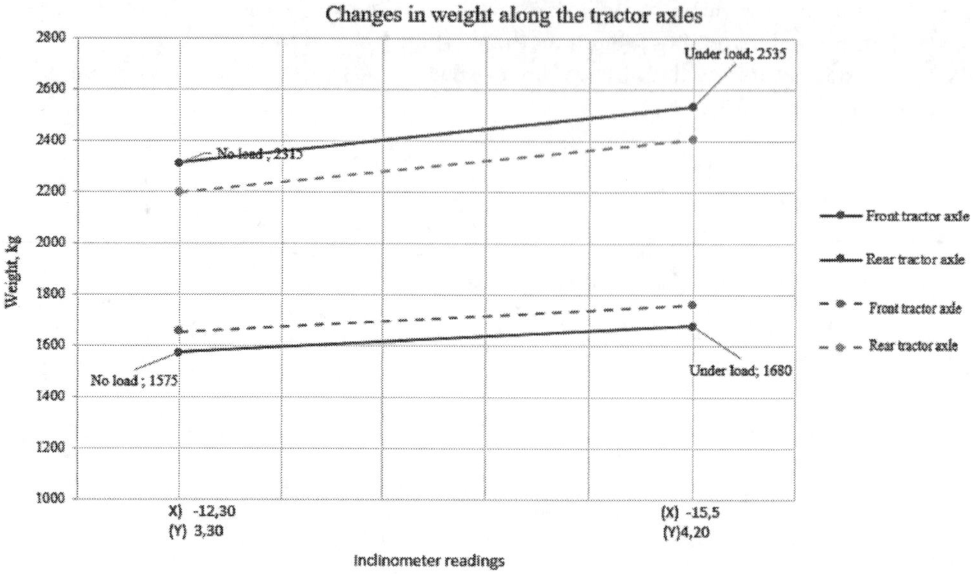

Fig. 4. Results of experimental studies on the redistribution of mass when the device is operating in the tractor reloading mode

Therefore, an increase in the vertical load on: the front axle of the tractor — by 105 kg; rear axle of the tractor — 220 kg.

Based on the results obtained above, it can be concluded that the redistributed vertical load makes it possible to increase the traction and coupling properties of the tractor, reduce the amount of slipping, increase the speed characteristics of the MTU and expand its technological characteristics.

As the studies have shown, the results of theoretical and experimental data are within the confidence interval of 3.5-5 %, which indicates the reliability of the theoretical justification.

4 Discussion

As a result of solving the compromise problem, the following acceptable values are justified, all other things being equal. At the maximum depth of soil treatment, the lowering angle of the hitch is 27-12 degrees, the width of the disc harrow is 2.4-2.8 meters, the speed of the unit is 9-10 km/h.

At the maximum lowering angle of the hitch, the processing depth is 18-20 cm, the speed of the unit is 4-7 km/h, the width of the disc harrow is 2.9-3.0 meters. At the optimum value (6-8 km/h) of the unit's speed, the movement depth is 13.5-17 cm, the width of the disc harrow is 2.4-3.0 meters, the angle of lowering of the lift is 27-16 degrees. With the maximum working width, the lowering angle is 27-17 degrees, the soil treatment depth is 13-17 cm, the speed of the unit is 11-12 km/h.

The comparative economic tests carried out showed that use of the MTZ-80 tractor and the BDT-3 harrow with an installed harrowing device made it possible to increase the productivity per hour of the main working time by 14.2 % and reduce the fuel consumption per unit of cultivated area by 8.6 % compared to with MTZ-80 and BDT-3 tractors operating in the serial version.

An analysis of the results of traction tests indicated that there was a redistribution of the components of the power balance in both the serial and experimental tractors.Therefore, a decrease in the power spent on slipping was recorded for an experimental tractor in comparison with a serial one from 15.18 % to 10.25 %. On the basis of the data obtained, it can be concluded that use of the offered device makes it possible to increase the traction and coupling properties in comparison with a serial tractor due to the redistribution of the coupling weight in the MTU scheme.

The fuel and energy assessment showed that use of 1.4 class tractor with a device for redistributing the MTU weight during harrowing allows reducing energy consumption by 32.93 MJ/ha compared to the serial version and obtaining an annual economic effect on harrowing in terms of the ruble equivalent in the amount of 38.6 rubles/ha.

5 Conclusion

Based on the above, in comparison with the results previously obtained by the authors [11, 12, 13, 14], it can be concluded that the offered device for redistributing the adhesion weight in the harrow system is a highly efficient design that implements original ideas and has a design novelty intended for increasing the efficiency of mechanization means in the agro-industrial complex. Study materials are widely used in the technology of plant growing, used in the Federal State Unitary Enterprise Sadovoe, SOYUZ LLC of the Seryshevsky District, KFH S.V. Kovaleva. Ivanovsky district, farm ZARECHNOE, Mikhailovsky district, farm A.V. Osipov . Blagoveshchensky district of the Amur region.

Implementation of the results obtained in production has made it possible to increase the efficiency when using wheeled tractors of the 1.4-2 class and the BDT-3 harrow to obtain significant economic benefits for the enterprise.

Based on the results of the work, the team of authors responsibly declares that there is no conflict of interest, information data and results were obtained as a result of the activities of the entire research group. Wherein, the spent material resources are not attracted funds and belong personally to the co-authors.

References

1. M. Gholkar, V. Salokhe, A. Keen, ASABE Meeting Presentation Paper Number, **096606**, 1-12 (2009)
2. B. Avarvarei, Medunarodni Simhjzij agronoma zbornikradova. Ostjek, **2**, 892-896 (2009)
3. I. Troyanovskaya, Procedia Engineering, **206**, 1540–1545 (2017)
4. V.I. Belyaev, M. Fruhauf, T. Mainel, Europa Regional, **1(4)**, 13-21 (2004)
5. R. Gopichandran, Mitigation-Adaptation Nexusfor Sustainability: Some ImportantCrosscutting and Emerging Considerations. Vulnerability of Agriculture, Water and Fisheries to Climate Change: toward Sustainable Adaptation Strategies (2014)
6. V. Crisan, Actual Tasks on Agricultural Engineering, **40**, 243-251 (2012)
7. S. Shchitov, P. Tikhonchuk, I. Bumbar, Revista ESPACIOS, 38(Nº 48), 35 (2017)
8. S. Shchitov, E. Kuznetsov, Journal of Advanced Research in Dynamical and Control Systems (JARDCS), 13, 707-714 (2018)
9. S.V. Shchitov, Z.F. Krivutsa, E. Kuznetsov, International Journal of Applied Engineering Research (IJAER), **14(24)**, 16747-16752 (2018)
10. S.V. Shchitov, E.E. Kuznetsov, E3S Web of Conferences, 262 (2021)
11. Li H., He J., Gao H., Zhang Z., Chen Y., Frontiers of Agricultural Science and Engineering, **2(2)**, 179-185 (2015)

12. D. Wulfsohn, T. Way, Advances in Soil Dynamics, 3, 209-252 (2009)
13. J. Bulinski, H. Niemczyk, Ann.Warsaw Agr. Univ. Agr., **50**, 3-8 (2007)
14. Xiujian Yang, Jian Xiong, Int. J. of Heavy Vehicle Systems, **20(3)**, 253-288 (2013)

Mechanized Placement of Drip Tape in Potato Comb

Boris M. Kizyaev[1], *Victor I.* Balabanov[2], and *Natalia B.* Martynova[2,*]

[1]All-Russian Research Institute of Hydraulic Engineering and Land Reclamation named after A.N. Kostyakov, 127550, Moscow, Russian Federation
[2]Russian State Agricultural University - MSHA named after K.A. Timiryazev, 127550, Moscow, Russian Federation

Abstract. Over the past two years, decrease in potato yield has been observed in the Moscow region. One of the main reasons is unfavorable weather conditions. To obtain a guaranteed yield, as well as to improve the quality of products, it is necessary to carry out reclamation works, including additional irrigation. Creation of the optimal water - air balance of the plant throughout the entire growing season will increase productivity, and improve the quality of agricultural products. Drip irrigation is the most economical method of irrigation, but its widespread use is hindered by the high proportion of manual labor when installing the drip tape. A mechanized method of laying the drip tape by placing the working equipment on the Grimme GF-75/4 comb former is offered. The existing methods for determining the irrigation rate do not consider the parameters of the humidification circuit, which leads to excessive irrigation and additional losses of irrigation water. The paper offers a method for determining the irrigation rate considering the shape of the humidification circuit and its geometric parameters.

1 Introduction

In the Moscow region, after the record potato harvest in 2019, which amounted to 460 tons, there has been a decline [1]. Last year, the region received less than due 36 tons of the planned harvest, and this year there are reasons to believe that the situation will repeat itself [2, 3]. The quality of the products has also noticeably decreased. For consumers, potatoes with a diameter of over 55 mm are most in demand, and until last year the share of such products intended for food purposes was about 80 % [4, 5]. Last year, this figure dropped to 50 %, and economy class chain stores were forced to make concessions, the rest switched to imports. This year, the region will also not receive part of the planned harvest, and imported products may appear on store shelves in the fall, and not in the spring, as in previous years.

Until 2020, there was a steady growth in this industry segment, the indicators improved due to the return of land to agricultural use during cultural and technical reclamation within

* Corresponding author: nmartinova@rgau-msha.ru

implementation of Agriculture of the Moscow Region State Program (Development of Reclamation of Agricultural Land subprogram) adopted in 2014 (Figure 1).

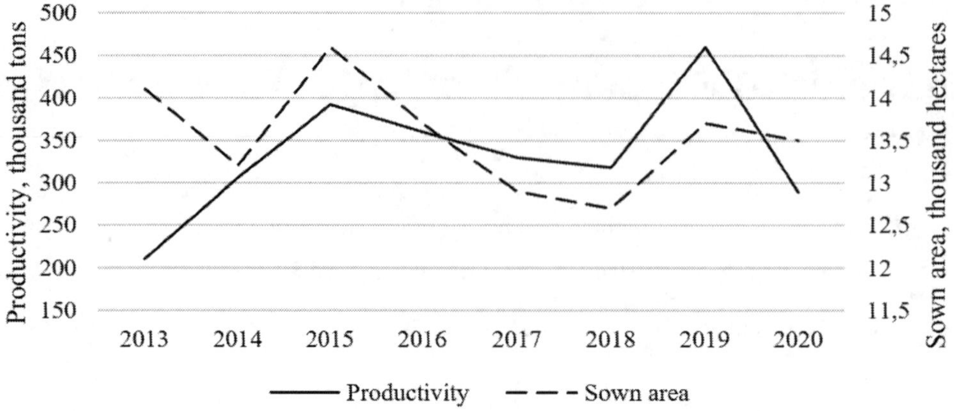

Fig. 1. Commissioning new areas and dynamics of potato yield.

Therefore, fluctuations in yield were compensated by introduction of new territories into agricultural circulation [6, 7]. Last year, implementation of the program was completed, and the last two years there has been a drop in yield, the main reason is the climatic features of recent years. In 2020, due to heavy rains at the beginning of summer, a significant part of the crop perished, while the harvested crop was significantly inferior to the indicators of previous years in terms of quality indicators [8]. In 2021, the negative trend worsened. This time, the reason was the almost complete absence of precipitation in the middle of summer, during the period when the daily water consumption of potatoes reaches its maximum values [9, 10]. The climatic feature of the Moscow region is fluctuations in precipitation within wide limits both during the growing season and throughout the year: high-water years alternate with moisture-deficient years (Figure 2).

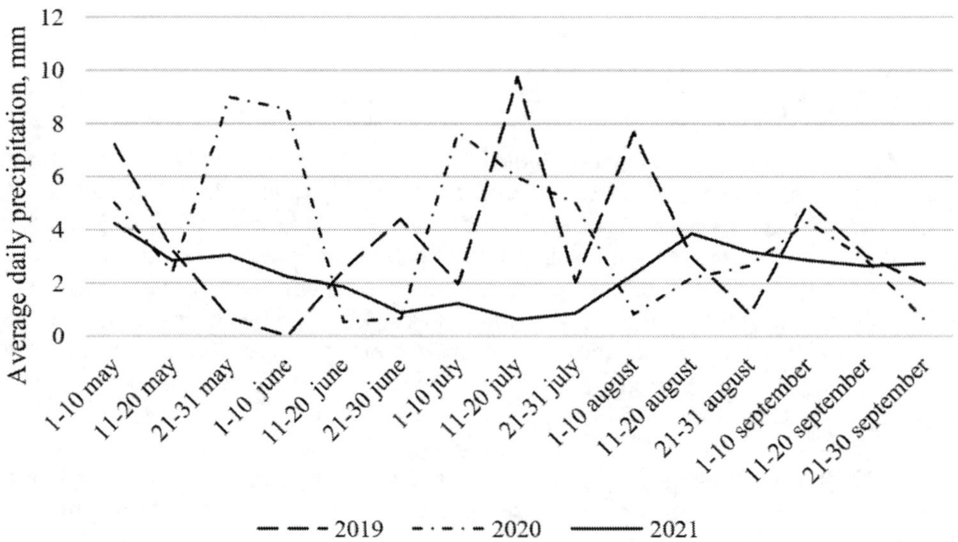

Fig. 2. The dynamics of precipitation within the growing season.

In such conditions, obtaining the stable harvest becomes a difficult task. The region, which ranks the 3rd in the Central Region in terms of gross harvest of potatoes, shall minimize weather risks. For the region with a widely developed network of processing enterprises with a total capacity of 170 thousand tons, obtaining a guaranteed yield that meets world quality standards is a task that requires an immediate solution.

Conducting the agro-reclamation measures will allow maintaining the optimal regime of water consumption throughout the entire growing season of plants. This will not only increase yields, but also improve the quality of combs. The total area of ameliorated land in the region is 395.7 thousand hectares, but most of it is in an unsatisfactory condition [11]. When carrying out irrigation measures, surface watering and sprinkling are used. More than half of the used sprinkler machines are beyond their service life, and new arrival — over 90 % are imported equipment. This runs counter to the import substitution program and the concept of the development of agricultural science developed by the distancing of agricultural sciences of the Russian Academy of Sciences that was supposed to increase the yield of crop production by 117 % compared to 2017 by introducing promising domestic machines into agriculture [12]. At present, the vast majority of promising domestic machine designs are at the stage of designing and testing prototypes. It is doubtful that by the specified date the basis of the park of sprinklers will be made up of mass-produced domestic equipment.

The way out of this situation could be an increase in the area occupied by drip irrigation. So far, this method of irrigation has not found widespread use in the region, its share is several percent. However, the advantages of this irrigation method are obvious: the delivery of moisture directly to the root zone makes it possible to almost completely eliminate the loss of water for evaporation and filtration into deep soil layers. For drip irrigation, the uneven terrain is not a problem, the complex configuration of the fields, in the spaces between rows, since they remain not watered, the growth of weeds is inhibited. Domestic enterprises are increasing the output of drip tape: in 2020, the output of the Russian - Israeli enterprise Polyplastic Rivulis Irrigation Pipeline Systems (St. Volzhsky, Volgograd Region) amounted to 200 million linear meters per year, New Age of Agrotechnology (St. Chaplygin, Lipetsk region) — 270 million linear meters per year. This made it possible to reduce the import of drip tape from 60 % to 30-40 %, but did not lead to a noticeable increase in the areas occupied by drip irrigation [13].

One of the reasons is the low degree of mechanization of the drip tape installation. The American companies Andros and Rain flo specialize in production of specialized equipment for laying drip tape. However, in their line of models there is no specialized machine for placing drip tape in a potato comb. Use of machines that are not intended for this type of work leads to uneven tension of the drip tape within installation process and, as a consequence, its twisting or breakage [14]. Creation of a special stacker of drip tape in a potato comb will expand application of this irrigation method.

2 Materials and Methods

When designing the working equipment of the drip tape stacker, the resistances arising during the laying process were determined [15]:

$$F_s = R_e + F_{pt} = R_e + F_{ft} + F_{ic} + F_{fc} =$$
$$= \mu \cdot G_s + k_p \cdot h \cdot b + \varepsilon \cdot h \cdot b \cdot v^2 + (f_a \cdot G_c \cdot d_a \cdot d_c^{-1} + 2 \cdot G_c \cdot l_c \cdot d_c^{-1}) \cdot (1 + e^{f_a \cdot \alpha}) \tag{1}$$

where: F_{pt} — resistance from pulling the tape, kN; F_{ft} — friction resistance of the tape in the stacker, kN; F_{ic} — force from unbalance of the coil, kN; F_{fc} — resistance to friction of the coil on the surface of the axis, kN; R_e — soil resistance to digging, kN; f_a — coefficient of friction of the axle in the bearings; G_c — weight of the tape coil, kN; d_a — axis diameter, m; d_c — coil diameter, m; l_c — displacement of the center of gravity of the coil from the axis of rotation, m; α — tape wrap angle; μ — coefficient of friction of the working body on the ground; G_s — stacker weight, kN; k_p — specific resistance to cutting, kN/m^2; ε — coefficient that takes into account the effect of cutting speed on the resistance to digging; h — development depth, m; b — working width, m, v — stacker speed, m/s.

When designing a drip humidification system, the irrigation rate shall be determined. The existing calculation methods do not take into account the shape of the humidification circuit, this leads to inaccurate results, mainly to excessive watering [16]. In the process of irrigation, water is delivered to the soil in small portions at regular intervals. The humidification circuit has a semi-elliptical shape [17]. Since gravitational and capillary forces act on a drop in the soil, the speed of moisture propagation in the horizontal and vertical directions is not the same. The vertical boundary of the circuit shall be determined [18]:

$$z = 2A^{-0,5}\sqrt{D_0 \cdot e^{\beta(W_m - W_0)} \cdot t} \tag{2}$$

where: A — correction factor; W_m — relative soil moisture; W_0 - initial soil moisture; t — time, s; β — parameter depending on soil and moisture content; D_0 — diffusivity coefficient at initial moisture content W_0.

The coefficients given in the formula can be determined by the formula [19]:

$$\beta = \frac{1}{W_1 - W_0} \ln\left(\frac{3(W_1 - W_m)^{3,5}}{W_1^2 (W_0 - W_m)^{3,5}\left(\frac{1}{W_0^2} + 2\frac{W_0}{W_1^3}\right)}\right) \tag{3}$$

The diffusion coefficient shall be determined from the expression:

$$D_0 = \frac{K_F \Psi_m W_m}{1 - \left(\frac{W_m}{W_1}\right)^3}\left(\frac{W_0 - W_m}{W_1 - W_m}\right)^{3,5} \cdot \left(\frac{1}{W_0^2} + 2\frac{W_0}{W_1^3}\right) \tag{4}$$

where: Ψ_m — value of soil moisture pressure at moisture W_m; W_1 — the limiting field moisture capacity of the soil; K_F — filtration coefficient.

To determine the horizontal boundary of the circuit, use the formula [20]:

$$x = \sqrt{\frac{2tD_0 e^{\beta(W - W_0)} \cdot \sqrt{A}}{\sqrt{A} - \beta(W_m - W_0)\sqrt{2D_0 e^{\beta(W_m - W_0)}}}} \tag{5}$$

Knowing the coordinates of the humidification circuit, the irrigation rate shall be determined as follows:

$$m = \frac{2 \cdot 10^{-2} \pi z x^2 (W_1 - W_0)}{3 \cdot l_d \cdot b_d} \tag{6}$$

where: l_d - distance between droppers, m; b_d — distance between adjacent combs, m.

Calculation of the irrigation rate considering the parameters of the humidification circuit will allow avoiding excessive irrigation and rational use of irrigation water.

3 Results and Discussion

Laboratory studies have confirmed that the traction resistance arising during the laying of the tape does not exceed 180 N at an estimated laying depth of 20 - 80 mm (Figure 3).

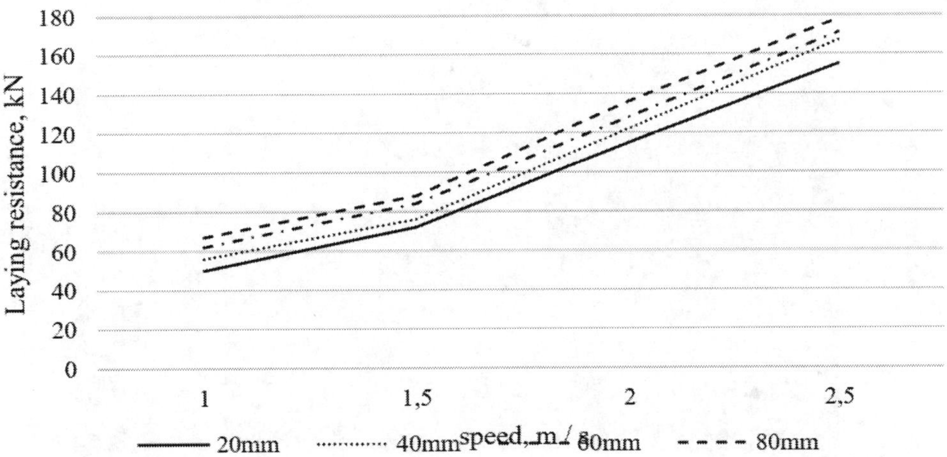

Fig. 3. Resistance to drip tape stacking.

Therefore, it is rational to place the working equipment of the tape stacker on the comb former, combining the process of forming the comb with the laying the drip tape. This will reduce energy costs and increase productivity. The working equipment of the drip tape stacker was installed on the Grimme GF-75/4 comb former and the drip tape was laid on the experimental site of the Field Experimental Station of the Russian State Agricultural University named after K.A. Timiryazev (Figure 4).

Fig. 4. Tests of the drip tape stacker.

In the process of further study, the boundaries of the humidification circuit within irrigation were determined, the irrigation rate was calculated and a drip tape with optimal parameters for a given crop was selected (Figure 5).

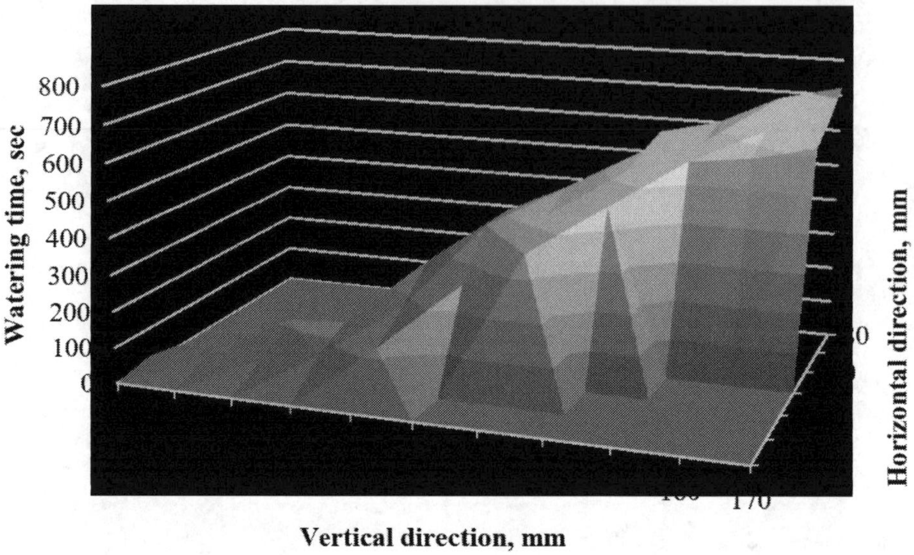

Fig. 5. The rate of the humidification circuit formation.

Watering was carried out using the calculated irrigation rate when the pre-irrigation moisture content reached 70 % of the maximum field moisture capacity. Considering the weather conditions, the time between waterings was adjusted (Figure 6).

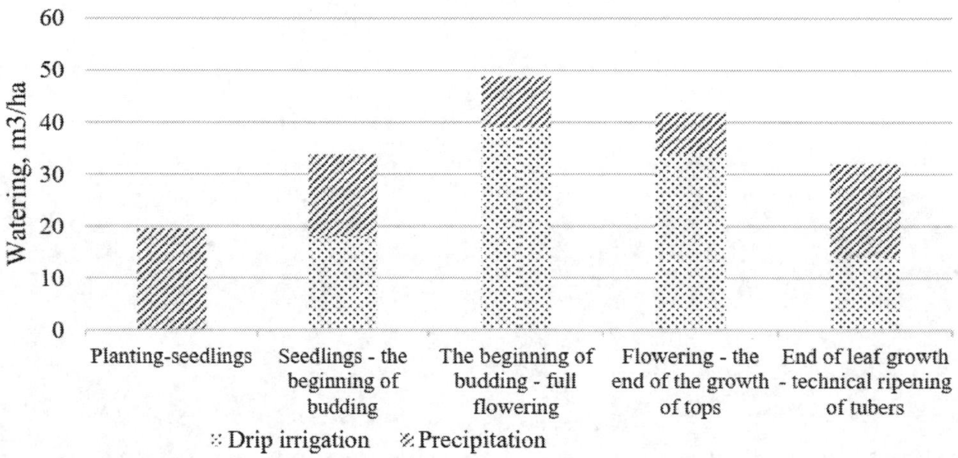

Fig. 6. Dynamics of daily water consumption.

4 Conclusion

In the Moscow region, in order to obtain a guaranteed potato crop, it is necessary to use agromeliorative measures. The advantage of drip irrigation is the ability to deliver irrigation water directly to the root-inhabited space, which practically eliminates the loss of water for evaporation and filtration into deep soil layers. Creation of the specialized machine for placing drip tape in the potato comb will expand the application of this irrigation method. Use of the drip tape stacker based on the Grimme GF-75/4 comb former made it possible to reduce labor costs by 22 - 28 %, and the combination of technological operations for forming potato combs and laying the drip tape — to increase work productivity by 32 - 36 % and reduce energy costs by 16 %. Calculation of the irrigation rate considering the parameters of the humidification circuit made it possible to reduce the irrigation water consumption by 12 %.

Acknowledgements

The article was prepared with the support of the Ministry of Education and Science of Russia within the framework of Agreement No. 075-15-2021-032 dated March 23, 2021 on provision of a grant in the form of a subsidy for creation and development of an engineering center on the basis of an educational institution of higher education and (or) a scientific organization within the framework of the federal project Development of Infrastructure for Research and Training of the national project Science and Universities.

References

1. A.S. Apatenko, Bulletin of the Federal State Educational Institution of Higher Professional Education "Moscow State Agroengineering University named after V. P. Goryachkin", **2(58)**, 23 (2013)
2. N.S. Sevrugina, A.S. Apatenko, E.V. Voytovich, Environmental management, **2**, 115 (2020)
3. V.A. Shevchenko, G.I. Bondareva., A.M. Solovjov, L.V. Kudrjavtzeva, Engineering for rural Development, 1065 (2020)
4. L.V. Kireycheva., L.A. Lentyaeva Influence of agricultural Production on Pollution of water Bodies. Environmental management, **5**, 18 (2020)
5. V.A. Shevchenko, V.A. Gubin V.K., L.V.Kudrjavtzeva, The Herald of the Moscow State Agricultural Engineering University named after V.P. Goryachkin, **88**, 27 (2018)
6. N.N. Dubenok, D.A.Bolotin, S.D Fomin., A.G. Bolotin, News of the Nizhnevolzhsky agro-university complex: science and higher professional education., **4**, 22 (2017)
7. V.N. Krasnoshchekov, D.G. Olgarenko, Prirodoobustroystvo, **4**, 51 (2016)
8. M.N.Makani, S.A. Sargent, L. Zotarelli D.J. Huber, C.A. Sims, Scientia Horticulturae., **197**, 428 (2015)
9. V.V. Borodychev, M.N. Lytov, A.S. Ovchinnikov A.S., V.S. Bocharnikov, Bulletin of the Nizhnevolzhsky agro-university complex: science and higher professional education., **4(40),** 21 (2015)
10. J. Reyes-Cabrera, L. Zotarelli, D.L. Rowland, M.D. Dukes, S.A. Sargent, American Journal of Potato Research, **91(5),** 504 (2015)
11. A.S. Ovchinnikov, A.A. Buber, Yu.P. Dobrachev, V.V. Borodychev, News of the Nizhnevolzhsky agro-university complex: science and higher professional education, **4**, 65 (2018)
12. N.B. Martynova, A.Yu. Korneev, Bulletin of the Federal State Educational Institution of Higher Professional Education "Moscow State Agroengineering University named after V.P. Goryachkin", **2(84),** 18 (2018)

13. E.V. Zhalnin, Bulletin of the Federal State Educational Institution of Higher Professional Education "Moscow State Agroengineering University named after V. P. Goryachkin", **6(82),** 10 (2017)
14. N.K.Telovov, Kh.A. Abdulmazhidov, The Herald of the Moscow State Agricultural Engineering University named after V.P. Goryachkin, **91(3),** 22 (2019)
15. O.A. Starovoitova, N.E. Shabanov, Bulletin of the Federal State Educational Institution of Higher Professional Education "Moscow State Agroengineering University named after V. P. Goryachkin", **4**, 34 (2016)
16. A.S. Shtanko, V.N.Shkura, Scientific journal of the Russian Research Institute of Melioration Problems, **3(31),** 39 (2018)
17. N.B. Martynova, G.I. Bondareva, S.K. Toygambaev, N.K. Telovov, Journal of Physics: Conference Series. Krasnoyarsk Science and Technology City Hall of the Russian Union of Scientific and Engineering Associations. Krasnoyarsk, Russian Federation, **42091** (2020)
18. L.V. Kireycheva, V.M.Yanshin, Melioration and water management, **5**, 31 (2019)
19. N.B. Martynova, S.K. Toygambaev, N.K. Telovov, K. Shavasov, Sh. Yusupov, 1st International Conference on Energetics, Civil and Agricultural Engineering. Tashkent, Usbekistan (2020)
20. V.P., Maksimenko, V.A. Shevchenko, V.K. Gubin, Bulletin of the International academy of Agricultural Education, **36**, 82 (2017)

BIO Web of Conferences **42**, 03005 (2022) https://doi.org/10.1051/bioconf/20224203005
SDGE 2021

Method for the influence design assessment of a wheeled agricultural vehicle on the soil

Ekaterina Balakina*, *Alexander* Konshin, and *Mikhail* Kochetov

Volgograd State Technical University, Volgograd, Russian Federation

Abstract. The impact of a wheeled agricultural vehicle on the deformable support surface determines the vehicle's ability to move, as well as soil compaction, which is not desirable in agriculture. The agricultural machine must not cause more pressure on the ground than is permissible. Therefore, in the tasks of design numerical modelling of the agricultural vehicle movement or trailer, it is required to calculate the specified parameter. It is impossible to calculate without knowledge of the geometric characteristics of the contact spot associated with the normal deformation of the tire under normal load. To calculate these characteristics, it is necessary to have universal dependencies for determining the normal stiffness of the tire. These are available for tires of various purposes. The elastic properties of ultra-low pressure tires are insufficiently studied. Experimental studies of the elastic properties of these tires have been carried out with the authors participation. However, there are currently no dependencies to describe them. This does not provide the possibility of a correct design calculation of the influence of such tires on the soil. The purpose of the work: to develop a universal method for calculating the influence on the soil of agricultural vehicle. A universal method for calculating the impact of a wheeled agricultural machine on the ground has been developed. Universal design-experimental dependence for determining the normal stiffness of ultra-low pressure tires is obtained. It takes into account tire pressure, normal load under specific conditions and geometric characteristics.

1 Introduction

Tires influence on a deformable support surface determines the vehicle's ability to move, as well as soil compaction, which is not desirable in agriculture [1-4]. The agricultural vehicle must not cause more soil pressure. Therefore, in the tasks of design numerical modelling of the agricultural machine movement, it is required to calculate the specified parameter. It cannot be calculated without knowledge of the geometric characteristics of the contact patch [5-8, 9-12] associated with the normal tire deformation under normal load. To calculate these characteristics, it is necessary to have universal dependencies to determine the normal tire stiffness. These are available for tires of various purposes [13-15]. The elastic properties of ultra-low-pressure tires are not well understood. Experimental studies of the elastic properties of these tires have been carried out with the authors participation.

* Corresponding author: fahrgestell2011@yandex.ru

However, there are currently no universal dependencies to describe them. This does not provide the possibility of a correct design calculation of the influence of such tires on the soil.

The aim of the research is the development a method for the influence design assessment of a wheeled agricultural vehicle on the soil.

2 Research methods

Has been investigated 7 ultra-low pressure tyres. In the course of the research, the load characteristics (the relationship between load and deformation) were measured at different tire pressures. They looked like pairs of experimental points (P_z; Z). They were approximated and experimental dependences $P_z = f(Z)$ were created. For example, the obtained dependencies are presented graphically in Figure 1.

Method is proposed for determination of coefficient of normal stiffness for ultra-low pressure tyre by initial experimental load characteristics $P_z = f(Z)$ of different tyres. Its stages are shown below.

A. Approximation for each ultra-low-pressure tire of the experimental loading characteristics $P_z = f(Z)$ by the same type functions for each pressure p_i. As a result, we have for each tire a set of functions in the form $P_z = a_1 \cdot Z^{b_i}$ for each pressure p_i. Here a_i, b_i – are constant coefficients.

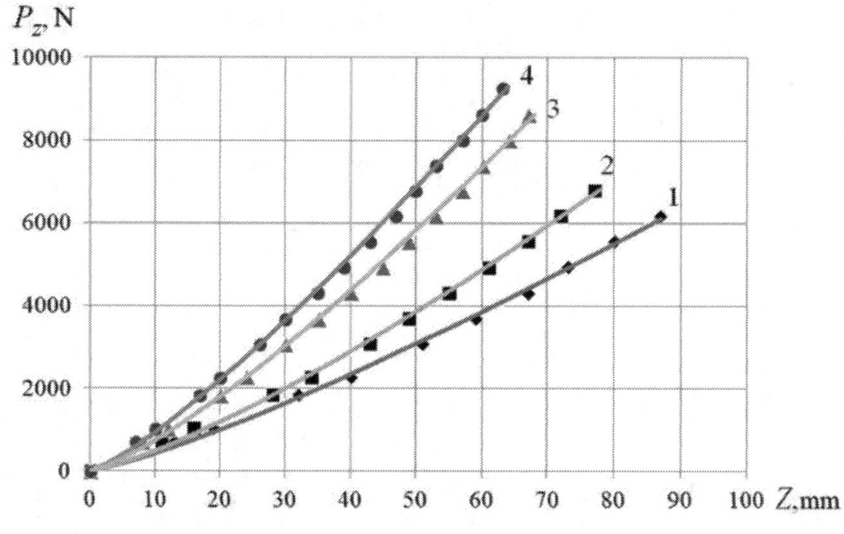

Fig. 1. Approximated experimental tire load characteristics 1300x600-533: 1 – p = 20 kPa; 2 – p = 30 kPa; 3 – p = 60 kPa; 4 – p = 70 kPa.

B. Approximation for each ultra-low pressure tire of the dependence for the obtained coefficients a and b on the pressure in the tire p. As a result, we get additional functions for each tire:

$$a = f(p) \text{ and } b = f(p) \quad (1)$$

They have the form:

$$a = k_1 \cdot p; b \approx const \quad (2)$$

As a result, for each ultra-low-pressure tire, the following expressions are obtained:

$$P_z = a \cdot Z^b \text{ или } P_z = k_1 \cdot p \cdot Z^b \quad (3)$$

C. Differentiating the normal load formula for normal deflection allows obtaining an expression for the normal stiffness of each ultra-low-pressure tire:

$$C_{tz} = \partial P_z / \partial Z = \partial (a \cdot Z^b) / \partial Z = a \cdot b \cdot Z^{b-1} = k_1 \cdot p \cdot b \cdot Z^{b-1} \quad (4)$$

Introduction of symbols $k_0 = k_1 \cdot b$; $d = b - 1$ gives a new expression:

$$C_{tz} = k_0 \cdot p \cdot Z^d \quad (5)$$

where $k_0 = k_1 \cdot b$; $d = b - 1$

D. Getting rid of the unknown value Z on the right side of equation (5) for each ultra-low-pressure tire:

$$C_{tz} = k_0 \cdot p \cdot (P_z / C_{tz})^d \quad (6)$$
$$C_{tz} = k_0^{1/(1+d)} \cdot p^{1/(1+d)} \cdot P_z^{d/(1+d)} \quad (7)$$

Denote:

$$k = k_0^{1/(1+d)} = (k_1 \cdot b)^{1/[1+(b-1)]} = (k_1 \cdot b)^{1/b} \quad (8)$$
$$n_1 = 1/(1+d) = 1/b;\ n_2 = d/(1+d) = (b-1)/b \quad (9)$$

As a result, expression (5) takes the general form:

$$C_{tz} = k \cdot p^{n1} \cdot P_z^{n2} \quad (10)$$

where p – tire pressure; P_z – normal tire load; k, n_1, n_2 – constant coefficients characterizing the tire design.

Thus, it was possible to get rid of the unknown value of the normal deflection Z in the formula (5) for calculating the coefficient of normal stiffness. Now this expression contains conditions: tire pressure p, normal tire load P_z and constant k, n_1, n_2, characterizing the tire design of this tire.

Table 1 shows the calculated constant coefficients for 7 models of ultra-low-pressure tires.

Table 1. Constant coefficients of expression (10)

Tire number	k_1	b	k_0	d	k	n_1	n_2
1	0.829	1.358	1.126	0.358	1.091	0.736	0.264
2	0.738	1.263	0.934	0.263	0.947	0.792	0.208
3	0.650	1.420	0.923	0.420	0.945	0.704	0.296
4	0.675	1.380	0.931	0.380	0.949	0.725	0.275
5	0.642	1.302	0.836	0.302	0.871	0.768	0.232
6	0.695	1.336	0.929	0.336	0.946	0.749	0.251
7	1.152	1.282	1.476	0.282	1.255	0.780	0.220

E. Obtaining a universal dependence for determining the normal (radial) stiffness coefficient of an ultra-low-pressure tire.

Table 2 shows the geometric characteristics of the considered ultra-low-pressure tires and the obtained constant coefficients of the expression for calculating normal (radial) stiffness.

Table 2. Tire characteristics and obtained coefficients

Tire number	Tire characteristics				Obtained coefficients			
	$[P_z]$, N	D_0, mm	H_t, mm	B_t, mm	H_t / B_t	k	n_1	n_2
1	4000	1300	380	530	0.717	1.091	0.736	0.264
2	6000	1300	380	600	0.633	0.947	0.792	0.208
3	8000	1300	345	700	0.493	0.945	0.704	0.296
4	8000	1300	345	700	0.493	0.949	0.725	0.275
5	6000	1200	330	600	0.550	0.871	0.768	0.232
6	8000	1300	345	700	0.493	0.946	0.749	0.251
7	8000	1300	345	700	0.493	1.255	0.780	0.220

Expressions are obtained

$$\begin{cases} k = 6.1 + 17.4 \cdot H_t / B_t - 14.6\,(H_t / B_t)^2 \\ n_1 = -0.67 + 4.82 \cdot H_t / B_t - 4\,(H_t / B_t)^2 \\ n_2 = 1.65 + 4.764 \cdot H_t / B_t - 3.94\,(H_t / B_t)^2 \end{cases} \quad (11)$$

where H_t – height of the tire profile in mm; B_t – width of the tire profile in mm.

Figure 2 shows the results of calculating the coefficients k, n_1, n_2 by expressions (11) in graphical form. Points show the according coefficients values for different tires from Table 2.

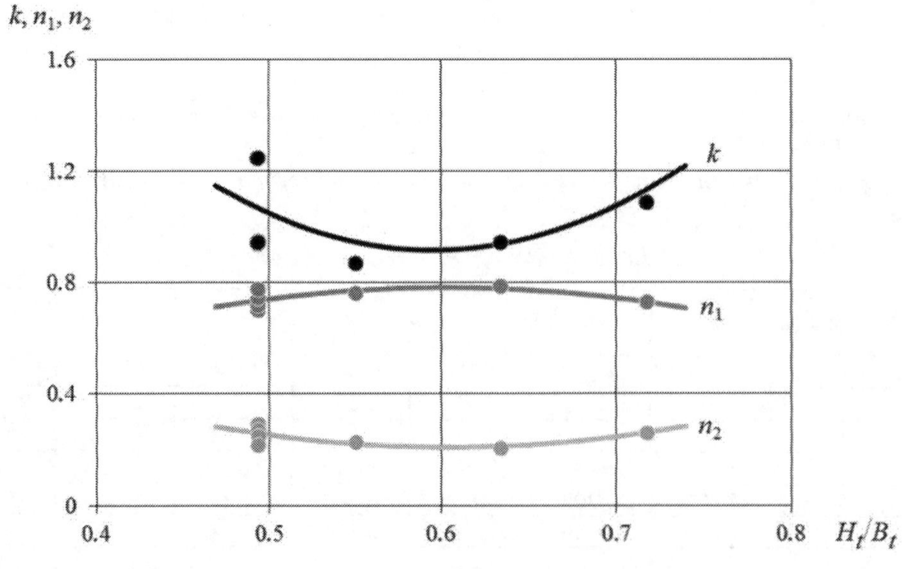

Fig. 2. Coefficients for calculating the radial stiffness of ultra-low-pressure tires: lines - calculation by approximated dependencies; points – experiment.

After calculating the coefficients k, n_1, n_2 for a tire with a given ratio H_t/B_t, it becomes possible to calculate the radial stiffness of an ultra-low-pressure tire from the expression $C_{tz} = k \cdot p^{n1} \cdot P_z^{n2}$.

3 Results and discussion

Thus universal calculation and experimental dependence (10) for ultra-low-pressure tires was obtained to determine the normal stiffness. It takes into account the tire pressure, the normal load under specific conditions and geometric characteristics. In the obtained dependence (10), the coefficients k, n_1, n_2 are functions of the section geometric characteristics of the profile (H_t/B_t). The indicated coefficients can be calculated by expression (11).

Using the obtained dependencies, it is possible to determine the normal tire deformation and the geometric characteristics of the contact patch of the ultra-low-pressure tire with the soil, as well as the pressure on the soil.

Soil pressure

$$p_c = P_z / F_c \tag{12}$$

where F_c – is the contour area of the contact spot.

According to the expressions obtained, it is possible to calculate the pressure on the soil – a particularly important parameter for these tires, because too much of its value may lead to obstruction of the agricultural vehicle movement or soil compaction, which is an undesirable factor in agriculture.

4 Conclusion

1. A universal method for calculating the influence on the soil of agricultural vehicle on ultra-low-pressure tires has been developed.
2. A universal calculated and experimental dependence is obtained to determine the normal stiffness of ultra-low-pressure tires. It takes into account the tire pressure, the normal load in the given conditions and the geometry. In the obtained dependence, three dimensionless coefficients are functions of the tire geometric characteristics: height and width.

References

1. V. I. Pryadkin, V. Y. Shapiro, Z. A. Godzhaev, Transport and technological machines on ultra-low-pressure tires (2019)
2. V. I. Kotlyarenko, S. V. Goncharenko, Z. A. Godzhaev, Tractors and agricultural machines, **2**, 17–21 (2014)
3. Z. A. Godzhaev, S. V. Goncharenko, A. V. Artyomov, Tractors and agricultural machines, **3**, 35–47 (2020)
4. Z. A. Godzhaev, A. Y. Izmailov, V. I. Pryadkin, Tractors and agricultural machines, **4**, 14–17 (2014)
5. E. V. Balakina, V. N. Zadvornov, D. S. Sarbaev, I. V. Sergienko, Y. N. Kozlov, International Conference on Innovations in Automotive and Aerospace Engineering, IOP, 27 May – 1 June 2019, Irkutsk, Russia (2019)

6. V. Y. Revenko, Z. A. Godzhaev, A. V. Rusanov, Tractors and agricultural machines, **5**, 48–54 (2019)
7. Cristian MINCA, Review of the Air Force Academy, **1**, 149–154 (2015)
8. S. Khaleghian, O. Ghasemalizadeh, S. Taheri, Tire Science and Technology, **44(4)**, 248–261 (2016)
9. F. Leiva-Villacorta, A. Vargas-Nordcbeck, J. P. Aguiar-Moya, LanammeUCR (2016)
10. P. Tomaraee, Spanish Journal of Agricultural Research, **13(2)** (2015)
11. V. Ivanov, Advances in Tribology (2010)
12. D. Woodward, 4th International Safer Roads Conference (2014)
13. E. V. Balakina, V. N. Zadvornov, M. S. Berezovsky, Automotive industry, **8**, 7–10 (2020)
14. V. I. Salnikov, A. A. Barashkov, V. N. Zadvornov, Automotive industry, **7**, 13–14 (2014)
15. V. N. Zadvornov, E. V. Balakina, N. A. Mishenkov, Journal of Friction and Wear, **41(4)**, 354–358 (2020)

Green technologies use in the smart grid construction in rural power supply systems

Igor Naumov[1,2], *Sergey* Podyachikh[1,*], *Marina* Polkovskaya[1], and *Aleksandr* Tretyakov[1]

[1]Irkutsk State Agrarian University Named After A.A. Ezhevsky, Irkutsk, Russia
[2]Irkutsk National Research Technical University, Irkutsk, Russia

Abstract. The article considers the using intelligent controls possibility in low-voltage rural electric networks to minimize the unbalance modes consequences. The proposed technology includes the digital data transmission compilation on the electrical energy parameters with a new balancing technical means the electrical network operating mode. Digital feedback is provided for changes the balancing device (BD) parameters by the unbalancing power consumption changing level. Based on the developed methods compilation, software for calculating unbalancing modes has been created, which makes it possible to assess the currents and voltages unbalancing effect on the power quality and its additional losses change. The "green" technology proposed version, which increases the economic and the electric energy environmental safety use in the rural electric power industry, contains a new constructive solution for the balancing device implementation. The proposed technology was tested on the measurement data basis in existing electrical networks. Based on the MALAB technologies use, changes studied indicators visualization in the before and after BD integration in the electrical network was carried out and its analysis was makes. Used on the "neural networks" MALAB technology, a preventive assessment of the unbalancing power consumption events development in the investigated operating electrical network is presented, as well as the proposed technology effectiveness assessment was carried out.

1 Introduction

Ensuring safe and reliable power supply to consumers is an essential requirement for the various economic orientation electric power systems. The intelligent networks introduction into modern power supply systems makes it possible to saturate them with new technical means that, without human intervention, can update these systems structure, increase their manageability. Thanks to its ability to control and regulate energy costs, as well as reduce the electrical equipment and its failures, smart grid technology makes it possible to improve the people's lives quality and thereby move towards a sustainable future. In today's deteriorating environmental situation, the automated and safe intellectual devices use in electrical networks has been called "green technologies". Their distinctive feature is the

* Corresponding author: psv78@yandex.ru

digital two-way feedback between the control tool and the processes occurring in the power supply system. Such communication is carried out through online software and services. Modern rural power supply systems are insufficiently equipped with control tools, they can be called "low observability" electric networks. In this regard, such "smart, green technologies" development is very relevant and allows us to implement the basic providing electricity requirements to consumers: power supply reliability, power quality, as well as compliance with economic and environmental safety measures. Let's call such technologies "GSG Technology". It has been stated repeatedly in numerous publications about the existing operating modes rural 0.38 kV electric networks that the objective these modes characteristic is the phase currents and voltages unbalance. At the same time, three-phase system unbalance leads to a significant change in quality indicators characterizing the unbalance level which values are set by Standards [1-3]. At the same time, the currents load unbalance not only reduces the power quality level and its attractiveness as a commodity, but also significantly increases its overspending, creating additional losses. Thus, the energy economic attributes saving and power transmission environmental requirements are violated, since the additional losses allocation creates conditions for the fire's occurrence [4-6].

The power quality indicators characterizing the three-phase system unbalance are the negative and zero sequences voltage unbalance coefficients, determined by the expressions:

$$K_{2u} = \frac{U_2}{U_1} \cdot 100\% \quad \text{and} \quad K_{0u} = \frac{U_0}{U_1} \cdot 100\% \tag{1}$$

In addition, the evaluating criterion slow voltage changes due to the currents unbalance load in the phases is also the steady-state voltage deviation, determined by the expression:

$$\delta U = \frac{U - U_{nom.}}{U_{nom.}} \cdot 100\% \tag{2}$$

In accordance with the Standard [1], this indicator value at the electricity transmission point to the consumer (0.4 kV buses) should not exceed ± 10%. If we are talking about the consumers connection point to the general-purpose electrical network (load node), then in this case, in the 95% measurement time interval, the indicator value should not exceed ± 5%.

At the same time, the symmetrical positive, negative and zero sequences currents components cause additional power losses, which can be characterized by a loss coefficient determined by the expression [7]:

$$K_P = 1 + K_{2i}^2 + K_{0i}^2 \cdot (r_1 + 3 \cdot r_N) \tag{3}$$

Where are the coefficients: $K_{2i}=I_2/I_1$, $K_{0i}=I_0/I_1$ which are determined based on the positive, negative and zero sequences currents values.

The article purpose is to develop a smart grid technology for assessing and minimizing the unbalancing consequences modes in rural electric networks. To achieve this goal, a tasks number have been formulated:

- based on the developed methods to create an algorithm and software "GSG Technology";
- to consider the using possibility a special balancing device with digital feedback;
- to test the proposed technology to assess the current unbalance mode operation level in the 0.38 kV electrical network;
- to analyze the unbalancing modes and prove the proposed model balancing device effectiveness;
- to carry out the results statistical processing obtained and to prognosis the development unbalancing power consumption events in the existing electrical network.

2 Materials and Methods

With unbalancing power consumption, the positive, negative and zero sequences voltages and currents components can be determined in accordance with the modular calculation method [7]:

$$U_1 = \frac{1}{6} \cdot \{[2U_A + \sqrt{3} \cdot (s + r) - (p + g)] + j \cdot [(s - r) + \sqrt{3} \cdot (p - g)]\};$$
$$U_2 = \frac{1}{6} \cdot \{[2U_A - \sqrt{3} \cdot (s + r) - (p + g)] + j \cdot [(s - r) - \sqrt{3} \cdot (p - g)]\};$$
$$U_0 = \frac{1}{3} \cdot [U_A + (p + g) - j \cdot (s - r)].$$
$$I_1 = \left(\frac{1}{6 \cdot I_A}\right) \cdot \{[3 \cdot I_A^2 - I_A \cdot (d + \sqrt{3} \cdot g) + 2 \cdot \sqrt{3} \cdot g] + j \cdot [\sqrt{3} \cdot I_A^2 - I_A \cdot (\sqrt{3} \cdot d - e) - 2 \cdot \sqrt{3} \cdot h]\}; \quad (4)$$
$$I_2 = \left(\frac{1}{6 \cdot I_A}\right) \cdot \{[3 \cdot I_A^2 - I_A \cdot (\sqrt{3} \cdot e - d) - 2 \cdot \sqrt{3} \cdot g] - j \cdot [\sqrt{3} \cdot I_A^2 - I_A \cdot (\sqrt{3} \cdot d + e) - 2 \cdot \sqrt{3} \cdot h]\};$$
$$I_0 = \frac{1}{3} \cdot (d - j \cdot e).$$

In these expressions: $p = \frac{U_A}{2} + \frac{U_B^2}{2U_A} - \frac{U_{AB}^2}{2U_A}$; $g = \frac{U_A}{2} + \frac{U_C^2}{2U_A} - \frac{U_{CA}^2}{2U_A}$; $s = \sqrt{U_B^2 - p^2}$; $r = \sqrt{U_C^2 - g^2}$; $g = b_1 \cdot c_1 + b_2 \cdot c_1$; $h = b_1 \cdot b_2 - c_1 \cdot c_2$; $c_1 = \sqrt{I_A^2 - b_1^2}$; $c_2 = \sqrt{I_B^2 - b_1^2}$; $e = \sqrt{I_N^2 - d}$.

The loss coefficient value (3), which is the energy losses ratio in the unbalancing mode to the corresponding losses caused by the positive sequence currents flow, allows to determine how much energy losses in the real unbalancing mode exceed those losses that can be obtained by taking appropriate measures to minimize the negative and zero sequences currents:

$$K_P = \frac{\Delta W_{UNB.}}{\Delta W_1} = \tau \cdot (I_1^2 \cdot r_1 + I_2^2 \cdot r_2 \cdot I_0^2 \cdot r_0)/I_1^2 \cdot r_1 \cdot \tau, \quad (5)$$

Where r_1, r_2, r_0 – active resistances of the positive, negative and zero sequences studied power line, respectively.

Thus, in order to evaluate the electrical network unbalance operation mode, it is necessary to know five currents: I_A, I_B, I_C, I_N, I_{BC} and six voltages: U_A, U_B, U_C, U_{AB}, U_{BC}, U_{CA}.

The I_N and I_{BC} values can be determined based on the corresponding vector diagrams construction as the currents projections geometric sum on the "X" and "Y" axes:

$$I_N = \sqrt{(I_{nx})^2 + (I_{ny})^2}; \; I_{nx} = I_{ax} + I_{bx} + I_{cx}; \; I_{ny} = I_{ay} + I_{by} + I_{cy}; \; I_{ax} = 0; \; I_{ay} = I_A; \; I_{bx} = I_B * 0.866025; \; I_{by} = -0.5 * I_B; \; I_{cx} = -0.866025 * I_C; \; I_{cy} = -0.5 * I_C; \; I_{BC} = \sqrt{(I_{BCX})^2 + (I_{BCY})^2}; \; I_{BCX} = I_{bx} + I_{cx}; \; I_{BCY} = I_{by} + I_{cy}. \quad (6)$$

In addition to these values, to evaluate the unbalanced mode, it is necessary to know the high voltage line parameters, supplying power to the 10/0.4 kV power transformer – the positive (negative) sequence complex resistance; the power transformer parameters – it's the positive (negative) and zero sequence complex resistance, as well as the 0.38 kV line parameters - the positive (negative) and zero sequence complex resistance. The studied indicators determination, characterizing the power quality and its additional losses with special balancing device inclusion into the electrical network is carried out in accordance with the method, described in detail in [8, 9].

The conducted studies devoted to the low-voltage electrical networks operating modes [11-18] have established that the most effective balancing means the 038 kV network operating mode is the special balancing devices use with zero-sequence currents minimal resistance. This device type is used mainly in electrical networks that electricity suppliers the utility load consumers (urban and rural).

Balancing devices described in [19, 20] can be used as such the BD example.

In this article, we propose to use a balancing device similar in its parameters to the devices [8, 9], but differing in that it is a compilation of the digital block and the electrical part block.

The International Electrotechnical Commission (IEC) published the IEC 61850 Standard "Networks and Communication Systems at substations" to solve the constructing problems the digital substations configuration for standardization purposes [21]. In accordance with this Standard, we propose a certain "GSG Technology" type, which functional scheme is shown in Figure 1. Figure 1 shows: 1. Combined optical current and voltage transformer. 2. Computer. 3. Software. 4. A balancing device. 5. The balancing device connection node.

Let's consider the above scheme in more detail.

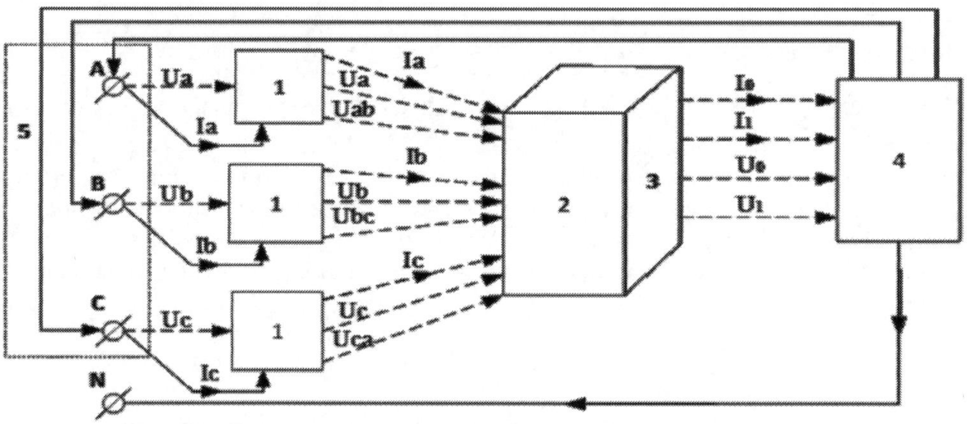

Fig. 1. The "GSG Technology" implementation functional scheme to assess and minimize the unbalancing mode consequences.

1 - Optical current (voltage) transformer. Their use in modern power supply systems is due to the fact that it is quite problematic to integrate analog TC and TV into a single monitoring and control network and the measurement accuracy no longer meets modern standards and requirements, since, according to our country general development the entire power system strategy, all substations should be digital format reconstructed pretty soon. At the same time, such devices advantages are obvious: no resonance, high linearity, no influence of the secondary circuits load, and most importantly - a very wide frequency range and increased safety. The such operation principal converters are based on the Faraday effect use, according to which, during the linearly polarized light propagation located in a magnetic field, the light polarization is plane observed rotation, that is, depending on the beam deflection angle, the electronics calculates the flowing current taking into account all existing errors. The optical sensor, located perpendicular to the current-flow tire, does not contact with it, and the signal is transmitted digitally via the shared bus to the electronics module *personal computer – 2.*

Thus, in accordance with IEC [21] the currents and voltages instantaneous values transfer occurs 80 times in the industrial frequency each period in digital form. At the receiving end (2 – computer), this information is collected and processed (filtering, digital processing, etc.) - it is supposed to collect 96 points for a period and transmit 2 measurements in one package.

3 – The calculation methods software implementation is based on the methods compilation [7-10]. The "Modulomethod" program algorithm is shown in Figure 2.

The digital data currents and voltages reception in from optical current transformers is supplemented by data on the electrical network parameters entered into the program: the high and low voltage lines full complex resistances and power transformer full complex resistances. In addition, it is possible to integrate data on three-phase symmetrical load into the program (asynchronous motors - if they are available in the power consumption system). The studied power quality indicators and its loss changes time diagrams visualization, in accordance with the calculation algorithm, is carried out on a computer monitor and its hard disk and can be showered at any required time. At the same time, the electrical network operation options are taken into account, both without balancing means, and when a balancing device is included in the any consumers concentration load studied node, as well as on the power transformer buses.

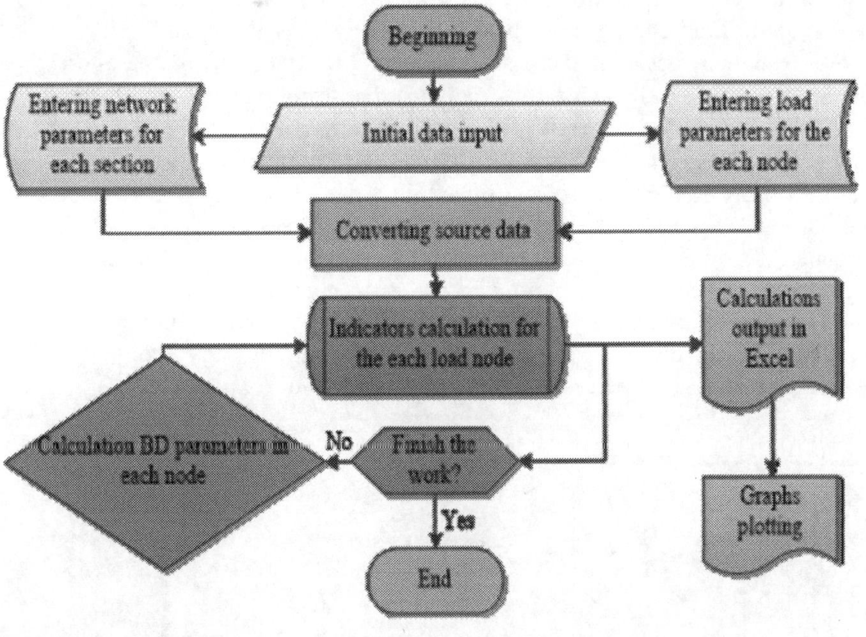

Fig. 2. The implementing calculations algorithm in the software "GSG Technology" using the program "Modulomethod".

4 – A balancing device. The traditional balancing means use is limited by their unregulated parameters constancy. This leads to additional losses in the device itself and significantly increases costs.

Therefore, the most optimal are BD, which change their power as the current unbalance changing level function. The proposed model prototype is an electromagnetic device described in [22]. The following an electromagnetic device model is proposed, which windings are connected according to the counter zigzag scheme, and are located on rigid fixed bases (frames) 7. Movable anchors 5, acting as a magnetic circuit, are fixed rigidly on one side on a spring bundle 2 (Figure 3).

The proposed device operation principle is based on the electromagnetism laws and is as follows. Before the device starts working, the steel cores (armature) 5, due to their own weight, enter the cylindrical windings 3 and 4, connected according to the counter zigzag scheme in each phase, at a dl_1 distance. At the same time, the all three cores gravity (F_{core}) will be balanced by the three sprin tensile force 2: $F_{core}=F_{spring}$. Since the windings connection according to the counter zigzag scheme reduces the windings resistance to zero-sequence currents by more than two magnitude orders, the total zero-sequence current is closed in the area between the device connection point and the unbalanced load. When the current passes through the coil, a magnetic field arises in it, which is uniform inside the cylindrical coil and is expressed by the: $B = \frac{\mu_0 \cdot I \cdot W}{l_к}$. The armature 5 will be affected by the electromagnetic force F_{em}, which at some point in time is balanced by the spring tensile force: $F_{spring} = F_{core} + F_{em}$. Since the current at each time moment in the three windings each is unchanged, the zero-sequence current also expends energy to increase the flow coupling in the $id\Psi$ amount, which is half the magnetic field energy. The other half is spent on covering the mechanical work $F_{em} \cdot dl$ in moving the core inside each phase winding. With an increase in the unbalanced power consumption, the zero sequence current increases, thereby increasing the electromagnetic field energy, cause an increase in the coil inductance. The cores 5 are drawn into each frame 7 and occupy the position dl_2. In this case, the magnetic field energy in each winding changes by the valued: $dW_н = 1/2 \cdot id\Psi$. When the zero sequence current decreases, the magnetic energy decreases, thereby reducing the inductance, i.e., the inductance self-regulates due to the force's equilibrium $F_{spring} = F_{core} + F_{em}$. It should be noted that in this case, the third-party forces energy is not used, and the work is carried out due to the zero-sequence current. It should be noted that in this case, the third-party energy forces is not used, and the work is carried out due to the zero sequence current. That is, as the flow coupling $d\Psi$ increases (decreases) in each winding, the inductance $dL=d\Psi/i$ increases (decreases) depending on the zero-sequence current magnitude. Thus, the cores movement inside the windings each will be determined by each phase zero sequence current magnitude. In other words, each phase inductors regulate their inductance depending on the current magnitude passing through them, therefore, their reactive conductivity Y_L is a zero-sequence current function$Y_L(I_0)$.

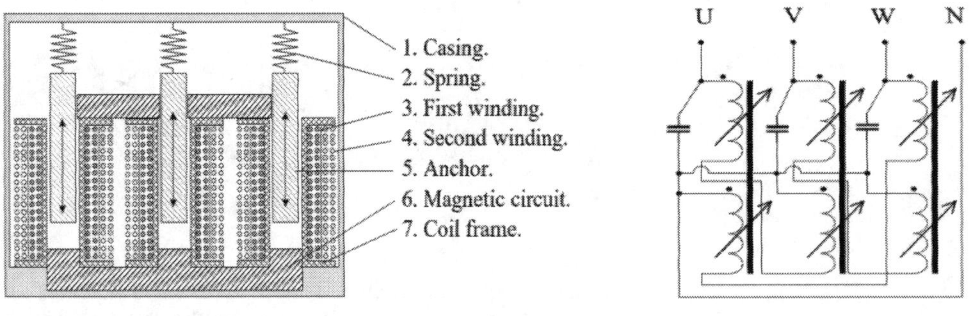

a) b)

Fig. 3. Electromagnetic balancing device: a) - design, b) - electrical circuit.

This balancing device advantage is that the inductors conductivity, and, consequently, the device power, is self-regulated by the zero-sequence current passing through it and, with a minimum (threshold) unbalancing power consumption level value, can be disconnected from the electrical network by a switching device means. Based on the above, the device parameters are adjusted precisely from the positive and zero sequence currents.

Based on the methods use described in [8-9], the balancing device parameters can be determined by the expressions:

$$Y_{BD1} = Y_{BD2} = \frac{1}{\left(2*\frac{U_1}{I_1}\right)*7.0714}; Y_{BD0} = 2*\left(\frac{3*I_0}{U_0}\right) \tag{7}$$

Were U_1, U_0, I_1, I_0 – the symmetrizing components of voltages and currents positive and zero sequences, determined, respectively, by expressions (4).

3 Results and discussion

3.1 Unbalancing operation modes investigation

To test the proposed technology, measurements were used in the current 0.38 kV network made in accordance with the CJSC IRMET Protocol (Electrical energy test report dated 02.27.2020). The tests were carried out by the JSC Irkutsk Electric Grid Company order (665709, Irkutsk region, Bratsk-9, s/b 786). The tests duration is 24 hours, the measurement interval is 1 minute. The measurements were carried out by a certified RESURS UF2M device (factory No. 2337) at the electricity transmission point to consumers on the TMG-250-10/0.4 power transformer 0.4 kV tires, which receives electricity via a 10 kV power transmission line made by AC70 wire, 1.259 km long. The device was installed on an outgoing 0.38 kV line, made with an insulated SIP-2A 3x25+1x35mm2 wire with a 0.270 km. length. Utility consumers (individual residential houses) with electric energy three-phase input are connected to the line. The electrical network under study has the following parameters.

The 10 kV line positive (negative) sequence total complexes resistances: $Z_{LH1}=Z_{LH2}=0,579+j0,346$ *Ohms*. The power transformer positive (negative) and zero sequence total complexes resistances are equal, respectively: $Z_{T1}=Z_{T2}=0,0094+j0,0272$ *Ohms*, $Z_{T0}=0,096+j0,2349$ *Ohms*. Similarly, for the low voltage lines: $Z_{LL1}=Z_{LL2}=0,324+j0,022358$ *Ohms*; $Z_{LL0}=0,3076+j0,0219$ *Ohms*.

Power transformer data: idling power P_{id}= 580 W; short circuit power $P_{sh,circ.}$= 4200 W; idling current $I_{id.}$= 1.2%; short circuit voltage $u_{sh.circ.}$ = 4.7%.

Based on the measurements made and the "Modulomethod" software use, changes time diagrams in currents and voltages in the studied electrical network were obtained (Figure 4). This figure analysis showed the following. As can be seen, the currents in the phases (Fig. 2, a) take different values. At the same time, phase "A" is the most loaded. The average current value in this phase during the measurement period was 45.28982 A. In the other two phases (B and C), the difference in current changes is not so clear, their values are on average 14.91498 and 17.7505 A, respectively. At the same time, the phase and phase-to-phase voltages averages values for the measurement period (Fig. 4) are, respectively: U_A=232.6257 V, U_B=235.7242 V, U_C=234.1277 V, U_{AB}=404.6016 V, U_{BC}=407.0769 V, U_{CA}=405.0429 V. As can be seen, three-phase power supply voltage system unbalance is not as clearly represented as it is typical for currents in phases.

 https://doi.org/10.1051/bioconf/20224203006

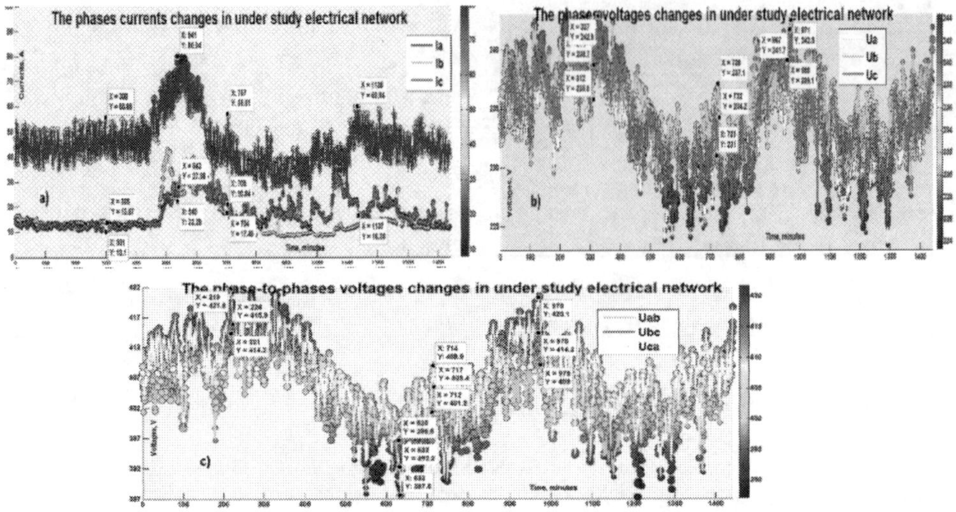

Fig. 4. The changes time diagrams in phase currents, phase and phase-to-phase voltages in the line under study.

To analyze the indicators determined by the expressions (1), (2) and (3) and the "GSG Technology" use presented in Figure 1, their calculation was performed and, based on the MATLAB graphics editor technologies use, the changes time diagrams in these indicators were constructed (Fig. 5-10). These diagrams clearly show the balancing device effect, which parameters, in accordance with Fig. 1, automatically change in accordance with the change in the unbalanced load power. The change in the negative sequence voltage unbalance coefficient is not considered in this article, since its value in the networks feeding the utility consumers load is minimal, and the balancing effect according to this coefficient is either weakly expressed or the opposite, is absent. It is since the balancing device parameters are calculated from the positive and zero sequences currents and voltages unbalancing indicators only.

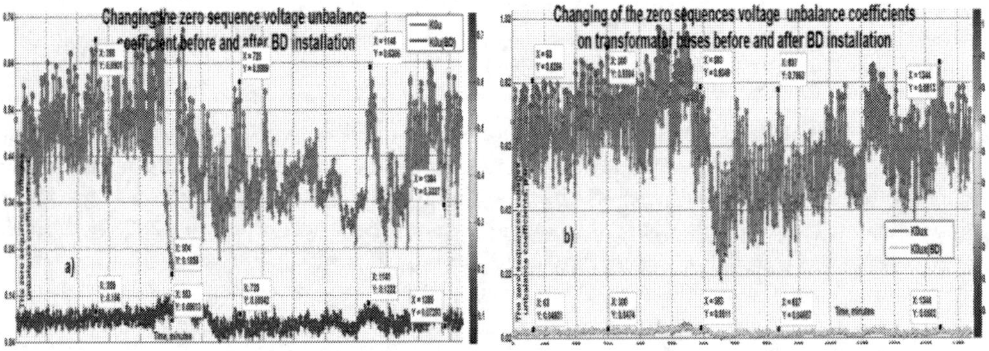

a) b)

Fig. 5. Changing time diagrams the zero-sequence voltage unbalance coefficient in the load node (a) and on the power transformer buses (b).

Figure 5 shows changes diagrams in the zero-sequence voltage unbalance coefficient in the power take-off node (at the 0.27 km distance from the power transformer 0.4 kV tires), as well as directly on the 0.4 kV tires. The changes analysis in these indicators showed the following. The power quality level according to the K_{0U} indicator meets the Standards requirements [1-3]. So, the average for the measurement period, this indicator value in the load node (Fig. 5, a) was 0.44%, on the transformer tires (Fig. 5, b) - 0.63%. The balancing device effect when it is connected is manifested in a significant decrease in K_{OU} - in the node to 0.087% (the balancing effect was 80.3%), on the tires - to 0.0377% (the balancing effect is 94%).

The power quality according on the voltage steady-state deviation does not meet the Standard's requirements. The δU indicator value at the electric energy transmission point (on transformer buses) should not exceed 10%. At the same time, at the general connection point consumers to the electrical network, this indicator value in the 95% measurement time interval should not exceed ± 5%. Figure 6 shows changes time diagrams in steady-state voltage deviations in the load node (the common connection point) when the absence and balancing device inclusion.

The Figure 6 analysis showed the following. In the BD absence, the average voltage deviation (δU) value during the measurement period was 6.73%, and the output beyond the established norm (5%) was observed in the 93.2% measurement interval time. Thus, the power quality according to the established voltage deviation does not meet the Standard's requirements. When BD was turned on in the load node, the average δU value decreased to 0.252% (balancing effect was 96.3%).

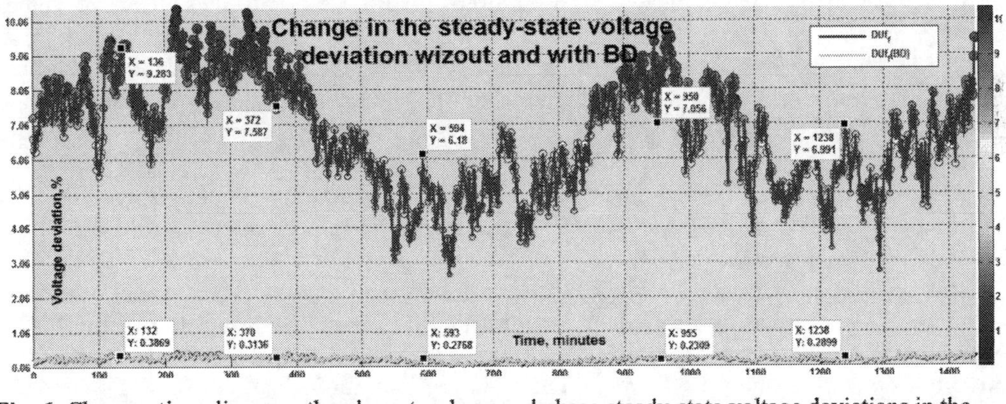

Fig. 6. Changes time diagrams the phase-to-phase and phase steady-state voltage deviations in the absence and BD inclusion.

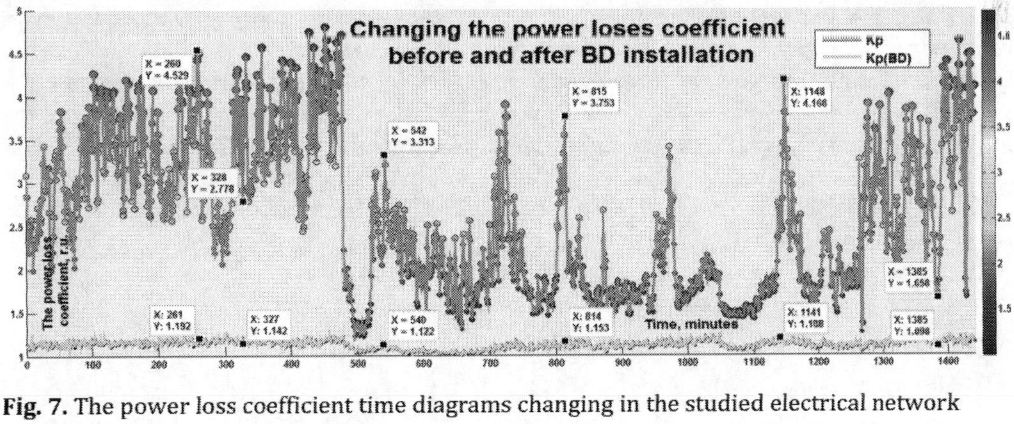

Fig. 7. The power loss coefficient time diagrams changing in the studied electrical network when the BD absence and its inclusion.

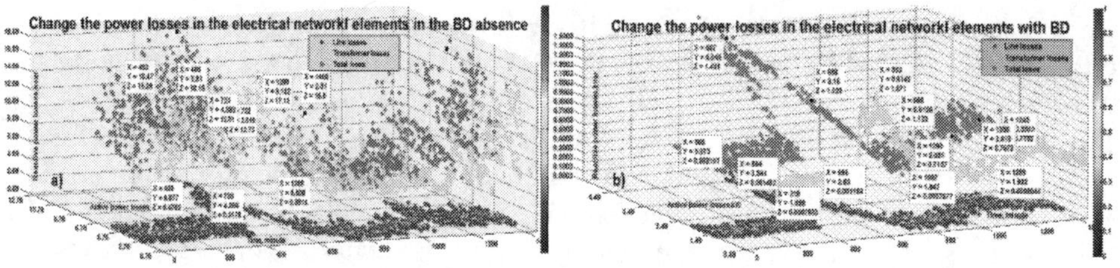

a) b)

Fig. 8. The BD symmetrizing effect visualization when power losses change in the electrical network elements: a) BD absence; b) BD is turned in the load node.

Let's consider how the power loss coefficient changes, which characterizes the increase in the unbalanced mode losses.

Figure 7 shows the change time diagrams in the power loss coefficient when the absence and BD inclusion in the load node.

Based on the Figure 7 analysis, it was found that in the BD absence, the power loss coefficient when it changes during the measurement time interval takes different values from the minimum (1.22) to the maximum (4.79). Its average measurement period value was 2.54. This means that the electrical energy losses caused by unbalancing power consumption are more than 2.5 times higher than those losses that could be caused by the only one positive sequence currents flow, that is, losses in a conditionally symmetric (balanced) mode.

When the balancing device is turned on, a significant balancing effect is observed in the load node. The average loss coefficient value during the measurement period was 1.12. That is, the balancing effect is 56%.

Let's analyze how the real electrical energy losses in the network under study change before the BD is turned on and integrated into the load node. Figure 8 shows the power losses change characteristics in the electrical network elements before and after BD switching on. As can be seen from Figure 8, in the BD absence in the 0.38 kV network under study, the active losses in the line during the measurement interval change every minute from 1.3 to 13.2 kW. The average these losses value during the measurement period is 4.377951 kW per minute. The line reactive power losses, varying from 0.09 to 0.9 kvar, averaged 0.3 kvar per minute. The line average full power losses are 4.388413 kVA. In a power transformer, active losses averaged 1.531783 kW over the period, reactive power – 9.771732 kvar, full power - 9.89115 kVA. When the balancing device was switched on, the line active, reactive and full power losses were, respectively: 1.924977 kW, 0.00079 kvar and 1.924977 kVA. Thus, the line active power loss balancing effect amounted to almost 66%, reactive power loss - almost 100%, and more than 56% of the total power loss. In the power transformer, the BD inclusion led to a reduction in active power losses by 65%, reactive power by 92.2%, and full power by 90.5%.

The change in total full power losses is shown in Figure 9.

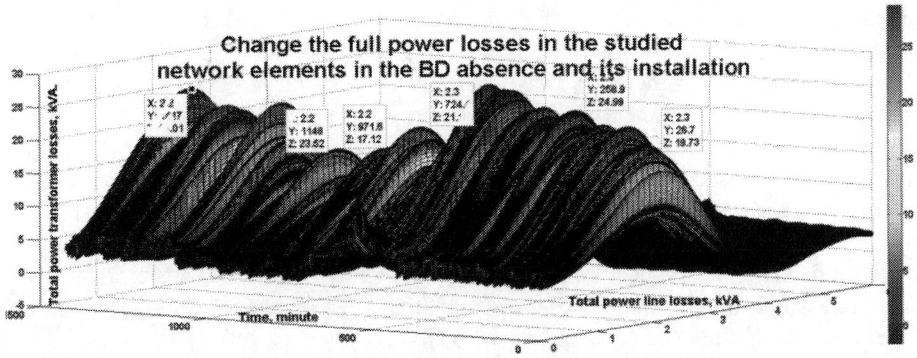

Fig. 9. The symmetrizing efficiency when changing the total full power in the electrical network elements.

The Figure 9 analysis showed the following. In the balancing device absence in the studied network, the average full power total loss per minute is 14.27956 kVA. The active power accounts share 5.909734 kW these losses, the reactive power share – 10.07459 kvar per minute. The balancing device inclusion leads to the average full total losses decrease on the 2.863912 kVA value. There are 2.46311 kW is the active power accounts of these full loses and 0.765389 kvar is for the reactive power. Thus, judging by the active power losses, the balancing device allows use us to save 3.53423 kW every minute. Under the investigated electrical network functioning condition without disconnections for one year, the electric energy savings will amount to 1857591.288 kWh. With the electric energy cost in the rural Irkutsk region areas equal to 0.861 rubles per 1 kWh [22], the saved losses cost will amount to 1599.386,098968 rubles. Considering that the creating cost technology in question is about 25-30% [16], the real using "GSG Technology" annual economic effect for the studied electrical network may amount to 479815.83 rubles. or 6771.32$. It should be noted that in this case, the improving power quality economic effect is not taken into account.

In addition, the balancing device costs described in [8] imply BD with the unregulated parameters use when its installed capacity is constantly used. In the regulated devise parameters use case described above, the BD power used and losses in it will be determined only by the current unbalance level, so the specified effect can be increased.

3.2 The studied indicators predictive evaluation mechanism

Common forecasting methods include statistical methods and Data Mining methods. Statistical methods include probability theory and mathematical statistics, regression analysis, maximum likelihood, time series extrapolation methods. Probability distribution laws are used to estimate many randomly varying parameters, among which we distinguish normal, gamma, Weibull, Pearson type III, three-parameter power law, etc. [24, 25]. One of the constructing conditions the probability distribution law is an autocorrelation relationship absence between the parameter under study values. To identify the relationship between the sample values or to refute this hypothesis, an autocorrelation coefficient calculated by the formula is used:

$$R_\tau = \frac{\sum_{i=1}^{n-\tau}(x_i-\overline{x}_i)\cdot(x_{i+\tau}-\overline{x}_{i+\tau})}{\sigma_i\sigma_{i+\tau}(n-\tau-1)} \tag{8}$$

Where n – is the sample size; τ – the shift order, varying from *0* to m (τ = 0, 1, 2, ..., m); x_i – the series values from x_1 to $x_{n-\tau}$; and σ_i – the average value and standard for the sample part from 1 to $n - \tau$; $x_{i+\tau}$ – the series values from $x_{1+\tau}$ to x_n; $\overline{x}_{i+\tau}$ and $\sigma_{i+\tau}$ – the average value and standard for a sample size from $1 + \tau$ to n.

The studied series parameters autocorrelation coefficients calculation showed that in most cases the intra-row connection is strong and significant (R_1>0.8). In this regard, the probabilistic laws construction is not possible. At the same time, autoregressive dependencies can be used to predict the studied indicator values in the short term, according to which the subsequent series values depends on the previous one. In other words, if the parameter value in question increased in the previous period (month, year), it will increase in the next, if there is a direct relationship between the values. In the case when the relationship is reversed (the autocorrelation coefficient is a negative value), the future value will decrease with an increase in the previous one.

The Fischer criterion was used to check the autoregressive model's quality shown in Table 1. The equations coefficients significance is estimated using the Student's criterion.

Table 1. The autoregressive models studied indicators constructed by minutes.

Indicator	Equation	Determination coefficient
	without a balancing device	
The zero-sequence voltage unbalance coefficient	$y_t = 0{,}88\ y_{t-1} + 0{,}054$	0,77
Power loss factor	$y_t = 0{,}94\ y_{t-1} + 0{,}15$	0,86
Total losses	$y_t = 0{,}90\ y_{t-1} + 1{,}40$	0,81
	with a balancing device	
The zero-sequence voltage unbalance coefficient	$y_t = 0{,}83\ y_{t-1} + 0{,}015$	0,68
Power loss factor	$y_t = 0{,}90\ y_{t-1} + 0{,}12$	0,81
Total losses	$y_t = 0{,}97\ y_{t-1} + 0{,}081$	0,94

The obtained equations can be used to predict the studied indicators in the short term.

Among the Data Mining methods, neural networks, decision trees, a genetic algorithm, the support vector method, and others are distinguished. We have adopted the neural network method as the basis, as the most commonly used in the preventive events assessment in electric power systems, the MATLAB modeling environment has been chosen as a creating means a neural network, since it is easy to use and has a high operation speed. In addition, there are a large methodological number materials on its use, which is very potential users valuable [26-28]. At the first stage, using autocorrelation coefficients, the analyzed indicators intra-row relationship was estimated, according to which there is a strong significant autocorrelation relationship in all variables, which makes it possible to use autoregressive forecasting models. To create a neural network, one of the design and neural networks training main MATLAB tools, Neural Time Series, was used.

A nonlinear regression is used as a tool (Fig. 10), which implements the simplest direct propagation network and allows predicting the effective feature value taking into account previous values with a different shift d. Let's consider the power loss coefficient *Kp*, like one of the three analyzed indicators.

To check and test data from 1440 this indicator values (the value corresponding to the minute measurement time interval), a 80:10:10 ratio percent was randomly selected, corresponding to the separation: network training - 80% (1152 target time intervals), approval - 10% (144 intervals) and testing - 10% (144 intervals) (Fig. 11).

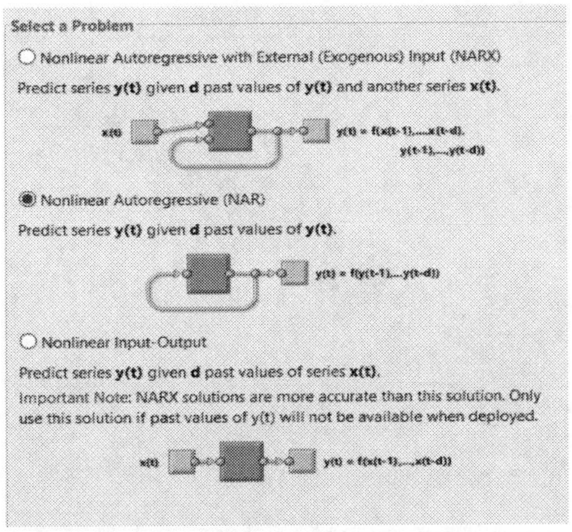

Fig. 10. Nonlinear autoregressive neural network model.

The creation and network training are carried out in an open form cycle, since training with an open loop (one-step prediction) is more efficient than with a closed loop (multi-stage prediction), and it allows us to provide the network with the correct feedback inputs, even when we train it to get the correct feedback outputs. After training, the network can be transformed into a closed loop form or into any other form required by the application. In this regard, the following architecture was adopted by random iteration: the hidden neurons number is 70, the delays number is 2 (Fig. 12). When working with the created neural network architecture (Fig. 12), its indicators were obtained (Fig. 13, 14), characterizing the using this model possibility to carry out the change's states preventive assessment in additional energy losses due to phase current unbalance.

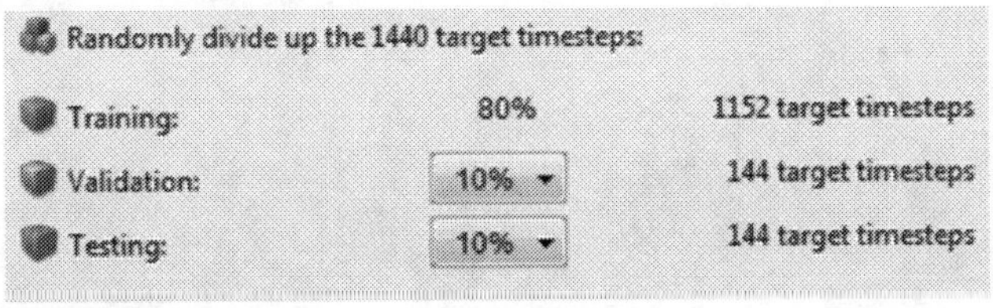

Fig. 11. Validation and Test Data.

The reference vectors values are laid out in the dotted line form, these corresponding parameters individual values are indicated by the "circle" type markers, the line shows all the studied model values correlation. The overall R score exceeds 0.8, which indicates a sufficiently high accuracy and constructed model adequacy.

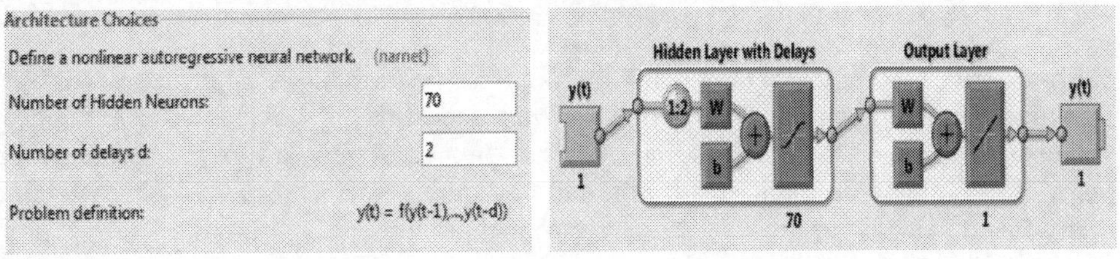

a) b)

Fig. 12. Building a neural network architecture: a) the neurons and delays number selection; b) the neural network configuration.

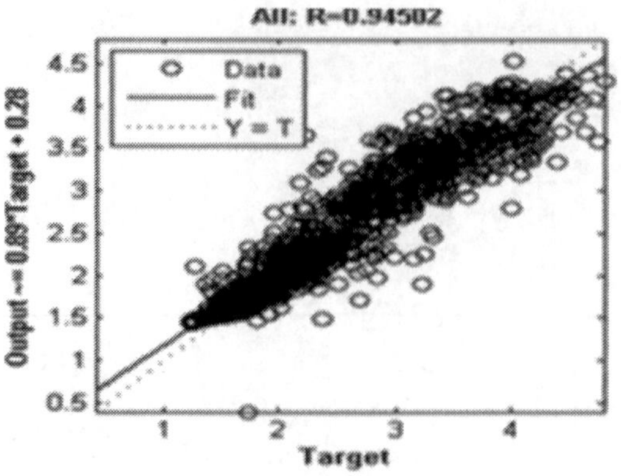

Fig. 13. Diagram of the neural network data general distribution.

Figure 14 shows the neural network training result, according to which the error during training has an insignificant value with some values exception "falling out" from the time series.

A selective retrospective the studied indicators forecast using a neural network showed that the actual values deviation from the forecast does not exceed 3%.

In addition, the loss coefficient retrospective prediction results comparison for the autoregressive equation and the neural network showed that the forecasts average deviation from the actual values is 1.9%, and for the neural network - 2.6% (Fig. 15). Thus, it can be concluded that linear autoregression describes the original series more accurately than the nonlinear one.

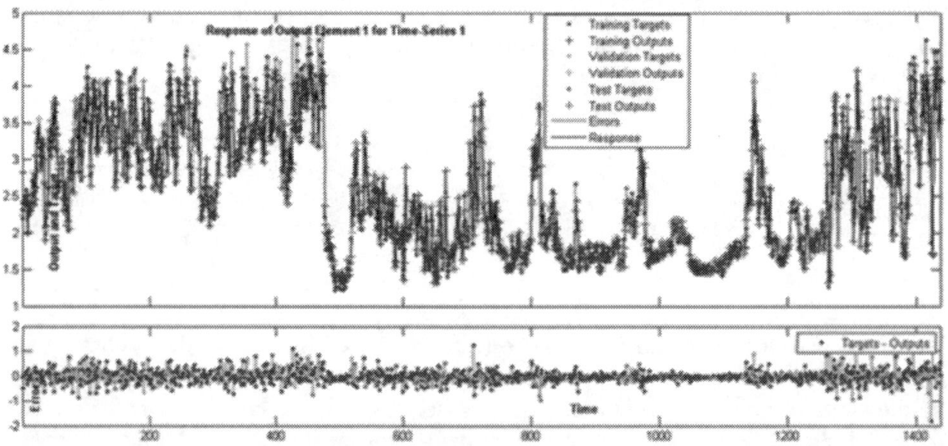

Fig. 14. Time series response in neural network training.

Retrospective forecast, obtained by autoregressive equations and neural networks showed that the actual values deviation from the forecast has a satisfactory value and does not exceed 10% (Table 2). It should be noted that the neural network gives a more accurate forecast. However, training and testing the network using various input indicators requires more time than calculating the autoregressive model parameters.

The graph comparing the initial data and the loss coefficient retrospective prediction data without a balancing device using the autoregressive equation and a neural network (Fig. 15) shows that the predicted values have practically no deviations from the actual ones.

Table 2. The average the power quality indicators forecast values deviation from the actual data in absolute value, %.

Indicator	The forecast value average deviation from the actual value in absolute value, %	
	according to the autoregressive model	by neural network
without a balancing device		
The zero-sequence voltage unbalance coefficient	7,96	7,71
Power loss factor	7,56	7,45
Total losses	9,67	9,44
with a balancing device		
The zero-sequence voltage unbalance coefficient	9,01	9,01
Power loss factor	1,43	1,41
Total losses	6,31	6,17

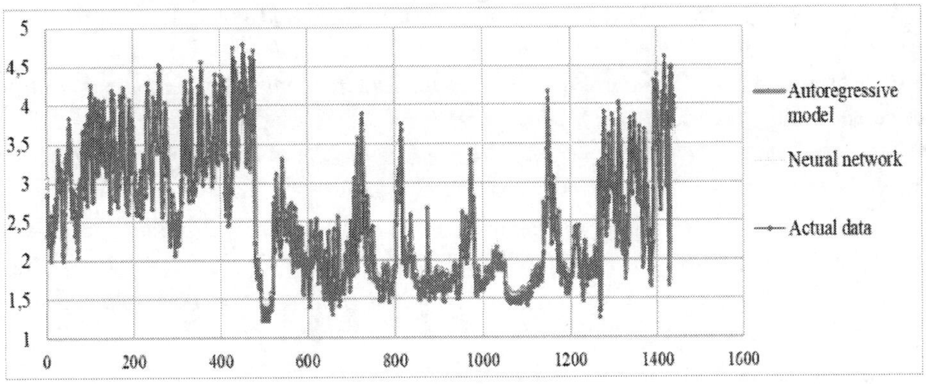

Fig. 15. Actual values and a retrospective forecast result based on an autoregressive model and a neural network.

4 Conclusions

Based on the conducted research, the following was established

- 0.38 kV voltage rural electric networks are characterized by the phase currents and voltages unbalance. At the same time, changes in the currents and voltages unbalance are always different on different the three-phase electrical network sections, which is due to the unbalancing power consumption probabilistic nature.
- A digital technology based on the measuring, calculation and decreas of the unbalancing indicators, characterizing the power quality and its loss minimization is proposed to evaluate the 0.38 kV electrical networks unbalancing modes operation.
- Developed: calculation software and a balancing device to minimize the power quality indicators and its losses caused by the currents and voltages unbalance.
- The proposed technology approbation was carried out on the currents and voltages measurements basis in the 0.38 kV electrical network. At the same time, the power quality and its additional losses analysis showed that the using efficiency a balancing device to reduce the power quality established indicators was more than 90%, and to reduce energy losses - close to 60%. The annual reducing energy losses economic effect for the studied electrical network amounted to 479816 rubles.

• Comparison of the studied indicators preventive assessment methods allowed to establish that the linear autoregression methods use is the most effective for the unbalancing regimes evaluation creating forecasts and minimizing them.

Reference

1. GOST 32144-2013 INTERSTATE STANDARD. Electric energy. Electromagnetic compatibility of technical equipment. Power quality limits in the public power supply systems Moscow. Standartinform (2014)
2. IEEE 1159-2019 - IEEE Recommended Practice for Monitoring Electric Power Quality (2019)
3. EN 50160:2010/A3:2019. Voltage characteristics of electricity supplied by public electricity networks (2019)
4. I.V. Naumov, N.V. Savina, I. V. Yamshchikova, M. V. Shevchenko, IPASJ International Journal of Electrical Engineering (IIJEE), **5(5)** (2017)
5. I. V. Naumov, International Science and Technology Conference "EastConf" INSPEC. Accession Number: 18724557 1-6. (2019)
6. I.V. Naumov, M.V. Shevchenko, P.P. Kozhushko, Journal Trends in the development of science and education, **30(2)**, 24-27 (2017)
7. F. D. Kosoukhov, *The current and voltage unbalance indicators calculation and analysis methods in rural distribution networks*, 42 (1984)
8. I. V. Naumov, *Methods and technical means the currents unbalance and electric energy losses reducing in rural distribution 0.38 kV networks*, 277 (1989)
9. I. Naumov, *Reducing losses and improving the quality of electric energy in rural distribution networks of 0.38 kV with the help of symmetric devices*, 387 (2002)
10. I. Naumov, E. Belousova, Vestnik Krasgau, **1**, 99-107 (2017)
11. A. S. Lukovenko, V. V. Kukartsev, V. S. Tynchenko, A. S. Mikhalev, V. A. Kukartsev, K. A. Bashmur, Journal of Physics: Conference Series. International Scientific Conference "Conference on Applied Physics, Information Technologies and Engineering - APITECH-2019", 55086 (2019)
12. Y. Xie, L. Wang, P. Li, T. Li, X. Su, Journal of Physics: Conference Series, 1549(5), 052053 (2020)
13. G. Bao, S. Ke, Energies, **12(15)**, 2842 (2019)
14. I. V. Naumov, S. V. Podyachikh, IOP Conf. Series: Earth and Environmental Science, **808**, 012006 (2021)
15. I. Naumov, S. Podyachikh, D. Ivanov, A. Tretyakov, A. Bastron, International Scientific Forum on Sustainable Development and Innovation (WFSDI 2021) E3S Web of Conferences, **295**, 02005 (2021)
16. I. V. Naumov, S. V. Podyachikh, D. A. Ivanov, SAHD 2021. SHS Web of Conferences **101**, 02012 (2021)
17. D. E. Dulepov, T. E. Kondranenkova, Bulletin of the NGIEI, **4(71)**, 28-35 (2017)
18. A. V. Vinogradov, A. V. Vinogradova, I. O. Golikov, V. E. Bolshev, International Journal of Emerging Electric Power Systems, **20(3)**, 3 (2019)
19. G. V. Lukina, I. V. Naumov, A. A. Lukin et al., RF Patent no. 26699 RF, MPK7 H 02 J 3/26 (2003)

20. I. V. Naumov, D. A. Ivanov, S. V. Podyachikh, et al., RF Patent no. 2490768 RF, MPK6 H 02 J 3/26 (2013)
21. GOST R 54835–2011/IEC/TR 61850–1:2003. Communication networks and systems in substations - Part 1: Introduction and overview (IDT), Moscow. Standartinform, 2012. The introduction date is 2012-09-01 (2012)
22. A.K. Shidlovskij, V.G. Kuznetsov, N.N. Kaplychnyj, et al., SU Patent no. 1206881 SU, МПК H 02 J 3/26, (1986)
23. Tariffs for Electric Energy for the Population of the Irkutsk Region and Consumers Equated to the Population. https://sbyt.irkutskenergo.ru/
24. S.V. Volkov, L.M. Rubakov, Energy problems, **1**, 84-89 (2004)
25. M.A. Undin, E.V. Rud, D.N. Golovakha, Collection of articles of the scientific and practical conference: Advanced achievements in the application of automation, robotics and electrical technologies in the agro-industrial complex, 190-195 (2019)
26. L.P. Bilgaeva, K.G. Vlasov, *He Study of Models of Neural Network Forecasting in Matlab,* **1(7)**, 11-19 (2017)
27. J. MacLennan, Z. Tang, B. Crivat, *Data Mining with Microsoft SQL Server 2008,* 675 (2009)
28. Oracle Cryslat Ball, http://www.hyperion.ru/

BIO Web of Conferences **42**, 03007 (2022) https://doi.org/10.1051/bioconf/20224203007
SDGE 2021

Ensuring sustainable spatial development from the perspective of land relations regulation

Dzhamilya Saralinova*, *Diba* Dokhkilgova

Kadyrov Chechen state University, A. Sheripova, 32, 364093 Grozny, Russia

Abstract. Solving new problems of sustainable development of society and territories requires improving the regulation of land relations to meet the managerial, production and technological needs of society. Improving the efficiency of regulating land relations and the use of lands of all categories in the context of their global degradation is one of the most important socio-economic problems in modern conditions. To solve the problems of land management, their rational use and protection, optimization of agricultural production, taking into account the environmental safety of products, an integrated approach to the regulation of land relations is needed, which is provided by the system of land management and land monitoring. The need to improve land relations management is obvious and indisputable, and the topic of the work seems relevant and timely. The purpose of the work is to develop practical recommendations for improving the mechanism of land relations management at the state and municipal level. Management decision-making should be based on the principles of strategic and tactical planning of innovation activities, taking into account the geopolitical, economic and socio-cultural specifics of the territory.

1 Introduction

One of the priorities of regional government bodies in recent years remains the purposeful coordination of the work of the sectoral executive authorities of the subject of the Russian Federation to optimize the composition of the property complex and bring it in line with the powers of the region. Priorities, goals, tasks of socio-economic development of the subject of the Russian Federation and the directions of activities for their implementation are approved by the resolution of the legislative body of the subject of the Russian Federation.

The existing distribution of the land fund by land categories, land users and land requires improvement. To do this, in the specific conditions of the regions, it is necessary to develop a methodology for monitoring land plots and land management aimed at a multifactorial assessment of land, taking into account the peculiarities of both natural and climatic factors and human economic activity in the region, the use of innovative technologies taking into account modern energy and environmental aspects. The impact of innovations on the sustainable development of society was noted in the innovative works of A. Smith, which describes the relationship between the theory of exchange and the theory

* Corresponding author: saralinova@list.ru

of production. Following him, D. Ricardo justified the sustainable development of the economy based on innovations in agriculture and resource conservation. Today, the global scarcity of resources and the presence of planetary environmental problems has become obvious. The obvious alternative is: a global catastrophe due to violations of global homeostasis, or a transition to a fundamentally different mode of civilization development.

As one of the main tools for regulating land relations, it is necessary to consider the mechanism for resolving issues of local importance through the powers assigned to the governing bodies at the municipal level. Considering the issues of land relations management, it should be remembered that most municipalities are rural territories. Evaluation of the efficiency of land use is an actual indicator of municipal management.
Identification and analysis of the direction of existing investment flows at the global, national and regional levels becomes a key element of strategic planning of the activities of government bodies at the regional and municipal levels. The subject of attention are such areas as: access to key markets; ownership of land plots, their availability and liquidity; efficiency of investments in land plots and taxation of real estate; issues of land legislation; problems of effective use of municipal lands and rural settlements; price level and other aspects.

2 Materials and Methods

On the basis of theoretical and methodological provisions and the regulatory framework of the Russian Federation, relying on a systematic approach, comparative analysis, informal modeling and comparison, systematization of data, the peculiarities of regulating issues of ownership and land relations, the disposal of land plots are investigated. As a result of the generalization of the analyzed materials, it was revealed that at all levels of land management, modern promising tools for the qualitative organizational and economic development of territories and rational disposal of land plots as a certain basis for their economic growth are not fully applied. This made it possible to identify problems in the implementation of powers to dispose of land plots at the municipal level and to justify the priority directions of their development.

State regulation of the restructuring of land plots is carried out with the help of anti-crisis management tools and measures within the framework of land policy. Thus, regulatory documents have been developed that contain criteria for the insolvency of tenants and owners and recommended management methods, such as a competitive mechanism and market sale of land, privatization of land plots. Spatial development of the economy is a state task, but its solution is achieved with the help of the state or municipal entity in determining the territorial location of the land plot intended for the placement of a specific object, in determining the purposes of using a specific site, its area, planned assistance to investors in providing infrastructure facilities, etc. In this regard, the approach of the Ministry of Economic Development of Russia seems justified, which determines the relationship between the purpose of a group of real estate objects and infrastructure components, land plots, premises of various profiles with the implementation of production and the creation of conditions for effective work.

3 Results and Discussion

The Russian Federation is a country with the largest territory and land area in the world, which refers to the factors of the "inner strength" of our state, along with other features and advantages: human, material, spiritual, natural. But, first of all, the territory of Russia requires a rational approach in management, both in rural settlements and in existing urbanized territories [1].

The management of the land fund of the Russian Federation is a separate area of public administration.

The land fund management bodies are integrated into an integrated system (**Figure 1**).

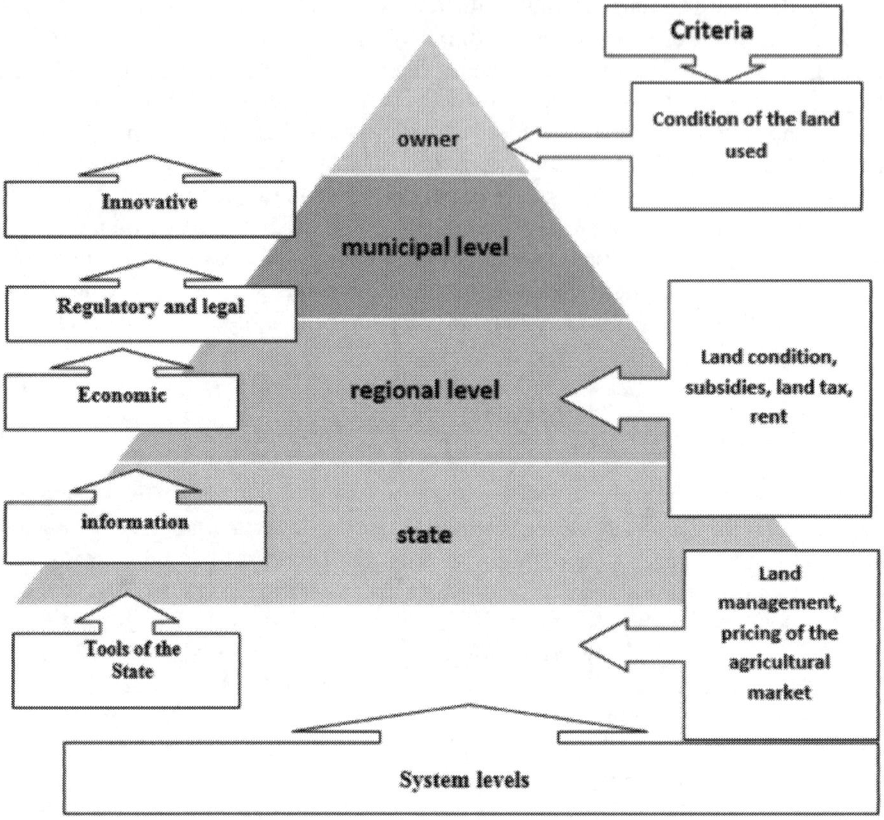

Fig. 1. Land Fund management system of Russia

The powers of Russia in the field of land relations include:

1)establishment of the foundations of federal policy in the field of regulation of land relations;

2)development and implementation of federal programs for the use and protection of land;

3)establishment of restrictions on the rights of owners of land plots, land users, landowners, tenants of land plots, as well as restrictions on the turnover of land plots;

4)management in the field of land monitoring, land supervision and land management;

5)establishment of the procedure for withdrawal, reservation of land and withdrawal of land plots for state and municipal needs, etc.

The main goal and direct competence of regional and municipal authorities are issues of the standard of living of the population. Regional ownership is one of the tools for solving problems in this area [2].

The subjects of the Russian Federation manage and dispose of land plots owned by the subjects of the Russian Federation.

One of the priorities of the regional government bodies in recent years remains the purposeful coordination of the work of the sectoral executive authorities of the subject of the Russian Federation to optimize the composition of the republican property complex and bring it in line with the powers of the region. The powers of the subjects of the Russian Federation include the development and implementation of regional programs for the use and protection of lands located within the borders of the subjects of the Russian Federation; reservation, withdrawal of land plots for the needs of the subjects of the Russian Federation; other powers that are not attributed to the powers of the Russian Federation or to the powers of local self-government bodies.

Priorities, goals, tasks of socio-economic development of the subject of the Russian Federation and the directions of activities for their implementation are approved by the resolution of the legislative body of the subject of the Russian Federation, which is reflected, as a rule, in strategic planning documents [3]. The main functions for their implementation are performed by the executive body of state power - the Ministry of Property and Land Relations. The actions and efforts of municipalities are aimed at completing the work on the cadastral registration of all lands and the delimitation of ownership rights to them. From these positions, the main emphasis is placed on the development of strategic and medium-term plans for the use of productive lands.

All participants in economic activity should be involved in achieving sustainable development. This can be realized thanks to the participation of the state, business and the financial sector in the process. At the same time, each of them plays a certain role – the state gives signals and creates conditions for conducting economic activity. The financial sector directs cash flows to the necessary activities, and the business implements the achievement of sustainable development in practice. Thus, the interaction of all three parties will allow achieving social and environmental results along with financial returns [4].

According to the norms for crediting tax revenues to the budget, indicated in the Budget Code of the Russian Federation, land tax and property tax of individuals are received in full to the municipal budget.

Currently, the main source of income for local budgets in the Russian Federation is property taxation. It should be noted that the tax revenues of local budgets are differentiated depending on the type of municipality. Thus, the budget of municipal districts receives these taxes levied only on inter-settlement territories.

At the municipal level, the authorities can take measures to optimize the established tax benefits for local taxes that are not associated with an increase in investments in the territory of the municipality or region, by assessing their effectiveness and including them in the system of assessing the quality of regional and municipal finance management. For municipalities, both direct payments from the federal budget and the budget of a constituent entity of the Russian Federation can be used, aimed at the introduction of these technologies by local governments (in the form of inter-budget transfers directed specifically to a specific municipality), and financial and legal incentives, when the amount of budget funds allocated from a higher budget increases depending on the timing of the introduction of technologies, the installation of specific programs, as well as the coverage of users involved and using the necessary technologies.

In addition, there is a need for management in the direction of limiting the concentration of land plots among large owners and land users. One of the most important conditions for the development of this sphere is a developed infrastructure of the land market and a low level of transaction costs.

Problems in the implementation of the powers to involve real estate objects in economic turnover by executive bodies of state power and local self-government bodies are:

– low investment attractiveness of objects;
– a large degree of wear of objects;
– territorial remoteness of objects from cities and district centers;
– lack of a source of financing for the involvement of objects in economic turnover;

– low quality of urban planning documentation;

– lack of a developed system of engineering communications.

As tools for involving unused or inefficiently used land plots in economic turnover, the following are used:

– transfer of unused land plots from one balance holder to another;

– gratuitous transfer of land plots to federal ownership, to the ownership of a subject of the Russian Federation;

– transfer of unused land plots for rent;

– giving consent to the sale of land plots;

– withdrawal of unused land plots in order to implement it through privatization;

– provision of land plots both at auction and without bidding (including for the implementation of investment projects)–

– other solutions (including attracting investors by concluding concession agreements, etc.).

The implementation of the above measures will contribute to the rational use of real estate objects and land plots owned by the subject of the Russian Federation, as a result of:

– identification and involvement in the economic turnover of unused real estate objects, including land plots;

– identification and elimination of inconsistencies between the layouts of existing technical passports and the actual location of premises in buildings;

– cadastral registration, state registration of the rights of unaccounted and unregistered land plots and/or real estate objects;

– identification and elimination of violations in the execution of contracts of gratuitous use, lease agreements of property of the subject of the Russian Federation.

On the other hand, the further development of various forms of activity largely determines the creation of significant financial and administrative conditions. Indeed, the spatial development of the economy is a state task, but its solution is achieved with the help of the state or municipal entity in determining the territorial location of the land plot intended for the placement of a specific object, in determining the purposes of using this site, its area, planned assistance to investors in providing infrastructure facilities, etc. The task of organizing, in particular, the management and development of the land management system allows us to determine the vector of the direction of progressive development of land relations.

The creation of a favorable investment climate is usually based on the desire of the authorities to implement the goals of national and regional economic development. The following approaches can be used:

– financial incentives (direct subsidies, grants and loans);

– tax instruments (tax holidays and reduced tax rates)–

– other approaches, including subsidies for the purchase of land, the use of infrastructure and various services.

It should be noted that experts consider it expedient to legally delimit the regulatory role of state bodies, on the one hand, and the areas of activity of owners, developers and business administrators, on the other hand.

It is obvious that the implementation of the intended goals is associated with the need to accumulate significant financial and administrative resources [5]. Diverse and rich experience the existing structure of the organization requires the definition and clarification of further directions of development.

The need to involve unused or inefficiently used land plots in economic turnover requires the involvement of certain instruments of influence, in particular, the transfer of unused land plots to lease or from one balance sheet holder to another is considered effective, as well as the withdrawal of unused land plots for the purpose of its implementation through privatization. It is also possible to sell land plots, including through bidding, and without bidding (for example, for the implementation of investment projects). In addition, it is possible to attract investors by concluding concession agreements and so on.

An important aspect of the disposal of land plots is their role as the basis for the organization of business activity, especially when conducting production activities, including by enterprises of the transport and communication structure of the region, energy, water supply and waste disposal systems [6].

Improving the mechanism of land management and relations at the municipal level is advisable in the following areas:

– develop strategic and medium-term plans for the use of productive lands;

– complete the cadastral registration of all lands and the delimitation of ownership rights to them;

– limit the concentration of land plots among large owners and land users;

– to develop the infrastructure of the land market and reduce the level of transaction costs.

To improve the current situation with land plots, it is proposed to carry out the following measures: to develop regulatory documentation on registration requirements, with the presence of penalties for non-compliance with the requirements; to include in the annual monitoring indicators such as land EC, DCT, private and comprehensive assessments of anthropogenic impact on the territory.

It is also necessary to carry out annual proper monitoring of:

– use of land plots;

– preservation of soil fertility and humus layer of soil;

– organization of optimal recreation of the land, reclamation and conservation of land;

– chemical treatment against pests and diseases in the areas;

– regular and timely implementation of the best available technologies in production.

In order to improve the system of land management in the region, it is advisable to:

– identify among the existing land plots those that are not properly formed;

– systematize the work on the formation of land plots and the establishment or modification of their permitted use;

– to form the missing information base and improve the existing system of urban planning activities.

It is necessary to develop a functional integrated approach to the regulation of land relations under the control of management bodies, which is more in line with the new production relations associated with the conditions of a market economy.

4 Conclusion

A pragmatic approach to determining the powers of authorities to dispose of land plots with an emphasis on the methodology of a comprehensive assessment of land use will increase the competitive advantages of the subjects of the Russian Federation, municipal and other administrative-territorial entities of the country within the framework of ensuring the effective use of land resources.

The state investment policy to support investment activities in each region has its own characteristics due to the innovation policy of the region, natural and climatic conditions, the availability of energy resources, the features of the infrastructure of investment activities, the existing production, personnel and innovation potentials, the geographical location of the region, the environmental and demographic situation, the investment attractiveness of the region. At the same time, the goals and objectives of the regional investment policy in the field of innovation, the forms and methods of its implementation may not coincide in different subjects of the Russian Federation [7].

The criterion of continuous development links regional sustainability with sustainable development (improving the lives of people around the world), and future research should focus on accelerating economic development through policy, allocation of funds, or simply creating a basis for the sustainability of all processes [8].

In order to increase the efficiency of land use of all categories in the conditions of their global degradation, all subjects of land relations need to accumulate information about the state and nature of land use, have an idea of changes in this area and see prospects for their further development.

World practice shows that effective innovative development is possible only in a managed innovation system. Management decision-making should be based on the principles of strategic and tactical planning of innovation activities, taking into account the geopolitical, economic and socio-cultural specifics of the territory. A comprehensive solution to the task offers comprehensive, timely and accurate information about the management object. Therefore, it is necessary to develop and organize monitoring of innovation processes in the region.

This approach eliminates the contradictions between the unsystematic use of land and the need for systematization of methods and technologies for regulating land relations as a complex complex, which is dictated by the process of widespread digitalization, the development of science and technology in general. The need for the early introduction of electronic document management in the field of land relations regulation in the digital economy implies the allocation of appropriate financial resources for this. Various incentives can be used for private subjects of financial and legal relations, for example, tax and legal incentives for the introduction and use of digital technologies. The issue of introducing digital technologies for municipalities is particularly acute, since, as a rule, they do not have their own financial resources to spend on this area. Another direction of the formation of the financial and legal status of municipalities is to expand the opportunities of citizens to participate in the financial activities of municipalities, including their involvement in the budget process, which, with well-structured work, contributes to increasing public confidence in local self-government bodies and the financial and legal policy pursued by the state. The study of the forms of such participation in the framework of financial law is just beginning and requires further development.

The application of geomonitoring methods as the basis of territorial development and management of territories is also justified. Thus, the solution of the problem is achievable on the basis of a complex of scientific research and practical work aimed at systematization and improvement of the system of regulation of land relations and sustainable development of territories during the digital transformation of the economy as a whole.

References

1. N.G. Zhavoronkova, G.V. Vypkhanova, Lex Russica, **2(171),** 50 (2021)
2. E.A. Korotaeva, Bulletin of the Udmurt University. Economics and Law Series, **27(4),** 104 (2017)
3. A.V. Denisova, Society and law, **4(41),** 95 (2012)

4. A. Beksultanova, T. Magomadova, and M. Barzaeva, Proceedings of the International Scientific and Practical Conference on Sustainable Development of Regional Infrastructure, 298 (2021)
5. Yu.Yu. Yurkova, Bulletin of the Voronezh State Agrarian University, **4(59),** 250 (2018)
6. O.B. Mezenina, Moscow Economic Journal, **9**, 1 (2020)
7. L.N. Roshchina, Finance and credit, **14(494),** 7 (2012)
8. M. Szabo, M.S. Csete, T. Palvolgyi, European Journal of Sustainable Development, **7(1),** 395 (2018)

The use of RS-485 and CAN interfaces for connecting sensors in climate control systems of greenhouse complexes

Victoria Petropavlovskaya[1*], *Alexander* Shishkov[2], *Sergey* Zamaraev[1], *Nikita* Pshenisnov[1], and *Maksim* Gort[1]

[1]South Ural State University, 454080, Chelyabinsk, Russia
[2]Moscow Polytechnic University, 107023, Moscow, Russia

Abstract. Purpose of this article is to demonstrate the features of the use of existing connection interfaces for the functioning of climate control systems of agricultural complexes. The article contains the author's classification of the characteristics of the connection interfaces. The two most suitable interfaces are considered. The article compares RS-485 and CAN protocols. Key differences between CAN and RS-485 for use in agricultural complexes are shown. A test sample of a system containing 3 sensors was presented.

1 Introduction

According to statistics, 4 billion people in the world (56%) live in cities [1]. A person consumes 1.5 kg of food per day. The total daily food intake reaches 6 million tons. Greenhouse complexes are being built near major cities to compensate for food losses and reduce the cost of supplies.

Year-round greenhouses create the ability to supply daily crops to consumer and silage grass to livestock farms. Plants need an optimal range of climatic conditions to maximize their yield. Yield conditions are temperatures above 14 C and illumination of at least 8 hours a day for most light-loving plants [2]. The optimum temperature for plants is maintained by a climate control system that includes a heating system, humidifiers, lamps and air condition sensors. All elements of the climate control system are connected to each other via an interface.

Modern agricultural complexes can cover an area of up to 30 hectares. This poses a challenge for the interface used, as the signal strength decreases with increasing cable length.

The problems of climate formation in greenhouses have been raised more than once. A.N. Zadeev in his article describes the calculation of the thermal regime of the greenhouse [3]. This article is necessary when setting up the interface of the heating system of a greenhouse complex. The main factors are the ambient air temperature and indoor

* Corresponding author: ViktoriaPetropavlovskaya@mail.ru

insolation. The amount of heat transmitted by sunlight throughout the year on the territory of Kazakhstan is shown in Figure 1.

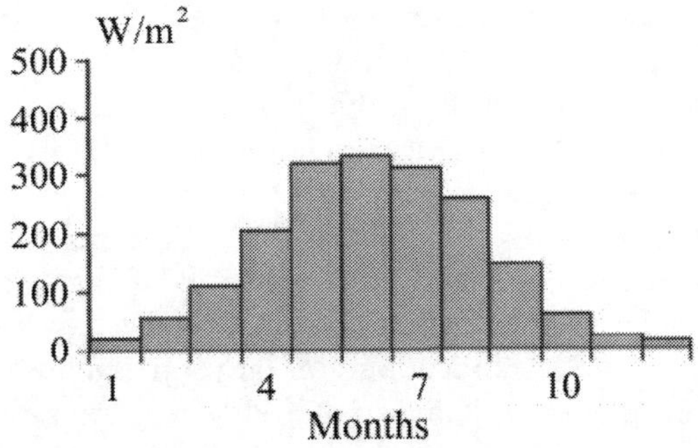

Fig. 1. Change of insolation during the year.

Agarkova A.M. and Iveshko G.G. raises the issue of rational consumption of energy resources when heating greenhouse premises. [4]. When operating a greenhouse complex in winter, it is necessary to take into account 2 important factors:

- Do not allow the temperature to drop below optimal.
- Avoid excessive consumption of fuel for space heating. This can lead to excessive evaporation and increased humidity in greenhouses.

The trend towards urbanization dictates the need to provide food for more and more people all year round. Efficient use of land in the suburban zone is possible with the competent organization of greenhouse complexes.

The purpose of this article is to demonstrate the features of the use of existing connection interfaces for the functioning of climate control systems of agricultural complexes.

2 Matherials and methods

Connection interfaces have not clear classification system yet. Each of them still has a set of characteristics that reveal different aspects such as:

- Data transfer rate;
- Coverage of the network;
- Simplicity of the transfer protocol;
- Energy saving;
- Safety;
- Reliability and performance;
- Market prevalence.

The specifics of working with greenhouse complexes involves changes in the microclimate of greenhouses within hours of time at considerable distances (up to 30 hectares). The transmitted data has no significant commercial value. A systematic failure in data transmission will be difficult to notice until it leads to crop losses.

All listed factors put forward the requirements for network coverage (distance and number of sensors), and neglect factors such as data transfer speed and network security. The personnel of the agricultural complex does not always have the quick network repair experience. Therefore, simplicity of the protocol and market availability are also desirable, but not paramount.

The interface used in most enterprises and ICS devices is RS-485. This interface has the ability to organize two-way data exchange with just one twisted pair of wires. It provides a communication length up to 1200 meters and assumes connection up to 32 devices (sensors) [5, 6].

The soil is saturated with fertilizers and moisture, which provoke cable corrosion. Therefore, the cable is laid above the ground. But the cable can be damaged by vehicles in this case. It is recommended to run the cable along the walls of the greenhouse complex. The pairing of RS-485 and power circuits should be avoided, especially in a common braid. The overhead cable can be replaced or repaired easily.

RS-485 interface has code combinations prohibited for transmission, which complicates its inner architecture. The RS-422 interface is devoid of this drawback. Its simplicity in comparison with RS-485 can be seen in Figure 2. However, it assumes the connection of only 10 devices (sensors). Its application for large greenhouse complexes will not be considered. An example of RS-485 connection is shown in Figure 2.

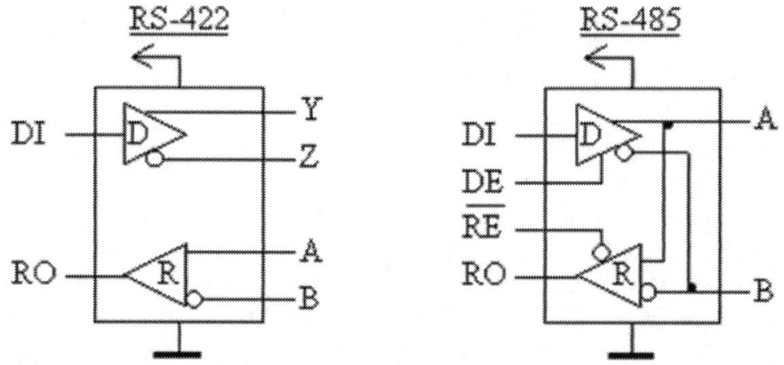

Fig. 2. Connection diagrams RS-422 and RS-485.

The CAN (Controller Area Network) interface can be a full-fledged analogue of RS-485 interface [7]. The CAN format is common in many areas. It was originally developed for the automotive industry. The CAN interface allows data transmission over 1000 m and up to 64 devices. The protocol provides for special methods of error detection, error signaling and self-monitoring to achieve maximum reliability of the transmitted data.

The following measures are used to detect errors:

- The transmitter compares each bit on the bus with the transmitted bit to confirm correct transmission at the bit level;
- Cyclic Redundancy Check (CRC) is performed;
- Bit stuffing is used;
- Check of each transmitted frame is used.

This increases the size of the overhead in the package and the complexity of the internal architecture, which complicates the repair. The authors find these security measures redundant for transmitting data on the state of the greenhouse complex.

3 Results and discussion

As a test sample, the authors assembled a temperature control system based on a greenhouse facility in Rudny, Kazakhstan. The RS-485 protocol was used (Figure 3).

Three sensors are installed: the first is at ground level, the second is at the ceiling level, the third is on the street. It is possible to monitor online what is the temperature in the greenhouse at a given time and change the temperature and humidity indicators.

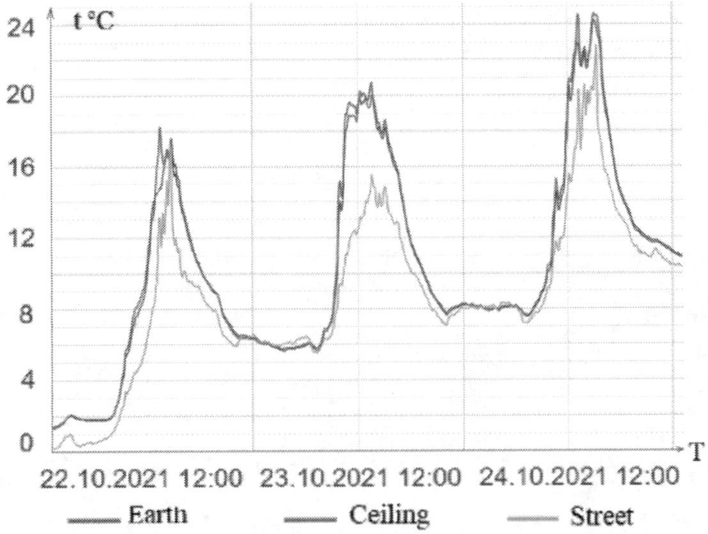

Fig. 3. Temperature control system performance.

Modernization of greenhouses, introduction of a climate control system and efficient use of resources allows increasing the yield of the agricultural complex by 20 percent [8].

Outside air temperature significantly affects the greenhouse yield due to large heat losses through the greenhouse fences. The light of the sun is biologically important for plants. However, insolation does not have a significant effect on the indoor temperature in the winter period.

Analysis shows that a large number of devices cannot be connected to a network with a simple architecture. This limits the choice of technical solutions to advanced interfaces. For large greenhouse complexes, for those who grow similar agrocultures, the RS-485 interface is suitable. It allows enough sensors to be placed within a 1200 meter radius around the main controller. For a dense accumulation of greenhouses for growing various crops, the CAN interface is suitable. It allows you to cover a circle with an area of 30% less, however the placement of sensors in the accessible area will be up to 3 times denser.

In multi-node systems at speeds above 1 Mbit/s, under harsh conditions, the advantage remains with CAN. The CAN interface has a wide temperature range. It is well suited for automotive and transportation applications. This is not required to transfer data from a greenhouse agricultural complex. The transfer of temperature and humidity parameters for greenhouses is better handled by the RS-485 interface. Excessive CAN productivity in this situation is not applicable [9, 10].

4 Conslusion

The article provides a detailed comparison of the RS-485 and CAN interface protocols. Key differences between CAN and RS-485 for use in agricultural complexes is shown. CAN interface provides 30% closer sensor coverage. It is designed for larger industries such as automotive and aerospace. RS-485 provides just enough sensor coverage and has a technological reserve of the recognition level. It also has 30% more coverage.

References

1. E. M. Shcherbakova, Demoscope Weekly, **645-646**, 9 (2014)
2. L. P. Seryakova, Meteorological conditions and plants, **77** (1971)
3. A. N. Zadeev, Industrial electric heating and electric heating, **4,** 34 (2013)

4. A. M. Agarkova, G. G. Iveshko, Rational use of fuel and energy resources during the operation of greenhouses, Budivelnik, Kiev, 120 (1985)
5. B. Perrin, Circuit Cellar, **108**, 2-9 (1999)
6. E.A. Ben, RS-485 for dummies, Mayak, Moscow, 21 (2003)
7. S. D. Erokhin, Electronic components, **2**, 42 (2018)
8. R. Gee, V. Retnyuk, Components and Technologies, **1(18)**, 60 (2019)
9. A. G. Garganeyev, Study of the energy efficiency of the microclimate of the greenhouse complex, National research Tomsk polytechnic university publishing, Tomsk, 103 (2019)
10. A. Lokotkov, Modern automation technologies, **3**, 110 (1997)

Development of the pneumatic system with an annular aspiration channel for vibrating and rotation grain separator

Vasiliy L. Andreev[1,*], *Vladimir V.* Shilin[2], *Valeriy V.* Ilichev[1], and *Tatjana S.* Holodenina[1]

[1]Nizhniy Novgorod state engineering-economic university, 22A, Oktyabrskaya street, Knyaginino, 606340, Russia
[2]Vyatka state agrarian-technological university, 133, Oktyabrsky avenue, Kirov, 610017, Russia

Abstract. The pneumatic system with a vertical annular aspiration channel for vibration-centered grain separators been developed. The rational values of the disk grain distributor are the length of the sectors 0.15 m, the angle of inclination of the sectors 60°. The rotation speed of the distributor in the seed mode (grain supply 5…15 t/h) should be 100…200 min^{-1} and in the food mode (grain supply 15…25 t/h) must be 100…150 min^{-1}. The effect of the release of light impurities in the developed aspiration system decreases from 45% to 20% with an increase in the supply of grain G from 5 to 25 t/h, herewith at the same time the loss of a full-fledged grain to waste decreases from 0.6...0.8% to 0.3...0.4%, while the effect of the release of light impurities in the aspiration system of the mass-produced vibration centrifugal grain separator decreases from 11% to 4% with the same losses of grain to waste.

1 Introduction

Larger agricultural enterprises have higher grain yields [1]. Therefore the use of highly efficient and capacity machines for seed and grain cleaning is important.

The most widespread in agro-industrial enterprises of the Russian Federation are the air and sieve grain cleaning machines [2]. To separate the source material into fractions these machines use a gravity field, so they have a less specific productivity of the sieves surface.

Centrifugal separators have higher values of specific productivity in which the exit of particles through the separating surface of the sieves occurs under the action of inertial forces far exceeding gravity [3].

Currently a large number of the vibration centrifugal grain separators of various brands are being mass-produced [4-7 and others]. For example the A1-BCS-100 vibrating grain separator [8] is designed to isolate impurities from grain material that differ in thickness, width and aerodynamic properties (figure 1).

This separator has four identical blocks. Each block consists of a housing 6, an annular aspiration channel 3, a rotating rotor with a sieve drum of three separate cylindrical sieves 5, 9, 10 and sieves cleaning devices 7, 13.

* Corresponding author: andreev.vas@mail.ru

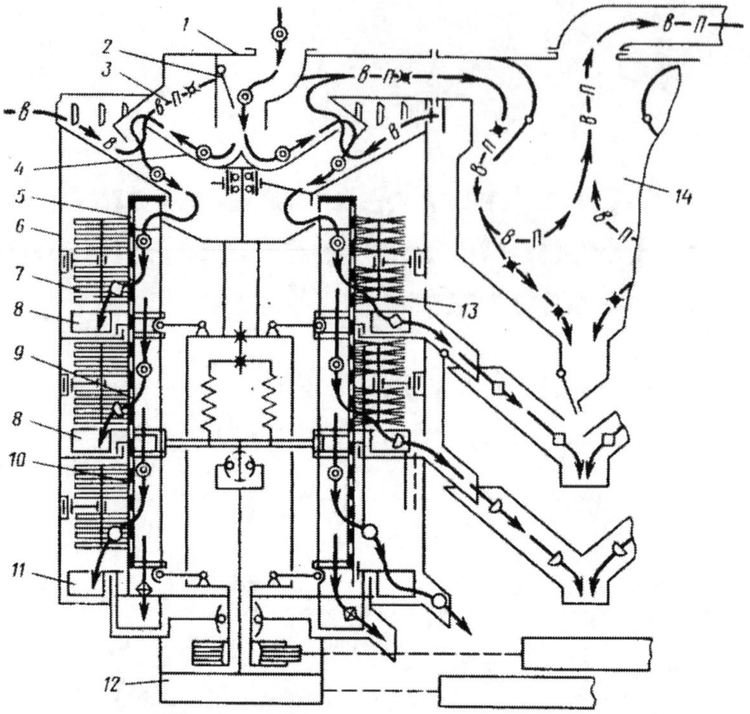

- dry unpeeled grain
- small impurities
- crushing (fine) grain
- refined grain
- large impurities
- light impurities
- grain dust
- air flow

Fig. 1. Technological scheme of one block of separator unit A1-BCS-100: 1 – dispenser; 2 – valve; 3 – aspiration channel; 4 – disc feeder; 5, 9, 10 – sieves; 6 – housing; 7 – rubber cleaner; 8, 11 – blades; 12 – vibrator; 13 – brush cleaner; 14 – sedimentary chamber

This separator has four identical blocks. Each block consists of a housing 6, an annular aspiration channel 3, a rotating rotor with a sieve drum of three separate cylindrical sieves 5, 9, 10 and sieves cleaning devices 7, 13.

The grain material through the dispenser 1 enters the disc feeder 4, which evenly feeds grain into the annular aspiration channel 3. Light impurities are released in the channel 3 and carried out with the air flow into the sedimentary chamber 14, where the bulk of them settles and is removed. At the same time the internal cavity of the sieve drum is aspirated. The remaining light impurities are released in the cyclone which has a high cleaning efficiency, so that the concentration of grain dust in the air of the working area does not exceed the permissible norms.

Next the grain material flows by gravity to a rotating distribution device that moves the material to the upper sieve 5 that releases small impurities. Due to the rotational and oscillatory motion of the sieve drum, the grain mixture moves in a uniform layer along the sieves from top to bottom. Small impurities pass through the sieve apertures 5, while the grain material is moved to grading sieve 9. The puny and part of the crushed grain passes through the holes of the sieve 9, the cleaned material with large impurities moves to the sieve 10, where the clean grain passes through the holes and is removed from the machine, and large impurities go down along the lower sieve 10.

However this separator has some disadvantages.

The grain material in the annular aspiration channel 3 is processed once at an increased speed, but for a short time, since the channel 3 formed by the disc feeder 4 and the feeder body has a short length, as a result of which light impurities not have time to separate from the cleaned material.

Also the channel has different hydraulic resistance along its length due to the location of the air intake, which is located on the side wall of the housing 6, which causes a large unevenness of the air velocity in the channel.

Too the device for feeding grain material into the annular aspiration channel has not been sufficiently studied.

2 Materials and methods

The pneumatic system with a vertical annular aspiration channel for the vibration centrifugal grain separator has been developed and its experimental sample has been manufactured (figure 2).

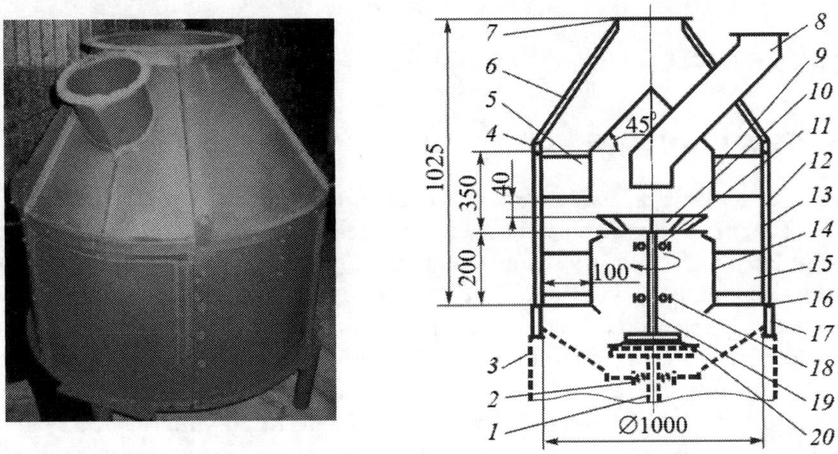

Fig. 2. Pneumatic system with a vertical annular aspiration channel and with a disk grain distributor with inclined sectors: a – general view; b – schema; 1 – the shaft of the sieve separator; 2, 11, 18 – bearing support; 3 – the housing of the sieve separator; 4 – the upper ring of the housing of the pneumatic system; 5 – the upper spacer plate; 6 – the housing of the pneumatic system; 7 – the mounting flange of the duct for removing exhaust air; 8 – the grain loading pipeline; 9, 14 – the inner wall of the aspiration channel; 10 – the disc grain distributor; 12 – the lower housing rack of pneumatic system; 13 – the outer wall of the aspiration channel; 15 – the lower spacer plate; 16 – the lower ring of the pneumatic system housing; 17 – the rack of the sieve separator housing; 19 – the shaft of the grain distributor disk; 20 – coupling

Through grain loading pipeline 8 the grain mixture is fed to the rotating disk grain distributor 10 with replaceable sectors. Under the action of centrifugal forces the cleaned material moves first along the horizontal surface and then along the inclined sectors of the distributor 10 enters the separation zone of the vertical annular aspiration channel formed by the outer 13 and inner 9, 14 walls. Here the cleaned material is blown by the air flow created by the fan, which is not shown in the figure. The light impurities trapped in the aspiration channel 8 are removed together with the exhaust air through an air duct connected to the flange 7. Spacer plates 5, 15 of the same length provide the same depth of the vertical annular aspiration channel along its length and height in the separation zone. Exhaust air intake is carried out from the upper part of the pneumatic system. The outer diameter of the vertical annular channel had a value of 1.0 m and its depth was 0.1 m.

Rotation frequency 10 grain disk distributor was values $n = 100...250$ min^{-1}, but changed using an electric motor and a DC source. The angle of inclination of the sectors had meanings $\alpha = 0...60°$, and their length was $l = 0...0{,}20$ m (fig. 3).

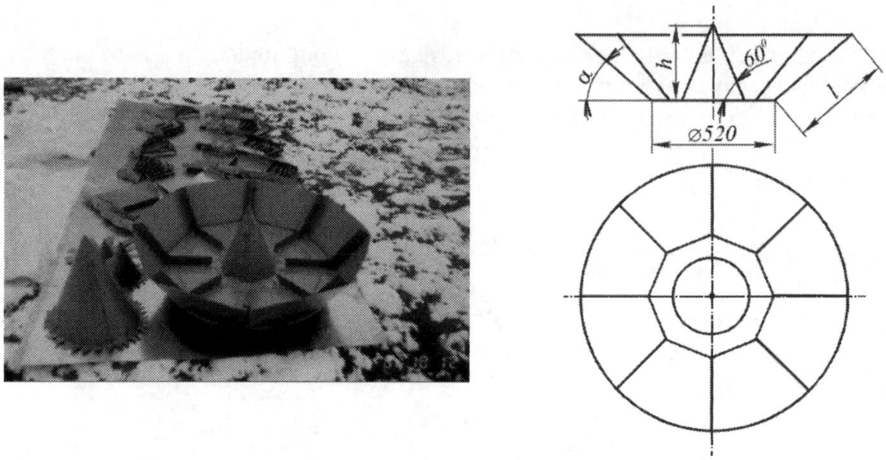

Fig. 3. The rotating disk grain distributor with replaceable sectors

3 Results and Discussion

To determine the rational parameters of the disk distributor when it is functioning in the aspiration system, a two-level central composite rotatable plane 2^{4-1} of the first-order experiment for four factors [9] is implemented. After the implementation of the experiment plan, regression models were obtained, but which do not adequately describe the real processes (used F-criterion of Fisher at the probability $p = 0{,}95$).

Therefore further implemented is a three-level plan for the second-order Box-Benkin experiment for four factors [9]. As the main culture, the barley seeds of the Abava variety with a thickness of more than 2.4 mm in an amount of 95% were used, and the impurity was small rye with a thickness of less than 2.0 mm in an amount of 5%. Factors and levels of their variation are shown in table 1.

Table 1. Factors and levels of their variation when implementing an experiment of Box-Benkin's plan of the second order for four factors

Coded value of factors	The name of the factors, their designation and unit of measurement	Factor levels					
		Seed mode			Food mode		
		-1	0	1	-1	0	1
x_1	Grain G supply, t / h	5	10	15	15	20	25
x_2	Frequency *n* of rotation of the distributor, min $^{-1}$	100	175	250	100	175	250
x_3	Lenght *l* of sector, m	0	0,075	0,150	0	0,075	0,150
x_4	Angle α of sector, degree	0	30	60	0	30	60

The optimization criteria adopted effects of selection of impurities from the grain mixture on the seed E_{SEED} and food E_{FOOD} (%) modes, as well as losses a_{SEED}, a_{FOOD} (%) of full grain in waste.

After the implementation of the experiment's plan, regression models were obtained, which adequately describe the actual processes (used F-criterion of Fisher at the probability $p = 0{,}95$):

$$E_{SEED} = 23{,}43 - 5{,}51 \cdot x_1 + 3{,}92 \cdot x_3 + 4{,}52 \cdot x_4 + 1{,}38 \cdot x_1^2 + 0{,}415 \cdot x_1 x_2 - 0{,}79 \cdot x_1 x_4 - 0{,}05 \cdot x_2^2 + 0{,}79 \cdot x_2 x_3 + 0{,}44 \cdot x_2 x_4 + 0{,}92 \cdot x_3^2 + 4{,}08 \cdot x_3 x_4 + 1{,}88 \cdot x_4^2; \quad (1)$$

$$a_{SEED} = 0{,}137 - 0{,}040 \cdot x_1 + 0{,}104 \cdot x_2 + 0{,}173 \cdot x_3 + 0{,}160 \cdot x_4 - 0{,}011 \cdot x_1^2 - 0{,}020 \cdot x_1 x_2 - 0{,}040 \cdot x_1 x_3 - 0{,}024 \cdot x_1 x_4 + 0{,}017 \cdot x_2^2 + 0{,}088 \cdot x_2 x_3 + 0{,}093 \cdot x_2 x_4 + 0{,}054 \cdot x_3^2 + 0{,}230 \cdot x_3 x_4 + 0{,}330 \cdot x_4^2; \quad (2)$$

$$E_{FOOD} = 15,76 - 2,17 \cdot x_1 - 0,21 \cdot x_2 + 2,80 \cdot x_3 + 3,45 \cdot x_4 + 0,40 \cdot x_1^2 - 0,31 \cdot x_1 x_3 - 1,10 \cdot x_1 x_4 + \\ + 0,06 \cdot x_2^2 + 0,25 \cdot x_2 x_3 + 1,07 \cdot x_2 x_4 - 0,84 \cdot x_3^2 + 2,82 \cdot x_3 x_4 + 0,18 \cdot x_4^2; \quad (3)$$

$$a_{FOOD} = 0,068 - 0,004 \cdot x_1 + 0,077 \cdot x_2 + 0,105 \cdot x_3 + 0,121 \cdot x_4 + 0,010 \cdot x_1^2 - 0,006 \cdot x_1 x_2 - \\ - 0,008 \cdot x_1 x_4 + 0,013 \cdot x_2^2 + 0,065 \cdot x_2 x_3 + 0,074 \cdot x_2 x_4 + 0,033 \cdot x_3^2 + 0,151 \cdot x_3 x_4 + 0,055 \cdot x_4^2. \quad (4)$$

Analysis of equations (1)...(4) showed that as on seed mode with changes in the pneumatic bandwidth G from 5 to 15 t/h, and on food regime when changes in the capacity of the pneumatic system G from 15 to 25 t/h rational values are the length of the sectors l = 0,15 м (x_3 = 1) and angle of inclination of sectors α = 60° (x_4 = 1), but the frequency of rotation of the grain disk distributor on seed mode (fig. 4, a) should be n = 100...200 min^{-1} (x_2 = -1...0,33), but on food mode (fig. 4, b) should be n = 100...150 min^{-1} (x_2 = -1...-0,33). Permissible losses of the main grain in waste with light impurities on seed mode $[a_{SEED}]$ = 1,0% and food mode $[a_{FOOD}]$ = 0,5%.

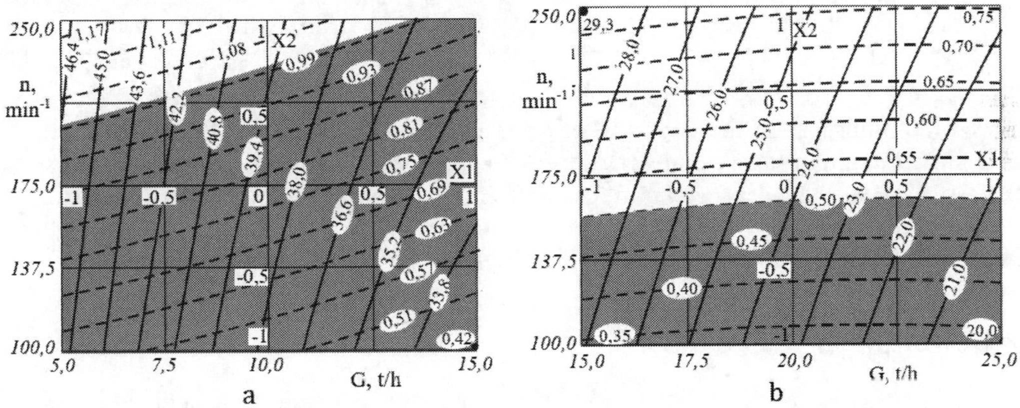

Fig. 4. The dependences of the effects of impurities and losses of full grain into waste during the functioning of the aspiration system from the supply G of grain and the frequency n of rotation of the distributor at the values of factors x_3 = 1 (the lenght of sectors l = 0,15 м) и x_4 = 1 (The angle of inclination of sectors α = 60°); a – on seed mode; b – on food mode; ——— – effects E_{SEED} and E_{FOOD} of selection of impurities (%); – – – – – losses a_{SEED} and a_{FOOD} basic grain in waste (%); ▒ – areas of permissible grain loss $[a_{SEED}]$ = 1,0% и $[a_{FOOD}]$ = 0,5%

The studies of the developed pneumatic system were carried out in comparison with the pneumatic system installed in the serial-produced vibration centrifugal separator MZP-25, having a bandwidth of 25 t / h. To estimate the effectiveness of impurities is passed through both pneumatic systems, 500 kg of prepared grain mixture (95% - the Abava barley seeds variety with a thickness of more than 2.4 mm and 5% - the small winter rye with a thickness of less than 2.0 mm). The air flow rate in the pneumatic channels during experiments was not changed and installed at the beginning of the removal of the full grain into waste at the lowest seeds of the grain mixture, taking into account the allowable losses of the main grain. All highlighted light fraction, including a small number of barley grains and the main part of winter rye grains, was shot in the tissue filter. The separation of the mixture was performed on solids with rectangular openings 2.2 × 20 mm. The experiments were carried out in three-time repetition.

The dependence of the effect E of removing of impurities in the serial-produced pneumatic system and in the developed pneumatic system with an annular aspiration channel from the supply *G* of grain with the same losses *a* of grain in waste shows at fig. 5.

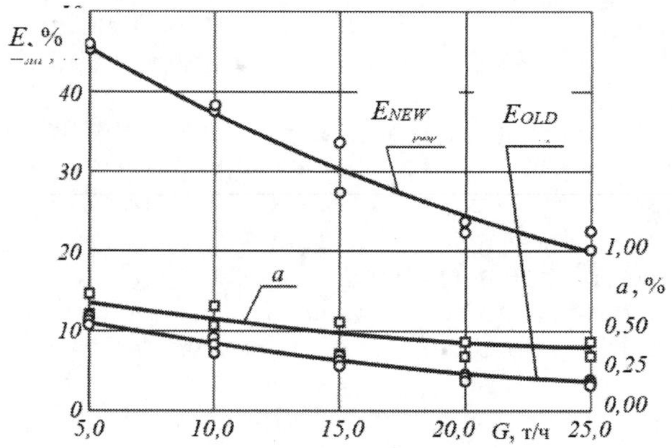

Fig. 5. The dependences of the effects of emission of lung impurities in the developed pneumatic system E_{NEW} (length of the sectors of the distributor l = 0,15 m, the angle of their tilt α = 60°, rotation frequency n = 115 min^{-1}) and serial produced pneumatic system E_{OLD} from the supply of *G* grain with the same losses *a* and the main grain in waste

With an increase in the supply *G* of grain in aspiration systems from 5 to 25 t / h and simultaneously decreasing losses *a* of a full grain in waste from 0,6...0,8% to 0,3...0,4% effect E_{NEW} of election of light impurities in the developed pneumatic system decreases from 45% to 20%, while the effect E_{OLD} of the separation of light impurities in the serial produced pneumatic system is reduced from 11% to 4%.

4 Conclusion

The developed pneumatic system with a vertical ring aspiration channel and a disc-distributor of grain with inclined sectors more efficiently releases impurities from seed and food grain compared with the pneumatic system of mass-produced vibration centrifugal grain separators MZP-25 with the same losses of the main grain in waste.

References

1. S. Petrova, D. Zjukin, *Assessment of factors for the increase in the productivity of grain crops*
2. V.I. Orobinsky, A.M. Gievsky, V.A. Gulevsky, *Prospects for improving the technology of post-harvest grain and seed processing and technical means to implement it*
3. N.E. Avdeev, Centrifugal separators for grain, Moscow, Kolos, 152 (1975)
4. Vibration separators A1-BCS-100, https:// sinref.ru/
5. Vibration separators R8-BC2S and R8-UCS-200, https:// mukomoll.ru/
6. Grain centrifugal separators BCS-25, BCS-50, BCS-100, https:// melagrosnab.ru/
7. Centrifugal grain vibrating separator B1-VCS Belogorie, https://nizhniynovgorod.selhoztehnika.net/
8. B.V. Lebedev, Industrial seed processing and storage, Moscow, Agropromizdat, 255 (1991)

9. S.V. Melnikov, V.R. Aljoshkin, P.M. Roshchin, Planning an experiment in agricultural process research, Leningrad, Kolos, 168 (1980)

Experimental determination of comparative sensitivity of eimeria to chemotherapeutic preparations and phytobiotics

Inessa Yakovleva

BSAU named after V.Ya. Gorin, Cand. Sc. (Biology), 308503 Maisky, Belgorod, Russia

Abstract. The sensitivity of the organism of broiler chickens artificially infested with a mixture of three types of eimeria to anti-eimeria drugs used in the Belgorod region was determined. For the experiment, synthetic coccidiostatics (decoquinate, maduramycin, nicarbazine, robenidin), ionophore antibiotics (lasalocid, monensin, narazine) and a complex phytobiotic "AdiCox AR" were taken. The smallest loss of live body weight of chickens was observed when feeding them in the diet of lasalocid, then decoquinate. Feeding the specialized feed additive "AdiCox AR" ensured the preservation of gains in live weight of chickens by 19.0% less than the most effective ionophore antibiotic lasalocid in our experiments. Then, in descending order, followed: narazine 10%, monensin 20%, nicarbazine 25%, maduramycin 1%, robenidin 6.6%. In connection with the obtained results, we consider it expedient to use lasalocid 15% at a dose of 500 g / t feed, decoquinate 6% at a dose of 500 g / t feed and AdiCox AR at a dose of 300 g / tons of feed. The use of narazine 10%, monensin 20%, nicarbazine 25%, maduramycin 1% should be carried out under constant monitoring of the sensitivity of the pathogen to the listed drugs. We consider the use of robinidin 6.6% inappropriate.

1 Introduction

Despite the use of rotary and shuttle programs in the pharmacological prevention of eimeriosis, the causative agents of this disease quickly develop resistance to many drugs. It has been proven that eimeria are capable of transmitting, at the genetic level, to the next generations of pathogens, the developed resistance to anticoccidial drugs, therefore it becomes more and more difficult to select therapeutically active compositions for the prevention of eimeriosis in poultry complexes [1, 2, 3]. Since over the past 10 years on the veterinary pharmaceutical market, only complex drugs have been offered, created on the basis of already known and selected according to the principle of synergy of their components, the sensitivity of eimeria to them, obviously, will not be as long as the manufacturers of these drugs would like [4, 5]. Therefore, it is worth paying the attention of veterinarians to phytobiotics - plant complexes that are devoid of negative side effects on the body of birds, such as, for example, the well-known synthetic drugs. In addition, they contain biologically active compounds that have an adaptive effect on the body of birds in biologically uncomfortable conditions of large poultry farms [6, 7].The effectiveness of the herb St. John's wort in the prevention of eimeriosis has been proven; Echinacea purpurea, fir coniferous flour, oregano oil, thyme extract - in increasing the general and specific resistance of the body of birds, which is a direct factor in the prevention of bacterial, viral and parasitological diseases, including eimeriosis [8-12]. Phytobiotics normalize impaired metabolism, stimulate the functions of immunocompetent organs, and have stress-protective and growth-stimulating effects [13-17]. The pharmaceutical market of veterinary drugs in Russia, unfortunately, is represented by a very small list of phytobiotics, while the plant raw materials in our country are more than enough for the development and production of this group of drugs.

The feed additive “AdiCox AR” (manufacturer: “AdiFeed Sp. z oo”, Poland), which has recently appeared on the pharmaceutical market, consists of micronized ground and root parts of the following plants: Capsicum annuum L. var. minimum (Miller) Heiser, Sinapis alba L., Saponaria officinalis L., Acorns calamus L. in the amount of 91.5-93.5% of the total weight, as well as auxiliary substances: iron oxide - 1%; a mixture of flavoring substances - 0.5%; hydrogenated vegetable oil - 5-7%. The process of superfilling (micronizing) particles is widely used in the manufacture of pharmaceuticals. Due to this process, a high bioavailability of active substances is achieved, a faster absorption and development of a therapeutic effect, a decrease in the risk of side effects [18, 19].

The aim of our study was to determine the sensitivity of the organism of broiler chickens, artificially infested with a mixture of three types of eimeria, to anti-coccidiosis drugs used in the Belgorod region. For the experiment, we took synthetic coccidiostatics (decoquinate, maduramycin, nicarbazine, robenidin), ionophore antibiotics (lasalocid, monensin, narazine) and the recently appeared on the market a complex phytobiotic (specialized feed additive) of Polish production "AdiCox AR".

2 Materials and methods

From the average sample of broiler chickens' droppings located in the production shops of one of the poultry farms of the Belgorod region, the culture of Eimeria was isolated and identified in accordance with the methodology [20, 21, 22]. The species composition of eimeria was determined by morphological characteristics and by the place of their localization in the intestines of broiler chickens after infection. Experimental broiler chickens were divided into groups of 6 heads each and kept in cages, taking into account the required technological norms of the area per head.

The first group of chickens was not infected with the Eimeria suspension, i.e. the bird was healthy. The second was drunk with a suspension of eimeria, i.e. artificially infected them with eimeriosis, but they received no treatment. The studied anti-coccidiosis drugs of various pharmacological groups were added to the compound feed in the doses recommended in the instructions. The day before the infection with Eimeria and another 9 days from the moment of infection, the prepared mixtures were fed to the chickens of the experimental groups; in total, the preparations were fed for 10 days in a row. The third group was fed a mixed feed mixture with 6% decoquinate; the fourth group - with lasalocid 15%; the fifth group - with maduramycin 1%; the sixth - with monensin 20%; the seventh - with narazine 10%; eighth - with nicarbazine 25%; the ninth - with robenidin 6.6%; the tenth - with a specialized feed additive "AdiCox AR".

At the beginning and at the end of the experiments, the chickens were weighed, the average weight of one head was determined, rounding off the indicator to grams, and then monitoring was carried out in groups, comparing the indicators with similar ones. The observation of broiler chickens was carried out within nine days from the moment of infection, taking into account the clinical manifestations of eimeriosis. A postmortem examination of the dead chickens was carried out, paying attention to the peculiarities of the lesions of the intestinal mucosa.

3 Results and discussion

A mixture of E. acervulina, E. maxima and E. Tenella cultures was isolated from a sample of broiler chickens, with a quantitative predominance of E. maxima. Their oocysts are shown in Fig. 1.

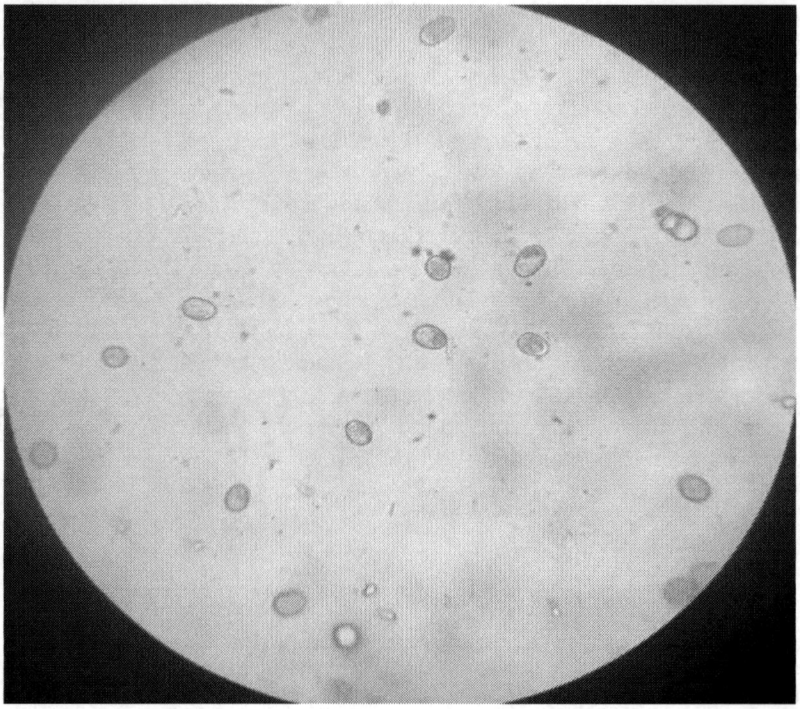

Fig. 1. Oocysts of E. acervulina, E. maxima and E. tenella isolated from broiler chicken droppings (eyepiece × 10, objective × 40).

Indicators of weight gain in experimental chickens and their safety are the main criteria for the effectiveness of the studied drugs. Indicators of the dose of the studied drugs, the safety and weight gain of the control and experimental chickens are presented in table 1.

Table 1. The results of determining the sensitivity of the field mixture of E. acervulina, E. maxima and E. tenella, isolated from broiler chicken droppings to the studied drugs.

Name		Drug dosage, g/t	Number of chicks, animal units	Dead, animal units	Live out, %	Gain, %
Group	Drug					
1	Control (healthy bird)	-	6	0	100,0	152
2	Control (infested bird)	-	6	3*	50,0	21
3	Decoquinate 6%	500	6	0	100,0	107

Table 1. Continued.

4	Lasalocid 15%	500	6	0	100,0	121
5	Maduramycin 1%	500	6	1**	83,3	74
6	Monensin 20%	500	6	1***	83,3	80
7	Narazine 10%	700	6	2***	66,6	96
8	Nicarbazine 25%	500	6	2***	66,6	77
9	Robenidin 6,6%	500	6	2**	66,6	63
10	Feed additive «AdiCox AR»	300	6	1***	83,3	102

* - multiple intestinal lesions of the dead bird, characteristic of coccidial invasion, were found;
** - insignificant lesions of the intestines of the dead bird were found
*** - death due to reasons not related to coccidiosis

As can be seen from the table, the doses of the drugs used were different: from 300 g / t of the specialized feed additive "AdiCox AR" to 700 g / t of the feed narazine 10%, all other drugs were fed at a dose of 500 g / t of feed (all of the above doses were taken from the instructions for use). In the second group (invasive control), the survival rate of chickens was only 50%. Postmortem examination of three dead chickens from this group showed the presence of characteristic signs of coccidiosis: multiple lesions of the intestinal mucosa, characteristic of infection mainly with E. Maxima. Autopsy of one head from the fifth experimental group receiving maduromycin 1% with food and two heads from the ninth experimental group receiving 6.6% robinidin with food revealed the presence of minor lesions of the intestinal caecum, characteristic of coccidiosis. An autopsy of all the dead chickens of the sixth, seventh, eighth and tenth experimental groups revealed the causes of the death that were not associated with coccidiosis. In the experimental groups receiving decoquinate and lasalocid, no deaths were recorded, i.e. 100% safety of chickens was observed. For the convenience of perceiving the results obtained in the experiment, the weight gain of the control and experimental chickens, we presented in the form of a diagram, on which we arranged the indicators obtained from the use of drugs in the order of decreasing their effectiveness. So in our experiment, the highest sensitivity of eimeria was noted to lasalocid 15%, then to decoquinate 5%, then to the specialized feed additive "AdiCox AR". The sensitivity of the pathogen to robenidin 6.6% turned out to be the lowest, which is obviously due to its frequent inclusion in the schemes of preventive treatments for eimeriosis of the poultry head of this complex.

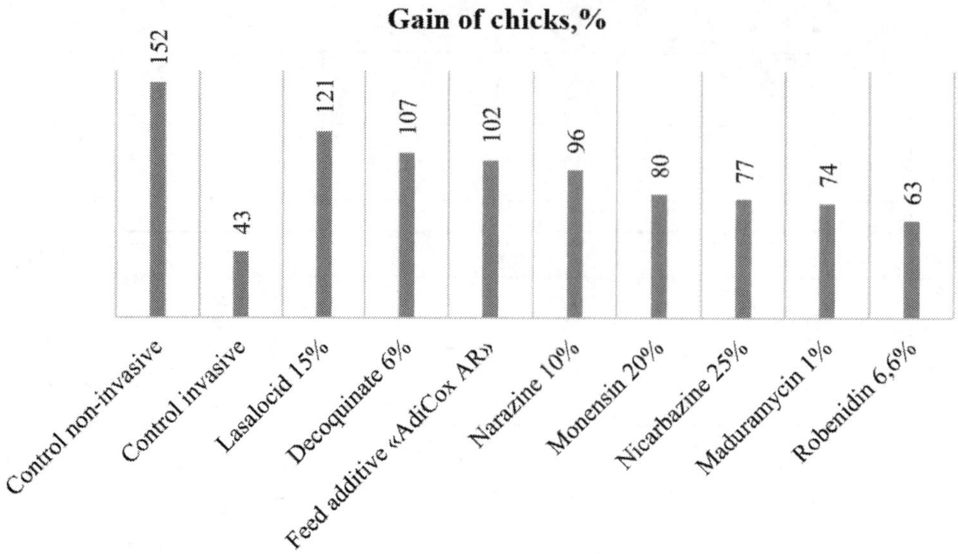

Fig. 2. Gain of chicks, %.

4 Conclusion

Thus, the smallest body weight loss during infestation of broiler chickens with a mixture of Eimeria cultures (E. acervulina, E. maxima, and E. Tenella) was observed when fed with lasalocid in the diet of 15% (-31% relative to the control non-invasive group), then decoquinate 6% (-45% relative to the control non-invasive group). Feeding a specialized feed additive belonging to the pharmacological group of phytobiotics and not containing chemical components ensured the preservation of gains in live weight of chickens by only 19.0% less than the most effective ionophore antibiotic lasalocid 15% in our experiments. Then, in descending order, follow: narazine 10%, monensin 20%, nicarbazine 25%, maduramycin 1%, robenidin 6.6%.

In connection with the obtained results, we consider it expedient to use lasalocid 15% at a dose of 500 g / t feed, decoquinate 6% at a dose of 500 g / t feed and a specialized feed additive "AdiCox AR" at a dose of 300 g as prophylactic anti-coccidiosis agents at poultry sites of a poultry farm in the Belgorod region. / t feed. The use of narazine 10%, monensin 20%, nicarbazine 25%, maduramycin 1% should be carried out under constant monitoring of the sensitivity of the pathogen to the listed drugs. The use of robinidin 6.6% at the poultry sites of this poultry farm is considered inappropriate.

References

1. E.O. Kachanova, R.T. Safiullin Rus. J. Parasitol. **4**, 97- 104 (2019)
2. V.I. Smolensky, A.L. Kiselev, T.G. Titova Poultry Press J. **1**, 50-52 (2018)
3. 3 R.T., Safiullin, T.G. Titova, T.A. Nurtdinova Rus. J. Parasitol. **3**, 288-298 (2017)
4. T.G. Titova, V.M. Razbitsky, I.M. Biryukov *Sensitivity of field isolates of eimeria chickens to combined anti-coccidial drugs*, in Proceedings of the conference "Actual issues and prospects for the development of agricultural sciences", 11 May, 2017, Omsk, Publishing house: Innovation center for development, education and science, 24-27 (2017)

5. L.M. Kashkovskaya, V.A. Orobets Vet. J. **8**, 31-33 (2019)
6. V.S. Buyarov et al. Agr. Sc. Bulletin, **3 (84)**, June, 2020, 44-59 (2020)
7. D.A. Latysheva, E.V. Ulrikh *History and current state of the production of phytobiotic feed additives in the world*, Materials of the VII international scientific and practical conference “Climate, ecology, agriculture of Eurasia”, Irkutsk, 2018, 125-136 (2018)
8. V.D. Avdachenok Uchenye zapiski UO VGAVM. **1**, 7-10 (2016)
9. A.D. Chernyshova *Influence of Echinacea purpurea on the growth of broiler chickens*, Collection of articles of the All-Russian scientific-practical conference of young scientists, 68-70 (2017)
10. V.I., Terentyev, T.I. Anikienko Bulletin of KrasGAU. **5**, 163-166 (2011)
11. N.V. Yunyaeva, K.V. Salandaev, A.V. Slyusar Poultry Sc. J. **8**, 43-45 (2016)
12. E.A. Kishnyaykina, K.V. Zhuchaev Vestnik NSAU. **4 (49)**. 74-80 (2018)
13. Z.N. Alekseeva et al. *Phytobiotics for growing pullets of chickens* in Proceedings of the scientific-practical conference for teachers, students, undergraduates and graduate students, dedicated to the 80th anniversary of the NSAU, 98-101 (2016)
14. N.N. Lantseva et al. Influence of functional properties of probiotics and phytobiotics on productivity indices of broilers, **2**, 1417-1423 (2015)
15. V.A. Fedotov et al. Poultry Press J. **8**, 33-37 (2018)
16. O. Trufanov Animal husbandry of Russia, 5-7 October (2016)
17. N.I.Petenko, D.A. Butskovsky, A.N. Gneush *The influence of phytobiotics on the zootechnical indicators of quails,* Modern aspects of production and processing of agricultural products: a collection of articles based on the materials of the V International scientific and practical conference dedicated to the 15th anniversary of the department of technology for storage and processing of livestock products of the KSAU, 247-251 (2019)
18. Pharmacokinetics and Pharmacodynamics, https://ru.wikipedia.org/
19. Yu.A. Selivanova Poultry Press J. **1** 37-40 (2018)
20. R.T. Safiullin L.A. Bondarenko Theory and practice of parasitic diseases of animals, 269-272 (2014)
21. M.V. Krylov Determinant of parasitic protozoa, 693(1996)
22. R.T. Safiullin, R.R. Murzaev, O.N. Andreyanov Theory and practice of parasitic diseases of animals, 135-139 (2011)

Range of goods in framework of agricultural enterprise management

Nikolay Sergeevich Bondarev, and *Galina Sergeevna* Bondareva

Kuzbass State Agricultural Academy, Markovtseva st., 5, 650056 Kemerovo, Russia

Abstract. The study of the product range of the enterprise, in particular, the agricultural ones, is one of the priority areas. Maintaining a stable demand for reduced goods is an obligatory condition for strategizing activities, a guarantee of development. The presented research reflects the theoretical foundations of enterprise management, in terms of building a strategy; the assortment features for both commodity and trade ones are disclosed; the authors highlighted the specifics of agricultural production through strategic management and its goal setting. In the main part of the work, there is a block focused on the study of the assortment within the framework of the agricultural enterprise management, based on the product range optimization. This is supported by the conclusions that this approach is a source of diversification of an agricultural enterprise production activity and takes into account both the existing market needs and the capabilities of the enterprise itself when creating the necessary conditions for its development, ensuring competitiveness not only in local markets, but also includes a possibility to expand its market zone. The assortment of goods was studied in a separate block on the example of a specific agricultural enterprise - the greenhouse complex JSC "Sukhovskiy", a member of the Siberian agricultural complex of the brand "Em Rodnoe". The assessment of the main economic indicators of the activities of JSC "Sukhovskiy" was carried out. For the purpose of practical testing of the developed directions of strategizing the product range of JSC "Sukhovskiy", there was proposed a variant of its expansion by introducing roses into the "line" of production as the most popular and demanded flowers in high demand among the population. In addition, the agrotechnology of this production in the conditions of a greenhouse economy is shown, possible sales channels for finished products are proposed, the costs of the "rose" assortment line are structured and the predicted economic effect from the implementation of the proposed measures is calculated.

1 Introduction

Greenhouse activity is one of the most labor-intensive branches of agricultural production. Despite the difficulties and the presence of a number of specific features, the presence of increased demand stimulates agricultural enterprises to develop in this area [1; 9].

The process of implementing measures for development and improving the assortment strategy is aimed at the creation and implementation of new business solutions and technologies [14].

Within the framework of the study, it is proposed to apply the accumulated experience in the development of the greenhouse business based on the expansion of its assortment positions in the management system of a specific agricultural enterprise.

The object of research is the production activities of JSC "Sukhovskiy".

The subject of the research is management and technological processes and decisions in the field of production strategy.

Expected results:

1. Creation of highly competitive production;
2. Meeting the needs of population;
3. Increasing the range of products.

2 Research methods

In this aspect of the study, strategic management is methodically investigated on the basis of the goal-setting processes, organic nature of resources, processes sequence, consistency of decisions and approaches, adequacy of management actions in the current situation and inherent capabilities. System management decisions can acquire their functionality only in the case of building a rational planning vector in the context of optimality of business processes [15]. The fundamental principle is the action "from the future to the present", and not "from the past to the future."

In this aspect of the study, strategic management is methodically investigated on the basis of goal-setting, the organic nature of resources, the sequence of processes, the consistency of decisions, the consistency of approaches and the adequacy of management actions in the current situation and inherent capabilities. System management decisions will acquire their functionality only in the case of building a rational planning vector in the optimality conditions of business processes [15]. The fundamental principle is the action "from the future to the present", and not "from the past to the future."

On a strategic basis, the elements of assortment management in an agricultural enterprise are investigated:

1. Setting the goal of assortment management;
2. Study of the environment for an agricultural enterprise functioning (external and internal). It should be determined assortment program in relation to the target segment and competitive environment based on the selection of potential competitors;
3. Development of a commodity strategy and the formation of a production program for an agricultural enterprise;
4. Formation of a product offer and subsequent entry into the local market;
5. Evaluation of the results of the selected assortment strategy implementation.

3 Data and primary analysis

3.1 Strategic enterprise management

External environment affects activities of an enterprise, incl. agricultural, significant impact and subject to constant change. The ongoing economic, political, social processes, and in the current period epidemiological one, as a result of the global pandemic COVID-19, require constant adaptation of business to building flexible management systems ready for these changes. All these conditions for the dynamism of processes, the increase in their complexity, transformation, proceeding often unpredictably, indicate the importance of identifying the existing patterns in retrospect and, therefore, building a model of the behavior of an economic entity at the current moment and a business strategy for the future.

So, strategic management is interpreted as a modern view of organization of the enterprise management system, aimed at developing strategic decisions in order to strengthen competitive positions, team building, providing admissibly possible management tools for the implementation of the desired strategic positions and their transformation and current production and economic plans.

Approaches to strategizing the production activities of an enterprise imply the development of an appropriate methodology and methodology for its implementation in the face of constant transformation of the external environment, the resource capabilities of the internal environment, the existing resource potential.

Changing conditions of the external environment and lack of readiness for these changes in the internal environment form together quite specific realities of the business environment. Its functioning is carried out in conditions of uncertainty, which accordingly affects business processes in the real sector of the economy, not so much on their perception, but on their uncontrollability.

The evolution of the "strategic management" concept comes from the 60s-70s of the twentieth century, the need for which arose because of the need to “reflect the difference between management carried out at the highest level and current management at the production level. The need to make this distinction was driven primarily by changes in the business environment. As the leading idea, reflecting the essence of the transition to strategic management from operational management, was the idea of the need to transfer the focus of senior management to the environment in order to respond appropriately and in a timely manner to the changes taking place in it, to respond in a timely manner to the challenge posed by the external environment” [2].

At the moment, the scientific literature presents a fairly significant number of scientific views, practical opinions regarding the interpretation of the concept of "strategic management", the content of which comes from various organizational aspects of the management process, its complexity and versatility.

The goal-setting of the management strategy of an agricultural enterprise is based on building key positions (Figure 1) [8].

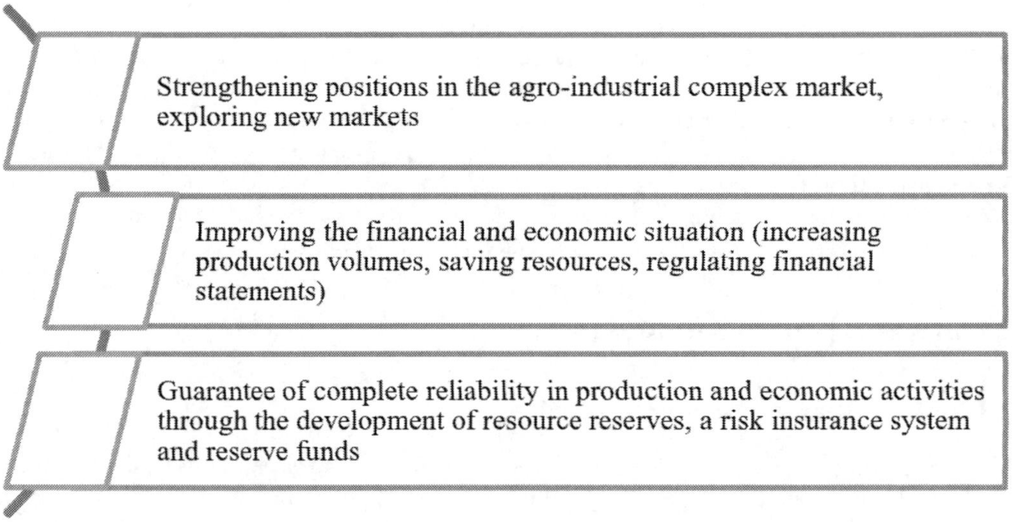

Fig. 1. Key positions for formation of the management strategy of an agricultural enterprise

The pointed key positions inherently reflect all organizational aspects of an agricultural enterprise with an emphasis on production activities as the most significant. Any enterprise, first of all, is a production system, the activity of which is aimed at the implementation of the main function - production.

In the process of the enterprise's functional activity, it is influenced by various factors, which significantly affects the approach to strategic management. It should be noted that agriculture is characterized by quite specific impact factors that are not inherent in other industries. Figure 2 clearly shows the factors influencing the organizational activity of an agricultural enterprise from the standpoint of strategic management [4].

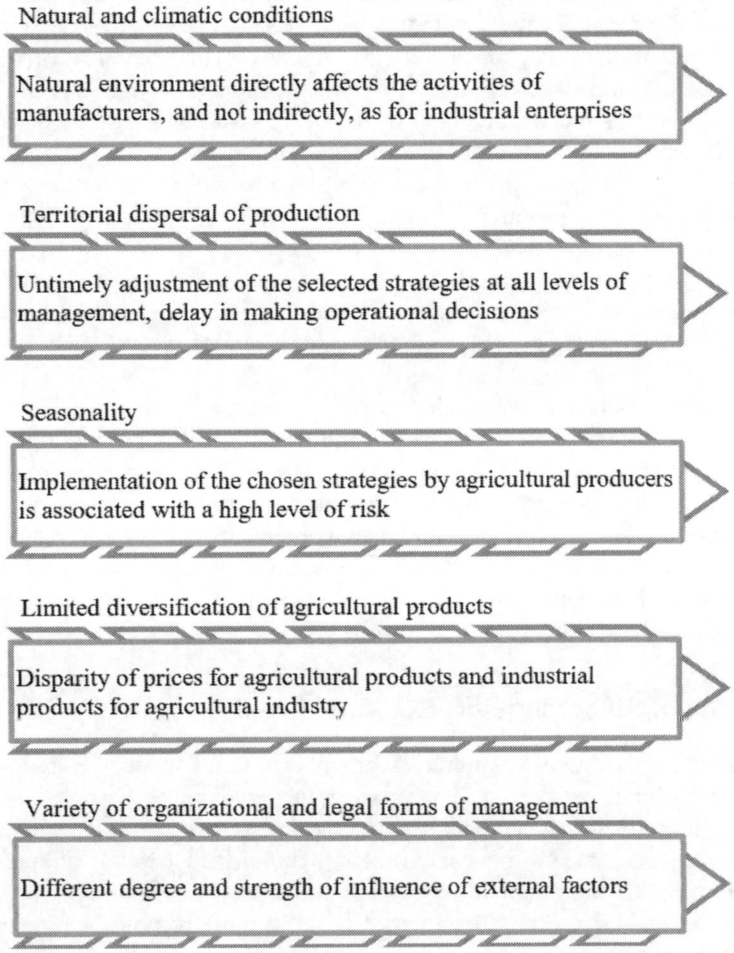

Fig. 2. Factors affecting organizational activities of an agricultural enterprise from the strategic management standpoint

3.2 Assortment within the framework of agricultural enterprise management

By origin (from French), "assortment" means "selection of products of various classes and designs" and means "a list of goods or services offered by the organization to meet specific human needs"; "a set or list of goods by type, name, purpose, grade and type of raw material used" [7]; "an element that connects the enterprise and the market, acting as an object of economic exchange and a means of satisfying existing needs" [5], etc.

This list of interpretations of interpretations of the term can be expanded and expanded, but it becomes clear that it is a certain species diversity that is common: a list, a list, a variety, the name of goods (works, services).

Thus, considering the assortment, a new category "product assortment" appears, which, without losing its classification features, is also endowed with production (industrial and commercial) ones.

In the first case, it is the sectoral affiliation of the produced goods' assortment or the sectoral affiliation of the enterprise that produces the given assortment of goods. In the second case, the emphasis is placed on a certain set of the commodity list, which is presented in wholesale or retail trade enterprises, i.e. in retail or wholesale trade networks, centralized places of sale of products, spontaneous markets. A feature of the organization of the trade assortment is the inclusion in its classification of the territory of production of products - a domestic product or a product of foreign production. In this connection, it can be argued about a greater species diversity of the trade assortment, in comparison with the industrial assortment.

It should be noted that the assortment management strategy in agricultural enterprises is based on optimizing the product assortment, since this direction is a source of diversification of production activities. Thus, it is necessary to take into account that the assortment management strategy should take into account both the needs of the market and the needs of the enterprise itself in creating conditions for creating development potential and ensuring its competitiveness. The complete interdependence of these processes should be provided.

In this connection, when forming an assortment strategy, it is necessary to take into account the fact that a commodity producer in the process of his production activity must operate precisely with the needs of the market and the prevailing market situation, and his product and trade assortment must satisfy market needs, both in quality and quantity.

Thus, "the private goals of assortment management are achieved by developing and implementing a set of measures for the assortment strategy, which gives an idea of what the main stages of achieving goals are, and how the existing financial, material and labor resources of the enterprise should be used for this" [6].

3.3 Study of the product range of JSC "Sukhovskiy"

JSC "Sukhovskiy" is the largest supplier of green crops and fresh vegetables of its own production in the Kemerovo region. The area of the greenhouse complex is 60,000 m^2. More than 300 million rubles were invested in fixed assets; more than 250 million were invested in current assets; 381 thousand rubles - authorized capital; about 400 million rubles - long-term loans that went to modernize the complex; about 400 million rubles - annual revenue; 31 million rubles - net assets; about 30 million rubles - annual profit from sales (but even this does not allow reaching a net profit - since the previously received losses are repaid, according to the dynamics - in 3 years the first net profit will be received) [3].

According to the official website of the greenhouse complex of JSC "Sukhovskiy", the history of which is almost 34 years old (date of creation: July 27, 1987), began with the production of cucumbers and tomatoes. A modern greenhouse complex also has the opportunity to offer the consumer more than a dozen types of greens - from dill with parsley to selected types of lettuce, as well as eggplants grown in their own greenhouses (with the mastered available resources and growing technologies, these products are excluded from the product range) [3].

The main economic indicators of JSC "Sukhovskiy" are presented in table 1.

Table 1. Dynamics of the main economic indicators of JSC "Sukhovskiy" for 2017-2019

Indicator	Unit of measurement	2017 г.	2018 г.	2019 г.
Area of the complex	Ths. m^2	60		
Yield	t/m^2	0,095	0,094	0,95
Gross production	t	5700	5640	5694
Average number of employees	people	229	223	225
Average monthly salary of 1 production worker	ths. rub.	28,8	27,7	28,2
Revenue	ths. rub.	439204	383151	374 796
Cost price	ths. rub.	391904	332604	335 937
Gross profit from sales	ths. rub.	47300	50547	38 859
Product profitability	%	12	15,2	11,6
Profit margin	%	2,5	4,2	1,4

In modern conditions of the development of greenhouse vegetable growing, detailed methodological recommendations have been developed that help to choose which products to grow, both in terms of yield and in terms of economic feasibility for sale at one time or another of the year in conditions of greenhouse farms for areas that have their own specific features for conducting agricultural activities [10; 11]. The correct selection of high-yielding and unpretentious varieties serves as the basis for the rational management of greenhouse vegetable growing, and also contributes to maintaining a competitive position in the current market environment.

In the conditions of modernization of the existing equipment of the greenhouse economy, active work is being carried out to improve agricultural technologies [12; 17]. For example, practical approaches to pest control are being developed aimed at increasing the yield of crops grown, incl. due to the quality of the produced product range [16].

In modern conditions, hybrids of domestic and foreign selection are cultivated at JSC "Sukhovskiy". The species composition of the produced product range and a description of its nutritional value are reflected in table 2 [3]:

Table 2. Product range of JSC "Sukhovskiy"

Name	Nutritional value per 100g of product (average values)
Basil (green)	Proteins - 2.5g; fats - 0.5g; carbohydrates - 2g; calorie content - 25 kcal/100kJ.
Melissa	Proteins - 2.5g; fats - 0.5g; carbohydrates - 2g; calorie content - 25 kcal/100kJ.
Cucumber	Proteins - 0.5g; fats - 0.1g; carbohydrates – 3.5g; calorie content - 15 kcal/70kJ.
Spring onion	Proteins – 1.5g; fats – 0.1g; carbohydrates – 3g; calorie content – 20 kcal/80kJ.
Sorrel	Proteins – 1.5g; fats – 0.3g; carbohydrates – 3g; calorie content – 20 kcal/85kJ.
Salanova salad	Proteins – 1.5g; fats – 0.2g; carbohydrates – 2g; calorie content – 15 kcal/65kJ.
Arugula salad	Proteins – 2.5g; fats – 0.5g; carbohydrates – 2g; calorie content – 25 kcal/100kJ.
Cilantro	Proteins – 2g; fats – 0.5g; carbohydrates – 3.5g; calorie content – 25 kcal/110kJ.
Dill	Proteins – 2.5g; fats – 0.5g; carbohydrates – 6.5g; calorie content – 40 kcal/170kJ.
Celery	Proteins – 1g; fats – 0.1g; carbohydrates – 2g; calorie content – 15 kcal/55kJ.

Table 2. Continued

Romaine lettuce	Proteins – 1g; fats – 0.3g; carbohydrates – 3.5g; calorie content – 20 kcal/90kJ.
Oak leaf salad	Proteins – 1.5g; fats – 0.3g; carbohydrates – 1.5g; calorie content – 15 kcal/60kJ.
Parsley	Proteins – 4g; fats – 0.4g; carbohydrates – 7.5g; calorie content – 50 kcal/210kJ.
Mint	Proteins – 4g; fats – 1g; carbohydrates – 7g; calorie content – 70 kcal/290kJ.
Purple Basil	Proteins – 3g; fats – 0.6g; carbohydrates – 2.5g; calorie content – 30 kcal/125kJ.
Frillice salad	Proteins – 1.5g; fats – 0.2g; carbohydrates – 2g; calorie content – 15 kcal/70kJ.
Afizion salad	Proteins – 1.5g; fats – 0.2g; carbohydrates – 2g; calorie content – 15 kcal/65kJ.
Tomato	Proteins – 0.7g; fats – 0.3g; carbohydrates – 4.3g; calorie content – 20 kcal/100kJ.

Product shelf life: 10 days at temperatures from 3 °C to 6 °C and relative humidity from 90% to 95%. For a fresh cucumber: 15 days at a temperature of 10 °C to 14 °C and a relative humidity of 85% to 90%.

Also JSC "Sukhovskiy" grows a seasonal product - tulips. Seasonally (spring) two varieties have been introduced into the product range: Alcatraz (cooling period: 16 weeks, Triumph group, red color, plant height 40 cm, flowering period IV-V month), Bombita (cooling period: 16 weeks, Triumph group, color red with yellow, plant height 45 cm, flowering period IV-V month, forcing time 20 days).

The product range of the greenhouse complex is only 60% sold within the Kemerovo Region-Kuzbass, and 40% is sold to other regions of the Siberian Federal District.

Gross production of manufactured products at JSC "Sukhovskiy" is distributed in the following proportion: cucumbers 82%; tomatoes 15%; lettuce commodity group and other grown crops 3%.

4 Results and discussion

Assortment management of an agricultural enterprise implies the development of a product assortment plan [13]. In this connection, it seems relevant to develop an event aimed at expanding the range of products manufactured by JSC "Sukhovskiy".

Scientific search for various information and research of market analysts indicate that in order to improve the strategy of managing the product range, the most popular and possible direction for expanding commodity positions is the cultivation of flowers, which will significantly expand and diversify the production capabilities of the greenhouse industry.

For Siberia regions, where due to natural and climatic conditions there is an inherent peculiarity of agricultural production (including a low temperature regime in the winter season and a lack of sunlight), the greenhouse business is a rather promising direction of activity.

Expenses on gas and light take a significant share in the cost of flower production, about 60%. Other expenses are overhead costs, salaries, transportation.

To cut costs and be sure of making a profit, you need to be clear about which flowers to grow. It is also necessary to take into account the seasonal demand for this product (Fig. 3).

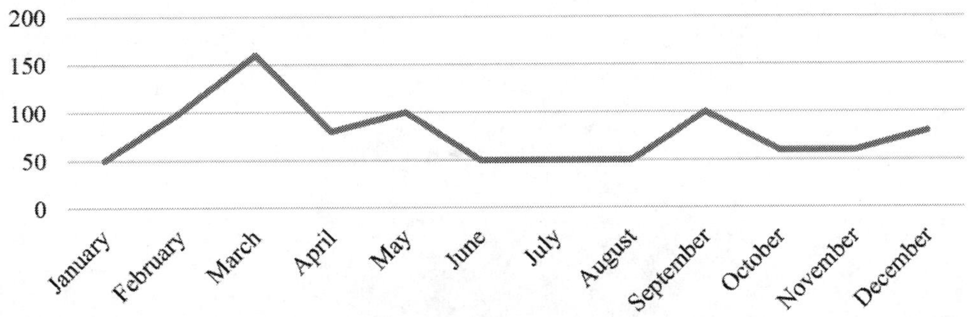

Fig. 3. Annual dynamics of seasonal demand for flowers

To expand the range of products manufactured by JSC "Sukhovskiy", it is proposed to grow roses. Since according to statistics, roses are the most popular and demanded flowers for bouquets and are always in high demand among the population.

Growing roses in greenhouses includes: *preparing soil for the crop* (humus (20kg), superphosphate (30g) and ammonium sulfate (30g), peat (no more than 10kg) per 1m^2 are added to the soil; the soil can also consist of mullein, peat, peat (1:5:4) with a soil thickness of 80cm; before use, the mixture is kept in a pile for about a year; before planting flowers, meat and bone meal (1kg per 1m^2) and mineral fertilizers are added). *Preparation of seedlings for planting* (immersion in water for 24 hours; dropping into the ground, leaving the grafting site in the air; correct and regular moistening (at the level of 70%) and spraying with a remedy containing copper from the development of diseases). *Planting roses* (scheme: 30*30 or 40*25, or there are no more than 12 units per 1m^2; good spillage; temperature control - should not exceed 10^{0}C before forcing, the optimal temperature to maintain the growth of roses is about 22^{0}C, and soil - not less than 12^{0}C).

Rhythm of the production process of growing flowers in greenhouse farms depends on a rationally organized process of selling products, its planning and dynamism through organized sales channels, both to local markets and to global ones [18]. Proposed sales channels for finished products from the greenhouse complex of JSC "Sukhovskiy": opening of its own flower shop, point of sale; making contracts with intermediaries; building contractual relationships with firms participating in the local flower market; sale through trade and retail networks; implementation via the Internet, "promoting" through popular social networks.

Forecasted cost of roses for the varieties *Ksenia* and *Proud* proposed for introduction: cost price: 50 rub. and 100 rub., selling price: 80 rub. and 120 rub., economic effect: 30 rub. and 20 rub., respectively.

The forecasted cost structure when expanding the product range of JSC "Sukhovskiy" with the introduction of roses into the "line" is shown in Figure 4.

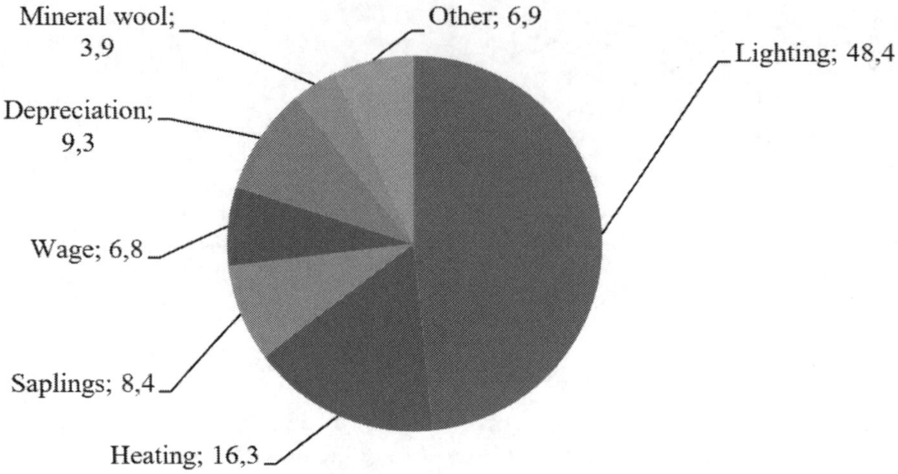

Fig. 4. Forecasted cost structure for the assortment line "roses", %

Based on the forecasted cost structure for the "rose" assortment presented in Figure 4, it was determined that the cost element "lighting" (i.e. energy resources) accounts for about 48%, three times lower, but no less significant for the "heating" element – 16.3%. To a lesser extent, costs are represented by the item "mineral wool" - 3.9%.

Figure 5 shows a graphical display of economic prospects for JSC "Sukhovskiy" when expanding the product range.

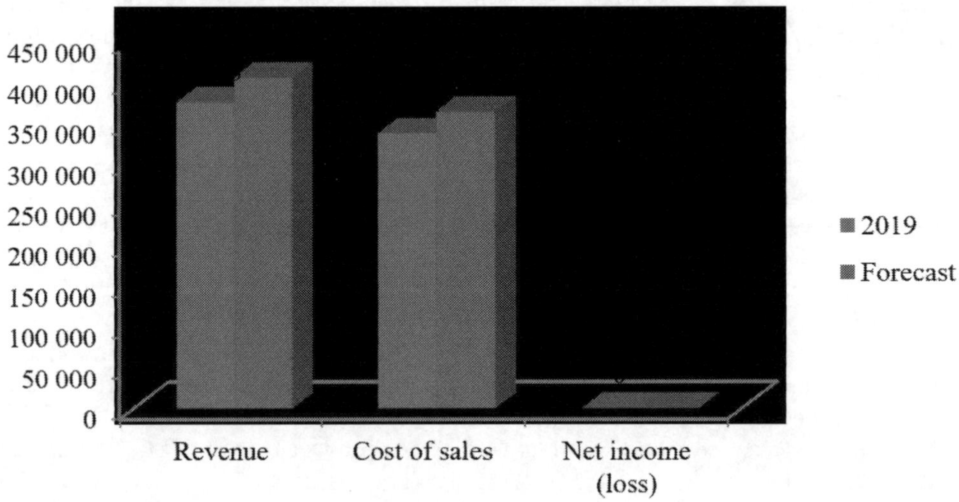

Fig. 5. Forecast economic effect of expanding the product range manufactured at JSC "Sukhovskiy", thousand rubles

5 Conclusion

Thus, the adoption of the proposed management decisions in the greenhouse complex of JSC "Sukhovskiy" to expand the range of products due to the new product "rose" will give opportunities to: 1) diversify production; 2) get a significant economic effect; 3) enter new sales markets; 4) increase the awareness of the brand *Em Native*; provide a trade assortment with a new type of local production (local analogs of an industrial scale are absent in the regional market); receive additional profit in the amount of 168 thousand rubles per year.

References

1. N.S. Bondarev, G.S. Bondareva, E.E. Khazieva, Modern S&T Equipments and Problems in Agriculture, **1-10** (2020)
2. H.O. Gaziev, Economics, **2(40)** (2019)
3. EmRodnoe: Official site of the combined agricultural complexes "Yemelyanovskiy" and "Sukhovskiy", http://emrodnoe.ru/company
4. E.L. Zolotareva, K.M. Konorev, Bulletin of the Kursk State Agricultural Academy, 4 (2010)
5. T.L. Koikova, Yu.A. Kopytina, E.L. Ralnikova, Concept, 10 (2017)
6. O.S. Bedskikh, *Comprehensive assortment management in the system of strategic planning of the activity of industrial enterprise structures*, 23 (2008)
7. L.V. Orlenko, *Assortment, merchandizing and examination of fur and fur products*, 272 (2019)
8. L.R. Bilalova, F.N. Mukhametgaliev, Economy Vector, **4(22)**, 67 (2018)
9. E. Appolloni, F. Orsini, N. Michelon, A. Pistillo, I. Paucek, G. Pennisi, G. Bazzocchi, G. Gianquinto, International Society for Horticultural Science, **1298**, 59-70 (2020)
10. W.A.P. Weerakkody, S.M.M.R. Mawalagedera, *Recent developments in vegetable production technologies in sri lanka* (2020)
11. A. Dudnyk, M. Hachkovska, N. Zaiets, T. Lendiel, I. Yakymenko, Technology Cente, **4(2-100)**, 72-78 (2019)
12. M. Eben-Chaime, A. Bechar, A. Baron, International Journal of Production Economics, **134(1)**, 246-254 (2011)
13. J.D. Hoskins, Journal of Retailing and Consumer Services, **57**, 102234 (2020)
14. A. Timonina-Farkas, A. Katsifou, R.W. Seifert, European Journal of Operational Research, **285(3)**, 1058-1076 (2020)
15. Y. Alan, M. Kurtuluş, C. Wang, Manufacturing and Service Operations Management, **21(3)**, 620-635 (2019)
16. J. Del Castillo Múnera, B. Belayneh, A. Ritsvey, E.E. Koivunen, J. Lea-Cox, C.L. Swett, Agricultural Water Management, **226**, 105737 (2019)
17. P. Diaz, R. Carrera, IEEE CHILEAN Conference on Electrical, Electronics Engineering, Information and Communication Technologies, , 8988049 (2019)
18. B.F. Cardoso, E. Giampietri, A. Finco, P.F.A. Shikida, Agris On-line Papers in Economics and Informatics, **9(2)**, 23-32 (2017)

Research of the factor dependence of egg and egg products consumption in the subjects of the Russian Federation

Galina Sergeevna Bondareva, and *Nikolay Sergeevich* Bondarev

FSBEI HE Kuzbass State Agricultural Academy, 650056, st. Markovtseva, 5, Kemerovo, Russia

Abstract. Importance of consummating eggs and egg products in the diet of the population is determined by their high nutritional value. An egg contains a complete animal protein, it has consumer accessibility (both economic and physical) and prevalence in food markets regardless of the territory. The widespread availability (both assortment and price) of eggs and egg products to some extent can compensate for the deficit in consumption of other types of livestock products, which are often not available for socially vulnerable groups of the population with a low level of income. This food group is one of the most valuable products in the structure of rational consumption that meets the modern requirements of a healthy diet, which is reflected in the relevant Recommendations of the Ministry of Health of the Russian Federation [1]. The conducted research is based on the analysis of the main parameters of eggs and egg products' consumption. The emphasis is made on the study of statistical information on the number of the living population; the prevailing average per capita money income; consumer average retail prices for the specified food group; the level of consumption, its compliance with a rational level that meets modern requirements for healthy nutrition in equal subjects of the Russian Federation, both in the region, territory, republic, autonomous okrug, autonomous region, and in cities of federal significance: Moscow, St. Petersburg, Sevastopol. The result of the study was confirmation of the direct dependence of the reduction in the consumption of eggs and egg products as a result of a decrease in the population's income. The territorial aspect also indicated the existence of such dependence. All this allows us to assert that the fundamental factor affecting the level of consumption of eggs and egg products is the economic affordability of the food group, which depends on the income of buyers [11; 13].

1 Introduction

In order to ensure the economic and physical accessibility of food, food security in the territories of the Russian Federation, it is definite that the price of an egg and other types of basic food products should vary depending on the average per capita monetary income of the population, its purchasing power. All this proceeds from a certain axiomatic relationship: the higher the income, the higher the prices for food, and the inverse relationship: the lower the income, the lower the price of food. Otherwise, there will be a territorial disparity in food supply.

The value of eggs and egg products in the diet of the population is explained due to the presence of balanced nutrients in them, in their properties, equal to milk and beef. So the energy value of 100g of egg mass averages 157 kcal, which is comparable to 72g of boneless beef, 44g of pork or 350g of milk [2; 10]. However, it should be noted that the costs of purchasing the specified set of products are not comparable and may exceed more than twice.

According to Russian Statistics Committee, for 2020 the average retail price of one egg in Russia was 6.38 rubles, while the amount of milk equivalent in energy was 23.4 rubles, boneless beef - 38.61 rubles, pork - 29.8 rubles [4]. In this connection, milk and products of its processing, meat and products of its processing, eggs, bread and bakery products, vegetables and fruits can be attributed to the main foodstuffs supporting the standard of living [3, p. 3]. The presence of such significance determines the relevance of the study.

2 Basic Methods

The research approach is implemented on the basis of statistical material of the constituent entities of the Russian Federation with the highest and lowest average per capita money incomes of the population in terms of the achieved values in terms of the following parameters: the average annual population, the level of per capita money income, the current level of average retail prices for eggs, the level of consumption of eggs and egg products, of the population with monetary incomes below the established value of the subsistence minimum, the level of consumption to the established rational norms of food consumption that meets modern requirements for a healthy diet. To confirm the importance of applying this approach, the authors methodically used the allocation of territories with significant opportunities for economic accessibility of food and, accordingly, territories where an inversely proportional situation is actually developing: in addition to the prevailing low average per capita money income, a high proportion of the population with money incomes below the subsistence minimum was noted.

As additional research methods in the presented article, the following are applied: economic and statistical, observation, graphic, synthesis, analysis methods. The empirical basis of the study was statistical information presented on the official website of the Federal State Statistics Service.

3 Data and Primary Analysis

From the standpoint of the designated study, the dynamics of the average annual number of population and the level of their incomes for equal subjects of Russia have been analyzed (Figure 1; 2) [4].

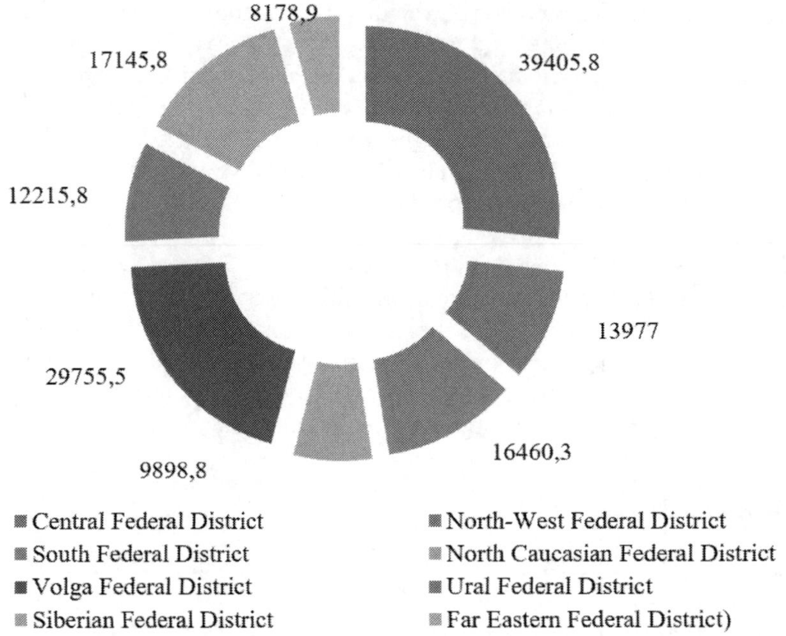

Fig. 1. Distribution of the average annual population of the Russian Federation by federal districts according to data for 2019.

Analyzing the average annual population in the context of federal districts, it was determined that the largest share of residents is noted in the Central Federal District (26.8%) and the Volga Federal District (20.3%), while the Far Eastern Federal District has the smallest number of population in the presence of one. an increase of 8,178.9 million people in 2019 (5.5% of the average annual population of Russia), against 6,174 million people in 2017.

The average per capita monetary income of the population, grouped by size (Figure 2), indicates that the most frequent distribution of the number series is concentrated within the amounts of 25,000-30,000 rubles; the peak of the distribution is 25,000 rubles, which is typical for 32 equal subjects of the Russian Federation out of 85.

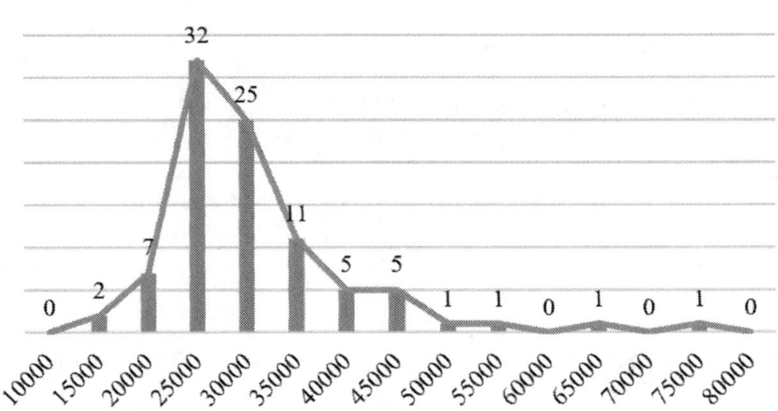

Fig. 2. Distribution of equal subjects of the Russian Federation by the level of per capita monetary income, units (01.01.2021).

At the same time, the territories with the highest level of per capita money income according to the data for 2020 include the subjects of the Central, Northwestern and Far Eastern federal districts with the average value of the indicator in the amount of 47,224 rubles, 38,903 rubles and 38,550 rubles, respectively.

From the standpoint of assessing the economic accessibility of basic types of food for equal subjects of the Russian Federation based on the level of per capita money income, namely, the republics of Ingushetia and Tyva, within the established range of no more than 18,000 rubles, it is necessary to take into account the ratio of the prevailing price level for the specified food group, i.e. to study the consumer market for the group of food products "eggs and egg products".

4 Results and discussion

As the general features of the functioning of the consumer market by groups of food products, it is customary to single out the features inherent in the territories (Figure 3) [5, p. 67-68].

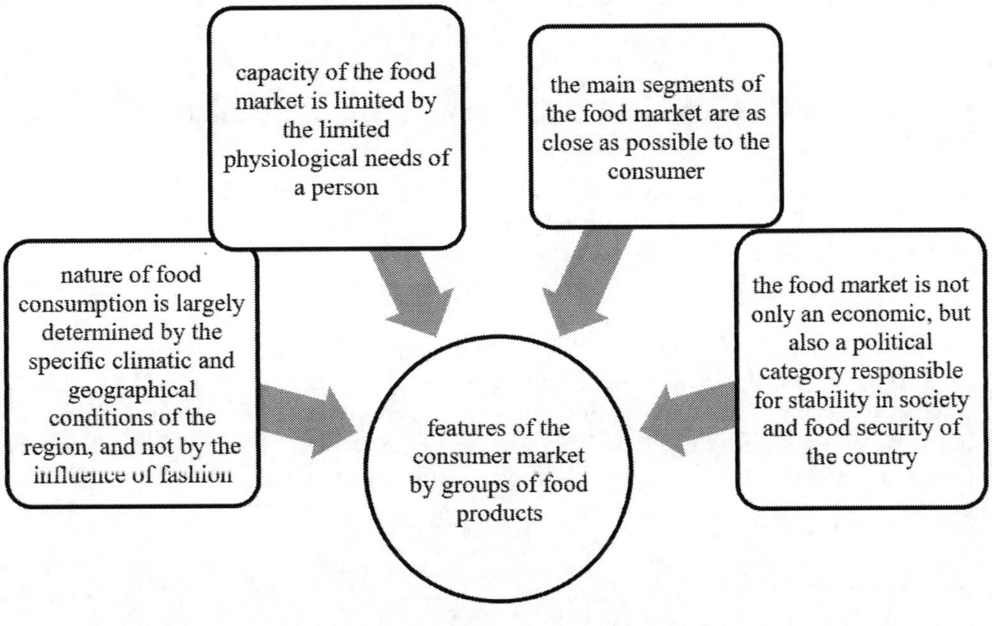

Fig. 3. Functioning features of consumer market by groups of food products.

Taking into account the specifics of the organization of activities in the production of eggs and poultry meat, it is necessary to focus on the importance of the natural and economic conditions of management. Often, the priority is precisely the territorial localization of production [6; 8]. The presence of a constant demand for this food group, year-round production capabilities with an organized sales market allow enterprises in this industry to almost completely eliminate the seasonality factor [7; 14].

Based on a statistical study, the subjects of the federation were identified with the lowest average per capita income (the Republic of Ingushetia (17,105 rubles) and the Republic of Tyva (17,582 rubles)) and with the highest level of average per capita monetary income (Chukotka Autonomous Okrug (89,059 rubles) and the Yamalo-Nenets Autonomous Okrug (RUB 88 579)). In other subjects of the federation with an average level of per capita money income, the values may vary, which makes it difficult to identify a trend. The median price level was determined as a reference indicator [9].

Analyzing the subjects of the federation designated as the object of research (with the lowest and highest level of per capita money income), it was determined that in the republics of Ingushetia and Tyva, the average retail prices for the food group "egg" amounted to 59.95 rubles and 63.56 rubles, in the Chukotka Autonomous Okrug and the Yamalo-Nenets Autonomous Okrug - 154.97 and 75.22 rubles, respectively (Figure 4).

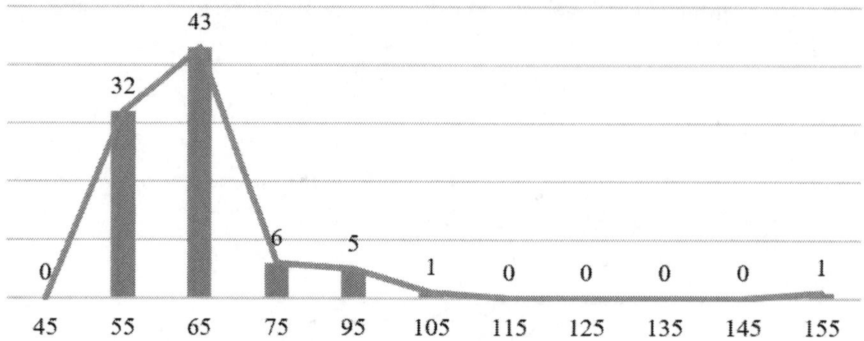

Fig. 4. Distribution of equal subjects of the Russian Federation based on the current level of average retail prices for an egg (as of 01/01/2020).

The existing price level on the food market for eggs, as a commodity group, indicates the establishment of the average retail price in January 2020 at 65.44 rubles per 10 pieces, with a minimum value of 46.37 rubles that was noted in the Republic of Mordovia.

Carrying out a similar study on the level of consumption of eggs by the population, the following data were obtained (Figure 5). The main groups are identified by the level of consumption: the population of 16 subjects of the Russian Federation consumes eggs at the level of 210 pieces, the population of 32 subjects - 230 pieces, the population of 18 subjects - 250 pieces per capita. The consumption rate of eggs, based on the recommended rational standards, is 260 eggs per year, i.e. in 71 constituent entities of the Russian Federation, the population under-consumes the specified food group in accordance with the established border.

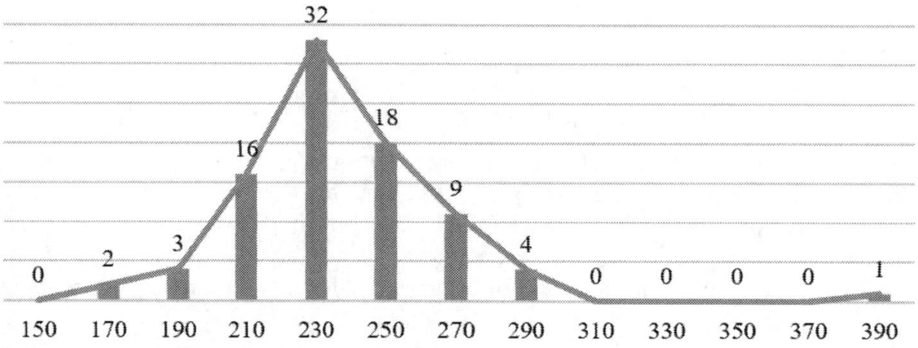

Fig. 5. Distribution of equal subjects of the Russian Federation by the level of egg and egg product consumption (as of 01/01/2020).

The population of the republics of Ingushetia and Tyva consume 199 and 172 pieces per a household member, respectively. In turn, in the Chukotka Autonomous Okrug and the Yamalo-Nenets Autonomous Okrug, the population consumes 184 and 294 pieces, respectively; whereas the median consumption of eggs is at the level of 230 eggs.

As an indicator that strengthens the presented study, the proportion of the population with monetary incomes below the established value of the subsistence minimum in the respective regions is taken into account. So in the Republic of Tyva, the subsistence level is determined by the border of 8,846 rubles, in the Republic of Ingushetia - 10,571 rubles, while in the subjects with the highest level of income as the Chukotka Autonomous Okrug and the Yamalo-Nenets Autonomous Okrug - 24,875 rubles and 16640 rubles per capita. Regarding the proportion of the population with monetary incomes below the established value of the subsistence minimum, the most problematic situation was again noted in the Republic of Tyva and the Republic of Ingushetia, where the proportion of this population is critically high - 34.7% and 30.5%, respectively. The statistics for the Chukotka Autonomous Okrug and the Yamalo-Nenets Autonomous Okrug are as follows: 8.7% and 5.6%, respectively, with an average Russian value of 12.3% according to data in 2020 [4].

Low incomes of the population, a high proportion of the population with monetary incomes below the established value of the subsistence minimum (the Republic of Tyva is more than 1.3 of the total population), the underdevelopment of the consumer market for food product groups negatively affects the purchasing power of the population, the economic accessibility of basic food groups their physical availability.

Low economic opportunities of the population, the prevailing price level, its growth affect the constant decline in purchasing power [12]. In 2019-2018 alone, the population of the Republic of Tuva in the “egg and egg products” food group “lost” the opportunity to purchase more than 80 eggs per year (3048 eggs versus 2963 eggs, respectively). In the Yamalo-Nenets Autonomous Okrug, the situation is directly related - 11,735 pieces and 10192 pieces, i.e. +154.3 dozen eggs per year.

5 Conclusion

The revealed structure indicates the following statements:

- in the constituent entities of the federation with the lowest level of per capita money income, the consumption of eggs and egg products is in the under-consumption zone from the established rational norms of food consumption that meet modern requirements for a healthy diet - 260 pcs/year/person. The value of consumption in the Republic of Ingushetia was 76.5% of the rational nutritional norm; Tyva - 66.1%, respectively.
- in the constituent entities of the federation with the highest level of per capita money income, the under-consumption of eggs and egg products is observed in the Chukotka Autonomous Okrug - 76 pcs. (70.7% of the rational consumption rate); while in the Yamalo-Nenets Autonomous Okrug, an over-consumption of eggs was noted, which amounted to 34 pcs. (113.1% of the rational consumption rate).

Thus, the revealed trend confirms the axiomatic statement about the presence of a direct dependence of the consumption of eggs and egg products on economic factors: the average retail price and income of the population.

References

1. Order of the Ministry of Health and Social Development of the Russian Federation of August 19, 2016 No. 593n «On the approval of recommendations for rational norms of

food consumption that meet modern requirements for healthy eating», https://www.rosminzdrav.ru/
2. Draft program "Development of poultry farming in the Russian Federation for 2010-2012 and for the period up to 2018-2020", http://www.mcx.ru/
3. T.V. Gomelko. *Theoretical foundations of food market formation: monograph*, 104 (2007).
4. Unified interdepartmental information and statistical system (EMISS): Federal State Statistics Service, www.fedstat.ru
5. O.P. Nikiforova, I.V. Antokhonova, Bullit. Of Buryat. state univ, **2,** 60 (2010)
6. C. Currie, M. Hammon, American Planning Association, **1,** 56 (2021)
7. M.F.a. Neves, A.W.b. Gray, C.E.c. Lourenço, F.A.d. Scott, Wageningen Academic Publishers, **1,** 138(2021).
8. M. Clark, D. Tilman, Environmental Research Letters, **12 (6),** 064016 (2017)
9. A. Mishra, A. Agarwal, Inderscience Publishers, **1,** 107 (2021)
10. Y. Bai, R. Alemu, S.A. Block, D. Headey, W.A. Masters, Elsevier Ltd, **99,** 101983 (2021)
11. S.D. Rathnayaka, S. Selvanathan, E.A. Selvanathan, Blackwell Publishing Ltd, **1,** 97 (2021)
12. G. Giraud, M. Grasselli, (March 2021). Elsevier B.V., **183,** 901 (2021)
13. N. Baral, B. Paria, B. Behera, P. Mishra, Elsevier Ltd, **21,** 100296 (2021)
14. O. Kosenchuk, O. Shumakova, A. Zinich, S. Shelkovnikov, A. Poltarykhin, Colloids and Surfaces A: Physicochemical and Engineering Aspects, **10(5),** 991 (2019)

On the question of the use of inter-settlement territories in the structure of urban agglomerations

Valerija Vylegzhanina[1,2,*], Ilgiz Giniyatov[2], Nadezhda Dobrotvorskaja[2,3], Viktor Timonov[4]

[1]Federal Service of State Registration, Cadastre and Cartography for Novosibirsk region, 630091, Derzhavina Str., 28, Novosibirsk, Russia
[2]Siberian State University of Geosystems and Technologies, Plakhotnogo Str., 10, Novosibirsk, Russia
[3]Siberian Federal Scientific Center of Agrobiotechnologies RAS, Novosibirsk Region, r. p. Krasnoobsk, 630501, Russia
[4]Novosibirsk State University of Architecture, Design and Arts named after A. D. Kryachkov, 630099, Krasny Prospekt, 38, Novosibirsk, Russia

Abstract. The article discusses the reasons that contribute to the formation of agglomerations on the territory of the Novosibirsk region, overpopulation of urban areas that make up the agglomerations, as well as the reasons for the decrease in rural inhabited space on the territory of municipalities located within the boundaries of the regions of the region, both within the agglomeration itself and on the territory districts of the region located outside the agglomeration. The results of the influence of urban agglomerations on the state of agricultural lands located within and outside the specified agglomerations are considered. An analogy is drawn between these processes in relation to agricultural lands with lands of the named category located on the territory of the Russian Federation. A possible solution to these problems is proposed, providing for the development of conceptual approaches, including the introduction of phased structural and investment measures.

1 Introduction

More than 20 years have passed since the beginning of the land reform in the territory of the previously existing Soviet state, the purpose of which was development of various forms of land management, the formation of a multi-structured economy by abolishing the state monopoly on land [1].

Since formation of the Russian Federation, land reform has been gaining momentum. A number of laws were adopted that allow citizens, within the framework of various organizational and legal associations, to carry out agricultural activities, which is based on the principle of preserving the intended use of land plots [2, 3].

The security function of the state is based on the regulation of land relations within the framework of the law, and supervisory measures are also carried out to correct these legal

* Corresponding author: pravo@uy.nsk.su

relations [4]. Wherein, the current situation picture in the field of rational use of agricultural land is not in favor of the designated category of land.

Over the years, the emerging difficult socio-economic situation in the country has led to the depopulation of rural areas, the outflow of the able-bodied population to cities and over-concentration of the population in large cities and adjacent settlements [5]. This process has led to formation of urban agglomerations, concentrating the scientific, cultural potential, personnel reserve, an overabundance of the working population [6]. The reverse side of this process is, respectively, the desertification of rural areas accompanied by degradation of rural settlements and, as a consequence, their abolition. Accumulation of agricultural production near more or less "live" settlements and near urban agglomerations is accompanied by decrease in acreage and decrease in the quality of agricultural land.

The purpose of this article is to analyze modern socio-economic phenomena in inter-settlement territories using the example of the Novosibirsk region. The objectives of the study were to study the influence of urban agglomerations on the state of agricultural land and rural settlements located in the territory of the Novosibirsk agglomeration.

2 Materials and Methods

The object of study is agricultural land located within the boundaries of the Toguchinsky district of the Novosibirsk region included in the Novosibirsk agglomeration. Within preparation and presentation of the study results, the data of the public cadastral map posted on the Federal Service of State Registration, Land Register and Mapping website; Earth remote sensing materials; official data published on the websites of Novosibirskstat, the Government of the Novosibirsk Region, the Ministry of Agriculture of the Novosibirsk Region; on the official website of the administration of the Toguchinsky district of the Novosibirsk region, the data obtained using the Google maps program, were used.

3 Results and Discussion

The Novosibirsk Region (NSO) is no exception to the general socio-economic picture summarized in the introduction of this article in relation to the entire Russian Federation.

Historically, the bulk of settlements in the territory of the Novosibirsk Region arose on the place of fortifications – military fortifications erected to protect immigrants from the European part of Russia in the 17th century during the development of Siberia. The main occupations of the population were arable farming, fishing, hunting and carriage. Favorable climate, large areas with fertile lands have led to the dominance of the agricultural industry throughout all time periods of development of the territory [7]. The rapid, poorly prepared implementation of the land reform of the 90s of the last century led to an unfavorable socio-economic situation in the agriculture of the country and the Novosibirsk region in particular, caused an outflow of the able-bodied population to large regional centers and cities of the Novosibirsk region, which in turn led to decrease in agricultural activity, increase in the area of uncultivated farmland and, as a result, degradation of agricultural land.

According to a number of decrees of the President of the Russian Federation, within the framework of regional legislative initiatives in order to stabilize the situation described above, the Government of the Novosibirsk Region has developed a strategy for the socio-economic development of the Novosibirsk Region for the period up to 2030 (Strategy) [8].

The main direction of the Strategy is formation of urban agglomerations and territories of advanced socio-economic development (TASED) in NSOs. The largest urban agglomeration is Novosibirsk one, which includes 7 districts of the region, 4 cities and 1 village. The Novosibirsk agglomeration is formed on the basis of the principle of the spatial organization of the region: the city of Novosibirsk and the cities of the region located near it, which are most susceptible to its influence [8]. In the territory of the agglomeration, two TASED are formed – this is industrial community Linevo of Iskitimsky district and industrial community Gorny of Toguchinsky district, which are monocities, within the boundaries of which it is planned to create new city-forming enterprises with an increase in the number of jobs [8].

The object of this study is agricultural land within the Toguchinsky district of the Novosibirsk region, located between the village of Lnikha and the village of Borovushka, which are part of the Repyevsky village council of the Toguchinsky district of the NSO. The village of Lnikha and the village of Borovushka were previously part of the large collective farm "60 years of October", which was engaged in production of agricultural products and was reorganized in 1993 into Lnikha association of peasant farms (Lnikha APF) [9]. As a result of the transformations, the peasants of the Lnikha APF became the owners of agricultural land with cadastral number 54:24: 035801:79 with a declared area of 27,711,792 sq.m. [10] on the basis of common shared ownership.

It shall be noted that the document — the basis for assigning the land of this category to the peasants was a state act certifying the ownership of land, life-long inherited possession, perpetual (permanent) use of land by citizens, enterprises, institutions, organizations or peasant (farmer) farms drawn up in accordance with Decree of the Council of Ministers of the RSFSR dated September 17, 1991 No. 493 [11]. As a rule, these are previously registered land plots, the registration of which was carried out before March 1, 2008, information about their boundaries is absent in the Unified State Register of Real Estate (USRN), the area is declared, that is, approximate [12]. The location of these agricultural lands is shown in Figure 1.

Fig. 1. Agricultural Lands of Lnikha APF.

If we refer to the public cadastral map of the Federal Service of State Registration, Land Register and Mapping [13], multiple inclusions in the considered agricultural lands of the forest fund lands are visually displayed. According to the soil map of the Novosibirsk region, these lands belong to the north-forest-steppe zone of the right bank of the Ob River. In geomorphological terms, the territory is an erosion-denudation upland, which is characterized by cone-shaped hills up to 390 m above sea level, steep deeply incised river valleys, slopes of more than 5°. From an agroecological point of view, these are highly erosional lands with a soil cover that includes dark gray and gray forest soils, podzolized chernozems, and meadow soils [14]. Specific features of natural conditions (rugged terrain, forested area) determine the rapid overgrowth of production fields and work sites with weeds and trees and shrubs in case of non-use of land for its intended purpose as a result of lack of agricultural activities [15, 16]. This process can be clearly traced on satellite images provided in the Google Maps program (Fig. 2).

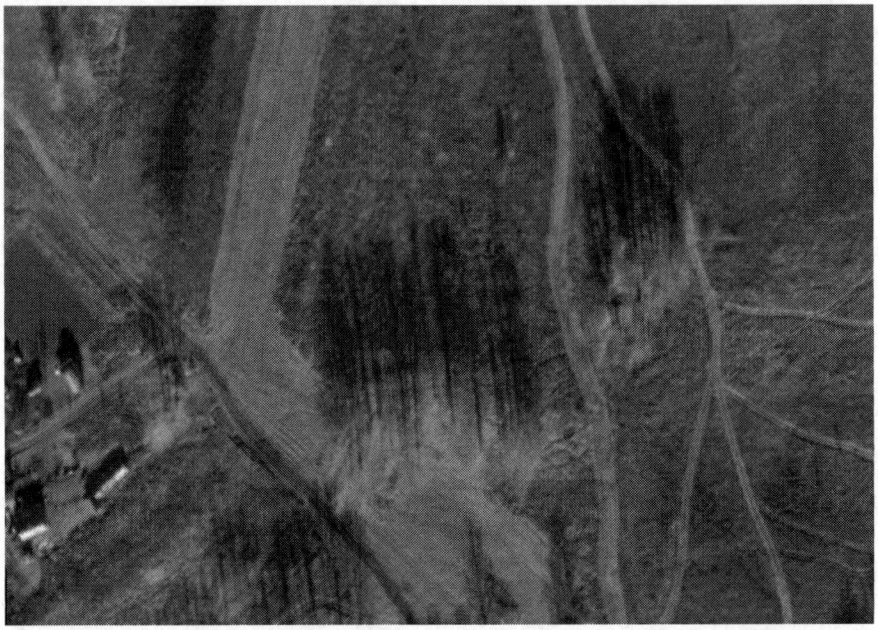

Fig. 2. Overgrowth of fields near village Borovushka, Toguchinsky district, Novosibirsk region (images of 2021, Google maps program, scale 1: 2500).

To identify the reasons for non-use of land, an analysis of the data of the All-Russian Population Census in different years was carried out (Table 1) [17, 18].

Table 1. Change in population in small settlements of Toguchinsky district of Novosibirsk region.

District	Name of the settlement	2002	2007	2010	2020
Toguchinsky	Borovushka	60	43	21	N/A
	Lnikha	546	542	589	N/A

The NSO government, within the framework of a comprehensive program for the socio-economic development of the Toguchinsky district in 2011-2015 and for the period until 2025, allocated funds for improvement and development of engineering infrastructure: the construction of a water supply system and the construction of antenna-mast structures for cellular communications in Borovushka and Lnikha, renovation of the house of culture and the Linikha library. In 2025, it is planned to perform reconstruction from the 4th to the 2nd category of the main road on the Bugotak - Lnikha - Repyevo section and to strengthen the railway tracks in order to establish protection against landslides [19]. Within the framework of the long-term target program "Development of gasification of the territory of settlements in the Novosibirsk region for 2012-2016", gasification schemes were developed for Lnikha village, Bugotak station, Bugotak village, and design estimates for gasification of Gorny settlement, Bugotak village, Bugotak station, Lnikha village in the amount of 25 million rubles. Wherein, there are no offers for investment activities in terms of revitalizing these settlements in the comprehensive program of socio-economic development of the Toguchinsky region in 2011 - 2015 and for the period until 2025 [19]. Currently, there is a decrease in the number of students in rural schools, although Lnikha village is considered a remote village in which educational activities were previously carried out [20]. According to the electronic websites of real estate agencies, summer residents are settling in the villages of Borovushka and Lnikha, buying up village houses for recreation and summer seasonal residence [21, 22]. Therefore, the village of Borovushka practically exists as a dacha society. In the village of Lnikha, rural life is still glimmering.

Now let's refer to more general official statistics characterizing the above quantitative components in the Toguchinsky district of the Novosibirsk region. According to the information of the administration of the Toguchinsky district of the Novosibirsk region, at present, in the territory of the Toguchinsky district of the NSO, the number of settlements of the district is 107 (villages, settlements, railway sidings). By 2002, the villages of Vyemka, Kazansky, Kazarma 132 km and the village of Agafonikha were abolished [9]. At the beginning of 2017, over 1.5 thousand settlements were registered throughout the Novosibirsk region. 56 of them are registered, but do not have a population. Since the 2010 census, more than 30 settlements have disappeared from the map of the Novosibirsk region as a result of the abolition [23].

The dynamics of the population in the regions under consideration in relation to the population of the city of Novosibirsk in the period from 2002 to 2020 [24] is reflected in Table 2.

Table 2. Change in the population in the Toguchinsky district of the NSO, as well as in the city of Novosibirsk in the period from 2002 to 2020.

Periods	Novosibirsk	Toguchinsky district
2015	1,425,508	65,515
2015	1,621,835	55,843
2015	1,625,631	55,427

These data characterize the growth and concentration of the population in Novosibirsk, at the same time, there is a constant trend of population decline in the considered districts of the region. In general, the total number of the population in the Novosibirsk region is constantly increasing. So in 2002 the population of the NSO was 2,692,251 people, in 2007 — 2,647,195 people, in 2019 — 2,795,777 people, in 2020 — 2,798,170 people. [24]. Therefore, with a constant increase in the total population in the territory of the Novosibirsk region, there is a process of migration of the rural population to cities and regional centers. These trends indicate an outflow of labor resources and a reduction in the habitable space in the territory of the region's districts. These circumstances contribute to reduction of the cultivated areas of agricultural crops, leading to degradation of the remaining uncultivated agricultural lands. According to the results of the All-Russian Agricultural Census of 2016 for the period from 2007 to 2019, the number of sown areas of agricultural crops is steadily decreasing in the territory of the NSO [25]. However, in 2020, the quantitative indicator of sown areas slightly increased compared to the previous year [26] (Table 3).

Table 3. Comparative table of the number of sown areas of agricultural crops by NSO for the period from 2007 to 2020.

Sown area	2007	2010	2014	2017	2018	2019	2020
Ths ha	2,419.0	2,320.1	2,376.4	2,380.0	2,226.6	2,225.7	2,272.0

Therefore, there is a trend of constant population growth in the Novosibirsk region: its number in 2020 compared to 2007 increased by 5.4 %. Wherein, the number of agricultural crops sown in the Novosibirsk region in 2020 decreased by 6.1 % compared to 2007. This is 147 thousand hectares of uncultivated agricultural land, in respect of which the process of degradation has begun.

4 Conclusions

When summing up the study, we believe that it is possible to draw the following conclusions.

Formation of the Novosibirsk agglomeration in the territory of the Novosibirsk region is due to the created socio-economic situation, which entailed the migration of the rural population to regional centers and cities located near the city of Novosibirsk, which are most susceptible to its influence. These circumstances contribute to the process of overpopulation of these cities located in the territory of the Novosibirsk agglomeration and, in turn, to decrease in the rural inhabited space in the territory of municipalities located within the boundaries of the region's districts within the Novosibirsk agglomeration itself. Along with these processes, there is a decrease in the number of sown areas of agricultural crops on agricultural lands located in the territory of the Novosibirsk agglomeration, as well as in the territory of the entire Novosibirsk region. These circumstances contribute to the degradation of uncultivated agricultural land. It shall be noted that the described processes and the obtained study results do not differ from the general picture emerging in relation to agricultural land located throughout the territory of the Russian Federation.

Elimination of these negative processes accompanying the creation of agglomerations, in our opinion, is possible when developing a conceptual direction within the framework of a state scientifically grounded food security program, for example, "Agrarian Russia", the purpose of which could be phased structural and investment measures aimed at development of domestic agricultural production, restoration of suburban farmland and consolidation of the population in the inhabited territories. Any form of economic activity is acceptable here — from agricultural holdings and farms, up to individual private farmsteads with elements of agritourism. Wherein, reducing the consequences of sanctions pressure at the international level and eliminating trade barriers will increase the export of agricultural products abroad, which will also serve as a powerful incentive for development of this industry.

References

1. Law of the RSFSR dated November 23, 1990 No. 374-1 "On Land Reform".
2. Federal Law dated June 11, 2003 No. 74-FZ "On the Peasant (Farm) Economy".
3. Federal Law dated July 24, 2002 No. 101-FZ On the Turnover of Agricultural Land".
4. "Land Code of the Russian Federation" dated October 25, 2001 No. 136-FZ.
5. T. G. Nefedova, N.V. Mkrtchyan, Moscow University Bulletin, Series 5, Geography, **5**, 58 (2017)
6. D. N. Baranov, *Modern trends in development of the economy of urban agglomerations: monograph,* 188 (2015) http://www.iprbookshop.ru/
7. P. A. Slovtsov, History of Siberia. From Ermak to Catherine II M.: Veche, 512 (2006)
8. Decree of the Government of the Novosibirsk Region dated March 19, 2019 No. 105-p "On the Strategy of Social and Economic Development of the Novosibirsk Region for the Period until 2030" https://www.nso.ru/
9. Wikipedia, https://ru.wikipedia.org.
10. Official website of the Federal Service of State Registration, Land Register and Mapping. Public cadastral map https://pkk.rosreestr.ru
11. Decree of the Council of Ministers of the RSFSR dated September 17, 1991 No. 493 "On the Approval of the Forms of the State Act on the Ownership of lAnd, Life-Long Inherited Possession, Unlimited (Permanent) Use of Land"
12. Official website of the Federal Service of State Registration, Land Register and Mapping, https://rosreestr.gov.ru
13. Official website of the Federal Service of State Registration, Land Register and Mapping. Public cadastral map, https://pkk.rosreestr.ru

14. Soil map of the Novosibirsk region. Scale 1:400000, 1987.
15. Decree of the Government of the Russian Federation dated September 18, 2020 No. 1482 "On Signs of Non-Use of Agricultural Land Plots for their Intended Purpose or Use in Violation of the Legislation of the Russian Federation."
16. N.I. Dobrotvorskaya, Bulletin of Novosibirsk State Agrarian University, **1(50),** 1 (2019)
17. All-Russian population census 2010. The population of urban and rural settlements of the Novosibirsk region.
18. All-Russian population census of 2002. Population by rural settlements of the Novosibirsk region.
19. Comprehensive program of socio-economic development of Toguchinsky region for 2011-2015 and for the period up to 2025, https://www.nso.ru
20. Electronic version of the regional regional newspaper Toguchinsky Bulletin, **24** (2016) http://www.toguchin.nso.ru
21. Real estate agency Onrealt, https://onrealt.ru
22. NGS real estate, https://novosibirsk.n1.ru
23. Register of settlements of the Novosibirsk region. Newspaper Soviet Siberia, **146** (2007)
24. Official site of Novosibirskstat, https://novosibstat.gks.ru/
25. Federal State Statistics Service. All-Russian Agricultural Census 2016: https://rosstat.gov.ru/
26. Official site of the Ministry of Agriculture of the Novosibirsk Region, http://mcx.nso.ru/page/

The problem of ensuring food security of the Russian Federation and sustainable development of the economy in the context of energy indicators and rates of human food consumption

K.A. Vorobev, *T.S.* Orlova, and *L.N.* Safiullin

Ural State University of Economics, Yekaterinburg. Russia

Abstract. In this article, the authors consider the problem of ensuring food security of territories as one of the areas of sustainable economic development. The authors' studies have shown that solution to this problem largely depends on the regulation at the state level of individually balanced human nutrition in order to ensure the optimal ratio of the components necessary for life.

1 Introduction

Considering the main problem of ensuring sustainable development of the Russian economy in terms of ensuring food security, it shall be noted that in order to accomplish this task, it is necessary to arrange provision of consumers with constant access to the rational food product. Human nutrition shall be balanced in the context of the optimal ratio of fats, proteins, carbohydrates, minerals, vitamins in the diet and the correspondence of the daily caloric content of a person's food to his/her individual energy consumption. This is necessary in order for the population of the country to have the opportunity to be economically active for as long as possible due to high-quality and balanced nutrition that ensures the continuation of life. Nevertheless, the problem of ensuring food security in Russia in the context of sustainable economic development has objective distinctive features and cannot be put on a par with the problem of hunger and food security in Asia and Africa.

2 Materials and methods

Table 1 provides the range of averaged rates of physiological nutritional requirements for able-bodied persons in the age of 18-59 is systematized on the basis of the basic nutritional elements indicated in the recommendations of Rospotrebnadzor MR 2.3.1.2432-08 [7] and other sources [9], since the human body in normal development requires more than 600 food components.

Table 1. Range of averaged daily rates of nutritional needs of a person in the age of 18-59

Basic nutritional elements, unit of measure	Daily rate
Human energy consumption per day, kcal	1800 - 4200
Proteins, g	58 – 117
of which, animal proteins, g	32 – 64
Carbohydrates, g	257 – 586
Fats, g	60 – 154
Minerals	
P, mg	1200
Ca , mg	800
Fe , mg	18
Zn , mg	15
I , mg	0,15
Vitamins	
A (retinol), mg	0,8 – 1
B1 (thiamine), mg	1,1 – 2,1
B2 (riboflavin), mg	1,3 – 2,4
B6 (peridoxin), mg	1,8 – 2,0
B9 (folacin), mg	0,2
B12 (cyanocobalamin), mg	0,003
C (ascorbic acid), mg	70 – 80
E (tocopherol), mg	8 - 10
PP (niacin), mg	14 – 28

Rationality of nutrition is determined by reimbursement of the body's energy consumption, which are expressed in calories, kilocalories (kcal), reflecting the energy value of food products. The average values of energy metabolism for an adult in Russia are summarized in Table 2.

Table 2. Average values of the daily rates of energy metabolism of an adult in Russia considering the gender, weight, and age

Daily energy exchange for men, kg / kcal		
Weight, kg	18-39 years, kcal	40 – 75 years, kcal
55	1430-1520	1240-1350
60	1500-1590	1300- 1410
65	1570-1670	1430-1550
70	1650-1750	1500-1620
75	1720-1830	1500-1620
80	1810-1920	1570-1700
85	1900-2010	1640-1780
90	1990-2110	1720-1870
Daily energy exchange of women, kg / kcal		
Weight, kg	18-39 years, kcal	40 – 75 years, kcal
45	1120-1150	1030-1080
50	1190-1230	1100-1160
55	1260-1300	1160-1220
60	1340-1380	1230-1300
65	1410-1450	1290-1370
70	1490-1530	1360-1440
75	1550-1600	1430-1510
80	1630-1680	1500-1580

From the above data, it follows that the signs of complete, partial starvation of a person are due to many different nutritional indicators of the diet and energy exchange, as well as many relevant indicators of gender, age, weight, the nature of a person's activity, his/her place of residence, traditions, and etc. Their severity for health, life a person, considering his/her daily food, is determined by the level of energy adequacy, the amount of energy reserve and necessary substances in the human body. The initial energy imbalance is usually covered by human glycogen in the amount of 1500-2500 kcal, which provides the energy requirement of a starving person per one day in conditions of relative rest.

According to the data of domestic studies, it was concluded that insufficient food leads to the decrease in the total energy requirements of the body and the corresponding weight of the person in the range of 30-35 %. The decrease in human body weight within 45-50 % of the normal (initial) value is considered incompatible with life.

3 Results and discussion

Based on the foregoing, with some convention due to the individual characteristics of the human diet, we conclude that the main energy indicators of food consumed by a person include:

1) consumption of food with an energy below the minimum (up to 1000 kcal) for 2-3 weeks entails the death of a person or irreversible consequences for his/her health;

2) consumption of food with an energy in a minimum scope (1000-1800 kcal) for 1 day or more means malnutrition, physical limitation of the human body and entails the corresponding human diseases;

3) consumption of food with an energy in the normal scope of 1800-3000 kcal means the body's work in a limited economical version, which may be accompanied by corresponding diseases;

4) consumption of food with an energy in the optimal volume of 3000-4000 kcal means that the body receives a balanced diet, which prolongs life, subject to the conditions of the quality and diet of this food.

The provided generalized energy indicators of a food product consumed by humans are shown in Figure 1.

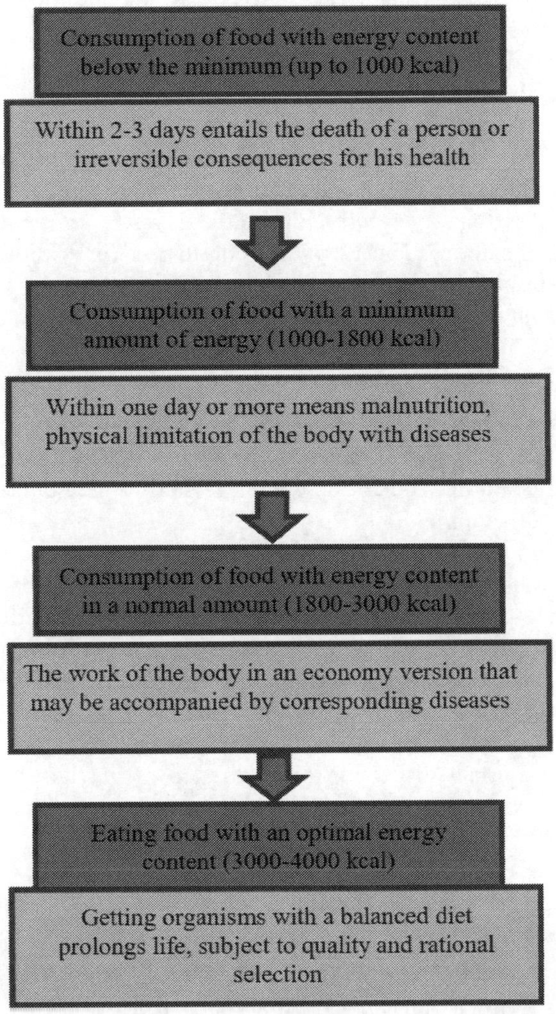

Fig. 1. Generalized norms of energy indicators of food consumed by humans

Wherein, it is worth focusing on the peculiarities of the territorial residence of a person, which are characterized by objective factors of ensuring food security and sustainable development of the Russian economy and play an important role in providing people with food.

Distinctive features of ensuring food security in Russia, in our opinion, can be defined as follows:

1) supplemented and refined elements of the official Food Security Doctrine that provide:

- establishment of the FS Doctrine measures by the norms of strategic planning, which form the corresponding directions of state policy;
- determination of the Russia's food security in the context of its food independence - self-sufficiency in the main types of Russian agricultural products, foodstuffs, and raw materials;
- designation of rational norms of food consumed by a person in accordance with the developed scientific modern principles of nutrition considering the range of averaged norms of daily food needs of a person, and the developed generalized norms of energy indicators of food consumed by a person;

- development in the FS Doctrine of strategic provisions of national security and generic economic security of Russia;
- the FS Doctrine shall consider the recommendations of the UN FAO only in relation to the maximum share of imported food and the volume of food reserves;
- due to the fact that based on the studies of domestic and foreign authors [2, 5], in Russia, in contrast to the African and Asian regions, the problem of hunger and malnutrition is not so acute, it is necessary to consider the list of national interests of the FS sphere characteristic of Russia, which reflect the measures to eliminate the identified main problems of ensuring the food security of the population (providing the population with constant access to rational, high-calorie food and pragmatic use, protection of the land resource as the main source of food production).

4 Results and discussion

There are objective general features of ensuring food security in Russia substantiated by the study that must be reflected in official documents related to the country's food security and sustainable economic development.

According to foreign researchers [1, 3, 4, 6], in Russia there is no threat of a population explosion, as noted in Asia and Africa, which predetermines the characteristic differences of the Russian FS system from general measures to ensure the FS in the context of global growth in the world population, which are reflected in various economic studies [11] and doctrinal documents of the Russian Federation on demographic policy [8].

The factors of increasing global competition in the context of the FS are complicated by the unresolved issues in Russia from the number of institutional social problems: significant social inequality and regional disproportion; presence of barriers to development of the agro-industrial complex; weak feedback between science, education and business; insufficient competition in some agricultural areas; poor development of social capital; depletion of traditional raw materials; the need to reform the health care system [10].

Therefore, the first main problem in ensuring the food security and sustainable development of the economy of the Russian Federation shall be clarified and reflected in the Doctrine of the Russian Federation in the form of a provision on providing the population with constant access to food considering its energy and functional content based on the objective characteristics of the territorial residence of a person (absence in Russia of starving people, the threat of a population explosion, and etc.), as well as from the individual characteristics of health and the human body.

References

1. D. Bazzana, B. Zaitchik, G. Gilioli, Structural Change and Economic Dynamics (2020)
2. F. Ceballos, S. Kannan, B. Kramer, B. World Development, **136,** 105069 (2020)
3. T. Hua, W. Zhao, S. Wang, B. Fu, P. Pereira, Conservation and Recycling, **163,** 105102 (2020).
4. L. Van Geffen, E. van Herpen, S. J. Sijtsema, Conservation and Recycling, **163,** 105041 (2020)
5. Z. M. Gitonga, M. Visser, C. Mulwa, World Development Perspectives, **20,** 100239. (2020)
6. Z. Xu, Z. Zhang, H. Liu, F. Zhong, J. Bai, S. Cheng, Food Policy, 101918 (2020).
7. Methodical recommendations MP 2.3.1.2432-08 "Norms of Physiological Needs for Energy and Nutrients for Various Groups of the Population of the Russian Federation"

(approved by Chief State Sanitary Doctor of the Russian Federation on December 18, 2008), https://base.garant.ru/2168105/

8. Decree of the President of the Russian Federation dated May 07, 2012 No. 606 (as amended on October 25, 2019) "On Measures to Implement the Demographic Policy of the Russian Federation", http://www.pravo.gov.ru
9. S.E. Bebinov, O.N. Krivoshchekova. Federal State Budgetary Educational Institution of Higher Vocational Education SibADI, (2016) https://www.elibrary.ru/
10. R.M. Nureyev, A. Inshakova, V. Sorokozherdieva, S*trategies for the modernization of Russia: problems of formation of the rule of law and an effective innovative economy* (2012)
11. L.N. Safiullin, A.F. Mingazova, Kazan Economic Bulletin, **1(39),** 71(2019)

BIO Web of Conferences **42**, 04005 (2022) https://doi.org/10.1051/bioconf/20224204005
SDGE 2021

Criteria for assessing food consumption patterns in the brics countries in accordance with sustainable development goals

Anna Diachkova[1,2,*], *Sergey* Tikhonov[2], *Olga* Tomyuk[3], and *Natalya* Tikhonova[2]

[1]Ural Federal University named after the first President of Russia B. N. Yeltsin, Ekaterinburg, Russia
[2]Ural State University of Economics, Ekaterinburg, Russia
[3]Ural State Law University, Ekaterinburg, Russia

Abstract. Achieving quality nutrition, food security, ending hunger, combating climate change and other challenges facing the world. These problems can be solved by building a sustainable food pattern. The aim of the article was to assess food patterns in accordance with sustainable development goals (2,3,12): the level of energy consumption and the structure of consumption (WHO recommendations gap), sustainable pattern gap, the share of imported food, the amount of harmful emissions. From the results obtained, the BRICS countries have not yet reached the goals of sustainability in the food system and require measures to change the food pattern: development of local farming, reducing the consumption of ultra-processed products, increasing organic products, reducing poverty and unemployment in the country, reducing the consumption of products that polluted the environment, reducing food waste and losses, buying rational quantities of products for eliminated the risk of product spoilage. These changes will help move towards sustainable food consumption.

1 Introduction

Achieving food security and improved nutrition - the second of the UN Sustainable Development Goals by 2030 - remains one of the most pressing development challenges in the world today [1]. Even before the pandemic, about 690 million people worldwide suffered from hunger, which is 8.9% of the world's population. This means that if the level of food security does not rise in the near future, then by 2050 the proportion of the population starving will increase several times. Daily energy consumption decreases below the norm, if vegetables and fruits do not become more accessible, then their consumption will also fall, as the population will prefer more high-calorie and affordable food (potatoes, cereals, sugar).

National policies and international programs aimed at providing quality and safe food necessary for a healthy and fulfilling lifestyle are becoming important [2,3]. In this regard, the focus of the study and subsequent adjustment is directed to the individual diet.

* Corresponding author: tihonov75@mail.ru

Researchers Kearney J., Lowe B., Fraser I., Souza-Monteiro D. Vasileska A., Rechkoska G identify patterns of nutrition in individual countries [4-6]. The level of development of the food industry and the income of the population are named as key determinants. Thus, it is found that low-income consumers make the decision to replace the lack of calories (energy) with cheaper and more affordable products. Substitutes are potatoes, bread or pasta, that is, foods high in carbohydrates. [7].

Some studies show that increasing income does not always lead to quality nutrition [2]. For example, in China, when economic reforms took place, as a result of which income increased by 4 times, consumption of foods high in fat increased along with income. Consumption of animal products (meat products) increased by 30%, while consumption of traditional products (rice, wheat, vegetables) decreased.

It should be noted that urbanization has led to the dominance of large supermarkets with a wider range of products, leading to the globalization of food consumption patterns. The activity of the urban population is on average 10-15% lower than in rural areas. High-calorie eating patterns are more common in cities than in rural areas. Climate is one of the most important determinants of the food basket. In every country in the world the climate is different and, even in one country in different parts of the weather conditions may vary. Countries can be dominated by droughts, floods, hurricanes, storms, fires, thunderstorms and more. As for the climate, in some countries it is cold, in some hot, somewhere humid, and in some arid. F. Karpena found out that dry shock reduces the level of calorie intake and, consequently, macronutrients: fats, proteins and carbohydrates. The dietary pattern becomes poorer during a drought, and the population does not consume enough calories. [8].

In accordance with the above determinants, we can observe the dependence of the diet model on agriculture, productivity, economic development and the level of the country's welfare. This leads to the formation of food patterns characteristic of individual countries of the world. It should be recognized that it is important not only to consider the characteristics of dietary patterns, but also to what extent they correspond to the goals of sustainable development.

The **aim** of the study: to determine the nutritional patterns of the BRICS countries based on the proposed assessment criteria that meet the goals of sustainable development.

2 Materials and methods

The theoretical basis of the work was formed by the main provisions of the theory of consumer choice, the theory of food systems, a resource-saving approach to food production, and modeling of food recipes. The research is based on the method of ranking, systematization, comparison, generalization and interpretation of the results. The information base of the study is presented by the official statistics of the World Health Organization, the Food and Agriculture Organization of the United Nations, and the Ministry of Health of the Russian Federation.

3 Results and discussion

Evaluating the nutrition models of the Brics, which differ in climate, the level of development of agriculture, the economy as a whole, cultural and historical traditions, we saw their differences and the dependence of the models on the identified nutritional determinants.

Sustainable development is becoming more and more a part of every country, business, and household every year. Deteriorating environmental conditions, depletion of natural resources, climate change, and declining biodiversity are forcing the whole world to change its daily behavior for the well-being of the world as a whole. This also applies to the nutrition of the population. The challenges of sustainability in the food system are also massive. The consequences of production and consumption affect the ecological, social and economic spheres of life.

A sustainable food system can be defined as a system that provides healthy food to meet current food needs while maintaining healthy ecosystems, that can also provide food for future generations with minimal negative impacts on the environment, and that rewards local production infrastructure and distribution, makes nutritious food available to all [9-12].

Thus, a sustainable eating model is a system that provides healthy food to meet current food needs while maintaining healthy ecosystems.

Let's single out the indicators according to which the power model can be considered sustainable: (1) a healthy and balanced diet; (2) affordable food; (3) local food, production and distribution infrastructure (short food supply chains); (4) minimal negative impact on the environment of the agricultural and food industries; (5) a healthy ecosystem for future generations.

It should be noted that a universal diet for each person is not called. The exact composition of a healthy, balanced, varied diet depends on individual characteristics such as lifestyle, level of physical activity, as well as on determinants (religion, income, climate, place of residence, etc.). All this influences the individual approach to the choice of the composition of the grocery basket.

The World Health Organization, the Ministry of Health of the Russian Federation recommend a balanced consumption structure bleaching of individual food items. A rational, sustainable model must be based on more than health-promoting recommendations. In connection with the aggravated situation related to pollution of water resources, climate change, deterioration of marine and terrestrial ecosystems, it is worth taking into account environmental problems. One of the solutions to the problem is organic (ecological) products.

Much of the work on sustainable models is focused on foods such as meat and dairy products. The main reason for the emission of greenhouse gases is due to natural causes. Namely, in the process of digestive activity of ruminants, methane is released. Therefore, special attention is paid to meat and dairy products. In earlier sources, meat is a food that should be limited for reasons of religion, health concerns, animal welfare concerns. Now, in more recent literature, the emphasis is increasingly shifting towards increasing sustainability by reducing the consumption of meat and dairy products. The authors also suggest reducing the consumption of these products in several ways. The first form is to reduce the portion size of meat and dairy products, the second form is to reduce the frequency of these foods by eating lean meals several times a week. To reduce your meat intake, you can replace it with other foods with unsaturated fats (fish, soy, lentils, beans, etc.). But there is also a risk that fish will replace meat in people's diets, thereby increasing pressure on fish stocks.

The food balance is a comprehensive picture of a country's food supply structure during a given reporting period. In the balance of food products for each type of food, sources of supply should be identified. The total quantity of food produced in a country, added to the total imported and adjusted for any changes in stocks that may have occurred since the beginning of the reporting period, gives the supply available during that period.

Thus, based on the above, we will propose key criteria for assessing a diet that meets the goals of sustainable development: 1) the level of energy consumption (calories) per day (comparison with the diet according to WHO recommendations); 2) consumption structure; 3) the gap in achieving a sustainable eating pattern; 4) the share of imported food products; 5) the amount of emissions of harmful substances.

Let's estimate the average amount of energy consumed by the population. The norm for the number of calories is 2800-3000. We will offer criteria for assessing the nutritional model on a scale from 0 to 1, where the highest score (1) is awarded to those countries that are included in the recommended norm. In accordance with this rating, the BRICS countries are assessed according to the first criterion, namely the level of energy consumption (see Table 1).

Table 1. Estimation of total energy consumption in BRICS countries [7,9,13-17]

Country	Energy consumption (ccal)	Rang
Brazil	3215	0,7
China	3199	0,8
India	2425	0,6
Russia	3212	0,7
South Africa	2838	1

According to the results of evaluating the first criterion, it should be said that South Africa is in the normal zone, China is in second place, whose indicator tends to the norm. Brazil and Russia are rated the same. And India has the worst rate of all. India is in a calorie deficit per person per day.

The second assessment criterion is the level of consumption of food groups, which are presented in the recommendations from the Ministry of Health. 11 food groups represent approximately 90% of the diet. Unfortunately, the recommendations do not say anything about such products as nuts and legumes, so these groups are not included in the analysis and will not affect the rating of the BRICS countries.

The grading scale is also presented from 0 to 1, where 1 is the best grade (awarded to those countries that are included in the recommended norm, that is, the difference from the norm in percentage terms does not exceed 5%) and 0 is the worst. The data obtained are presented in table 2.

Table 2. The difference in food consumption from the sustainable model [1,13,14,18-21]

Product	Recommended kg / year / person	Sustainable pattern gap (%)				
		Brazil	China	India	Russia	South Africa
Wheat products	96	42,4	33	35,7	36,4	38,3
Cereals	29	172,4	345,8	303,7	40,4	318,4
Potato	90	45,5	29,7	66,5	12,3	64,0
Sugar	24	167,9	66,2	47,8	217,5	71,9
Butter	7	194,1	20,1	42,1	176,1	85,9
Vegetables	140	63,3	159,6	37,0	26,5	71,7
Fruits	100	0,3	0,1	39,1	37,4	77,3
Meat	76	33,7	12,5	94,2	5,0	8,7
Eggs	16	33,1	23,4	79,6	3,3	63,3
Milk	340	58,3	93,2	68,5	57,5	85,3
A fish	24	62,5	61,1	71,5	16,3	73,5

Comparing the countries, we can see that Brazil, China, India and South Africa are similar in that they consume a large amount of cereals, especially rice and corn. Russia stands out among other countries for the excessive consumption of flour products and potatoes. At first glance, China has the healthiest model among the BRICS countries due to the minimum consumption of sugar and butter, but at the same time consuming enough or even more vegetables. Russia and Brazil are different in that they consume more oil, sugar and alcohol than other countries. India is characterized by the fact that it consumes in excess only those foods that really should be limited, namely, butter, sugar and cereals.

If countries are ranked according to the rating reflecting the quality of food, then the following conclusions can be drawn. China seems to be in the most advantageous position with an adequate consumption of vegetables and fruits, as well as a low consumption of butter and sugar. The second place is Russia, despite the high consumption of flour products and potatoes, Russia has good indicators of consumption of fruits, vegetables and dairy products. Then South Africa and Brazil, but Brazil loses in that it consumes more butter and sugar than South Africa (SA). India is short of the norm for almost all the necessary food products, at the same time, it consumes in excess of sugar, cereals and oil. These facts suggest that India has the worst nutritional model.

Table 3 presents calculations reflecting the difference between real consumption and the recommended rate (in percent), and regardless of whether the product is consumed in deficit or surplus, in any case the difference has a negative effect.

Table 3. Assessment of food supply for food groups in BRICS countries

Food partten	Brazil	China	India	Russia	SA
Ccal/day	0,7	0,8	0,6	0,7	1
Wheat products	0,6	0,7	0,6	0,6	0,6
Cereals	0	0	0	0,6	0
Potato	0,5	0,7	0,3	0,9	0,4
Sugar	0	0,3	0,5	0	0,3
Butter	0	0,8	0,6	0	0,1
Vegetables	0,4	1	0,6	0,7	0,3
Fruits	1	1	0,6	0,6	0,2
Meat	0,7	0,9	0,1	1	0,9
Eggs	0,7	0,8	0,2	1	0,9
Milk	0,4	0,1	0,3	0,4	0,2
A fish	0,4	0,4	0,3	0,8	0,3
Total	5.4 (0.45)	7.5 (0.63)	4.7 (0.39)	7.3 (0.61)	5.2 (0.43)

To understand whether the populations of countries can afford to eat sustainably, it is necessary to calculate the costs of real diets and compare them with sustainable ones. It is worth noting that if a population eats within an amount that is equal to or close to the sum of the rational model, then it can be concluded that a sustainable diet is available for a given country.

By calculating the cost of the diet for each country, it is possible to determine not only how much money is needed for a rational sustainable diet, but also to compare in which country food is more affordable (see Table 4).

Table 4. Comparative table of the cost of food models in the BRICS countries [13-14]

Product	Russia	South Africa	China	Brazil	India
Wheat	20,79	16,75	21,93	14,38	57,42
Eggs	19,06	9,78	26,02	22,68	0,62
Potato	16,48	7,24	20,44	26,69	12,67
Vegetables	63,66	40,89	359,66	30,92	46,13
Fruits	59,44	38,30	108,46	62,79	67,45
Cereals	3,11	17,13	56,41	17,43	84,28

Table 4. Continued

Meat	147,49	194,64	202,23	180,22	14,76
Milk	55,57	16,50	11,61	47,08	73,95
Sugar	88,38	1,41	1,76	1,32	
Cost of real food model	475,07	342,65	808,52	403,5	357,28
Cost of a rational food model	535,21	717,68	748,95	509,61	804,42
Accessibility assessment	0,9	0,5	1	0,8	0,4

Comparing the cost of a rational food model and the current food model, it became clear that the model is absolutely affordable for China, less available for Russia and Brazil. Estimates for South Africa and India show that among the BRICS countries, a sustainable diet is virtually unavailable.

Further, to assess the level of imported food products, data on the share of food imports were used. The amount of emissions of harmful substances is estimated by the volume of emissions of CH4, CO2, NO2. These emissions are calculated from the following activities: growing crops, using synthetic fertilizers, introducing manure into the soil, manure residues on pasture and burning plant residues. Since these indicators do not have a specific rational norm, they will be calculated from the maximum number among the BRICS countries. That is, the largest number of imports and emissions will be estimated at 0 points.

Table 5. Assessment of imports and emissions of the BRICS countries [13-14,18-25]

Country	Food imports (%)	Estimated level of imports	Emissions	Estimated level of emissions
Brazil	5,8	0,5	155221,4	0,6
Russia	11,7	0	25020,18	0,9
India	3,9	0,7	304327,9	0,3
China	6,64	0,4	404692,3	0
South Africa	7,23	0,4	12806,59	1

After evaluating all the criteria defined earlier, we get the following results. The total is also displayed on a scale of 0 to 1 based on averages. The Russian model turned out to be the best model of food consumption and was estimated at 0.60. Brazil (0.59) and South Africa (0.58) are practically on a par with Russia. Then China (0.51) and India (0.45), which are also close to each other in terms of nutrition. As the assessment shows, India has the worst performance among the BRICS countries.

Table 6. Total assessment of food patterns accordance with sustainable pattern

Criteria for evaluating food pattern	Rang				
	Brazil	China	India	Russia	SA
Calories per day	0.45	0.63	0.39	0.61	0.43
Sustainable pattern gap	0,8	1	0,4	0,9	0,5
Import level	0,5	0,4	0,7	0	0,4
Emission level	0,6	0	0,3	0,9	1
Final grade	0,59	0,51	0,45	0,60	0,58

Thus, after analyzing the food indicators of each of the BRICS countries, it was possible to assess the current situation with regard to the nutritional status of the population [26]. Weaknesses have been identified, namely food groups that are more detrimental to the diet and thereby reduce the level of sustainability of the food system. These foods include vegetables and fruits [27]. Also, it turned out to assess the availability of food in the BRICS countries, the level of imported food and the level of emissions from agriculture.

4 Conclusion

Achieving quality nutrition, food security, ending hunger, combating climate change and many other challenges facing the UN and the world. Part of these goals can be solved by creating a sustainable dietary pattern. Diet changes are essential for a sustainable future.

Thus, based on data on food consumption in the BRICS countries, it was possible to draw conclusions about the prevailing patterns of food consumption. It became clear that Brazil, China, India and South Africa consume excessive amounts of cereals such as rice and corn, while Russia attaches great importance to wheat products and potatoes in its model. All countries, except China, do not consume enough vegetables, thereby reducing the level of the country's nutritional model. China stands out among the rest for the minimum consumption of butter and sugar. And India, in comparison with other BRICS countries, consumes the smallest amount of energy compared to the recommended amount.

From the results obtained, it is clear that the BRICS countries have not yet reached the goals of sustainability in the food system and require significant measures to change the nutritional model. Such as the development of local farming, reducing the consumption of ultra-processed products, policies aimed at reducing poverty and unemployment in the country, reducing the consumption of products that have a burden on the environment, or completely abandoning them in the direction of organic products, it is also worth reducing the level of food waste and losses, that is, to reduce waste when transporting food, and consumers to buy rational quantities of products in order to reduce the risk of product spoilage [28]. All of these changes will help move towards sustainable food consumption.

References

1. WHO. Malnutrition (2020). https://www.who.int/
2. WHO. Healthy eating (2018). https://www.who.int/
3. A. Prosekov, S. Ivanova, Geoforum, **91**, 73-77 (2018)
4. J. Kearney, *Food consumption trends and drivers, Philosophical Transactions of the Royal Society*, 365 (2010)
5. A. Kolesnyak, I. Kolesnyak, IOP Conference Series: Earth and Environmental Science, 315 (2019)
6. ROSPOTREBNADZOR. Recommendations for citizens: healthy eating (2019). https://www.rospotrebnadzor.ru/
7. L. R. G. Guilhermea, A. S. Lopes, A. P. Corguinha, *Challenges and opportunities for a sustainable agriculture in Brazil* (2018)
8. J. Guthrie, L. Mancino, C. Lin, Psychology & Marketing, **32**, 501-511 (2015)
9. Ministry of Health. Information on the permissible norms for the consumption of alcoholic beverages (2017). https://minzdrav.gov.ru/

10. T. Allen, P. Prosperi. Modeling Sustainable Food Systems, Environmental Management, **57**, 956-975 (2016)
11. S. Coderoni, M. Perito, Journal of Cleaner Production, 252 (2020)
12. A. Azzurra, A. Massimiliano, M. Angela, Sustainable Production and Consumption, **17**, 95-107 (2019)
13. FAO and WHO. Sustainable Healthy Eating. Guidelines (2020)
14. FAO. http://www.fao.org/
15. Yu. G. Bendersky, I. V. Varfolomeev, A. P. Lopatin, *Problems of the Economic Assessment of the Natural Resource Potential of the Krasnoyarsk Territory* (2001)
16. M. Verain, H. Dagevos, G. Antonides, Appetite, **91**, 375-384 (2015)
17. X. Jiao, N. Mongoi, Journal of Integrative Agriculture, **17**, 755-764 (2018)
18. X. Liu, S. Zhang, J. Bae, Applied Energy, 204, 489-496 (2017)
19. P. N. Lomakin, Ensuring food security in Russia: internal and international aspects (2017)
20. L. F. Pinto, R. Hajjar, P. Newton, A. Agrawal, D. Adshead, D. Bini, M. Bogaerts, L. Cirhigiri, V. Maguire-Rajpaul, A. Chaves, C. Dermott, J. Milder, P. Pinho, I. Robinson, M. Rodkin, E. Wollenberg. *Transitioning to more sustainable, low emissions agriculture in Brazil, Climate Change, Agriculture and Food Security* (2016)
21. P. Priyadarshini, P. Abhilash, Land Use Policy, **96** (2016)
22. L. Ronquest-Ross, N. Vink, G. Sigge, South African Journal of Science (2015)
23. L. Reisch, U. Eberle, S. Lorek, Sustainability: Science, Practice and Policy, **9**, 7-25 (2013)
24. C. Tobler, V. Visschers, M. Siegrist, Appetite, **57**, 674-682 (2011)
25. J. Aertsens, W. Verbeke, K. Mondelaers, G. Huylenbroeck, British Food Journal, **111**, 1140-1167 (2009)
26. A. Vasileska, G. Rechkoska, Procedia – Social and Behavioral Sciences, **44**, 363-369 (2012)
27. K. Rask, N. Rask, Food Policy, **36**, 186-196 (2011)
28. B. Lowe, I. Fraser, D. Souza-Monteiro, Psychology & Marketing, **32**, 481-485 (2015)
29. S. Haggblade, N. Ne-Nsope, J. Staatz, Food Policy, **71**, 27-38 (2017)

Role of livestock in sustainable agriculture for food security

Maria V. Karpova[1,*], *Nina V.* Roznina[1], *Aliya U.* Esembekova[1], *Alla V.* Shulgina[2], and *Valentina M.* Flakina[2]

[1]Kurgan State Agricultural Academy by T.S. Maltsev, Kurgan, Russia
[2]Kurgan State University, Sovetskaya Str., 63, 4, Kurgan, Russia

Abstract. The livestock sector plays a special role in the formation of the rural economy and the life of society, performing a wide range of different functions. Agriculture, as a result of its activities, provides new jobs, in connection with which unemployment is reduced, there is also a decrease in rural migration, there is a change in the competitive environment, as a result of which the needs of the food market are being satisfied. A comprehensive assessment of the impact of animal husbandry on the development of the territory of the Kurgan region is the goal of our study. The article examines the development trends of the industry on the example of the Kurgan region. The influence of the development of animal husbandry on the provision of food security in the region has been studied. Such indicators as the volume of agricultural products for all types of categories, livestock, production of basic livestock products by categories of farms in the Kurgan region for 2016-2020 were considered and analyzed.

1 Introduction

The livestock sector is a strategically important production activity that provides the necessary food resources in solving food security problems. Trends in livestock production have a significant impact on how well these problems are addressed. In modern conditions in animal husbandry, agricultural producers are an important element of the national agri-food system, making up a certain part of the production of all livestock products. Subjects of all categories of agriculture help in solving the problems of rural development of a social nature. Livestock farming is an integral part of the agri-food system of the Kurgan region, it affects the food and economic security of the country and its regions, preserves the quality of the environment, maintains landscapes, provides employment for the population, socio-economic development of rural areas, and etc. Therefore, a comprehensive analysis and definition of the role of animal husbandry in the development of agriculture is relevant.

* Corresponding author: mdusheva@rambler.ru

2 Materials and methods

The study used statistical data from the Federal State Statistics Service. All the main branches of animal husbandry are considered; the dynamics of the livestock population, the volume of agricultural production, and the production of basic livestock products by categories of farms were considered as indicators reflecting the development of livestock industries.

In the process of writing the paper, various methods of scientific study were used, such as observation and collection of facts, analysis and synthesis, graphic.

3 Results and discussion

Support for small businesses has proven its relevance and will continue in the Kurgan region. Unprecedented attention is paid to development of small businesses in the region. In total, in 2020, 38 grants were issued to novice farmers, 11 grants to family farms, 3 grants to agricultural consumer cooperatives, for the first time 33 regional grants were provided for formation and improvement of peasant (farm) farms. So that farms can successfully develop further, the Department of AIC is developing a mechanism that allows them to sell products, incl. enterprises that have taken advantage of state support [1-2].

In the region, as of January 1, 2021, the following were recorded: 1138 peasant (farming) households (according to statistics); 27 agricultural consumer cooperatives, of which 4 - credit, 13 - processing, 8 - supply and marketing, 2 - service; 167.8 thousand families with personal subsidiary plots [3-4].

The main task of animal husbandry is to meet the needs of the population for food resources. Agricultural production represents data on the volume of crop and livestock production of all agricultural producers, including private sector farms (household farms, peasant (farm) households and individual entrepreneurs), in a cost estimate at actual prices (Table 1).

Table 1. The volume of agricultural products by categories of farms, mln.r.

	2016	2017	2018	2019	2020
Farms of all categories					
Production volume - total	38094.0	38576.7	39511.4	46268.9	42823.4
including:					
crop industry	21890.1	23452.7	22248.2	29732.6	26494.5
livestock industry	16203.9	15124.0	17263.2	16536.3	16328.8
Households of the population					
Production volume - total	16868.4	15777.2	17471.8	17615.2	15377.5
including:					
crop industry	4819.3	5356.7	5215.6	5928.1	4373.9
livestock industry	12049.1	10420.5	12256.3	11687.1	11003.6
Peasant (farming) households and individual entrepreneurs					
Production volume - total	6684.4	6921.1	6907.7	9614.6	9074.5
including:					
crop industry	6169.9	6356.5	6053.1	8759.3	8110.2
livestock industry	514.5	564.6	854.6	855.3	964.3

* according to the data of the Federal State Statistics Service

The volume of production of individual agricultural entrepreneurs in 2020 and households of the total value of agricultural production in the region amounted to about 57 % (Table 1, Figure 1), while the structure of production changes insignificantly over this period.

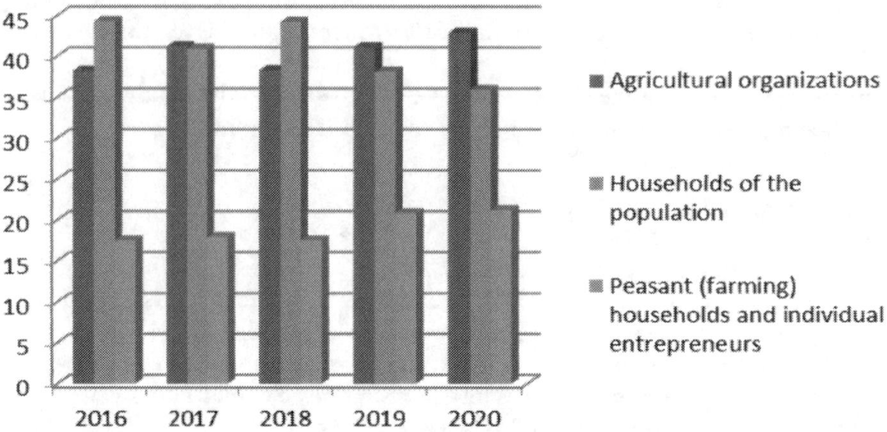

Fig. 1. Percentage of agricultural products by categories of farms, %

The largest share of the volume of production, 44.3 % in 2016, was obtained in household farms, the smallest share of 17.5 % in peasant farms. During the period under review, this structure has changed and in 2020 the largest share of agricultural production was received in agricultural organizations and amounted to 42.9 %, the least 21.2 % also in peasant farms. The volume of output for the analyzed period increased in agricultural organizations by 4.7 %, in peasant farms by 3.7 %, in households, this indicator decreased by 8.4 %, during this period a decline in production is observed in households, both for crop production and livestock production, which is associated with a decrease in acreage in households.

The composition of livestock products includes the cost of raw products obtained as a result of growing and economic use of farm animals and poultry (milk, wool, eggs, and etc.), the cost of growing (offspring, growth, weight gain) of livestock and poultry according to the data of the analyzed period, and etc. [5.6].

One of the main indicators reflecting the level and dynamics of the development of animal husbandry are indicators of the number and composition of the livestock of farm animals (Table 2).

Table 2. Livestock of farm animals in the Kurgan region, thousand heads (2016-2020)

Name	2016	2017	2018	2019	2020
Farms of all categories					
Cattle	117.5	117.0	122.0	120.8	120.5
Cows	49.5	48.4	50.3	50.0	50.7
Pigs	87.8	81.2	77.9	77.2	69.5
Sheep and goats	125.5	118.2	114.4	111.8	108.7
Horses	10.8	10.7	15.1	16.2	15.3
Households of the population					
Cattle	67.8	66.0	65.6	63.0	63.1
Cows	29.1	28.4	28.3	27.5	27.3
Pigs	44.1	33.7	34.4	32.4	27.8
Sheep and goats	113.3	107.5	106.3	100.2	96.2
Horses	6.9	6.8	10.7	11.6	10.7

Table 2. Continued

Peasant (farming) households and individual entrepreneurs					
Cattle	8.4	10.3	16.8	18.9	21.7
Cows	4.1	4.8	7.8	8.8	10.1
Pigs	15.7	18.8	15.4	15.9	16.7
Sheep and goats	10.2	9.4	6.8	9.3	9.3
Horses	1.0	1.1	1.7	2.0	2.4

*according to the data of the Federal State Statistics Service

In households for the period from 2016 to 2020, there is a decrease in the number of livestock, by 7.0 %, and in farms of all categories and in peasant (farm) farms (K(F)Kh), the number of cattle, including cows increased in organizations of all categories by 2.5 %, in individual entrepreneurship farms by — 2.5 times. During the same period, in all forms of farming, the number of pigs, sheep and goats decreased, but there is a positive dynamics of increase in the number of horses for the period under review.

With regard to the volume of production of farm animals, a similar trend can be traced (Table 3).

Table 3. Production of basic livestock products by category of farms in the Kurgan region for 2016-2020

	2016	2017	2018	2019	2020
Farms of all categories					
Livestock and poultry produced thousand tons	67.1	68.5	70.6	68.4	68.5
Milk, thousand tons	216.1	212.8	202.1	195.5	189.9
Eggs, mln. pcs.	102.0	96.6	94.6	91.8	91.5
Households of the population					
Livestock and poultry produced thousand tons	42.6	41.7	41.8	41.3	40.1
Milk, thousand tons	143.0	139.9	127.9	125.5	122.9
Eggs, mln. pcs.	84.3	80.6	78.7	78.1	77.5
Peasant (farming) households and individual entrepreneurs					
Livestock and poultry produced thousand tons	2.8	4.1	5.4	5.1	5.5
Milk, thousand tons	7.8	7.8	11.1	12.8	12.9
Eggs, mln. pcs.	4.6	3.1	3.1	1.5	1.4

* according to the data of the Federal State Statistics Service

In 2020, in small forms of individual entrepreneurship, the production of livestock and poultry meat (in live weight) increased by 2.0 times, in farms of all categories — by 2.0 %. In the households of the population there was a decrease in this indicator — by 5.9 %. For the years under consideration, in these forms of management, there is a decrease in milk yield and egg collection, — by 14.0 % and 8.1 %, respectively. In small forms of individual entrepreneurship, milk production increased by 65.3 %, but in this form of farming, egg production decreased by 70 %.

The volume of sales of agricultural products includes the sale of products by agricultural organizations through all channels: organizations that purchase for state and municipal needs, processing organizations, wholesale trade organizations, in markets, stock exchanges, auctions, through their own distribution network and public catering organizations, in the order wages, barter, and etc. (Figure 2).

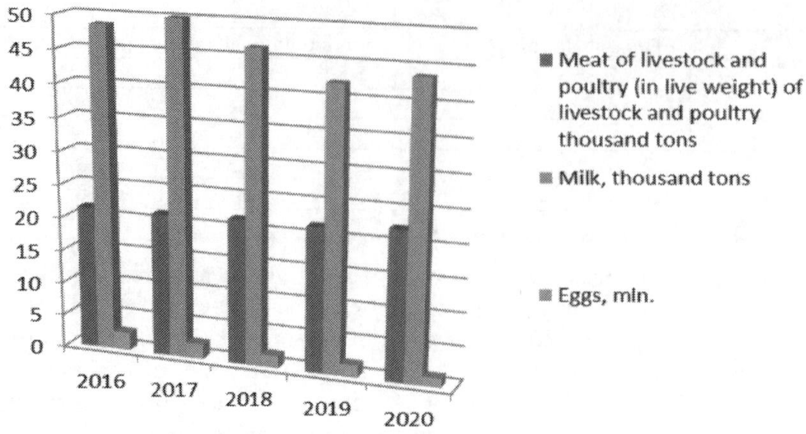

Fig. 2. Sales of livestock products by agricultural organizations

Based on the data in Figure 3, it can be seen that the sale of agricultural products of all organizations, there is a slight increase in the sale of cattle meat by 4.9 % in 2020 compared to 2016, as well as an increase in the sale of eggs by 37.5 %; during the period under review, the volume of milk sales decreased by 11 % [7-8].

Analyzing the positive changes in the increase in the production of livestock products of agricultural organizations, it can be assumed that the livestock industry in the region has a future, if the government pays attention to this, allocating material assistance and other support measures for livestock farming [9-10].

4 Conclusions

There are positive changes in the development and activities of the agricultural sector of the Kurgan region in the livestock industry, but at the same time there are problems that hinder the development of this industry:

1. High cost of purchasing feed and low purchasing price of livestock products (milk and meat).
2. Low economic indicators were obtained in dairy cattle breeding, which is an obstacle to attracting investments.
3. High degree of depreciation of fixed assets.
4. Lack of financial resources for technical re-equipment of farms in the agricultural sector.
5. Low provision of highly qualified personnel [6].

In general, the high risk of innovative projects, unequal competition conditions in the domestic food market, the underdevelopment of the insurance and guarantee system in the implementation of projects, all these conditions, hinder the development in the agricultural sector of the Kurgan region, including in the livestock industry.

References

1. S. A. Zotov, Bulletin of the Saratov State Socio-Economic University, **2(46)**, 47-50 (2013)
2. A. Bogoviz, S. Shkodinsky, E. Kondratieva, S. Lobova, A. Bazanova, Agro-industrial complex: Economics, Management, **8**, 4–14 (2017)

3. N.D. Gushchenskaya, D.V. Paliy, M.A. Sumarokova, A.U. Yesembekova, A.Yu. Anfalova, M.V. Pavlutskikh, International Transaction Journal of Engineering. Management. & Applied Sciences & Technologies, **11(14)**, 143-152 (2011)
4. T.I. Mazaeva, Bulletin of the University, **1**, 59–63 (2012)
5. N.D. Gushchenskaya, D.V. Paliy, M.A. Sumarokova, M.V. Pavlutskikh, Ecological-Socio-Economic Systems: Models of Competition and Cooperation (ESES 2019), Advances in Social Science, Education and Humanities Research, 392, 357-360 (2020)
6. S.Yu. Kornekova, E.D. Abdulhamidov, Society. Wednesday. Development, **4(29)**, 31–34 (2013) N. Yu. Novikova, Agro-food policy of Russia, **4(16)**, 19-21 (2013)
7. Nina Roznina, Maria Karpova, Alla Shulgina, Valentina Borovinskikh, Alevtina Volkova, Agricultural development of the Kurgan region in modern conditions, 10001 (2021)
8. Nina V. Roznina, Maria V. Karpova, Yevgenia S. Sokolova, Valentina A. Borovinskikh, Valentina M Flakin, The European Proceedings of Social and Behavioural Sciences EpSBS (2021)
9. L. R. Slepneva, I. V. Badmaeva, Bulletin of the Trans-Baikal State University, **12**, 154-159 (2014)
10. N. V. Yakovenko, Almanac of modern Science and education, **3**, 205-207 (2009)

BIO Web of Conferences **42**, 04007 (2022) https://doi.org/10.1051/bioconf/20224204007
SDGE 2021

Modern ecological and hygienic problems of the agricultural economy associated with development of pond fish farming

Yury Eliseev, *Anna* Vojtovich*, *Yuliya* Eliseeva, *Nina* Pichugina, and *Dmitry* Eliseev

Saratov State Medical University n.a. V.I. Razumovsky, 410012, Bolshaya Kazachya street, 112, Saratov, Russia

Abstract. The materials on the study of the content of heavy metals and organochlorine pesticides in fish are presented. It has been established that the level of heavy metals and pesticides in fish is influenced by the territorial location of the pond production. Therefore, the content of lead, cadmium and mercury in the liver and gonads of carp was higher than the permissible values for fish growing in ponds located in the city limits near highways. The pesticides DDT and HCH were detected in the organs of fish in the territories of pond fish farming of farms. Residual concentrations of pesticides contained in fish did not exceed the maximum permissible levels. In order to safely consume fish grown near highways by the population, we recommend the removal of internal organs (liver and gonads) from fish products.

1 Introduction

One of the priority tasks of the state policy of the Russian Federation is the development of fisheries in inland waters. Wherein, the share of pond fish farming, as one of the main sectors of commercial aquaculture, must constantly develop. Fish farming is a very promising direction of agricultural activity, since these products are not only in high demand among the population, but also among entrepreneurs who sell them. In real conditions, even in the well-known fish farms of the Astrakhan and Rostov regions, fish production is extensive, due to sea and river fisheries, and not the development of pond fish farming. In general, in Russia, 60 % of the volume of aquaculture production falls on the south of the country (Rostov, Astrakhan and Volgograd regions, as well as Krasnodar and Stavropol territories) [1]. However, if we compare the economic efficiency of fish production with other sectors of agriculture, it shall be stated that the product of fish production when sold from one hectare of land will be several times more economically feasible than selling the amount of grain crops from the same area.

Moreover, pond fish is a food product necessary for humans, containing complete proteins that are easily digestible at any age, as well as essential (not synthesized in the body): fatty acids of the omega-6 group (linoleic and arachidonic) and omega-3, vitamins

* Corresponding author: voitovich.88@mail.ru

(A, D, E, K), mineral (calcium, phosphorus, magnesium, potassium, sulfur, chlorine) substances [2-4]. Considering the above, in recent years, there has been an increase in pond fish farming in Russia. So, in the territory of the middle Volga, namely, in Saratov region, practically in the urban area, near the bypass road, ponds for joint water use have appeared for fish farming. These include: a pond near the village of Kleshchevka; a cascade of three ponds located behind the Volzhsky district of the city, near the Elshanka river; two ponds located behind the Zavodskoy district of the city, one near the bypass road, on the Latryk river, near the village of Novoaleksandrovka, another pond near the village of Eremeevka. A number of large fish farming ponds are now being developed on the agricultural lands of the agro-industrial centers of the Saratov region in the Engels, Yekaterinovsky, Rtishchevsky, Pugachevsky, Ivanteevsky districts. All of the above reservoirs are provided with a convenient year-round access to the territory of pond fish farming.

The filling of ponds is mostly natural. The ponds are mostly stocked with carp, sazan, silver carp, crucian carp and tench. We selected for study the fish-breeding enterprises of the Saratov region, which switched to extensive forms of pond carp breeding, by growing carp in polyculture with carp [1].

In the territories of pond fish farming, trade pavilions for the sale of fresh fish have been installed, places for paid fishing have been organized.

In the territories of pond fishing, thanks to introduction of new technologies, all conditions have been created for the wintering of fish in the pond and the absence of unpleasant consequences in the form of suffocation. For this purpose, oxygen supply (aeration) is organized in a winter reservoir covered with ice with the help of electric air compressors. Also, by organizing holes, or openings on the surface of ice-covered ponds, toxic gases (ammonia and hydrogen sulfide) are removed from the water, which are formed in silty organic sediments from the waste products of fish and plants [5].

Wherein, there are numerous publications indicating the presence in water bodies of a colossal amount of various chemical pollutants capable of accumulating in aquatic ecosystems and, especially, in bottom sediments [6-7]. However, even if the concentration of ecotoxicants does not go beyond the maximum permissible values, commercial fish have the ability to deposit pollutants in the liver, kidneys, muscles, gonads and gill leaves [8]. Wherein, if data on the pollution of environmental objects by ecotoxicants (atmospheric air, water of reservoirs, soil), food products are widely presented in the available references [9], then information on the content of pollutants in the organs and tissues of the body of fish grown in pond fish farming are very limited and controversial, which served as the basis for this study.

2 Materials and Methods

For carrying out field studies, two reservoirs were selected: a pond in the area of the city limits near the village of Kleshchevka and a fish pond in the Ivanteevsky district of agro-industrial farms. In both reservoirs, pond carp breeding was carried out by growing carp in polyculture with carp. At least three to five specimens of each fish species (carp and sazan) were taken from each pond. The total combined weight of a particular test fish species from a particular pond, as a rule, did not exceed 3.0 kg.

Sampling of pond fish for laboratory tests and study was carried out taking into account the recommendations drawn up by MosMR 2.3.2.006-03. The recommended sample mass during the examination of the sample for the chemical investigation of the content of heavy metals (HM) and organochlorine pesticides (OCP) was at least 500 g.

To compile a combined sample from each specimen of a specific species of pond fish (carp and carp), in order to further study, the concentration of chemical contaminants in tissues and organs, three point samples were cut out from the cut fish: muscle tissue of the sidewall carcass, liver and milk. Therefore, each spot sample represented a minimum amount of production (about 3-5 grams) taken from a specific location of a specific fish species with the aim of further compiling a combined research average. Each of the pooled average product spot samples was packed in a sealed thermal container with a lock, which was labeled and the contents of the bag described. The labeled sample was transported to the certified analytical laboratory of the branch of CEP Ecomir LLC in a chilled form in a refrigerator bag. The marking contained information about the serial number of the sample, the type of fish, the place of taking a spot sample from the opened fresh fish, as well as the specific place of catch - the collection of products (pond, area).

In the research laboratory, the delivered samples were passed through a meat grinder, ground in a homogenizer and ground in a mortar until smooth. Chemical analysis of the samples taken was aimed at investigating the content of four HMs of lead (Pb), cadmium (Cd), mercury (Hg) and arsenic (As). Determination of the content of metals Pb, Cd, As in organs and tissues of fish was carried out using adsorption stripping voltammetry on an IVA-5 analyzer. The Hg content in the studied fish samples was determined by the method of flameless atomic absorption spectrometry of cold vapor with preliminary mineralization of the sample under pressure on a RA 915 M analyzer with a RP-92 pyrolytic attachment. The content of residual amounts of organochlorine pesticides DDT, HCH in homogenized fish samples was analyzed by gas-liquid chromatography on a Chromatek Crystal gas chromatograph with an electron capture detector in the measurement range of 0.005 - 0.5 mg/kg. The assessment of the permissible levels of safety for the detection of heavy metals and pesticides was carried out according to the Unified Sanitary and Epidemiological and Hygienic Requirements for Products (Goods) Subject to Sanitary and Epidemiological Supervision from 2010. Toxic elements were determined by standard methods: lead — according to GOST 26932, cadmium — according to GOST 26933, zinc — according to GOST 26934, copper — according to GOST 26931, arsenic — according to GOST 26930, mercury — according to GOST 26927 and MU 5178.

3 Results and Discussion

It is widely known that one of the main and most dangerous anthropogenic pollutants of water bodies today are heavy metals and organochlorine pesticides [10-22]. Through ecological trophic chains, ecotoxicants first enter the body of the inhabitants of water bodies, and then they can pose a significant danger to human health [23].

The studies carried out (see Tables 1 and 2) found that the accumulation of TM and KhOP more intensively took place in the organs (liver and milk) of fish than in muscle tissues. Wherein, the content of metals Pb, Cd, Hg in the organs of carp (Table 1) living in ponds located in the city limits near the highways near the village of Kleshchevka was 1.5-2.5 times higher than the maximum permissible values. Similar data were previously obtained by I.N. Bedritskaya, 2000 [24] in the study of fish toxicosis, taking into account the habitat and the nature of the diet when growing carp on the warm waste waters of the Kirishskaya GRES-19 and Kirishskaya NPP. The intake, followed by the accumulation of TM in the liver and gonads of carp, was apparently associated with the cumulating ability of benthophages (chironomids, oligochaetes), which are the main natural food for this fish species. Wherein, it shall be noted that in the bodies of carp inhabiting the ponds of the Ivanteevsky district, located on farmland, far from major highways, the detected concentrations of heavy metals were significantly lower than the regulated AEL. Moreover, the metal arsenic not present in the emissions of road transport, regardless of the place where pond fish was grown, was detected in organs (liver and gonads) in background concentrations, 100-1000 times lower than permissible levels, and in muscle tissue samples of carp and sazan was absent. The studies carried out revealed significant differences in the content of heavy metals and organochlorine pesticides in the organs and tissues of the organism of farmed fish. So, if the concentrations of heavy metals were present in quantities exceeding the permissible levels in the organs (liver and milk) of carp and carp living in ponds near major highways of the city, then organochlorine pesticides, on the contrary, were found only in the body of fish grown in artificial reservoirs in the territories of therural farming agglomerations. Wherein, residual concentrations of organochlorine pesticides (DDT and HCH) were either not detected, which was typical for fish muscle tissue, or were found in carp organs at the level of permissible concentration values. Similar results were obtained by us in the study of the content of heavy metals and pesticides in the tissues and organs of the carp, as well as the places of their breeding (Table 2).

Table 1. Distribution of HM content in tissues and organs of carp grown in ponds near the village of Kleshchevka and Ivanteevsky district of Saratov region

Heavy metals (AEL for fish content)	Concentrations of heavy metals (mg/kg) in tissues and organs of carp grown in fish ponds					
	near the village of Kleshchevka			Ivanteevsky district		
	liver	milk	muscle	liver	milk	muscle
Pb (1.0 mg/kg)	2.5±0.3	1.1±0.02	0.1±0.03	0.9±0.2	0.2±0.01	No
Hg (0.6 mg/kg)	0.9±0.07	0.7±0.05	0.4±0.05	0.3±0.05	0.1±0.01	No
Cd (0.2 mg/kg)	0.3±0.06	0.3±0.06	0.1±0.02	0.1±0.02	0.05±0.001	No
As (1.0 mg/kg)	0.01±0.002	0.001±0.0003	No	0.01±0.002	0.001±0.0003	No

Table 2. Distribution of the content of organochlorine pesticides in the tissues and organs of carp grown in ponds near the village. Kleshchevka and Ivanteevsky district of Saratov region

Names of organochlorine pesticides (AEL for fish content)	Concentrations of organochlorine pesticides (mg/kg) in tissues and organs of carp grown in fish ponds					
	near the village of Kleshchevka			Ivanteevsky district		
	liver	milk	muscle	liver	milk	muscle
DDT and its metabolites (0.3 mg/kg)	0.01±0.002	No	No	0.3±0.02	0.2±0.01	No
HCH (α, β, γ isomers) (0.03 mg/kg)	0.007±0.07	No	No	0.03±0.004	0.03±0.004	No

4 Conclusion

Therefore, the studies carried out have established that the greatest threat to public health can be posed by fish that inhabit ponds near major highways of the city limits, which are sources of heavy metal emissions. For the safe use of such products, the removal of internal organs (liver and milk) from fish products shall be recommended. It shall also be recommended to limit the work associated with creation of ponds within the city limits and along major highways.

References

1. Yu. P. Mamontov, V. Ya. Sklyarov, N. V. Stetsko, *Pond fish farming*, 216 (2010)
2. O. E. Zhigileva, N. I. Bezukh, *Assessment of the state of the ide population from the Ob river in the oil production area, Problems of immunology, pathology and health*

protection of fish and other aquatic organisms, Materials of the scientific - practical conference Borok, Moscow, Russian Agricultural Academy, 161 (2007)

3. Yu. V. Kleshchina, Yu. Yu. Eliseev, Hyg and sanit, **92-1**, 81 (2013)
4. T. Kennedy, Techol. Irel. Apr., 43-44 (1986)
5. N. F. Burdina, Young Scientist, **25(263),** 69 (2019)
6. S. Musaev, Yu. Eliseev, I. Lutsevich, Bulletin of the Samara Scient Center of the Rus Acad of Sci, **13(1-8),** 1914 (2011)
7. S. Musaev, Yu. Eliseev, I. Lutsevich , Hyg and sanit, **91-5**, 101 (2012)
8. I. L. Golovanov, V.T. Komov, Voprosy ichth, **45-5**, 695 (2005)
9. Yu. Eliseev, V.Spirin, S. Chechomov, Yu. Eliseeva, Hyg and sanit, **100-5**, 482 (2021)
10. S. R. Ishbulatova, N. M. Kazachkova, International Student Scientific Bulletin, **4-3** (2017)
11. P. A. Popov, *Heavy metals in fish of the Ob River*, 373 (2000)
12. M. F. Zaranyica, Environ. Sci and health, **33-7,** 1895 (1997)
13. N. Yu. Stepanova, V.Z Latypova, A. M. Mukhametshin, TOFEA Bulletin, **4**, 44 (2005)
14. D. Hinton, Mar. Environ. Res., **28(1-4),** 411 (1989)
15. L. Jorgesen, B. Pedersen, Marine pollution Bulletein, **28-4**, 235 (1994)
16. K. Grah, P. Franfe, R. Halcbach, Symposia Biologia Hungarica, **29**, 357 (1985)
17. D. Mallins, B. McCain, D. Brown, Env Sci and Techn, **18-9**, 705 (1984)
18. S. Adams, K. Shepard, M. Greely, Marine Envir Res, **28-1-4**, 459 (1989)
19. R Bradley, J. Morris, Water, Air and Soil pollut, **27**, 341 (1986)
20. M.M. Tyaptirgyanov, V.M. Tyaptirgyanova, Yakutsk med jour, **3(47)**, 66 (2014)
21. M.M. Tyaptirgyanov, V.M. Tyaptirgyanova, Yakutsk Med Jour, **3(47)**, 69 (2014)
22. M.M. Tyaptirgyanov, V.M. Tyaptigyanova, Yakutsk Medi Jour, **1(49)**, 34 (2015)
23. E. Sergeeva, Yu. Eliseev Saratov jour of med scient research, **4-4**, 18 (2008)
24. I. N. Bedritskaya, *Influence of heavy metals on the organism of fish reared in waste waters of power plants, Dis ... Cand. biol. Sciences, Troitsk,* 185 (2000)

Advanced digital transformations for food security

Elmira Amirova[1,*], *Ilnur* Safiullin[1], *Olga* Bakhareva[1], and *Amina* Sakhbieva[2]

[1]Kazan State Agrarian University, Kazan, 420015, Russia
[2]Kazan (Volga region) Federal University, 420008, Russia

Abstract. The issues of digital transformation of the modern agricultural economy and the impact of innovative processes on its effective and sustainable functioning in order to ensure food security are analyzed in this article. The key directions for the development of digital production are shown, with the help of which the real prospects for the development of the agrarian sector of the national economy are determined. The current state of digital transformation of the agricultural sector of the economy is analyzed using modeling of the product impact and process innovations on the development of the agricultural economy. The influence of knowledge spreading in the Internet era, the development of network technologies and the use of digital platforms significantly expand the possibilities of both modern network business and business in the agricultural sector of the economy in solving problems related to the implementation of research and development in the field of digital transformation of the agricultural sector of the economy. In the conclusion, it is concluded that the development of rural areas and ensuring of food security in Russia will largely depend on the level of application of information technologies.

1 Introduction

Almost all spheres of the national economy of Russia are subject to the penetration of progressive technologies that contribute to the integration of digital technologies into the agricultural sector of the economy. Since the end of the last century, the consequences of the use of digital technologies in agriculture have led to innovative changes in soil-agrotechnical and organizational-territorial conditions, which can significantly increase crop yields, animal productivity, reduce production costs and other types of costs, protect the environment, preserve soil fertility [1].

Modernization of all spheres of the national economy is carried out with the implementation of digital technologies which help to discover promising opportunities for effective business and public administration [2, 3]. Transforming the digital space, it becomes possible to organize the production and sale of competitive agricultural products, provide high-quality services and get the best administrative solutions. Elements of computer modeling make it possible to reduce the time and funds required for the development and creation of new products and services, starting with the breeding of plants and animals,

* Corresponding author: elmira_amirova@mail.ru

ending with complex solutions for digital transformation of both individual agribusiness units and the region as a whole [4, 5].

In the modern world, the digital transformation of the agricultural economy is proceeding at a high pace. It has been determined that the number of users of the latest IT solutions grows tenfold annually in the agricultural sector in numerical terms. [6]. Ensuring food independence of our state directly correlates with the pace of implementation of artificial intelligence and advanced solutions in the agricultural sector of the economy. Development and creation of a national digital platform for the management and coordination of all participants in agricultural markets, which still continue to belong to an area with undiscovered potential due to the fact that the quality of communication networks in the Russian Federation does not meet the requirements of the proposed breakthrough strategies for the modernization of agricultural production [7].

The goals of the digitalization processes in the agricultural sector in Russia are the following: achievement of a large increasing in the efficiency of work at enterprises; transformation of state management processes in the field of agricultural economics; ensuring the effectiveness and efficiency of taken decisions; increasing of investments in the Russian economy and export proceeds; development, production control and integration of large amounts of data to create the latest technologies; achieving the maximum profitability of organizations. [8].

Necessary condition for the digital transformation of the agricultural economy is the creation of social institutions in the form of laws, standards, norms and rules. All digital data must be legally and officially enshrined in the international and state acts [9].

Despite all the obviousness of the need for digital transformation, a full-fledged transition of the agricultural economy to management based on the use of progressive digital technologies will be feasible only if this need is used for the entire system of economic relations. Optimization of all structural elements of economic activity is impossible without a developed digital infrastructure [10, 11]. Consequently, at present, the study of various aspects of the digital agrarian sphere transformation of the economy and the processes of its development is very relevant from a scientific and practical point of view.

Nowadays, digital technologies affect most of the areas. This number also includes agriculture, which is a strategically important area for the Russian Federation. The Ministry of Agriculture of the Russian Federation implements a truly significant project "Digital Agriculture" having the ambitious task to increase the productivity of agricultural enterprises twice by 2024 with the help of digital technologies [12].

2 Materials and Methods

The scientific works of Russian scientists and economists on the digital transformation of the economy are the theoretical and methodological basis of this article. [13, 14]. Modern conditions of development predetermine the need to study the issues of digital transformation of the industry, which consists in developing roadmaps for a strategy for the digital transformation of the agrarian economy and adapting the activities of agricultural enterprises in order to form a digital strategy.

The object of the research is the agricultural economy in connection with the infrastructural aspects of its digitalization. Abstract-logical, economic-statistical and graphic research methods were used in the paper.

3 Results and Discussion

The agricultural sector of the economy differs from other sectors of the economy in its conservatism; digital products are not developed here. However, the digital transformation of the industry is necessary in terms of production process management and logistics operations. An example of such actions is the use of digital maps in the cultivation of fields, referred to as "digital farming" technology.

When using this technology, the state of the land and the determination of weather conditions is carried out using regularly uploaded photographs to the monitoring system using satellites, as a result, the artificial intelligence makes technologically important decisions on carrying out the necessary field work. Further, on the basis of mathematical models using Big Data technologies and the use of artificial intelligence in the entire chain of the production cycle, it is possible to forecast of the harvest volume and indicators of demand for finished products.

An idea of the impact of the widespread introduction of innovations in the agricultural sector of the economy on the rate of economic growth can be obtained by analyzing their specifics and features in different periods. Note that the innovations of the second half of the XX century were mainly of a grocery nature. At this time, one could observe the emergence and introduction into production of space rockets and nuclear power plants, computers of various generations and mobile phones, the improvement of production technology (from the production of chemical fibers and fabrics to the production of aircraft, cars), etc.

Expansion of knowledge in various fields made it possible to develop the current agricultural production and constantly produce better products that acted as competitive goods and stimulated production, increasing of production volumes, providing increase of income and profits of private business and growth of the economy of different countries and the globe as a whole.

The main feature of modern process innovations is that their use in the agricultural business primarily provides a reduction or savings in costs for a particular type of activity, allows you to process large amounts of information, transmitting them over long distances in real time. But process innovation does not necessarily lead to the growth and expansion of output in every such business, and therefore is not a direct and immediate factor of GDP growth. Process innovations provide network businesses with the ability to distribute and redistribute information, provide services of an intangible nature that do not create cost. In this sense, they can lead to a reduction or slowdown in economic growth, expressed by the rate of GDP growth. The relationship of product and process innovations with the development of the economy is shown in the figure 1 [8, 18-20].

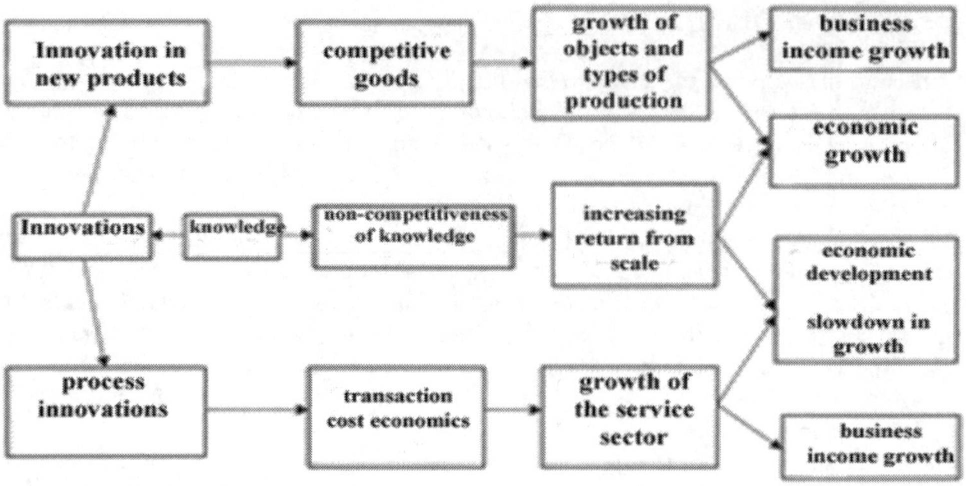

Fig. 1. Impact of product and process innovations on the economic development.

It's necessary to keep in mind that at present the product innovations continue to develop such as robotics in the agricultural sector, 3D printers, new materials, digital platforms, etc. Such developments require significant capital expenditures, the use of highly qualified specialists, significant investments associated with the production of goods and their promotion to the market. The creation of competitive goods can still serve as a definite factor in economic growth. E-commerce, cloud technologies, Internet services are developing at an increasing pace since they often do not imply large capital investments.

In accordance with the Decree of the President of the Russian Federation dated May 7, 2018 No. 204 "On national goals and strategic objectives of the development of the Russian Federation for the period up to 2024", a national program "Digital economy of the Russian Federation" has been formed. One of the objectives of the program is to create a system of legal regulation of the digital economy based on a flexible approach in each area. The implementation of this task is dedicated to the federal project "Normative Regulation of the Digital Environment", which is supervised by the Ministry of Economic Development of the Russian Federation [9].

The following key directions for the development of digital production are distinguished: the formation of technological groundwork [15, 21-23] and research competencies; personnel and education [16, 17]; normative regulation; information security; information infrastructure.

Now the use of information technology in the agricultural sector involves the use of not only computers. Digital technologies make it possible to control the full cycle of both livestock and crop production. So modern devices measure and then provide the parameters of the local climate, soil, plants, etc. All these data from various types of technical means are analyzed by special programs. Online applications, as well as mobile applications, help agronomists and farmers to calculate the fertilization schedule, forecast the harvest, determine the best time for sowing and harvesting, etc.

Modern technologies in the field of agriculture for quite a long time have been actively using in such countries as Europe, Canada and the United States. The demand from domestic farmers for modern digital technologies is increasing, even though they are far from the indicators of the above countries. If you rely on the opinions of experts, then it is the digitalization process that will help make a powerful leap in the development of the national agro-industrial complex.

The main goal of the project for the digital transformation of agriculture, implemented by the Ministry of Agriculture, is the digital transformation of the agricultural sector of the economy by introducing platform solutions and digital technologies into this industry in order to achieve an increase in the efficiency of the functioning of digital enterprises in the agricultural sector. [14, 15]. The initial step of this project was the creation of a national platform - Digital Agriculture. The order regulating the formation of the platform was issued back in February last year. Rostec takes an active part in the transition of the country's agricultural industry to digital technologies. So, the Ministry of Agriculture of the Russian Federation and the Russian state corporation in April 2020 signed an agreement on cooperation in the field of implementation of digital technologies in the agro-industrial complex.

4 Conclusion

Thus, the current state of digital transformation of the agrarian sector of the economy has been analyzed using modeling the impact of product and process innovations on the development of the agrarian economy. The impact of the dissemination of knowledge in the Internet era, the development of network technologies and the use of digital platforms significantly expand the possibilities of both modern network business and business in the agricultural sector of the economy in solving problems related to the implementation of research and development in the field of digital transformation of the agricultural sector of the economy in order to provide food security of the country. Summing up, it is important to note that the use of digitalization in agriculture is still in the early stages of formation. The development of agricultural territories will largely depend on the level of application of information technologies. The basic directions of the digitalization process in the future should be the use of smart combines, tractors, unmanned aerial vehicles and other equipment, the development of various services and, of course, distance learning for workers. It is also important to note that it is necessary to properly implement state support for the agricultural sector of the economy, since the uncontrolled introduction of digitalization can ultimately lead to negative consequences for local agricultural producers, due to increasing of imports on the market.

References

1. E.F. Amirova, Bulletin of Kazan State Agrarian University, **13(1(48))**, 147-151 (2018)
2. A. Zakirova, G. Klychova, O. Doroshina, et al., E3S Web of Conferences: 2018 International Science Conference on Business Technologies for Sustainable Urban Development, SPbWOSCE 2018, St. Petersburg (2018)
3. O.V. Kirillova, E.F. Amirova, M.G. Kuznetsov, G.A. Valeeva, G.P. Zakharova, Innovative d irections of agricultural development aimed at ensuring food security in Russia, BIO Web of Conferences, 00068 (2020)
4. E.F. Amirova, O.V. Kirillova, M.G. Kuznetsov, Sh.M. Gazetdinov, Internet of things as a digital tool for the development of agricultural economy, BIO Web of Conferences, 00050 (2020)
5. Sh.M. Gazetdinov, M.Kh. Gazetdinov, O.S. Semicheva, F.F. Gatina, Reserves for improving the efficiency of integrated formations, BIO Web of Conferences, 00026 (2020)
6. O.V. Bakhareva, Journal of Advanced Research in Law and Economics, **10(8(46))**, 2262–2270 (2019)
7. M.S. Faskhutdinova, E.F. Amirova, I.N. Safiullin, L.G. Ibragimov, Human resources in the context of digitalization of agriculture, BIO Web of Conferences, 27, 00020 (2020)

8. S.S. Vinokurov, P.A. Guryanov, Spatial Economics, **4**, 93-111 (2016)
9. E.F. Amirova, M.G. Kuznetsov, E.Khakimova, A.Tolmacheva, Integrated development of digital agribusiness platform to support import substitution of food products, BIO Web of Conferences, 27, 00055 (2020)
10. A.L. Zolkin, A.N.Losev, D.V.Gridina, T.G.Aygumov, Research of problems of computer networks expert systems IOP Conference Series: Materials Science and Engineering, 1047(1), 012106 (2021)
11. A. Zakirova, G. Klychova, G., Ostaev et al, Organizational and methodological approach to managing financial flows of agricultural enterprises, E3S Web of Conferences, 164, 10009 (2020)
12. A.K. Subaeva, M.M. Nizamutdinov, L.M. Mavlieva, M.N. Kalimullin, BIO Web of Conferences, International Scientific-Practical Conference “Agriculture and Food Security: Technology, Innovation, Markets, Human Resources” (FIES 2019), 00226 (2020)
13. A.K. Subaeva, M.M. Nizamutdinov, L.M. Mavlieva, BIO Web of Conferences. International Scientific-Practical Conference “Agriculture and Food Security: Technology, Innovation, Markets, Human Resources” (FIES 2019), 00178 (2020)
14. A.U. Mentsiev, E.F. Amirova, N.V. Afanasev, Digitalization and mechanization in agriculture industry, IOP Conference Series: Earth and Environmental Science: III International Scientific Conference: AGRITECH-III-2020: Agribusiness, Environmental Engineering and Biotechnologies, Volgograd, Krasnoyarsk, Krasnoyarsk Science and Technology City Hall of the Russian Union of Scientific and Engineering Associations. – Volgograd, Krasnoyarsk: Institute of Physics and IOP Publishing Limited, 32031 (2020)
15. A.U. Mentsiev, A. U. Mentsiev, E. F. Amirova, IoT and mechanization in agriculture: problems, solutions, and prospects, IOP Conference Series: Earth and Environmental Science: III International Scientific Conference: AGRITECH-III-2020: Agribusiness, Environmental Engineering and Biotechnologies, Volgograd, Krasnoyarsk, Krasnoyarsk Science and Technology City Hall of the Russian Union of Scientific and Engineering Associations. – Volgograd, Krasnoyarsk: Institute of Physics and IOP Publishing Limited, 32035 (2020)
16. N.A. Safiullin, G.R.Valieva, D.I. Faizrakhmanov, L.N. Savushkina, Ch.M. Kurakova. Quality assessment of electronic state and municipal services using the example of the ministry of agriculture of the Russian Federation, BIO Web of Conferences, 00143 (2020)
17. V. Timiryanova, K. Grishin, D. Krasnoselskaya, Economies, **8(4)**, 87 (2020)
18. S. Kosnikov, T. Aygumov, F. Khamkhoeva, D. Krasnoselskaya, Amazonia Investiga, **9(28)**, 519-527 (2020)
19. K. Yusupov, K. Grishin, V. Timiryanova, D. Krasnoselskaya, Amazonia Investiga, **8(24)**, 452-463 (2019)
20. A.U. Mentsiev, Z. Gerikhanov, A. Isaev, APITECH-2019 Journal of Physics: Conference Series, 1399, 1-6 (2019)
21. R.V. Fattakhov, M.M. Nizamutdinov, V.V. Oreshnikov, Economy of Region, **15(2)**, 436-450 (2019)
22. A.V. Shmidt, T.A. Khudyakova, Proceedings of the 26th International Business Information Management Association Conference - Innovation Management and Sustainable Economic Competitive Advantage: From Regional Development to Global Growth, IBIMA 2015, 1617–1625 (2015)

23. V.V. Zhuravlyov, N.Y. Varkova, T.A. Khudyakova, S.V. Aliukov, Proceedings of the 32nd International Business Information Management Association Conference, IBIMA 2018 - Vision 2020: Sustainable Economic Development and Application of Innovation Management from Regional expansion to Global Growth, 550–561 (2018)

https://doi.org/10.1051/bioconf/20224204009

Modern aspects of ensuring the security of the agro-industrial complex

Larisa Bryantseva[*], *Natalia* Shishkina, and *Tatiana* Sabetova

Voronezh State Agrarian University named after Emperor Peter the Great, 394087, Michurina, 1, Voronezh, Russia

Abstract. The article substantiates the urgent need for the formation of import-substituting clusters in the agro-industrial complex to ensure food security. In modern reality, carrying out business activities, agro-industrial organizations are already using the advantages of integration to a certain extent, but at the same time, the key interests and guidelines of allied participants in entrepreneurial activities are insufficiently balanced. The authors have identified the methodological features of the formation of the subject of integration (agri-food polycluster formation), using the cluster tools of intersectoral reengineering. The basis for the introduction of the subjects of integration - allied participants in entrepreneurial activity - was their initial assessment and analysis of the factors and conditions of business activity, where the negative results of import substitution, which is an effective threat to food security, can be traced as much as possible. Evaluation of the target pairing through the integration of related business participants allows one to quantitatively assess the predominance of the target points of integration subjects - related business participants using information at the macro and micro levels.

1 Introduction

At present, the safety of the industry related to food production should be considered as an attributive part of the national food security, since there is no doubt that the solution of other strategic tasks of the development of the agro-industrial complex is problematic without domestic food production.

Thus, each entrepreneurial structure as a key business goal puts forward the achievement of a stably balanced and most effective business development today, as well as obtaining significant opportunities in the future.

In this regard, organizations of the agro-industrial complex are experiencing difficulties that threaten them with a decrease in the performance indicators of their business activities, which may arise due to the low security of the company's business units responsible for personnel, raw materials, technologies and technical potential from the aggressive financial and economic policy of the state, which affects both external and internal business relations of entrepreneurial structures.

[*] Corresponding author: blv2466@mail.ru

2 Materials and methods

Analysis of factorial features and variable changes in business of allied participants in entrepreneurial activity in the agro-industrial sector of the economy suggests the need to create import-substituting subjects of integration in the agro-industrial complex to ensure food security, since Russian agriculture depends not so much on imports of food products as on resource potential and technological components necessary for food and auxiliary industries [2;4].

The ultimate goal of business relations of the subjects of the agro-industrial complex is to ensure industrial safety, the optimal level of which automatically ensures the food security of the country.

The level of industrial safety is formed under the influence of factors of internal (underutilized opportunities) and external (diplomatic component) nature, food safety - by the level of industrial safety and the import of food products of a specific type. The developed methodology "Algorithm for indicating business relations in the agro-industrial complex" (using the example of sugar beet production) allows at each separately taken stage of the production business cycle: to determine the degree of underutilization of opportunities; to judge the optimality of business relationships; to determine the degree of influence of threats of non-optimal business relations on the level of industrial safety.

The value of the industrial safety coefficient corresponds to the lowest level of all indicators of business relations. The algorithm of the developed methodology for indicating business relations is presented in Table 1.

Table 1. Algorithm for indicating business relations in the agro-industrial complex (based on the example of sugar beet production).

Indicators	Potential values of the indicators	Actual values of the indicators
I_1	$\frac{M_{seeds.f}*(P_{seeds}*B_{seeds})_{norm}}{V_{seeds\ are\ necessary}}*100\%$	$\frac{V_{domestic\ seeds}*(\frac{U}{P}*B_{sugar})_f}{V_{seeds\ are\ necessary}*(\frac{Y}{P}*B_{sugar})_{norm}}*100\%$
I_2	$MIN(K_{ob.f}, K_{total\ area\ f})*100\%$	$\frac{(V_{domestic\ beetroot}*B_{sugar})_f}{V_{beets\ are\ necessary}*B_{sugar}}*100\%$
I_3	$\frac{M_{sugar.f}*(P_{sugar}*B_{sugar})_{norm}}{V_{sugar\ is\ necessary}}*100\%$	$\frac{V_{sugar\ beet.f}}{V_{sugar\ is\ necessary}}*100\%$
K_{ind}	$\underset{1}{\overset{3}{MIN}}\ И_{ni}$	$\underset{1}{\overset{3}{MIN}}\ И_{фi}$

I_1, I_2, I_3 – industrial security indicators;

K_{ind} – industrial security index, units;

$M_{seeds.f}, M_{sugar.f}$ - actual capacity of seed and sugar plants of the Russian Federation, t/day;

$K_{ob.f}, K_{total\ area\ f}$ - use factor of beet-harvesting machinery, crop acreage, units;

P_{seeds}, P_{sugar} - duration of the productive season for seed and sugar plants, days;

B_{seeds}, B_{sugar} - finished product output of seed and sugar plants, %;

$V_{\text{seeds are necessary}}$ - required amount of sugar for provision of the physiological need of the population, ths tons;

$V_{\text{seeds are necessary}}$, $V_{\text{beets are necessary}}$ - required amount of seed and sugar beet, taking into account

$V_{\text{sugar is necessary}}$, tons, ths tons;

$V_{\text{domestic seeds } f}$ - sugar beet seeds amount produced using domestic components only, tons;

U - sugar beet yield per 1 hectare, т/га;

P - seed material consumption per 1 hectare, kg;

$P_{\text{beets}.n}$ - sugar beet crop areas, ths ha;

$P_{\text{domestic beetroot}}$ - sugar beet crop areas planted with the seeds of Russian selection, ths ha;

$V_{\text{sugar beet}}$ - sugar produced from sugar beet, ths tons.

The use of such an approach in determining the required amount of sugar to meet the needs of the country's population makes it possible to judge the appropriateness of certain areas of development of business relations in the sugar beet complex.

Actual research of our time, analyzing the reasons and factors for the safe conduct of business by processing entrepreneurial structures, including sugar production, differ in species diversity: researching the nature of economic threats, researching the nature of industrial threats and researching the nature of food threats. At the same time, the greatest interest is shown in equilibrium (balanced) development management in organizations of the agro-industrial complex using cluster tools.

Potential threats to balanced business relations and the need to eliminate the dangers of their transformation into a negative synergistic effect for the company's business activities are the reason why it is necessary to find out the danger of this socio-economic phenomenon, on the one hand, and find out ways to level them, on the other hand.

At the same time, we designate two foundations of economic activity, leading to the threat of falling competitive business development:

(i) industrial and production aspects of business activities of entrepreneurial structures;

(ii) logistic relationships of interdependent entrepreneurial business units.

Analyzing the modern aspects of increasing the level of safety in the activities of organizations in the agro-industrial complex, it is necessary, first, to focus on issues of state support for such enterprises. At the same time, one should not underestimate the importance of documentation support for conducting effective business activities, including by organizations of sugar beet production (standards, regulations) that allow the agricultural sector of the economy to ensure the country's food security.

3 Discussion of results

In modern reality, the management of organizations increasingly uses such management methods that are associated with the benefits of integrating adjacent business participants [7]. Such integration, which makes it possible to reduce the threat to food security, creates the following advantages for its participants: concentration of subjects on a territorial basis; competition within the cluster; stability of economic and social ties; polycentricity; unidirectional co-development of the subjects of integration; positive synergy effect. Taking into account the above aspects of allied participants in entrepreneurial activity, we propose to use the methodological features of the formation of the subject of integration (agro-food polycluster formation), using the cluster tools of inter-sectoral reengineering (Table 2).

The basis for the creation of an APF is an expert factorial analysis of the functioning of allied participants in entrepreneurial activity; the negative consequences of import substitution that threaten the food security of the country are clearly visible [6].

Table 2. Processes of the implementation of the cross-sectoral reengineering project during the formation of the APF.

Inter-sectoral reengineering algorithm		Processes of creating an agro-food poly-cluster formation		Results
Stage	Name	Stage	Stage's essential contents	
1	Creating the future image	1.1	Balancing the key interests of regional business participants: the state; cluster; region; organizations, etc.	Identification of key landmarks of the cluster
		1.2	Cluster development strategy development: focusing on internal needs; export orientation; orientation towards import substitution	
2	Reverse reengineering	2.1	Assessment of the potential of the region	Determination of the contribution of the cluster to ensuring industrial and production security
		2.2	Development of a motivation system for cluster subjects	Regional (federal) program of support for the subjects of cluster formations
		2.3	Identification of existing external relations among the subjects of the cluster	P-model
		2.4	Identification of internal connections between the subjects of the cluster and the subjects of other clusters	O-model
3	Direct reengineering	3.1	Identification of opportunities for self-development of the cluster for a long-term period	P-model
		3.2	Identification of opportunities for self-development of competing clusters for a long-term period	Virtual O-model
		3.3	Revealing the causes, factors and conditions of intra-cluster and inter-cluster business relations	P-model
		3.4	Adjustment of goals, strategy, relationships, etc.	Actual O-model

We have considered the fundamental aspects and significant stages of the APF project implementation using the example of sugar beet production, since this industrial-production subsystem of the agro-industrial complex is currently a set of organizational and economic relations that most significantly reduce the country's food security.

Assessment of the conditions and trends in the development of the main production subsystems of the Russian agro-industrial complex makes it possible to state the need for an import-substituting cluster orientation to ensure industrial and production safety, since the intervention of imported products is due to the supply of not so much food products as resource and technological components that are constantly required by food production, as well as auxiliary and providing types of economic activity.

The conditions and factors for ensuring the conduct of a safe business by organizations of the agro-industrial complex should be assessed, in our opinion, taking into account the scale of their occurrence and the close relationship between the levels of safety of various types. From our point of view, the basis (basic platform) is industrial safety, the provision of which is possible through the highly efficient use of the resources and production capacities of organizations at the disposal of entrepreneurial structures, taking into account the invariably significant value of economic security. Neglect of underutilized opportunities and lost benefits by entrepreneurial agro-industrial structures will permanently lead to their loss of organizational and economic independence and stagnation of economic development at the macroeconomic level.

Applying the methodological features of the formation of the subject of integration for entrepreneurial agro-industrial structures, we recommend using the matrix of target pairing of subjects of integration-adjacent participants in entrepreneurial activity: the state, an integrated formation, processing organizations, agricultural organizations, research institutions, educational institutions, and other organizations. The essence of the matrix reflects the content of the dual advantages of such integrated formations: the activation of a specific target of a group of organizations-subjects of integration and the accumulation of professional knowledge, which ultimately helps to protect the advantages of competition of the agro-industrial structure at the inter-cluster level.

The methodological approach to the formation of a cluster is demonstrated along the diagonal of the matrix structure: firstly, the existing intra-cluster competition activates the vector of self-development and conjugation of the key tasks of the cluster; secondly, inter-cluster competition contributes to the formation of the vector of target pairing of the subjects of integration - adjacent participants in entrepreneurial activity [5].

The authors revealed that in the formation of such a subject of integration as APF (agro-industrial polycluster formation), there are all the necessary conditions for its integral self-development and the implementation of effective rivalry. As a result, at this stage it becomes realistic to determine the possible subjects of integration, capable of ensuring compliance with the principle of polycentricity of adjacent business participants, as well as quantifying the competitive advantages of the cluster at the inter-cluster level.

For the entrepreneurial structures of the agro-industrial complex, food security is of the greatest importance, the assessment of which, in our opinion, is advisable to be carried out in structural subordination (hierarchically), observing the sequence: analysis of the possibilities of ensuring the subject of integration of food security; analysis of the contribution of the subject of integration to ensuring industrial, production and food security. Such an approach, from our point of view, contributes to the conjugation of the targets of allied participants in entrepreneurial activity, which, as a result, will have a positive effect on the general state of the Russian agro-industrial complex [1;3].

On a planetary scale, it is generally accepted that state food security is ensured while maintaining the internal volume of consumption by 80% of its own food. About 20% is replenished by the state through food imports [8]. This proportion provides a certain state independence from the import of strategic food products. The situation in Russia is such that food security has given way to industrial security in the hierarchical structure of security, i.e. at present, the importance of the latter is higher. In fact, sugar produced from raw sugar is domestic, and when it is exported by other countries, importers are perceived and accounted for as Russian sugar. A paradoxical situation arises: imported white sugar, with its full consumption, would make up 5% of the total sugar consumption in Russia. Consequently, in fact, Russia exceeds the threshold of food independence by 15%, but the domestic sugar industry, on average, by 75% depends on imported raw materials and other components necessary for sugar production.

The logical connection, therefore, between the industrial, food and economic security of the country is breached. Food security, it turns out, falls out of this chain in a certain way, since it does not reflect the actual state of affairs in the sugar industry and the country as a whole.

4 Conclusion

Summarizing the above, we note that in the course of analyzing the essence of management as a system, considering security issues, including food security, we based on the following provisions. Firstly, the target of allied business participants is to obtain maximum economic benefits in the course of carrying out business activities, and secondly, business structures are obliged to try to make only economically justified costs and expenses that allow them to obtain optimal economic benefits, or to reduce the estimated costs in the future.

At the same time, the economic state of the Russian Federation can be considered safe if at least five conditions are met: living conditions and personal development that are acceptable for the majority of the population; stable socio-economic situation; military and political stability of society; the integrity of the state; the ability to resist the influence of internal and external threats.

One of the most significant elements of the government strategy, in our opinion, is the identification and monitoring of factors that reduce the stability of the socio-economic system of the state and the formation of economic policy that eliminate or mitigate their impact, the characteristics of threats to economic protection from security threats, the localization of which should be the first turn is focused on the activities of federal government bodies, focuses on the deformation of the structure of the Russian economy. Including due to such factors as the low competitiveness of products, the curtailment of production in vital industries, the conquest of the Russian domestic market by foreign firms for many types of consumer goods, etc.

References

1. A.V. Agibalov, L.A. Zaporozhtseva, V.B. Malitskaya, Z.P. Medelyaeva, Yu.V. Tkacheva, Agricultural companies as a russian regions' economic driver, 35th International Business Information Management Association Conference (IBIMA). Proceedings of the 35th International Business Information Management Association Conference (IBIMA) **1623** (2020)
2. A.I. Altukhov, Food Security of the Russian Federation: Assessment Methodology Issues Economics of Agricultural and Processing Enterprises, **7,** 2 (2016)
3. L.A. Zaporoztseva, The problem of reproduction in agriculture, 36th International Business Information Management Association Conference (IBIMA). Proceedings of the 36th International Business Information Management Association Conference (IBIMA) **1808** (2020)
4. V.N. Ivanov, S.N. Seregin, Agro-industrial policy of the EAEU: ensuring food security Suga, **10,** 14 (2015)
5. L.V. Bryantseva, Assessment of the potential of ensuring industrial and production safety on the basis of creation of polycluster formations in the aic Bulletin of the Saratov State Agrarian University. N.I. Vavilov., **10,** 70 (2009)
6. E.A. Vysotskaya, L.V. Briantseva, Coherence degree characteristic of production, industrial and food security of agro-industrial complex subsystems IOP Conference

Series: Earth and Environmental Science. 6th International Conference on Agriproducts Processing and Farming. **012102** (2020)

7. L.O. Makarevich, A.V. Ulezko, The mechanism of integrational interactions of economic subjects: methodological aspects, The Challenge of Sustainability in Agricultural Systems. Сер. "Lecture Notes in Networks and Systems" Heidelberg 23 (2021)
8. I.G. Ushachev, Import substitution in the agro-industrial complex of Russia: trends, problems, ways of development, Economy of agricultural and processing enterprises, **1,** 2 (2016)

Strategic analysis on the integration of agricultural holding assets

Ekaterina Ogorodnikova*, *Maria* Khokholush, and *Konstantin* Rostovtsev

Ural State University of Economics, 620144, 8 Marta Str, 62, Ekaterinburg, Russia

Abstract. The topic of integration of agricultural and food enterprises into holdings, the strategy of their cooperation and integration is of great importance for the food security of the country. The article presents the results of a study of the integration of the agricultural holding Rusagro, based on the use of a matrix of dependence of the company's profitability indicators on its relative value in the market. Approbation of this method allows to reasonably make management decisions regarding the development of an agricultural holding.

1 Introduction

Agriculture is the most subsidized area of doing business in the Russian Federation, and the state emphasizes the strategic importance of supporting this industry to achieve food security and social stability of the state [1]. At the same time, large agricultural producers are trying to find mechanisms to independently increase the stability and profitability of their business. One of these mechanisms is the process of integration and association. Agricultural markets and food markets are volatile. The state of demand for products is influenced by such factors as yield, market saturation, the ability to carry out export deliveries and government policy in relation to the import of products. The indicated influences create significant risks for the enterprises of the agro-industrial holding, which in turn causes the desire of agricultural holdings to minimize risks by diversifying business and compensating for losses in one area of activity due to successes in another [2,3]. The question arises, what structure of the holding should be chosen, what types of activities and enterprises should be included in the holding and how to bring the organization to the necessary criteria for strengthening its position in the market and increasing financial performance. The topic of combining enterprises into holdings, the strategy of their cooperation and integration is of great importance for agricultural holdings, certain aspects of this issue were considered in works [4] and [5]. The topic of institutional influence on the integration processes of agricultural producers and the formation of synergistic effects in the course of such integration is also widely discussed [6,7].

* Corresponding author: cmb_8@mail.ru

2 Materials and Methods

As a research method, a matrix of dependence of the company's profitability indicators on its relative value in the market is used. The value of the company's assets located on the abscissa (X-axis) reflects the desire to increase its presence in the market, due to the growth of profitability in a particular business division. On the ordinate (Y-axis) value of profit indicates the level of profitability of a particular business division [8].

Interpretation of business strategies using a matrix of dependence of the company's profitability indicators on its relative value in the market involves the allocation of three zones. The first of them is the zone of increased profitability of the agricultural holding, which signals the preference for the strategy of increasing the size of the company. The second zone - a stable level of the company's profitability indicates the absence of a threat of a reduction in the size of the company's assets and the absence of prerequisites for an increase in the size of the company. The red zone indicates a low level of profitability and the threat of a decrease in the presence on the presented market. Possible strategies fit into three directions, each of which has specific solutions.

The first is a growth strategy by activating business divisions that show the highest profitability. The main efforts in the implementation of this strategy are aimed at promoting the existing product on the market.

Second, the asset optimization strategy is used to consolidate resources in order to expand to promising market segments. The main solution is to retain the most profitable assets that generate the bulk of the profit and remove the less profitable business units from the portfolio. The result is an avalanche-like dynamics of an increase in the profitability of individual business units with a general decrease in their number.

The third is the strategy of reducing the size of the company, in which a systematic decrease in the value of assets is carried out in proportion to the decrease in profits. The strategy can be used by an agricultural holding if it is necessary to reduce one of the business divisions in order to obtain funds to strengthen another more promising direction. The strategy of reducing the size of the company is characterized by the fact that the most liquid assets are subject to sale, but the condition of ensuring an acceptable level of profit during the liquidation of the business remains.

3 Results and Discussion

As an indicator of the size of the company, the characteristic of the size of the company's non-current assets is used. The gradation is made in three levels: low, medium, high. The maximum value of the characteristic of non-current assets is taken to be the largest value of assets, relative to this value, the values of the characteristics are determined: high - at least 70% of the maximum value of the characteristic; medium - in the range of 30% - 70% of the maximum value of the characteristic; low - less than 30% of the maximum value of the characteristic.

As an indicator of the company's profit, the characteristic of the profitability of the company's non-current assets is used. The gradation is made in three levels: low, medium, high. High profitability is considered to be over 25%; medium - in the range of 5% - 25%; low - less than 5%. Next, we present information for compiling a matrix, table 1.

Table 1. Initial data for compiling a matrix of dependence of the company's profitability indicators on its relative market value for the Rusagro agricultural holding, 2013-2017

Holding business unit	2014		2015		2016		2017	
	Profitability, %	Assets, RUB bln	Profitability, %	Assets, RUB bln	Profitability, %	Assets, RUB bln	Profitability, %	Assets, RUB bln
Fat and oil products	14,46	17,3	14,46	17,3	4,35	19,9	4,35	4,55
Agriculture	2,07	10,7	2,07	10,7	2,56	20,21	2,27	18,8
Meat processing	4,52	18,1	4,52	18,1	3,97	17,9	6,02	20,5
Sugar production	14,49	22,5	14,49	22,5	13,86	37,2	12,19	30,4

Next, we present the matrix for the portfolio of business units of the Rusagro agricultural holding, Figure 1.

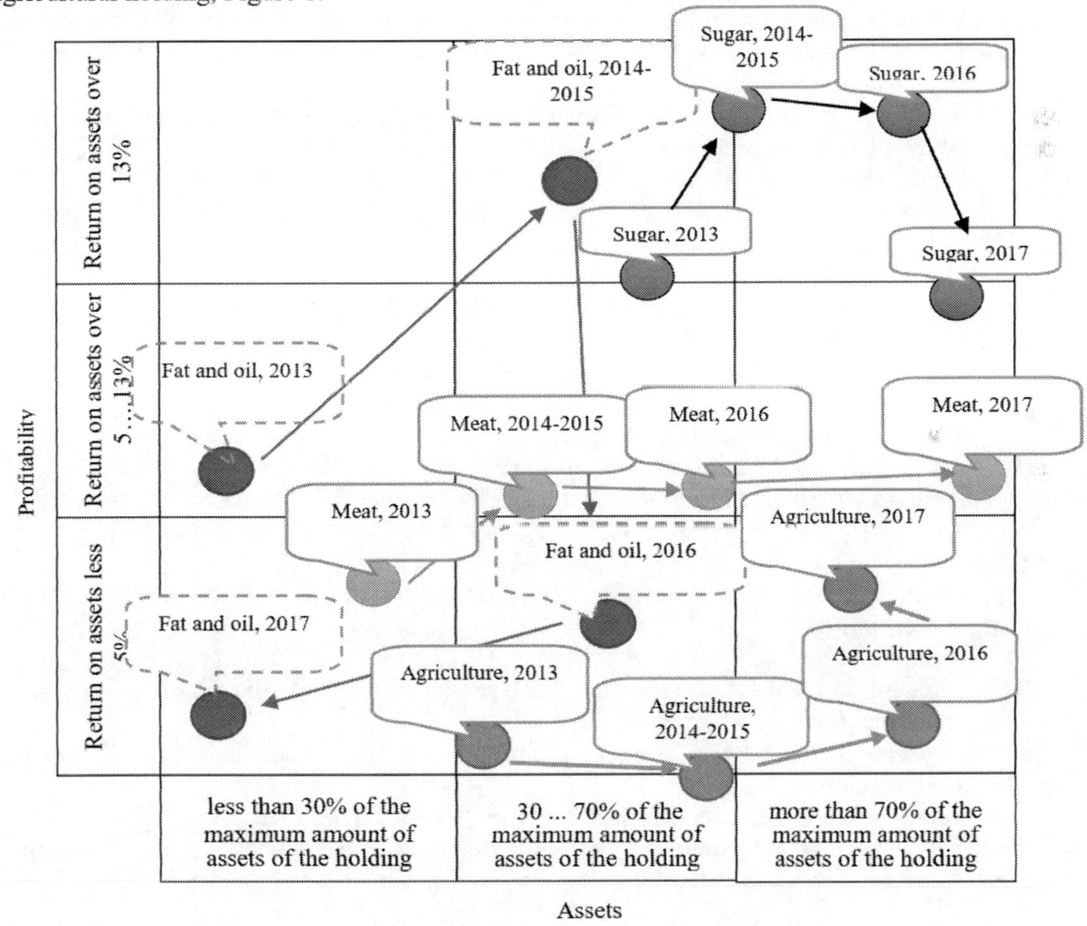

Fig. 1. Matrix of the relationship between company size indicators and its profit for 2013-2017.

The analysis allows us to conclude that 2017 was characterized by a dramatic decline in indicators in all business areas, except for the fat and oil segment. In the fat-and-oil segment, management errors were identified and eliminated, while simultaneously with the acquisition of assets and the adjustment of the strategy towards increasing the profitability of assets, the segment was successfully brought to the zone of high sustainable profitability indicators. In the meat segment, the downward trend in profitability, noted in 2016, continued to develop and, simultaneously with the acquisition of additional assets, brought the business into a zone of ineffective return on assets. According to the company's comments, the situation is not critical and despite the decline in the profitability of non-current assets, EBITDA showed growth due to an increase in the volume of these assets. In the current situation, according to the methodology of the applied analysis matrix, the company has two development alternatives: increasing the profitability of assets or removing them from the portfolio of business areas. In the sugar segment, the reason for the collapse in profitability is the pressure on prices, supported by the low world price, which led to a fall in the market in monetary terms. In 2017, sales prices for the company's sugar products fell by 23% to RUB 30.6 per kilogram excluding VAT. As a result, the profitability of the business has decreased. In the agricultural segment, there was also a landslide drop in prices for manufactured products due to record agricultural production in Russia and around the world. In addition to this, costs increased, in particular for fuel and lubricants, wages and land rent. The current state of affairs requires adjusting the company's strategy.

As a result of the study, the following main directions for the implementation of the agricultural holding's strategy can be identified:

In the segment of sugar production:

- increasing the productivity of factories in the Kursk region up to 5.0 thousand tons per day;
- reconstruction of the Krivets-Sugar plant;
- commissioning of a sugar silo with a capacity of 60 thousand tons.

In the segment of meat processing:

- increase in production up to 204 thousand tons of pork in live weight;
- management, control and reduction of production costs.

In the segment of agricultural production:

- an increase in the land bank by 20 thousand hectares: in the Tambov region - by 10 thousand hectares, in the Primorsky Territory - by 10 thousand hectares;
- technical re-equipment of the existing fleet of equipment, as well as providing new lands with equipment;
- business automation and launch of a breeding center for the development of improved sugar beet hybrids.

In the fat and oil segment:

- modernization of the existing production site of the oil extraction plant;
- construction of a site for obtaining lecithin;
- construction of a warehouse for finished products, designed to store the weekly production volume.

Thus, the use of matrix analysis tools in the long term will allow the agricultural holding to reach an acceptable level of profitability and will exclude from the development process a significant dependence on financing in the form of government support.

4 Conclusion

The strategic analysis tool presented in the article is used for the process of merging and developing enterprises, including agricultural holdings. The matrix of dependence of the company's profitability indicators on its relative value in the market allows to assess the risks that an enterprise may face without applying the correct strategy. The ways and methods of choosing the direction of development are considered. The analysis of the enterprise strategy is presented on the example of the agro-industrial holding "Rusagro" Group of Companies. The obtained research result is expressed in the form of a conceptual vision of the agricultural holding's strategy for future periods. The relationship between the value of non-current assets and the levels of profitability has been revealed. The investigated approaches make it possible to predict the market situation, analyze its processes and develop measures for the company's timely response to external changes. The results obtained in the work will allow developing the methodological aspects of the formation of holding strategies. The process of integration and merger of enterprises plays an important role for the country as a whole and can become an instrument for ensuring food security of the state, improving the situation of Russian enterprises, strengthening the country's economy as a whole.

References

1. V. Uzun, N. Shagaida, Z. Lerman, Land use policy, **83** (2019)
2. O. Kozlova, E. Lubkova, A. Shilova. IOP Conference Series: Earth and Environmental Science. **459**, 6 (2020)
3. R. E. Hoskisson, T. A. Turk, Academy of management Review, **15,** 3 (1990)
4. M.Graubner, I. Ostapchuk, T. Gagalyuk. European Review of Agricultural Economics, **48**,1 (2021)
5. O.Skydan, O. Budnik, L. Sus. Agricultural and Resource Economics: International Scientific E-Journal, **7** (2021)
6. A. Plakhin, E.Ogorodnikova, M. Khokholush, T.Guseva, M. Selezneva, Revista Inclusiones, **20** (2020)
7. A.Plakhin, E. Ogorodnikova, T. Kochergina, T. Guseva, M. Selezneva Amazonia Investiga, **9**, 26 (2020)
8. N. Morozko, N. Morozko, V. Didenko Journal of Reviews on Global Economics, **7** (2018)

A healthy eating plan for employees engaged in oil and gas exploration and production in the Far North

Andrey Vekovtsev, *Elena* Vovk, *Valery* Poznyakovsky, *Boisjoni* Tokhiriyon*, and *Valentina* Lapina

Ural State University of Economics, Ulitsa 8 Marta, 62/45, 620144 Ekaterinburg, Russia

Abstract. Proper nutrition is undoubtedly one of the most important factors of good health. Currently, the priority field of study for the Russian nutritionists is the development of well-balanced nutrition plans which can help to prevent or alleviate the most common medical conditions and work-related diseases. Extensive research has been carried out for the state-led program for the Development of Hydrocarbon Resources in the Northern Regions of Russia, including the Far North, where the huge oil and gas fields are located, and the specialists of Art-Life scientific production association have developed a healthy eating plan to help people working in the Far North maintain general well-being.

1 Introduction

The extreme weather conditions of the Far North are well-known, and because of the climate-related health threats this climatic zone in Russia is regarded as highly unfavourable for living. Employees who are brought to the Far North from other Russian regions experience an extended strain arising from both extreme cold and difficult work conditions. Research has demonstrated a phospholipids deficiency which leads to the exhaustion of the body's compensatory mechanisms, occupational diseases and pathological processes. Therefore, in order to improve the living conditions, it is primarily important to develop well-balanced diets with all the necessary micronutrients and macronutrients included. At the same time, the costs should also be taken into the account, as a very limited range of foods can be produced in the Far North; most foodstuffs have to be delivered from elsewhere [1,6-15].

In addition to the unfavourable weather conditions, employees engaged in oil and gas exploration and production suffer from chemical, physical and other harmful factors of the industrial environment. These negative factors disrupt metabolism and the immune system, and consequently affect general well-being, work performance and life expectancy. Therefore, there are a number of factors to be taken into account when developing diets aimed at replenishing nutrients and improving the general well-being of people living and working in the Far North [1-5].

* Corresponding author: tohiriyoni@gmail.com

2 Materials and methods

The materials used are natural raw materials of plant and animal origin, semi-finished products, laboratory and experimental samples of the food supplement. Standardized and modified test methods have been used to assess the quality characteristics of the product

3 Results and discussion

Although, nowadays Russian authorities are looking into ways to promote healthy dietary practices, the recent data on food consumption indicate the overall decrease in the consumption of fruit and vegetables, meat, dairy and fish products as well as the increase in sugar consumption. These changes inevitably result in different noncommunicable diseases and conditions. Comparing the nutritional guidelines for different Russian regions, we can notice certain differences. The Nutritional guidelines for the regions of the temperate climate zone and the Far North developed by the Federal State Budgetary Scientific Institution - the Institute of Nutrition, Biotechnology and Food Safety are presented in Table 1.

Table 1. Daily Nutritional Guidelines

Foods	Temperate climate zone (gram)	Far North (gram)
Whole grains	64.0	345.2
Fruit and vegetables	302.0	315.0
potatoes	275 .0	300 .0
Meat	167 .0	250 .1
Dairy	486 .0	1747 .1
Fats	40 .0	85 .0
Sugar	86 .0	115 .0
Fish	51 .0	54 .8

As can be seen from the table, the differences are significant, as energy intake has to be in balance with energy expenditure. Although other individual characteristics like gender, lifestyle, and physical activity have to be considered, climate is still a very important one. Thus, the energy intake for people living in the Far North is one and a half times as large as the energy intake for people living in the temperate climate zone.

Because of the chronic cold exposure people working and living in the Far North have to adjust to cold temperatures, but cold acclimatization takes time and while people are adapting to cold, iron and fluorine deficiencies cause iron deficiency anaemia, increased dental caries, hypovitaminosis and unsettle the immune system. For these reasons, foods fortified with minerals and vitamins have to be included in the diets of people in the Far North to help maintain healthy dietary practices.

One of the possible solutions, which can successfully cater for employees engaged in oil and gas exploration and production in the Far North, provide all the essential micronutrients and allow to develop healthy food environments, is the use of dietary supplements.

The specialists of Art-Life scientific production association have developed their healthy eating plan to help people working in the Far North maintain a healthy diet and prevent work-related diseases. The plan includes the following dietary supplements:

- Soft drinks concentrates (a vitamin fortified beverage with a choice of flavours)

- Fruit and berry jelly drinks fortified with vitamins and calcium (a wide choice of flavours)
- Protein shakes with vitamins and amino acids
- Protein bars

In addition, Omega 3 dietary supplement is recommended to be consumed. The complex is a source of essential fatty acids, vitamin E, selenium, lipoic acid, and dihydroquercetin. The data on the healthy eating plan are presented in Table 2 and the recommended healthy eating plan is shown in Figure 1.

Table 2. The Proposed Healthy Eating Plan: Contents

Drinks	Recommended daily amount (prepared drink in ml)	Contents		% RDI per drink/bar
Made from soft drinks concentrate fortified with vitamins	200 (once a day)	Energy value, kcal	56.4	2.0
		Carbohydrates, g	10.51	3.0
		Organic acids, g	0.4	-
		Vitamin A, mg	0.5	62.5
		Vitamin C, mg	100	167
		Vitamin B_1, mg	1.1	79
		Vitamin B_2, mg	1.1	69
		Vitamin PP, mg	10	56
		Tannin, mg	11.25	3.8
		Caffeine, mg	7.5	15
		Silybin, mg	30	100
Fruit and berry jelly drinks fortified with vitamins and calcium	200 (once a day)	Energy value, kcal	49.05	2.0
		Carbohydrates, g	13.62	4.0
		Organic acids, g	174	-
		Vitamin A, mg	0.5	63
		Vitamin E, mg	15	150
		Vitamin B_1, mg	1.9	136
		Vitamin B_2, mg	3.0	188
		Vitamin B_6, mg	7.3	365
		Vitamin PP, mg	9.0	50
		Vitamin B_5, mg	6.5	108
		Vitamin B_{12}, mcg	6.5	650
		Vitamin B_9, mg	0.6	300
		Vitamin H, mcg	48	96
		Vitamin C, mg	30	50
		Calcium, mg	130	13
Protein shakes with vitamins and amino acids	100 (three times a day)	Energy value, kcal	123	15
		Proteins, g	8.12	32
		L-carnitine, g	0.13	129
		L-taurine, g	0.13	99
		Glutaminic acid, g	1.20	26
		Isoleucine, g	0.39	58
		Methionine + cystine, g	0.42	70
		Valine, g	0.42	50
		Tryptophan, g	0.13	49
		Phenylalanine + tyrosine, g	0.62	42
		Threonine, g	0.33	41
		Leucine, g	0.59	38
		Lysine, g	0.49	36
		Arginine, g	0.56	27
		Proline, g	0.41	27
		Histidine, g	0.18	26
		Glycine, g	0.26	22
		Aspartic acid, g	0.76	18
		Alanin, g	0.34	15

Table 2. Continued

		Serine, g	0.36	13
		Carbohydrates, g	12.2	10
		Lactulose, g	1.0	150
		Dietary soluble fiber, g	0.265	40
		Fats, g	4.65	17
		Medium chain triglycerides, g	0.39	-
		Vitamin B_1, mg	0.31	22
		Vitamin B_2, mg	0.37	23
		Vitamin B_3, mg	3.67	20
		Vitamin B_5, mg	0.92	15
		Vitamin B_6, mg	0.37	19
		Vitamin B_9, mg	0.074	37
		Vitamin B_{12}, mg	0.83	83
		Vitamin C, mg	12.8	21
		Vitamin E, mg	1.38	14
		Calcium, mg	144	15
		Phosphorus, mg	149	19
		Iodine, mcg	23.7	16
Protein bars	1 bar (60 gram)	Energy value, kcal	173.45	7
		Proteins, g	20.34	27
		Fats, g	2.89	5
		Carbohydrates, g	16.52	5
		Calcium, mg	40.5	4
		Phosphorus, mg	33.75	4
		Iodine, mcg	6.75	5
		Soluble dietary fiber, g	3.49	12
		L-carnitine, g	0.11	37
		Caffeine, mg	11	22
		Vitamin B_1, mg	0.42	30
		Vitamin B_2, mg	0.48	30
		Vitamin B_3, mg	5.4	30
		Vitamin B_5, mg	1.8	30
		Vitamin B_6, mg	0.6	30
		Vitamin B_9, mg	0.06	30
		Vitamin B_{12}, mcg	0.0003	30
		Vitamin C, mg	18.0	30
		Vitamin D, mcg	1.5	30
		Vitamin A, mg	0.24	30
		Vitamin E, mg	3.0	30
Omega 3 complex	1 capsule	Eicosapentaenoic acid	90	35
		Docosahexaenoic acid	60	
		Alpha linolenic acid	200	
		Lipoic acid	9.0	30
		Dihydroquercetin	7.5	30
		Vitamin E	3.0	30
		Selenium	0.01	15

* Recommended daily intake in accordance with Technical regulations of the Eurasian Economic Union 022/2011.

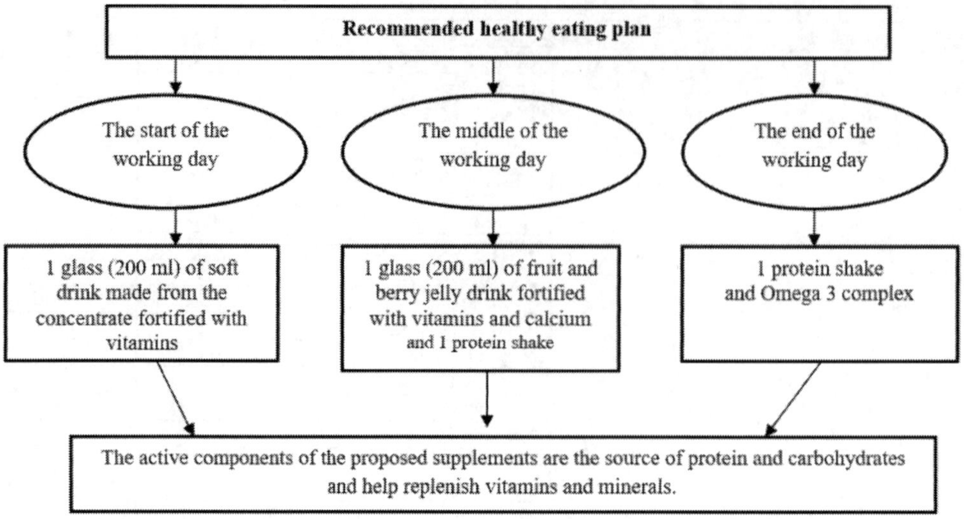

Fig. 1. Recommended healthy eating plan.

4 Conclusion

The proposed healthy eating plan helps counteract the effects of stress, improve digestion, and has a positive effect on gastric secretion. The active components improve liver function, regulate the viscosity and pH of blood and, thus, the heart rate and heart activity; strengthen metabolic processes, and increase efficiency. In addition, the active components make a favourable impact on the central nervous system, improving coordination of movements.

Vitamins perform catalytic functions and stimulate chemical reactions in the body. They assist in forming and functioning of enzymes, help absorb nutrients, enhance cell growth and overall strengthening of the body.

References

1. N.Kh. Amirov, Occupational medicine and industrial ecology, **4** (2003)
2. V.M. Akhmetov, Occupational medicine and industrial ecology, **5** (2002)
3. E.N. Belyaev, Nutrition issues, **3** (1996)
4. A.V. Shabrov, *Biochemical basis of the action of food micronutrients* (2003)
5. V.A. Tutelyan, Your food, **4** (2000)
6. B. Tokhiriyon, V. M. Poznyakovsky, S. Andrievskikh, Carpathian journal of food science and technology **12(1)**, 52–60 (2020)
7. B. Tokhiriyon, V. M. Poznyakovsky, N. M. Beliaev, International Journal of Pharmaceutical Research & Allied Sciences, **8(1)**, 115–122 (2019)
8. B. Tokhiriyon, E. V. Vyalyh, V. M. Poznyakovsky, Innovative technology of natural raw materials processing and biologically active complexes with systemic effect development, International Scientific and Practical Conference "Digital agriculture – development strategy" 167 (International Scientific and Practical, Ajman, 2019), 455–458 (2019)

9. A. A. Utebaeva, M. A. Burmasova, M. A. Sysoeva, Applied Chemistry and Biology, **6 (19)**, 100–109 (2016)
10. V. L. Fulgoni, D. R. Keast, R. L. Bailey, Journal of nutrition, **141**, 1847–1854 (2011)
11. H. Knopf, G. Sarganas, D. Grams, Bundesgesundheitsblatt – Gesundheitsforschung – Gesundheitsschutz, **62** (2019)
12. N. S. Karamnova, CardioSomatics, **8:1** (2017)
13. E. V. Averyanova, V. N. Khmelev, S. N. Tsyganok, V. A. Shakura, 18th International Conference of Young Specialists on Micro/Nanotechnologies and Electron Devices EDM 2017: Conference Proceedings, 255–259 (2017)
14. J. Sacco, V. Tarasuk, Procedia Food Science, **2**, 203–210 (2013)
15. M. F. San Martin, G. V. Barbosa Canovas, B. G. Swanson, Critical Reviews in Food Science and Nutrition, **42**, 627–645 (2002)

Ways to improve poultry productivity

Yulia Kolina[1], *Nadezhda* Momot[1], *Igor* Kamliya[1], *Aleksey* Trebukhov[2], and *Elena* Shaganova[2]

[1]Primorye State Agricultural Academy, 692510, Ussuriysk, Russian Federation
[2]Altai State Agricultural University, 656049, Barnaul, Russian Federation

Abstract. The article draws attention to the fact that the vitamin feed additive Introvit A+WS is an effective complex in its composition, consisting of 12 fat-and water-soluble vitamins, 18 essential and replaceable amino acids and 9 macro-and microelements that affect not only stability to various infectious diseases, but also to increase the meat productivity of young poultry. In recent years, industrially produced vitamins have become increasingly widespread, as well as feed additives used as a means of increasing animal productivity, reducing the cost of protein feed and increasing the efficiency of using feed nutrients. The vitamin feed additive Introvit A+WS in the indicated doses under production conditions had a positive effect on the weight gain of broilers and on the preservation of the poultry population, therefore, it shall be widely used in industrial poultry farming.

1 Introduction

A promising direction in increasing the productivity of poultry is the preservation of natural defense mechanisms inherent in poultry from nature, allowing in industrial conditions to keep adequately without losing productivity to respond to stress factors, to a lesser extent to be exposed to diseases of various etiologies [1-3]. Wherein, various feed additives increase the body's resistance to technological stress. They remove the negative immunosuppressive effect of vaccine drugs [4-8]. So, for example, the food additive Neokolistin – a water-soluble powder containing colistin, neomycin, tylosin and a set of vitamins, is used for therapeutic and therapeutic purposes for colibacillosis, pasteurellosis, salmonellosis, respiratory and other infections, the causative agents of which are sensitive to the components of the drug. The food additive Altisal-tm 1000 is prescribed for the treatment of poultry and pigs with gastrointestinal and respiratory bacterial infections caused by pathogens sensitive to tylosin. Water-soluble powder Ultravit AD_3E is prescribed for the treatment and prevention of a lack of vitamins A, D_3 and E in farm animals, including poultry. The feed additive Intromin Oral is used to balance the diet for missing minerals, to increase the digestibility and assimilation of nutrients.

Vitrocil is an antibacterial drug in the form of a solution for oral administration, intended for the treatment of pigs and poultry with gastrointestinal, respiratory and

The quantity and quality of food products, especially of animal origin, are of paramount importance in formation and maintenance of human health and maintenance of the adaptive capacity of his/her body to the environment. The quality of such products, namely, is determined by their microelement composition.

The purpose of this paper was to study the feasibility of using feed additives in poultry farming for the growth activity of young poultry.

To achieve this goal, the following tasks were identified:

- to study the high-quality vitamin composition of the feed additive Introvit A+WS;
- to investigate the amino acid composition of the feed mixture offered for poultry farming;
- to identify the presence of mineral elements in the composition of the additive used;
- to substantiate the production result with use of a feed additive, as well as without its use, on the growth activity of broiler chickens.

2 Materials and Study Methods.

The object of the study was broiler chickens of the Ross-308 breed, kept at the poultry farm Agroptitsa LLC in the Primorsky Territory. During the experiment on broiler chickens, two groups of birds, 50 heads each, were formed, selected according to the principle of analogues: experimental and control.

The conditions of keeping and feeding in all groups were similar, the chickens were housed on the floor throughout the entire period of rearing.

Access to water for broiler chickens was unimpeded. In addition to the main diet, the experimental group of young birds included a water-soluble vitamin additive Introvit A+WS in doses in accordance with the instructions for its use.

During the study, the following methods were used: clinical examination, palpation, thermometry, biometrics.

To determine the weight indicators, the chickens were weighed daily in the control and experimental groups. The control figures of the results obtained are shown on the 7th, 14th, 21st, 28th, 35th and 45th days of growing.

3 Results and Discussion

The vitamin additive Introvit A+WS used by us in the experimental group is a water-soluble powder containing a complex of vitamins, amino acids and microelements. It is used not only for the treatment and prevention of amino acid deficiency in farm animals, in stressful situations. In the experimental group, where the vitamin feed additive Introvit A+WS was used in young poultry, significant increases in the live weight of broiler chickens were revealed compared to the live weight of chickens in the control group on the 7th, 14th, 21st, 28th th, 35th and 45th days of growing. Introvit A+WS contains a complex of vitamins, both fat-soluble and water-soluble (Table 1). In addition, the feed mixture contains a wide range of nonessential and irreplaceable amino acids, as well as micro-macroelements (Table 2, 3).

Vitamins are known to be low molecular weight organic compounds belonging to various classes of organic substances. They are necessary for the vital activity of the body, as they participate in biochemical reactions, physiological processes. Most of the vitamins enter the body from the outside, with food or are added to the diet in addition.

Table 1. Vitamin composition INTROVIT A+WS

No.	List of vitamins	Name	Dose in 1 g of powder
		Fat soluble	
1.	Vitamin A	retinol acetate	20,000 IU
2	Vitamin D_3	cholecalciferol	6,000 IU
3.	Vitamin E	alpha-tocopherol acetate	60 mg
4.	Vitamin K_3	menadione	4 mg
		Water soluble	
5.	Vitamin B_1	thiamine hydrochloride	6 mg
6.	Vitamin B_2	riboflavin	12 mg
7.	Vitamin B_6	pyridoxine hydrochloride	8 mg
8.	Vitamin B_{12}	cyanocobalamin	60 mg
9.	Vitamin C	antiscorbutic vitamin	40 mg
10.	Vitamin B_9	folic acid	2 mg
11.	Vitamin H	biotin	300 mcg
12.	Vitamin remedy	niacin amide	80 mg

Along with proteins, fats and carbohydrates, which are a building, energy material, vitamins constantly, but in small quantities, must enter the body. A sufficient amount of vitamins has a positive effect on all organ systems of poultry, reduces susceptibility to infectious diseases. Hypovitaminosis – functional disorders are the cause of protein, fat, carbohydrate, and mineral metabolism disorders.

Table 2. Amino acid1 composition of INTROVIT A+WS

No.	List of amino acids	Group, replaceable/irreplaceable	Dose in 1 g of powder
1.	Alanin	Replaceable	2.24 mg
2	Arginine	Irreplaceable	1.24 mg
3.	Aspartic acid	Replaceable	5.10 mg
4.	Cysteine	Replaceable	1.14 mg
5.	Glutamic acid	Replaceable	8.80 mg
6.	Glycine	Replaceable	21.02 mg
7.	Histidine	Irreplaceable	1.12 mg
8.	Isoleucine	Irreplaceable	2.98 mg
9.	Leucine	Irreplaceable	4.74 mg
10.	Lysine	Irreplaceable	54.24 mg
11.	Methionine	Irreplaceable	50.82 mg
12.	Phenylalanine	Irreplaceable	1.32 mg
13.	Proline	Replaceable	2.68 mg
14.	Serine	Replaceable	2.72 mg
15.	Threonine	Irreplaceable	3.36 mg
16.	Tryptophan	Irreplaceable	0.84 mg
17.	Tyrosine	Replaceable	1.06 mg
18.	Valine	Irreplaceable	2.84 mg

Protein monomers are heterofunctional compounds - amino acids. Amino acids as constituent elements of proteins are involved in all vital processes and enter the body with food. The need for amino acids directly depends on the functional state of the bird and determines the metabolic rate, as well as the level of productivity.

Each protein in the body has its own primary structure, orientation in space. Synthesized from amino acids in the cytoplasm, the required protein chains are formed in sufficient quantities if amino acids are present in the required amount. THerefore, it is amino acids that are the most important nutrients.

Table 3. Composition of mineral elements INTROVIT A+WS

No.	List of minerals	Macro/microelements	Dose in 1 g of powder
1.	Choline chloride	Macroelement	5 mg
2	Iron sulfate	Microelement	12 mg
3.	Magnesium sulfate	Macroelement	12 mg
4.	Manganese sulfate	Microelement	12 mg
5.	Zinc sulfate	Microelement	12 mg
6.	Sodium chloride	Macroelement	100 mg
7.	Sodium sulfate	Macroelement	100 mg
8.	Potassium chloride	Macroelement	25 mg
9.	Calcium pantothenate	Macroelement	20 mg
	Total: 9 macro-microelements		

Organization of full-fledged feeding of animals is possible provided that all nutrients, including minerals, are provided in the diets in optimal quantities and ratios.

Mineral substances play an important and varied role in the animal organism. They determine osmotic pressure, affect energy, protein, carbohydrate and lipid metabolism, are a plastic material in formation of tissues and organs.

Mineral substances have no energy value, but their physiological role in the body is great. In the tissues of an animal organism, about 40 mineral elements are constantly found, but the physiological necessity has been proved so far only for 15, possibly necessary include fluorine, bromine, barium, strontium, calcium and phosphorus [9, 10].

Mineral elements, depending on their quantitative content in the body of the animal, are usually divided into two groups; the first includes the so-called macroelements – calcium, phosphorus, potassium, sodium, chlorine, magnesium and sulfur; the second one includes microelements – iron, zinc, copper, manganese, iodine, cobalt, molybdenum and selenium [9, 10].

4 Conclusions

Therefore, the vitamin feed additive Introvit A+WS is an effective complex in its composition, consisting of 12 fat- and water-soluble vitamins, 18 essential and irreplaceable amino acids and 9 macro-and microelements that affect not only resistance to various infectious diseases, but also to increase the meat productivity of young poultry. In recent years, industrially produced vitamins have become increasingly widespread, as well as feed additives used as a means of increasing animal productivity, reducing the cost of protein feed and increasing the efficiency of using feed nutrients. The vitamin feed additive Introvit A+WS in the indicated doses under production conditions had a positive effect on the weight gain of broilers and on the preservation of the poultry population, therefore, it shall be widely used in industrial poultry farming.

References

1. Yu. A. Kolina, Materials of the IV National (All-Russian) scientific and practical conference. In 4 parts, Ussuriysk, 312 (2020)
2. N. V. Momot, Hippology and Veterinary Medicine, **1(39),** 142 (2021)

3. N. V. Momot, Materials of the IV National (All-Russian) scientific and practical conference, Ussuriysk, 45 (2020)
4. Yu. O, Stoyanskaya. Proceedings of the 56th All-Russian Scientific Student Conference, Ussuriysk, 116 (2020)
5. V. E. Vorobyov, Innovations and modern technologies in agriculture: a collection of scientific articles based on the materials of the international scientific and practical Internet conference, Stavropol, 49 (2015)
6. E. I. Kolyvanov, Collection of scientific articles based on the materials of the IX International Scientific and Practical Conference dedicated to the 85th anniversary of the Faculty of Technological Management, Stavropol, 247 (2014)
7. G. Rusakova, Ya. Dergilev, M. Rusakova, materials of the international scientific and practical conference, Volgograd, 123 (2011)
8. N. Momot, Yu. Kolina, I. Kamliqa, A. Trebukhov, E. Shaganova, BIO Web of Conferences, **36** (2021)
9. J. Novotny, J. Pistl, D. Elias, H. Seidl, G. Kovac, R. Bobcek, Bull. Veter. Inst. in Pulawy, **47(2),** 559 (2003)
10. A. A. Spasov, Experimental and clinical pharmacology, **70(1),** 17 (2007)

Assessment of food security of the regions of the Arctic zone of the Russian Federation

A.N. Pestryakov, *N.V.* Sbrodova, *M.A.* Albycheva, and *E.A.* Reutova*

Ural State University of Economics, Yekaterinburg, Russia

Abstract. Ensuring food security and promoting the development of agriculture are included in the priority goals of sustainable development of the Arctic zone of the Russian Federation until 2035. The structural complexity and multidimensionality of food security in the region requires a clear methodology for its assessment. The article presents the results of a study of food security in the regions of the Arctic zone of the Russian Federation. This study is based on an analysis of the advantages and disadvantages of existing methods for assessing food security in the regions of the Arctic zone of the Russian Federation. The authors proposed a methodology for assessing the food security of the regions of the Arctic zone of the Russian Federation, which makes it possible to translate a phased multivariate analysis (the methodology was tested on the example of the Chukotka Autonomous Region).

1 Introduction

In 2010, according to the Decree of the President of the Russian Federation, the vector of the country's development in the field of agriculture was designated [1]. For Russia, the priority policy area has become food security, enshrined in the Food Security Doctrine. However, this document consolidated the provisions only for the territory of the entire country and did not have regional features. In this regard, on the territory of many constituent entities of the Russian Federation, their own normative legal acts were created to regulate relations to ensure food security. In these legal documents, systems of indicators for assessing food security are enshrined.

2 Materials and Methods

Currently, on the territory of Russia there is no unified system for assessing the food security of the region [6]. In the early studies of the authors, the current assessment methods enshrined in regulatory legal acts (including the Food Security Doctrine, the Law of the Murmansk Region "On ensuring the food security of the region", as well as the Resolution of the Government of the Yamal-Nenets Autonomous Region "On Yamalo-Nenets Autonomous Region"), analysis of security measurement tools was carried out.

The main disadvantages in the developed methods are the following provisions:

* Corresponding author: kata565722@yandex.ru

Lack of clear thresholds

The assessment tools do not cover all areas that form the food security of the region. The development of agricultural infrastructure, which provides storage, processing and transportation of manufactured products, remains not included in the methodology.

The methodologies are based on individual indicators that do not allow for an aggregate assessment.

The authors propose a methodology consisting of several sequential stages (table 1). At stage 1, an overall assessment of the level of self-sufficiency is given, a high indicator of which is the main task according to the Doctrine of the country's food security. Stage 2, including 3 indicators, establishes the main factors for the formation of a low level of food security in the region (low production, low consumption, low food quality). Stage 3 reveals the infrastructural features of the development of food security in the region, as the most important in ensuring food security in the regions of the Arctic zone of the Russian Federation.

Table 1. Methodology for assessing food security in the regions of the Arctic zone of the Russian Federation.

Stage	Rating parameter name	Calculation method	Parameter threshold
Stage 1. Food security	Self-sufficiency level	$\frac{\text{volume of food products produced in the region}}{\text{the volume of all food products in the region}}$	20%*
Stage 2. Food security indicators	Assessment of the physical availability of food	$\frac{\text{volume of food products produced}}{\text{rational consumption rates} * \text{population size}}$	100%
	Assessment of the economic availability of food	$\frac{\text{actual consumption food per capita}}{\text{rational consumption of food}}$	100%
	Assessment of food quality and safety	$\frac{\text{volume of low} - \text{quality and dangerous food products produced in the region}}{\text{the volume of all food products produced in the region}}$	0%
Stage 3. Scorecard.	Indicator 1. Consumer price index for the main socially significant food products	Index	30%**
	Indicator 2. Share of losses in the sale of food products	$1 - \frac{\text{the volume of all food products produced in the region}}{\text{production volume}}$	meat - 0.5%, bread - 0.5%, potatoes 0.6%, milk - 0.05%, vegetables 1.5% ***
	Indicator 3. Share of food stocks in total consumption by the population	$\frac{\text{food stocks}}{\text{actual food consumption per capita} * \text{population}}$	20%*

Table 1. Continued.

Stage 3. Scorecard.	Indicator 4. Availability of capacities for storage and processing of crop products.	If statistics are available, this is either a separate indicator or the amount of possible storage in warehouses / volume of products produced	100%
	Indicator 5. Volume of state social assistance to low-income families and low-income citizens living alone	$\frac{\text{amount of state aid}}{\text{the amount of funds required to provide for those in need}}$	100%
	Indicator 6. Volume of state support for legal entities and individuals engaged in the production of agricultural products per ruble of products sold	$\frac{\text{volume of state aid to agricultural producers}}{\text{agricultural products sold}}$	twenty%
	Indicator 7. Volume of state support for legal entities and individuals engaged in storage and processing of agricultural products per ruble of products sold	$\frac{\text{volume of state assistance to persons engaged in storage and processing of agricultural products}}{\text{agricultural products sold}}$	twenty%
	Indicator 8. Availability of areas for trade and catering per 1000 people	Separate indicator	208 m^2 ****

*The indicator is borrowed from [2].
**The indicator is borrowed from [3].
***The indicator is borrowed from [4].
****The indicator is borrowed from [5].

Based on the proposed system for assessing the state of food security in the Arctic zone regions of the Russian Federation, it is possible to comprehensively characterize the current position of the analyzed territory, as well as identify the factors of the greatest influence on the level of food security of the regions.

3 Results and Discussion

As part of approbating this methodology, it is proposed to analyze the food security of the Murmansk region, which is part of the Arctic zone of the Russian Federation. It is important to note that approbation is not possible for all indicators and (or) food products. This is due to the lack of statistical data both in the analyzed region and in most of the territories of the Arctic zone of the Russian Federation.

Stage 1. Food safety criterion.

The level of self-sufficiency for individual food products in the region is below the threshold values (Table 2). This is due to the lack of food production capacity. On the territory of the Murmansk region, there are only 7 enterprises operating in the animal husbandry industry, and 2 enterprises in the plant growing industry. These enterprises, in total, do not produce enough food to meet the demand of the region's population.

Table 2. The level of self-sufficiency in the Murmansk region for 2015-2020, in %.

	2015	2016	2017	2018	2019	2020
Meat and meat products	-27.23%	n / a	-304.32%	-337.18%	-340.10%	-391.09%
Milk and dairy products	99.79%	99.71%	98.85%	98.18%	99.43%	99.31%

Stage 2. Food security indicators.

A) Assessment of the physical availability of food reflects a negative trend towards low production of vital food products, which in general is the main negative factor that forms a low level of food security (Table 3).

Table 3. Physical availability of food in the Murmansk region for 2015-2020, in %.

	2015	**2016**	**2017**	**2018**	**2019**	**2020**
Meat and meat products	84.02%	21.57%	23.36%	23.51%	23.80%	20.32%
Milk and dairy products	7.22%	6.75%	7.41%	7.45%	6.88%	6.86%
Potato	72.93%	76.53%	68.08%	68.49%	71.29%	87.67%
Vegetables	3.82%	2.81%	4.59%	4.62%	4.68%	6.36%

B) The assessment of the economic availability of food products is at a low level due to the lack of food in the region and because of the consumption specifics of certain types of food. Due to the high demand for meat products, the region imports them from the neighboring regions and from abroad (Table 4).

Table 4. Economic availability of food in the Murmansk region for 2015-2020, in%

	2015	2016	2017	2018	2019	2020
Bread products	71.25%	74.69%	74.48%	72.60%	68.96%	68.44%
Potato	56.56%	59.56%	57.89%	55.44%	54.67%	53.22%
Vegetables and melons	66.07%	71.14%	66.29%	66.64%	64.79%	66.57%
Meat products	116.03%	120.14%	123.56%	122.60%	116.30%	120.27%
Eggs	87.31%	95.00%	98.08%	94.23%	92.31%	94.62%
Milk	82.50%	85.09%	85.12%	84.71%	81.76%	82.82%

C) The assessment of the quality and safety of food products [7] is at a low level due to food spoilage at the stages of its implementation, which is caused by the underdevelopment of infrastructure and logistics in the Murmansk region (Table 5).

Table 5. The share of hazardous products of the total food supply, percentage recorded on the territory of the Murmansk region by the supervisory authorities

	2015	2016	2017	2018	2019	2020
Meat	1.6	4.1	4.5	5.0	6.4	6.6
Bakery products	2.0	3.8	4.1	4.1	2.6	3.0
Milk and dairy products	7.4	6.8	7.4	14.3	3.5	2.1
Eggs	11.1	8.8	13.8	5.4	8.2	2.5
Vegetables	1.4	0	0.3	0.2	2.0	0.7
Potato	0	0	0	0	0	0

Stage 3. Scorecard.

Indicator 1. The consumer price index for the main socially significant food products as a whole did not exceed the threshold values in quarterly periods (Table 6).

Table 6. Consumer price index for basic food products in the Murmansk region for 2015-2020, in percentage

		Potato	Milk	Bird	Eggs
2015	I	101	103	100	117
	II	107	104	100	95
	III	88	89	100	100
	IV	120	103	101	119
2016	I	74	108	n / a	90
	II	127	104	n / a	81
	III	77	99	n / a	99
	IV	119	104	n / a	118
2017	I	86	102	99	98
	II	124	99	102	94
	III	85	101	100	97
	IV	107	105	97	118
2018	I	88	101	96	105
	II	114	98	97	95
	III	84	94	107	99
	IV	101	107	107	123
2019	I	122	101	98	101
	II	80	101	93	96
	III	150	102	98	93
	IV	77	98	107	105
2020	I	111	101	101	105
	II	142	101	97	108
	III	76	99	98	95
	IV	74	99	108	90

Indicator 2. The underdeveloped storage facilities, food processing and transportation in the region (table 7) also cause the share of losses in the sold food products.

Table 7. Share of losses in the sale of food products.

	2015	2016	2017	2019	2020
Potato	19%	19%	ten%	eight%	5%
Milk and dairy products	1%	1%	1%	1%	1%
Vegetables and melons	100%	77%	67%	thirty%	twenty%
Eggs	n / a	2%	3%	n / a	n / a

Indicator 3. The share of food stocks in the total food consumption is below the norm also due to the lack of storage facilities for the produced foods (Table 8).

Table 8. Food stocks in the Murmansk region for 2018-2020, percentage of consumption.

	2018	2019	2020
Corn	0.51%	0.00%	0.00%
Milk	4.49%	2.82%	2.64%
Meat	3.65%	3.27%	3.27%

Indicator 4. There are no statistics on the availability of storage and processing facilities in the region. However, the media and speeches of local authorities indicate the presence of a significant problem in the development of infrastructure, which is the main factor shaping food security threats.

Indicator 5. The volume of state social assistance to low-income families and low-income citizens living alone significantly exceeds the required norms, which, on the one hand, has a positive effect on the social strata of the population and their level of consumption as well as on the economic availability of food. On the other hand, a disproportionate amount of allocated funds can lead to incorrect spending of the budget, which can also negatively affect the region in the future. At this stage, the authors recommend reallocating funds in the region (Table 9).

Table 9. The volume of state social assistance to low-income families.

	2015	2016	2017	2018	2019
The volume of state social assistance to low-income families	306%	363%	628%	931%	811%

Indicator 6. The volume of state support for legal entities and individuals engaged in the production of agricultural products per ruble of products sold is presented in Table 10. This indicator is within the normal range, which generally has a positive effect on agricultural enterprises in the production of products, but does not create the so-called "Donor needle" in the Murmansk region (table 11).

Table 10. The volume of state support for legal entities and individuals engaged in the production of agricultural products in the Murmansk region for 2015-2020 per ruble of products sold, %.

	2015	2016	2017	2018	2019
Murmansk region	6.43%	4.67%	4.74%	3.19%	3.98%

Indicator 7. Volume of state support for legal entities and individuals engaged in storage and processing of agricultural products per ruble of products sold. Currently, there is no state support in this direction, which poses the main threat to the food security of the entire region.

Indicator 8. The provision of areas for trade and catering per 1000 people has a positive trend towards growth, with the exception of a decrease in 2020. This decline was driven by the Covid-19 pandemic. In the authors' opinion, the growth of indicators will be able to continue and reach an acceptable level only if restrictions are reduced both for the local population and for the trade industry.

Table 11. Provision of areas for trade and catering per 1000 people, m^2

	2015	2016	2017	2018	2019	2020
Provision of areas	146.9	168.4	197.4	201	209.4	173.5

Summarizing the analysis carried out in accordance with our methodology, let us compare the data obtained in Table 12.

4 Conclusion

Thus, the level of food security of the Murmansk region in 2020 is assessed as low with high risks for this region in terms of self-sufficiency in food. At the same time, in general, the growth of indicators in dynamics is not significant, except for the improvement in food prices.

It is important to note that the main factor behind the backwardness of agricultural producers and forming the main threat to food security is the underdeveloped infrastructure that provides storage, processing and transportation of manufactured products. This lag is also due to the lack of investment in the development of sub-sectors from both the state and the region.

Table 12. Final Food Security Assessment.

Stage	Assessment parameter title	Assessment parameter	Score obtained in 2015	Score obtained in 2020
Stage 1. Food security criterion	Self-sufficiency level	4 (2 for each type of product)	2	2
		Total for stage 1:	**50%**	**50%**
Stage 2. Food security indicators	Assessment of the physical availability of food	8 (2 for each type of product)	0	0
	Assessment of the economic availability of food	12 (2 for each type of product)	2	2
	Assessment of food quality and safety	12 (2 for each type of product)	2	2
		Total for stage 2:	**12.5%**	**12.5%**
Stage 3. Scorecard	Indicator 1. Consumer price index for the main socially significant food products	32 (2 points for each quarter for each type of product)	20	25
	Indicator 2. Share of losses in the sale of food products	6 (2 for each type of product)	0	0
	Indicator 3. Share of food stocks in total consumption by the population	6 (2 for each type of product)	0	0
	Indicator 4. Availability of capacities for storage and processing of crop products.	0 (no statistics available)	0	0
	Indicator 5. Volume of state social assistance to low-income families and low-income citizens living alone	2	2	2
	Indicator 6. Volume of state support for legal entities and individuals engaged in the production of agricultural products per ruble of products sold	2	2	2
	Indicator 7. Volume of state support for legal entities and individuals engaged in storage and processing of agricultural products per ruble of products sold	2	0	0
	Indicator 8. Availability of areas for trade and catering per 1000 people	2	0	0
		Total for stage 3:	**46.2%**	**55.8%**

References

1. Decree of the President of the Russian Federation of 30.01.2010 N 120 "On approval of the Doctrine of food security of the Russian Federation".
2. Food and Agriculture Organization of the United Nations. Assessment of food security and nutrition. https://www.fao.org.
3. Law of the Sverdlovsk region of January 31, 2012 N 6-OZ "On ensuring food security of the Sverdlovsk region".

4. Order of the Ministry of Industry and Trade of Russia dated 01.03.2013 N 252 "On approval of the norms of natural loss of food products in the field of trade and public catering" (Registered in the Ministry of Justice of Russia 05.04.2013 N 27999).
5. Government of the Murmansk region Resolution of December 10, 2010 No. 549-PP "On approval of standards for the minimum provision of the population with the area of retail facilities in the Murmansk region".
6. A. Pestryakov, E. Reutova, N. Sbrodova and A. Titovets. International Scientific and Practical Conference "Fundamental and Applied Research in Biology and Agriculture: Current Issues, Achievements and Innovations" (FARBA 2021), E3S Web of Conferences, **254** (2021)
7. Materials for the state report "On the state of sanitary and epidemiological well-being of the population in the Murmansk region in 2020" Department of the Federal Service for Supervision of Consumer Rights Protection and Human Well-being in the Murmansk Region.

Assessment of the sustainability of multipliers as the core of the region's agroeconomic system

V.V. Kalitskaya[1], *A.A.* Pustuev[2,*], *O.A.* Rykalin[1], *O.V.* Mustafina[1], and *I.M.* Perminova[1]

[1]Ural State Economic University, Ekaterinburg, Russia
[2]Ural State Law University, Ekaterinburg, Russia

Abstract. This article examines the role of multipliers – subsectors of agriculture, as the core of the agroeconomic system of any region, using the example of the Ural Federal Okrug, and also evaluates their sustainability based on materials from 2010-2015. Three basic subsectors are analyzed as multipliers: grain production, dairy farming and poultry farming. As a result of the author's calculations using the appropriate methods, a conclusion was drawn about the dependence of the basic industries on each other, as well as on the market model in a particular territory.

1 Introduction

In the agroeconomic system, as a set of technologically interconnected industries and types of entrepreneurial activity aimed through market relations to solve the problem of food self-sufficiency with a predominant focus on import substitution, three basic sub-sectors can be distinguished as elements of this system: grain production, dairy farming and poultry farming, that is, which are the basis for solving the above problem. Wherein, as a backbone, as the "core" of the region's agroeconomic system (RAES), one can call its own grain production, without which it is impossible to solve such a problem as the problem of food self-sufficiency in the region.

2 Results and Discussion

The backbone side of grain production, in our opinion, is that this sub-industry, being the "core" of the RAES and closely interacting with its other elements, is capable of forming this system under favorable conditions for integration. Moreover, a sufficiently high mechanized basis of grain production provides certain guarantees for achieving sustainable functioning of the entire RAES.

Integration of grain production with the poultry industry can involve other potential applicants of the RAES in the development process, that is, become a kind of multiplier of

* Corresponding author: pustuev@mail.ru

its development. As one of the elements, one can name compound feed production, which cannot work without grain, and poultry farming — without compound feed.

Let us analyze the state of sustainability of the functioning of the basic industries of the Ural Federal Okrug in terms of their suitability to act as multipliers of the RAES development.

Let us assess the sustainability of the poultry industry for the main types of poultry products at all levels: federal, district and regional ones presented in Tables 1, 2, 3.

Table 1. Dynamics of changes in production volumes of the main types of poultry products in the Russian Federation [1, 2]

Years	Production volume (Q)	
	poultry meat, thousand tons	eggs, billion pcs.
2005	1,388	37.8
2006	1,632	38.2
2007	1,925	38.2
2008	2,217	38.1
2009	2,555	39.4
2010	2,678	40.8
2011	2,897	41.1
2012	3,318	41.2
2013	3,832	41.3
2014	4,088	41.3
$\bar{Q} = \sum_{i=1}^{n} Q_i/t$	2,653	39.7
$K_y = \bar{Q}/Q_{max}$	0.65	0.96

Despite the stable increase in the volume of poultry meat production, the stability coefficient for this product is lower than that for eggs, which is caused by the uneven growth of this product. For example, from 2009 to 2010, the increase amounted to 123 thousand tons, and from 2012 to 2013 – 514 thousand tons, which indicates an underestimation of this indicator, especially since a sharp increase in production volumes is not consistent with a decrease in the effective demand of the population and low exports.

Despite the steady growth of these indicators, own resources in the poultry products market are less stable, which can be explained by the variability of demand due to presence of imported products (poultry meat) and imported products from adjacent regions (egg products) on the market, for which prices, as a rule, are lower than for domestic agricultural products.

Table 2. Dynamics of poultry production in the Russian Federation and Ural Federal Okrug with the calculation of the sustainability coefficient [1, 2]

Years	RF		Kurgan		Sverdlovsk		Chelyabinsk		Tyumen	
	meat in carcass weight, thousand tons	eggs, mln pcs.	meat in carcass weight, thousand tons	eggs, mln pcs.	meat in carcass weight, thousand tons	eggs, mln pcs.	meat in carcass weight, thousand tons	eggs, mln pcs.	meat in carcass weight, thousand tons	eggs, mln pcs.
2000	1,035	34,085	7.9	167.5	67	1,381.6	64	945.1	18.9	1,031.3
2001	1,098	35,242	8.2	225.7	70	1,344	67	987	21.4	1,058
2002	1,118	36,378	9.6	215.5	72	1,379	69	1,034	22.8	1,109.2

Table 2. Continued

2003	1,104	36,625	10.9	248.3	75	1,395	71	1,108	23.7	1,173.5
2004	1,350	35,901	10.1	233.3	78	1,375	76	1,139	24.6	1,196.5
2005	1,388	37,140	9.7	192.3	80	1,379	88.5	1,187.7	26	1,255.8
2006	1,632	38,216	9.4	119	83	1,312	91.4	1,213	27	1,268
2007	1,925	38,208	10.8	120.1	89	1,201	110.7	1,284	27.5	1,246
2008	2,217	38,058	10.5	120.9	96	1,237	122.2	1,315	26.9	1,281
2009	2,555	39,429	10.5	126.3	101	1,259	138.7	1,361.1	27.1	1,323.2
2010	2,830	40,600	12.5	119.5	105	1,278	159.8	1,305.6	29	1,359.3
2011	3,204.2	41,113	12.7	127.9	110	1,293	173.5	1,322.2	31.2	1,398.7
2012	3,624.8	42,033	12.7	132	112	1,321	196.5	1,354.9	33	1,414.9
2013	3,817	41,286	12.9	113.8	114	1,387.9	220.3	1,403.9	33.2	1,425.9
2014	4,217	41,860	11.1	107	106	1,387.5	234.3	1,477	34.5	1,445.1
2015	4,535.5	42,571.7	12.2	102	108	1,442.9	277.5	1,498	31.9	1,458.3
On average	2,345.5	38,673.2	10.73	154.4	91.6	1,249.4	135	1,254.8	27.4	1,277.8
K_u	0.517	0.908	0.83	0.622	0.81	0.866	0.486	0.83	0.794	0.87

Although the oligopolistic nature of the poultry market and increase in sales of it to large retail chains are gradually stabilizing not only sales volumes, but also price ratios between domestic poultry organizations.

The most unstable nature of the dynamics of growth of poultry resources in the food market was noted in 2010: in the second quarter, in comparison with the first one, the increase amounted to 29 thousand tons, in the third one, in comparison with the second one, – 59 thousand tons and in the fourth one – 383 thousand tons (on average 157 thousand tons).

Table 3. Dynamics of quarterly changes in production-export-import elements in the poultry market in 2013-2014 [1, 2]

Years and periods	Meat			Eggs		
	production, thousand tons	import, thousand tons	export, thousand tons	production, billion pcs.	import, bln pcs.	export, billion pcs.
2013 Total:	3,831.6	550.2	53.8	41,286.0	8,292.0	290.0
incl. I quarter	903.7	109.0	10.6	9,795.4	163.7	50.3
II quarter	918.8	131.6	13.5	11,217.2	388.9	72.4
III quarter	934.7	153.2	13.9	10,607.4	223.6	84.0
IV quarter	1,074.4	156.4	15.8	9,666.0	216.7	83.3
K_u	0.89	0.88	0.85	0.92	0.63	0.86
2014 Total:	4,088.0	467.2	61.8	41,263.3	915.2	185.0
incl. I quarter	936.1	88.3	15.6	9,547.6	242.2	44.2
II quarter	980.5	137.1	15.9	10,900.9	228.9	52.8
III quarter	998.8	107.6	15.0	10,933.8	219.1	56.5

Table 3. Continued

IV quarter	1,172.6	134.2	16.3	9,881.0	225.0	31.5
K_u	0.87	0.85	0.96	0.94	0.95	0.82
Average K_u	0.88	0.865	0.905	0.93	0.79	0.84

The average quarterly increase in imports in 2009 amounted to 55.6 thousand tons, in 2010 – 86 thousand tons, which indicates a constant increase in this indicator and an increase in selling prices.

Let us analyze the nature of the change in these prices for the period 2010-2015 using the example of the Ural Federal Okrug (UFO). It is very difficult to solve this problem on average in the country, since each region and even a constituent entity of the Russian Federation has its own price level, which depends on such indicators as the degree of competition in the poultry market in terms of the share of each poultry organization in it, the effective demand of the population, and climatic conditions.

The data in Table 4 indicate a stable growth in prices for poultry meat and a somewhat uneven change in prices for egg products.

Table 4. Dynamics of change in selling prices for poultry products in the Ural Federal Okrug for 2010-2015

Products	2010	2011	2012	2013	2014	2015	On average for 2010-2015	K_u	2015 to 2010, %
Poultry meat	85.8	89.2	92.0	91.2	90.6	108.7	92.9	0.854	126.7
eggs	30.4	27.6	33.8	33.1	40.1	48.0	33.5	0.74	157.9

An increase in the cost of material and technical resources has a negative impact on the economy of poultry organizations. Considering the high costs of compound feed and a decrease in stability in income generation, some of these agricultural structures are trying to compensate for the risk of managing their core activities by developing diversification activities, producing other types of agricultural products, animal husbandry and grain production.

Wherein, grain can be used for the production of compound feed, which is more profitable in the context of integration with compound feed factories.

The next basic industry (sub-industry) is grain production, which is unstable due to unstable yield, which directly depends on natural and climatic conditions (Table 20). Some regions suffer from drought, others from excessive moisture, and often from aridity during the growing season of plants and moisture during harvesting.

According to average statistical data and the results of study by a number of scientists, especially the "pre-market" period, biological, technological and "man-made" grain losses amount to 15-20 million tons [3, 4, 5, 6].

For example, as a result of the harvesting period prolonged in 2015 due to rainy weather, about 30 % of grain crops remained unharvested in the Sverdlovsk region, some of which remained under the snow [7, 8, 9].

The high seasonal load on the harvesting area on the grain harvester and other reasons do not allow completing the harvesting process within the time recommended by science and practice of 10-12 days. For each day following the above-mentioned period, losses increase almost exponentially.

In this regard, there is a significant difference even between the biological yield of grain crops and the "bunker" weight of threshed bread [3, 10, 11, 12, 13].

The next system-forming element of the RAES is dairy farming, which in terms of the production of milk and dairy products in the region amounts to an average of 6.4 % of the all-Russian level. The dynamics of milk production in the region and the country is quite stable, K_u = 0.912 (in the Russian Federation – 0.95), in the context of the regions of the

Ural Federal Okrug K_u was: 0.784 – Kurgan region; 0.934 – Sverdlovsk region; 0.879 – Chelyabinsk region; 0.91 – Tyumen region.

The results of calculating K_u for the backbone subsectors are given in Table 5.

Table 5. The values of the coefficient of sustainability of production of products of backbone subsectors in farms of all categories of subjects of the Ural Federal Okrug[1)]

Product type	Objects for calculating K_u	The value of K_u by UFO subjects			
		Kurgan region	Sverdlovsk Region	Chelyabinsk region	Tyumen Region
Grain	Sown area	0.89	0.83	0.897	0.935
	Yield	0.63	0.75	0.683	0.76
	Volume of production	0.545	0.812	0.66	0.701
Milk	Volume of production	0.784	0.934	0.879	0.91
Poultry meat	Volume of production	0.83	0.81	0.486	0.794
Eggs	Volume of production	0.622	0.866	0.83	0.87

It shall be noted that the above presented systemic industries (poultry farming, grain production and dairy farming) are directly related to feed production, therefore, this industry can also be connected to the systemic ones. However, the instability of economic relations in the system of "grain-feed" is a limiting factor in the connection of this industry to the multiplier of the development of other industries, especially to grain production. The instability of price ratios between grain and compound feed producers is forcing agricultural organizations to produce compound feed on their own resources. This tendency is typical, for example, for the Sverdlovsk region, where its feed mills use fodder grain from farms in the Orenburg and Kurgan regions as raw materials. The main reason is the disagreement of Sverdlovsk grain producers to supply grain to feed mills at a reduced price.

Presence of price disagreements can be explained by different market models in which "grain growers" and compound feed producers are located. The former operate in a model close to perfect competition, and the latter – to an oligopoly.

3 Conclusion

However, not all measures for development of the poultry industry in the Russian Federation outlined in the program are being implemented. So, for example, in section 4.2. The (Market Regulation) program outlined the following measures:

1. To develop national standards for poultry products, regulatory legal acts and methodological documents on price regulation.
2. Improvement of legal regulation, namely, the definition of product quality criteria and the level of trade margins.
3. To create the infrastructure of specialized wholesale markets for poultry products with the allocation of land plots on priority terms in order to develop trading activities, regulate supply and demand, reduce the cost of selling products, increase the role of exhibitions and fairs.
4. To consider the issue of including the industry in the system of purchases of eggs, meat and meat products from poultry meat for federal and regional state needs.
5. To expand the network of branded stores of poultry products, increase advertising support.

It can be noted that clauses 3 and 5 have not yet been fulfilled, since their own wholesale and retail system in the poultry market has not been created in the interests of agricultural producers. In fact, it is captured by large private retail chains, where it mainly sells its products to poultry farms.

In conclusion, we note that in modern crisis conditions, justification of the stability of economic indicators can be carried out only within certain (limited) limits, since it is practically impossible to consider all risk situations of a bifurcation nature. Difficulties in the calculation process also arise due to the need to consider the social side of sustainability of agroeconomic - territorial systems, the effective management of the development of which depends on social and motivational factors.

References

1. On the situation on the poultry market in 2005-2014 (2015)
2. Information bulletins of the Rospitsesoyuz, 80-88
3. Regions of the UFO: the results of socio-economic development (2010)
4. V. Veklenko, E. Zolotareva, APK: economics, management, **10**, 49-56 (2002)
5. A.V. Golubev, A.A. Golubeva, Economy of agricultural and processing enterprises, **3**, 12-14 (2014)
6. P.M. Ivanov, Economics and Mathematical Methods, **42**, 51-59 (2006)
7. A.N. Kashtanov, *Sustainability of farming: ways to improvement*, 230 (1983)
8. Agriculture, hunting and hunting farms, forestry in Russia (official publication), Stat. collection: Federal State Statistics Service (2015)
9. V.K. Chashin, A.L. Pustuev, *Monitoring in the food security system*, 439 (2004)
10. A.A. Chernyaev, D.V. Serdobintsev, Economy of agricultural and processing enterprises, **6**, 27-31 (2013)
11. M.M. Yuzbashev, O.V. Popova, Bulletin of statistics, **9** (1980)
12. E.V. Stovba, Economics and Management, **5**, 52-55 (2003)
13. A.N. Semin, *Methodological support of scientific research of economic problems of development of the agro-industrial complex of Russia*, 318-331 (2016)

Factors of development of children's sports schools in rural areas

Maksim Kocheryan, *Irina* Vashlyaeva, and *Tatiana* Volovik

Ural State University of Economics, 620144, 8 Marta Str, 62, Ekaterinburg, Russia

Abstract. The article presents an analysis of the external and internal environment of children's sports schools in rural areas in the form of a SWOT matrix. Children's sports schools in rural areas are distinguished by a huge set of sports disciplines, many their own facilities and a low pricing policy. Among the most significant disadvantages are the lack of active promotion of schools and a small percentage of paid services compared to budget ones. To improve the situation of children's sports schools in rural areas, it is proposed to create a promotion program that will help increase the number of students and affect the growth of the organization's income.

1 Introduction

Sports organization a legal entity, regardless of its organizational and legal form, carrying out activities in the field of physical culture and sports as the main activity. The market for physical education and sports in the Russian Federation is mainly represented by state and municipal organizations. The goal of the effective management of the state physical education and sports organization is a constant desire to meet the needs and sports interests of various categories of the population in physical education and sports services. This goal is only possible based on a comprehensive assessment of customer satisfaction.

One of the types of physical education and sports organizations are children's sports schools. Such organizations are municipal budgetary educational institutions whose main goal is the implementation of educational activities on additional general educational programs for children under 18 years of age and the implementation of services in the interests of the individual, society, and the state.

Children's sports schools are non-profit sports organizations since the purpose of their activities is not to make a profit. At the same time, non-profit organizations can carry out business activities, but only to achieve the goals for which they were created. As noted in the works [1-3], activities in the field of physical culture are considered in the concept of social marketing, which requires considering the interests of consumers on the one hand, and the interests of society on the other [4,5]. This means that customer needs come to the fore, as well as social indicators and the associated pedagogical effect. In the marketing of sports services Kahle [6] and Ferrand [7], the main criterion should be end-user inquiries [8]. Accordingly, the authors Shannon [9] and Roy [10] propose the use of a wide range of analytical tools survey methods.

2 Methodology

The research methodology includes the combination of survey technology and SWOT-analysis for the formation of solutions in the management of the municipal sports organization.

As part of an independent quality assessment, in September-October 2021, a survey was conducted of citizens - recipients of services in the field of physical culture and sports. A total of 119 people were interviewed. Respondents were adult beneficiaries or parents of minors. The respondents' data is shown in Figure 1.

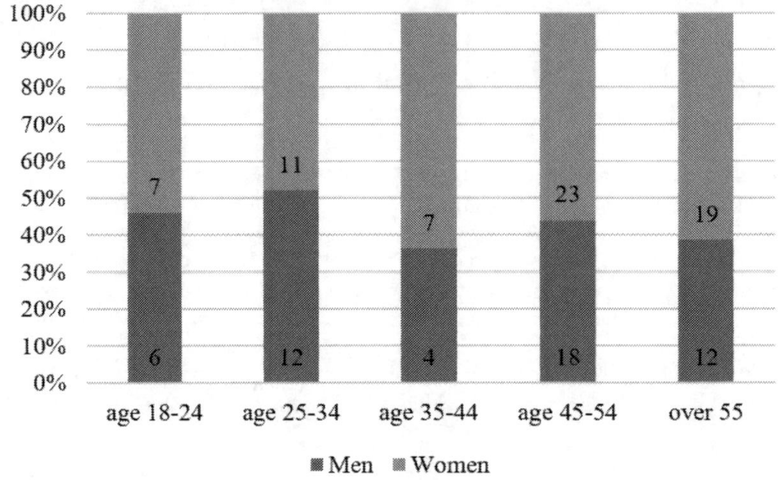

Fig. 1. Database of survey respondents

After assessing the quality of services by questionnaire methods, a generalization of the factors of a municipal sports organization's activities was carried out based on an analysis of its internal and external environment, including a strategic analysis of activities, a SWOT analysis and competitiveness analysis using a competitiveness polygon.

As a rule, the analysis considers the following factors of the internal environment that affect the activities of a children's sports school: organizational structure, sources of funding for activities, material, and technical base, labour resources, scientific and information resources. The external factors affecting the activities of municipal sports organizations include: the number of municipal subsidies, sports legislation, taxes, the economic situation in the country, the transfer of property rights to property and so on.

Figures and tables, as originals of good quality and well contrasted, are to be in their final form, ready for reproduction, pasted in the appropriate place in the text. Try to ensure that the size of the text in your figures is approximately the same size as the main text (10 point). Try to ensure that lines are no thinner than 0.25 point.

3 Results and Discussion

Next, we present the results of the study. Table 1 presents the results of a survey of recipients of municipal services in children's sports schools in rural areas.

Table 1. Quality assessment of the services provision in children's sports schools in rural areas, %

Quality indicators	Evaluation of the quality of services in %			
	Satisfied	Partially satisfied	Not satisfied	Difficult to answer
Building condition (premises)	68,3	27,5	4,2	-
Inventory status	89,4	10,6	-	-
The ability to get background information about the services provided	52,2	36,1	8,5	3,2
Satisfaction with personal interaction with the coaching staff	55,4	34	10,6	-
Service waiting conditions (conditions of waiting for children until the end of classes in the parents section)	55,7	34	10,3	-
Satisfaction with the opening hours	89,4	4,2	6,4	-
Satisfaction with employee competence	89,4	0	4,2	6,4
To what extent are you satisfied with the activity of the children's sports school?	59,6	23,4	-	17

As can be seen from the table, most consumers are satisfied with the activities of the municipal sports organization, however, it is necessary to note the partial satisfaction of consumers with personal interaction with the coaching staff and the opportunity to get background information about the services provided. These service parameters need to be improved.

According to the results of the analysis was formed a generalized matrix of the SWOT-analysis, presented in Table 2.

Table 2. The matrix of SWOT analysis of the activities of sports schools in rural areas

Strengths	Weaknesses
1. High qualification of the teaching staff 2. Minor staff turnover 3. The high level of quality of budget services 4. Low pricing policy 5. A large number of sports departments 6. A large number of infrastructure in fairly good condition 7. A large number of athletes studying on a budgetary basis 8. Good reputation due to the large number of athletes who have won various titles	1. Lack of time for the implementation of paid services 2. Low income from paid activity 3. Dependence on state authorities 4. Sports facilities are located in different places 5. Incomplete and sometimes untimely provision of equipment, inventory and outfit 6. The lack of own accounting (only central in the Office of Sports of the Sverdlovsk region) 7. Poor development of scientific and information resources, incl. advertising 8. The lack of young and promising coaches
Opportunities	**Threats**
1. State policy in the field of physical culture and sports will attract more people 2. Changes in the legislation of the Russian Federation, which will have a positive impact on municipal educational institutions 3. Transfer to use and management of a new high-tech sports facility 4. Attracting new consumer groups (people with disabilities) 5. The introduction of innovative educational technologies and changing the usual system of training 6. The emergence of new paid services	1. Reduction or invariance of the amount of municipal subsidies from the budget of the Sverdlovsk region 2. Inflation 3. The increase in the volume of municipal tasks 4. The emergence of new taxes and an increase in tax rates 5. The increase in the cost of utility payments; 6. Reducing the standard of living of the population

Children's sports schools in rural areas are distinguished by a huge set of sports disciplines, many their own facilities and a low pricing policy. Among the most significant shortcomings can be called the lack of active promotion of the school and a small percentage of paid services compared with the budget.

As for the external environment of the organization, the Russian economy is in an unstable state, so the incomes of the population are falling and the municipal budget also does not increase, which significantly affects the number of people involved in sports. In addition, the state policy now cannot be called favorable for the activities of such organizations, since new taxes constantly appear and utility payments increase, which occupy a significant part of expenses. However, currently, the state is actively creating strategic programs for the development and support of sports in the Russian Federation. Consequently, soon may be followed by favorable changes in sports legislation and in the system of training athletes.

4 Conclusion

Thus, because of the conducted research, the strengths, and weaknesses of the development of children's sports schools in rural areas were identified. The most vulnerable point is the lack of an effective promotion strategy, a strong link in comparison with competitors is the range of services and the price for them.

To improve the situation of children's sports schools in rural areas, it is proposed to create a promotion program that will help increase the number of students and affect the growth of the organization's income. Considering that the number of Internet users is growing every day, it will be effective to place your ads on various portals, forums, sites, on those where people read the news, because these sites are usually the most visited, as well as to actively develop pages in social networks such as Vkontakte and Instagram.

References

1. A. Buhler, G. Nufer *Relationship marketing in sports* (2012)
2. J. Burnett, A. Menon, D. T. Smart, Journal of Advertising Research, **33**, 5 (1993)
3. J. H. Donnelly, W. George *How consumers evaluation processes differ between goods and services* (1981)
4. S. B. G. Eysenck, D. H. Saklofske, Canadian Journal of Behavioural Science/Revue canadienne des sciences du comportement, **15**, 2 (1983)
5. G.R. Foxall, Journal of Marketing Management, **8**, 4, (1992)
6. L.R. Kahle, C. Riley. *Sports marketing and the psychology of marketing communication* (2004)
7. A. Ferrand, S. McCarthy, *Marketing the sports organisation: Building networks and relationships* (2008).
8. M. Mitre, P. Ordóñez de Pablos, International Journal of Sport Management and Marketing, **5**,3 (2009)
9. J. R. Shannon, Journal of services marketing, **13**, 6 (1999)
10. D.P. Roy, T.R. Graeff, Sport Marketing Quarterly, **12,** 3 (2003)

BIO Web of Conferences **42**, 05001 (2022) https://doi.org/10.1051/bioconf/20224205001
SDGE 2021

Process approach to organization of wild processing business

Galina Sergeevna Bondareva, and *Nikolay Sergeevich* Bondarev

FSBEI HE Kuzbass State Agricultural Academy, 650056, Markovtseva st., 5, Kemerovo, Russia

Abstract. The study examines the organization of a new business for the harvesting and processing of wild plants based on the development of traditional industries, trades and crafts of rural residents. As process solutions, a block of design justifications is presented, aimed at developing practical measures on the stated topic. In particular, the emphasis is placed on the economic justification and social significance when organizing a business for processing wild plants. In the substantive part of the study, a model for calculating capital investments is presented; sources of financing for a business project are indicated. Separately, it is worth highlighting the element of practical implementation, including the technical modernization of the production process of processing raw materials, namely, the improvement of the technological process through the use of microwave vacuum drying technology. The calculation of the need for personnel is performed; the possible qualification requirements for employees are indicated. The research of the market of similar products based on information search was carried out. An overview of the average retail prices is made according to the Internet portal "Russian agro-industrial server Agroserver.ru", on the basis of which the forecasted amounts of proceeds from sales and the expected profit from organizing a business for processing wild plants are calculated. Significant expected results are: an increase in the standard of living of the population collecting and harvesting wild plants; providing employment in rural areas; saturation of the market with products of wild plants; strengthening of competitive positions; saturation of the market with locally produced products; increasing the profitability of regional and local budgets. It is assumed that at the initial stage, the manufactured products will be sold mainly in the region. The sale will be carried out through trade networks, as well as in markets, stationary retail outlets, fairs, exhibitions, the main consumers of the products will be the population of the Kemerovo region.

1 Introduction

Recently, there has been great interest and a high level of demand for environmentally friendly products such as mushrooms, berries, medicinal herbs [1; 5; 10]. Due to the changing lifestyle of people, the popularity of this type of product is growing at an enormous pace.

Wild plants are understood to be renewable, natural, environmentally friendly natural products: herbs, mushrooms, berries [5; 10]. It should be noted that such products have no substitutes and have a fairly wide consumer demand, especially recently, due to the increased consumer demand for natural, environmentally friendly healthy food products [9; 12].

The scope of activities of the enterprises proposed for the organization: harvesting of natural non-woody plant raw materials (wild plants), preparing it for sale, subsequent processing and sale of wild plants and products of their processing to the end consumer.

As a result of the implementation of proposals for organizing a business for the processing of wild plants, it is planned to provide residents of the Kemerovo region, a number of regions of the Russian Federation and Siberia with high-quality competitive products of wild plants; ensuring employment of the rural population and, as a consequence, improving the quality of their life.

It is assumed that at the initial stage, the manufactured products will be sold mainly in the region. The sale will be carried out through trade networks, as well as in markets, stationary retail outlets, fairs, exhibitions, the main consumers of the products will be the population of the Kemerovo region.

2 Materials and Methods

To achieve the objectives of the study, the organization of business processes is presented on the basis of a process approach; all production is presented and described in the form of separate processes and stages. The theoretical provisions are based on the use of the monographic method, systematization, analysis of information; business economics is calculated based on the methods of economic and financial analysis, the forecast sales plan is built on the basis of expert judgment. Reflection of the final conclusions is based on an abstract-logical research method.

3 Results

This section proposes to consider the organization of business in key areas.

First of all, the organization of any business requires financial resources. To achieve the set goals, it is necessary to carry out technical modernization of the production process [8], to ensure its rhythm, to provide a quality component of manufactured products that is competitive in the market (local and external) not only in price, but also in quality.

It should be noted that there is a steady and growing demand for natural, and especially for organic food.

To carry out production for processing wild plants using microwave vacuum drying technology, the enterprise will need the following capital investments (Table 1):

Table 1. Model for calculating capital investments.

Category of costs	Average cost, rub.
Purchase, installation, adjustment of production equipment (microwave vacuum drying plant of the series "*Monsoon*")	550000
Vacuum packing machine	10000
Certification	50000
Rent	60000
Warehouse equipment	20000
Other expenses (packing materials - containers, special clothes, stationery, communication)	40000
Brand development and marketing strategy	20000
Total	750000

The total estimated cost of organizing a business for the processing of wild plants is 750 thousand rubles, including:

- investments in fixed assets and design work - 550 thousand rubles;
- investments in working capital - 40 thousand rubles.

Projects are supposed to be financed from two main sources of funds:

- funds in the amount of 500 thousand rubles, which can be attracted within the framework of grant support;
- own financing in the amount of 200 thousand rubles.

If the amount of own investments is much less or more, then problems arise in obtaining additional grant funds.

A registered owner is required, one person is enough. As a result of the implementation of a business for processing wild plants using microwave vacuum drying technology, on average, 3 full-time workplaces are created (Table 2).

Table 2. Average need for personnel for organizing the business of harvesting and processing of wild plants.

Employee position / Number, people	Salary
Head –Owner / 1	Due to the fact that the activity of the head is year-round, the salary has a fixed part and can be supplemented by a bonus part (depending on the availability of the bonus fund, its volume, performance results, etc.)
Dryer operator /1	Consists of a guaranteed salary and a bonus, depending on the volume of production, because the work of an employee directly depends on the workload of the equipment
Packing operator / 1	

In addition to the employees hired to work in the organization, outsourcing costs must be taken into account to support other business processes. In this model, the calculation of personnel costs was made in terms of the transfer of the implementation of the accounting and equipment maintenance functions. The main argument in favor of outsourcing these processes is the optimization of labor costs in terms of transferring insurance contributions to social funds. In addition, by establishing contractual relationships with specialists of the required profile, the enterprise relieves itself of the duty of a tax agent in terms of transferring personal income tax (Table 3).

Table 3. Qualification requirements for employees.

Position	Education	Qualities required	Work experience
Head	Higher	honesty, decency, conscientiousness, good knowledge of the computer, ability to manage	At least 1 year
Dryer operator	Vocational secondary	honesty, decency, conscientiousness, knowledge of the technological process, computer skills	At least 1 year
Packing operator	Vocational secondary	honesty, decency, conscientiousness, knowledge of the technological process, computer skills	At least 1 year

An enterprise with a simultaneous parallel processing of various raw materials (for example, drying of wild plants and packaging of finished products) requires a larger number of employees. At the same time, in some cases, the same personnel in different seasons can work in different areas.

The bulk of the personnel will be seasonal workers, whose wages will be paid only during the period of their main activity under the piecework wage system.

Next, the costs are determined (a list of fixed assets with an indication of the cost for each unit, rent calculation) for organizing a business (Table 4)

Table 4. Composition and average cost of organizing a business for the harvesting and processing of wild plants (data of 2020).

Type of expenses	Amount, rub.
Microwave vacuum drying plant of the series "*Monsoon*"	550000
Vacuum packing machine	10000
Certification	50000
Rent	60000 (5 thousand rubles/month * 12 months)
Warehouse equipment	20000
Cost of the technological process of drying wild plants	137000 (2 kWh * 12,500 kg * 3 h * 3.34) 3 h – average operating time of the dryer, 2 kWh/kg - average energy consumption of the dryer, 3,34 – coefficient of conversion of green raw materials, 12500 – final dry product
Other expenses (packing materials, containers, special clothing, stationery, communications)	40000
Brand development and marketing strategy	20000

The cost of the purchased equipment includes the total set of costs for the acquisition, installation, and adjustment of equipment, i.e. bringing it to full operation - "ready-to-operate". Due to the fact that production involves the creation of a full cycle from independent harvesting (collection of wild plants) and subsequent processing, bringing to the sale of processed products of wild plants to end consumers, there is a need to purchase a vacuum packer.

In the process of processing wild plants, it is permissible and quite rational to use the functional zoning of the rented production facility. This event will save money on the rental of additional premises, reduce the labor intensity of staff, by saving time [8; 13].

In addition to the economic feasibility of zoning, it is necessary to take into account the fact that the risk of contamination of raw materials, materials, semi-finished products and finished products cannot be excluded in the production process. In this connection, the production workshop must be divided into the following functional zones "clean" and "dirty", "wet" and "dry".

In this connection, the composition and amount of expenses includes the costs of equipping the premises: division into zones, creation of a rack system, technological passages, wiring.

Other expenses include: packing materials for packaging finished products (paper souvenir and birch bark packaging, polymer metallized vacuum packaging), containers, special clothing, stationery, communication services.

Having determined the expenditure components of the business, then the income component is determined on the basis of average retail prices for products (Table 5).

Table 5. Planning product prices

Product	Input parameters
MUSHROOMS dried, units – kg	
Porcini mushrooms 1 grade	3500.00 rub.
Porcini mushrooms 2 grade	2500.00 rub.
Aspen boletus	1700.00 rub.
Boletus	2000.00 rub.
Chanterelle	2200.00 rub.
Butterlets	1800.00 rub.
Honey mushrooms	1200.00 rub.
BERRIES, units – kg	
Blueberries (dried)	1300.00 rub.
Lingonberry (dried)	1350.00 rub.
Dried strawberries (wild) without stalks	3000.00 rub.
Hackberry (dried)	450.00 rub.
Rowan tree berries (dried)	300.00 rub.
Siberian rosehip	230.00 rub.
Hawthorn "red" fruit	250.00 rub.
Viburnum fruits (dried)	380.00 rub.
Dried buckthornberry (vacuum drying)	1200.00 rub.
Herbal mixture for bath steamers (phyto-collections, assorted) units – g	30 g. - 30 rub.\pc. 50 g. - 50 rub.\pc. 100 g. - 95 rub.\pc. 300 g. - 250 rub.\pc. 500 g. - 400 rub.\pc.
Siberian herbs (oregano, thyme, St. John's wort, tansy, etc.)	units – kg 50.00 – 2000.00 rub.

Reference: [3].

The presented prices for manufactured products are average market prices prevailing on the wild plants market in Siberia. Note that the price of food products is free and unregulated, depending on demand and market conditions [6].

Further, proceeding from the prices, the volume of sales was determined (Table 6).

Table 6. Forecasted sales volume (units).

Product	Quantity, kg/year
Dried berry	2500
Dried mushrooms	1250
Herbs	2500
Total:	6250

The quantitative data of the forecasted sales volume is determined by the minimum volume of production and subsequent sales in order to ensure a break-even business for the processing of wild plants, i.e. production volumes correspond to sales volumes. The real conditions are such that in good years it is possible to increase the volume of production by at least 2 times.

Based on prices and quantities, the amount of sales was determined (Table 7).

Table 7. Forecasted amounts of proceeds from sales (thousand rubles).

Product	Annual sales, thousand rubles
Dried berry	2250
Dried mushrooms	600
Herbs	1850
Total:	4700

Sales proceeds are calculated based on the sales price of the product. The prices for the products are set taking into account the prevailing prices on the market of wild plants in Siberia (Tomsk Region, Novosibirsk Region, and Altai Territory). In accordance with the marketing strategy (ensuring the competitiveness of goods), it is permissible to reduce the price of manufactured products by 10-15% below the market average [2; 7; 14].

4 Discussions

The profitability of organizing a wild plant processing business is a determining factor. In a market economy, the goal setting of any commercial enterprise is profit and its growth while minimizing costs. A steady growth in the results of financial activities is the basis for the development of the enterprise, the expansion of the scale of its activities, and the strengthening of competitive positions.

The following is the calculation of the expected result from the implementation of the business (Table 8).

Table 8. Forecast calculation of the expected profit from the organization of a wild plant processing business.

Indicator	Values
Net profit, thousand rubles	487
Return on sales, %	10,4
Payback of subsidies, months	13
Budgetary payback, months	17
Social efficiency of the project, workplaces	3

As a result of the calculations, it was determined that according to the results of the forecast period, the business will receive a net profit of 487 thousand rubles, while the return on sales will be 10.4%. The total payback of the received subsidy is 13 months, the budget payback is 17 months, and the social efficiency of the project is manifested in the creation of new jobs in the amount of 3 units.

A forecast sales plan is a kind of a specific program of activities aimed at selling finished products.

In the process of forecasting the sales plan, a research method of expert judgment based on knowledge and entrepreneurial experience was used (Table 9).

Table 9. Annual forecast sales plan, thousand rubles.

Type of products	Month										Year
	3	4	5	6	7	8	9	10	11	12	
Dried berry	Preparatory stage			75	225	425	525	450	300	250	2250
Dried mushrooms				50	100	100	100	100	100	50	600
Herbs			50	100	500	650	750	600	550	500	1850
Total:			50	225	825	1175	1375	1150	950	800	4700

At the initial stage of business organization, the acquisition, transportation, installation, adjustment of equipment, preparation of the rented premises, its zoning, conclusion of contracts for the sale of finished products, continuous search for potential sales markets will be carried out [2; 13; 15].

With the beginning of the spring-summer period, work begins related to the purchase of raw materials for subsequent processing, using equipment - microwave vacuum drying.

Thus, throughout the entire forecast period, the proceeds from sales for all types of products will be 4,700 thousand rubles.

The most affordable way to reduce the cost of paying taxes for persons engaged in the procurement and processing of wild-growing raw materials is to use the taxation system for agricultural producers (unified agricultural tax - UAT [4]).

At the initial stage of determining tax payments to budgets of different levels, the cost of production was predicted (Table 10).

Table 10. Forecast cost of production in 2021, thousand rubles.

Indicator	2021 год
Purchase cost of products, total, including:	1700
berry	750
mushrooms	200
herbs	750
Cost of the technological process of drying raw materials	137
Equipment acquisition and installation costs	550
Remuneration for labor with deductions	1410
Premises for rent	60
Material costs	150
Other expenses	125
TOTAL	**4182**

Thus, the purchase cost of raw materials by type of product: berries, mushrooms and herbs will be 750, 200 and 750 thousand rubles, respectively. The costs for the technological process of drying will amount to 137 thousand rubles. The forecasted total costs of personnel remuneration, including deductions, will amount to 4182 thousand rubles

The calculations made it possible to predict the final financial results of activities and calculate the amount of tax payments (Table 11).

Table 11. Forecast calculation of financial results and tax payments for 2021, thousand rubles.

Tax type	Tax base	Tax rate, %	Tax amount per year, thousand rubles
Contributions to extrabudgetary funds (pension fund, compulsory health insurance fund, social insurance fund)	1085	30	325
SAT value	4700-4182=518	6	31
Total tax payments			356
Net income			487

It is necessary to take into account that in addition to taxes paid by a legal entity as a tax agent (tax on personal income) and tax deductions from the wage fund, it is necessary to take into account the amount of payments under the special tax regime (SAT - Single Agricultural Tax) with a tax base "income-expenses" and a tax rate of 6% [4].

Consequently, the total amount of tax revenues to the budgets of different levels will amount to 356 thousand rubles.

5 Conclusions

The presented calculations allow drawing the following conclusions:

1. The process approach to organizing a wild plant processing business requires significant initial financial resources;
2. Even in the first season, income from the business commercialization allows recovering capital investments and operating expenses for the first year of implementation;
3. Subsidy payback and budgetary payback "come" in the second season of business implementation;
4. Despite the pronounced seasonality of income generation, the business organization for the integrated processing of wild-growing raw materials can reduce the risks of crop failure for each individual species of wild plants;
5. Organization of a business for processing wild plants allows creating more jobs, attracting seasonal workers (rural residents) in an amount of at least 250 people;
6. There is an opportunity to saturate the market of the Kemerovo region with natural, environmentally friendly products of healthy food.

References

1. T. G. Skurikhina, *Business model of organization the complex harvesting of wild-growing raw materials in the territory of the Tomsk region: study guide*, 134 (2015)
2. A. Osterwalder, I. Pigne, *Building business models: Handbook of the strategist and innovator*, 288 (2011)
3. Internet portal: Russian agro-industrial server Agroserver.ru, www.AgroServer.ru
4. Tax Code of the Russian Federation (part two), **117-FZ** (2000)
5. V. Kolerova, Business magazine, **3(239)** (2016)
6. A. A. Stepanova, Agriculture, **1,** 33 (2020) https://nbpublish.com
7. L. Argote, P. Ingram, Organizational Behavior and Human Decision Processes, **82(1),** 150 (2020)
8. M. Aurisicchio, A. Z. Van Der Laan, M. Tennant, Sustainable Production, Life Cycle Engineering and Management, 89 (2021)
9. M. Dewi, I. N. P. Aryantha, M. Kandar, International Journal of Recent Technology and Engineering, **8(2),** 81 (2019)
10. Y. Dogan, I. Ugulu, N. Durkan, Pakistan Journal of Botany, **45 (SPL.ISS),** 177 (2013)
11. F. Fusté-Forné, Tourism Geographies (2019)
12. R. Karousou, S. Deirmentzoglou, Journal of Ethnopharmacology, **133(1),** 191 (2011)
13. C. M. Lopes, A. Scavarda, L. F. Hofmeister, A. M. T. Thomé, G. L. R. Vaccaro, Journal of Cleaner Production, **1(142),** 476 (2017)
14. Ł. Łuczaj, V. Lamxay, K. Tongchan, K. Xayphakatsa, K. Phimmakong, S. Radavanh, V. Kanyasone, M. Pietras, M. Karbarz, Journal of Ethnobiology and Ethnomedicine, **17(1)** (2021)
15. L. Pereira, J. Santos, Á. Dias, R. Costa, International Journal of Knowledge Management, **17(1),** 1 (2021)

Agroecological cluster as a system integrator of the Baikal region

Evdokia Dugina, *Elena* Dorzhieva*, *Victor* Belomestnov, and *Luydmila* Garmaeva

East Siberia State Technical University, Department of Production Economics, Organization and Administration, 670013, 40 v, Klyuchevskaya st., Ulan-Ude, Russia

Abstract. The Baikal Region, despite its favorable geopolitical position (between Europe and Asia) and the most famous lake on the planet, located in its territory, is peripheral and depressive. The lag in socio-economic development is due to severe environmental restrictions imposed on economic activities. New business models and process solutions are necessary to enable the transition to a digital economy without harming the environment. Development of a agroecological cluster based on a digital platform will help increase the connectivity of the macroregional space, its integration and the rational use of the resources of the Baikal ecosystem. Due to generation of a synergistic effect that manifests itself in the socio-economic and environmental spheres the agroecological cluster acts as an integrator uniting the subjects of the "state – society – business" triad on the basis of mutually beneficial partnership.

1 Introduction

From the point of view of spatial development, the Baikal Region, which is a bridge between Europe and Asia and the area of intersection of the most important world axes (Christianity - Buddhism, West - East) [1] is of great geopolitical importance for Russia.

Revealing its integrating potential requires a transition to a targeted socio-economic partnership of three constituent entities of the Russian Federation - the Republic of Buryatia (RB), Irkutsk Oblast (IO), Trans-Baikal Territory (ZK) - and presence of a unique Lake Baikal postulates the need for close cooperation in the field of ecology to maintain stability of the Baikal ecosystem.

To transform the so-called "Baikal factor" from a restrictive barrier that impedes development of the regional economy into its competitive advantage, joint efforts of the RB, IO and ZK are required. Active actions of the "regional authorities - society - business" triad to expand and deepen interregional ties and establish effective interaction can launch the process of transformation of three neighboring regions of the Russian Federation into a single Baikal socio-ecological and economic system, which benefits from use of the Baikal brand within the country and for its outside.

* Corresponding author: elendorg@mail.ru

An inter-territorial agroecological cluster (AEC) can act as an integrator, the network structure of which allows it to flexibly integrate into modern production networks and digital industrial systems. AEC unites enterprises of the agricultural sector, companies from the supply, service, infrastructure sectors, public organizations, and etc.; their cooperative and integration interaction makes it possible to arrange the full-cycle agricultural production, which brings a synergistic effect to both the AEC participants and the socio-ecological-economic system as a whole through use of innovative agrobiotechnologies, digital solutions and modern business models focused on the balanced development of all stakeholders and maintaining the stability of ecosystems.

2 Materials and Methods

The theoretical and methodological basis of the study was the fundamental provisions of the theory of regional and spatial economics, general systems theory and the theory of competition. Within the study, an integrated approach was used, including a range of different methods and tools: categorical method, logical-structural analysis, synthesis, general scientific research methods, situational analysis, and etc.

3 Results and Discussion

The prospects of the Baikal Region entirely depend on how long-term vectors of development of the three constituent entities of the Russian Federation (RB, IO, ZK) correspond to the vector of the well-known and recognizable brand of Lake Baikal.

However, different visions of the mission and goals in the strategies of socio-economic development of these regions hinder the integration and effective interaction of regional authorities. The strategic priorities of the RB are closest to the principles of a green economy; IO focuses on the development of human capital, consolidation and attraction of the population, education of regional patriotism; ZK – to improve the level and quality of life of the population as a result of sustainable economic growth. The IO priorities mention the need to preserve the unique ecosystem of the region; for the regional authorities of the ZK, environmental problems, despite their obvious severity, are far from the foreground.

We agree with V.G. Belomestnov, B.B. Sharaldaev, Ch.D. Dashitsyrenov, I.A. Belomestnova, that one of the main mechanisms of influence of regional administrations on the subjects, ensuring the coordination of goals and strategies of the territories, is the organization of economic relations in forms that allow the interests of the subjects to be coordinated with a common mission and strategic priorities for development of the macroregion [2]. This mechanism can be formation of inter-territorial economic clusters, which will ensure coordinated spatial socio-economic development of regions. The authors offer introduction of innovation and integration network clusters (IINC) as the main tool for formation of such clusters. It is believed that production clusters and network regions provide the national economy with the strongest competitive position in the global market, since the synergistic effect of network mobilization makes it possible to respond more quickly to changes in the system of global exchanges.

Wherein, the authors note that the effect of clustering for leveling the level of development of regions is limited, because the complete elimination of socio-economic differentiation is, in principle, impossible: the goal is to increase the economic activity in the regions as much as possible and, on this basis, to increase their social attractiveness.

T.A. Shibaeva, A.V. Glotko adhere to a similar point of view: the primary task is to find an effective mechanism for managing the region, considering the economy clustering. An effective method is cluster-network management, in which a region with a predominantly cluster structure of the economy is viewed as a constantly forming and changing cluster-network space [3]. Cluster subjects form network connections and interactions around clusters: intra-cluster, inter-cluster, external. Development of network relationships (permanent and temporary) occurs when arranging production and expanding the scope of clusters. Since networkization is becoming a form of integration of interactions, we can talk about emergence of specific hybrid forms of economic entities - cluster-network [4].

Networked cooperation presupposes integration of the interests of all subjects of cluster-network associations and contributes to formation of their overall competitiveness. When managing such associations, the speed of building and saturation of network structures with network interactions at the interregional and international levels is of particular importance.

As one of the promising forms of network structures in the agro-industrial complex, T.A. Shibaeva, A.V. Glotko called the agro-industrial cluster characterized by systemic synergistic interaction of agricultural producers, food and processing enterprises, scientific and educational institutions, regional authorities. Traditionally, clusters are formed within the borders of the constituent entities of the Russian Federation, however, if it is necessary to expand the raw material zone and use the innovative potential of other regions, it is advisable to create interregional integration structures, the functioning of which requires a high level of development of ICT and transport and logistics infrastructure [3].

Considering the environmental risks and threats, in the Baikal region it is advisable to stimulate those activities that cause minimal damage to the Baikal ecosystem and contribute to increase in the level and quality of life of the population, i.e. bring a synergistic effect: in the economic, social and environmental spheres. It is necessary to abandon traditional industrial policies and actively introduce new business models that can integrate into the digital economy. As already noted, clusters, due to their flexibility, adaptability, informality, combination of cooperation and competition, stimulate the processes of business digitalization. So, an interregional AEC, created on the basis of a digital platform, can perform the following functions (Fig. 1).

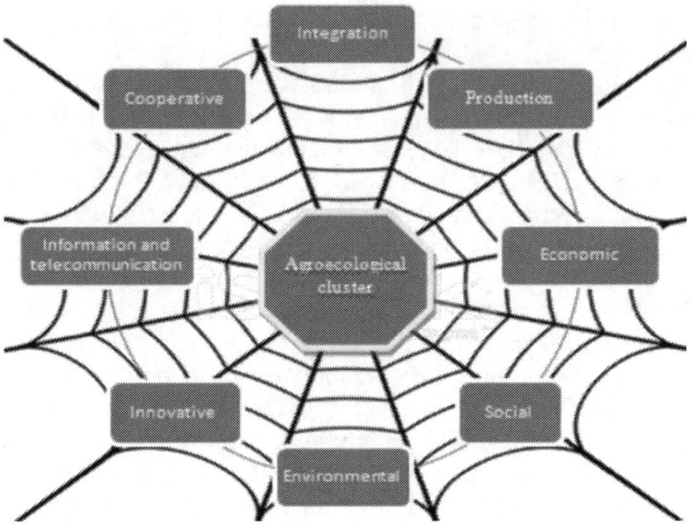

Fig. 1. The role and functions of the AEC in the Baikal Region.

Creation of an inter-territorial AEC will contribute to the gradual transformation of the Baikal Region into the Baikal socio-ecological-economic system based on increasing the connectivity of space as a result of the cluster performing the following functions:

1. Production – building value chains, including suppliers (feed, plant and animal protection products, equipment and technologies, etc.); agricultural producers (rural micro-clusters and clusters of small businesses (MFKh), specializing in production of farm and organic products for the healthy lifestyle market and proper nutrition; agricultural holdings and large agricultural organizations (SKhO) specializing in food for mass consumption); food and processing enterprises (whose main task is to increase the depth of processing, expand the range, develop the production of ethnic cuisine products); organizations providing transport and logistics, trade and intermediary services, ensuring the distribution and sale of products; enterprises engaged in waste processing. The expected result is an increase in production of livestock products by increasing the number of producers, increasing agricultural productivity with introduction of innovative technologies, and a more efficient organization of the sales system based on a digital platform.

2. Economic – at the micro level, saving on transaction costs, increasing profits, profitability, improving the financial condition of cluster members; at the meso-level, growth of GRP, employment, average per capita income, budget revenues at all levels, improvement of the business climate. At the level of enterprises and regions – an increase in competitiveness and economic efficiency. Development of economic activity in rural areas, diversification of the rural economy, business expansion opportunities using Internet communications and CRM systems (for example, Megaplan, Bitrix 24, Simple Business CRM, Microsoft Dynamics, Mango CRM, and etc.).

3. Social – sustainable development of rural areas, consolidation of the population in rural areas, preservation of the traditional way of life, traditions, culture, customs. Development of rural infrastructure by combining efforts of regional and municipal authorities, business, rural community, improving the demographic situation and the quality of life of the rural population. Accumulation of social capital (including using the Internet communities, groups, social networks), professional development of personnel (distance training, continuous self-education using materials posted on the

Internet, consultations, webinars, round tables in an online format). Increasing food security, improving public health while consuming high-quality and safe food (using a digital platform, software and hardware sensors and IoT technology will automatically download information on supplies and track the full path of food through supply chains. Buyers will be able to get all the information they need by scanning the QR code on the product packaging) [4].

4. Environmental – reducing the level of environmental pollution through use of innovative agrobiotechnologies in agriculture, the transition of a part of rural micro-clusters and clusters of small farms to organic livestock farming, waste processing, introduction of innovations and digital solutions in the agri-food complex, environmental education of the population (delivery of information through social networks, arrangement of environmental marathons, competitions for "green projects", and etc.) [5, 6].

5. Innovative – diffusion of innovations, development, distribution and implementation of innovative technologies in production and business processes, development of a regional innovation system in order to modernize the agri-food complex, digitalize the regional economy, implement joint innovative projects aimed at increasing productivity and productivity, resource conservation, rational organization of agricultural landscapes, preservation of the stability of the Baikal ecosystem, transition to an ecologically oriented innovation economy.

6. Information and communication – development of the ICT in the process of digitalization of the agro-industrial complex, increasing the availability of information, arranging the activities of companies on the basis of a digital platform, using viral marketing, promoting the AEC brand, entering the national and world food markets. Growth of network relationships, interaction with partners from other regions and countries, expansion of production and diversification of business due to erasure of boundaries and the ability to work with contractors from anywhere in the world [7, 8].

7. Cooperative – development of horizontal relationships between small, medium and large agribusiness, arrangement of networking between cluster members, regional authorities and society, joint implementation of infrastructure and other projects, attracting investors who prefer to invest in large-scale projects, formation of cluster-network systems as a result of the intersection of "key interests", development of intersectoral interaction, blurring the intersectoral boundaries in formation of the national digital ecosystem. Increase in the number of cluster members, intensive development of relationships, accumulation of social capital and diffusion of innovations lead to an explosive growth in the number of startups and proliferation of network structures.

8. Integration – formation and development of the AEC requires joint efforts of regional and municipal authorities, business and society. Integration of the Irkutsk Region, the Republic of Buryatia and the Trans-Baikal Territory into the Baikal socio-ecological-economic system becomes possible subject to formation of a unified vision of the prospects for development of the Baikal Region, implementation of interregional projects, and achievement of a balance of socio-ecological and economic interests. AEC functioning will contribute to the greening of the agro-industrial complex and transition to an environmentally oriented innovative development of the macroregion by ensuring a synergistic socio-ecological and economic effect at the micro- and meso-levels of the economic complex and localization of restrictions associated with action of the "Baikal factor".

This will undoubtedly serve as an incentive for deepening integration: the cluster can become a pilot project for working out mechanisms for implementing inter-territorial cluster forms of arranging the production and business processes aimed at increasing the connectivity of the Baikal Region. Nevertheless, the AEC is not the only and far from universal instrument for enhancing interregional interaction; its effectiveness largely depends on the level of development of the digital economy and readiness of society for large-scale implementation of fundamentally new technologies in Industry 4.0.

4. Conclusion

The future of the Baikal Region largely depends on the successful cooperation of the three constituent entities of the Russian Federation, in the territory of which the catchment basin of Lake Baikal is located. However, a different vision of the mission and goals of long-term development among regional elites is a disintegrating factor, the influence of which can be overcome through introduction of new business models focused on network interaction [9].

The role of a system integrator can be played by an agroecological cluster, which is characterized by a number of properties common to all clusters (network structure, hybridity, flexibility, adaptability, competition and cooperation, knowledge diffusion, and etc.), as well as capable of generating a synergistic socio-ecological and economic effect on all levels and areas, namely:

a) in the field of economics:

- for actors — profit, profitability, growth of competitiveness;

- for the region — growth of business activity, employment of the population, budget revenues, improvement of food supply, improvement of the system of inter-territorial food ties, development of infrastructural organizations;

b) in the social sphere:

- for the village — providing jobs, incomes, guarantees in the role of a backbone enterprise; fixing people in traditional places of residence;

- for the agro-industrial complex — recovery of the prestige of the labor of agrarians, establishment of inter-sectoral relations, introduction of innovative agricultural production;

c) in the field of ecology:

- for the environment — use of innovative agrobiotechnologies and digital solutions that minimize environmental damage;

- for biocenosis — biodiversity conservation, monitoring the environmental risks, implementation of an environmental safety system.

In the conditions of the Baikal Region, which suffers enormous economic losses due to the current strict protection regimes, the most urgent issues are environmental-oriented economic development, therefore, formation of an inter-territorial agroecological cluster can become a promising area of regional specialization. Development of organic livestock farming, use of digital and biotechnologies will allow inter-territorial clusters to develop intensively, stimulating entrepreneurial activity in the Baikal Region and reducing the anthropogenic load on the Baikal ecosystem. The so-called "Baikal factor", which is currently considered a limiting factor that reduces the economic attractiveness of the region, can be further regarded as a preference for development.

Acknowledgements

This work was made within the grant "Young Researchers of the East Siberia State University of Technology and Management".

References

1. V.V. Mantatov, *Baikal Region on the Way to Sustainable Development: Problems and Prospects*, Ulan-Ude, ESSTU (2005)
2. V.G. Belomestnov, B.B. Sharaldaev, Ch.D. Dashitsyrenov, I.A. Belomestnova, *Spatial socio-economic development of the region*, Ulan-Ude, ESSTU (2013)
3. T.A. Shibaeva, A.V. Glotko, VESTNIK ORELGIET, **2(52)**, 86 (2020)
4. R. Akhmadeev, A. Redkin, N. Glubokova, O. Bykanova, L. Malakhova, A. Rogov, Entrepreneurship and Sustainability, **7(2)**, 1149 (2019)
5. G. Daley, *Assessing our earth: Economics, ecology, ethics*. H. Daly, K. Townsend (Eds.), The MIT Press, Cambridge, MA. Russian Branch of ISEE, Moscow (1993)
6. A.V. Glotko, V.S. Balabanov, A.V. Balabanova, O.B. Repkina, L.A. Molchanova, R.A. Shichiyakh, Iranian Journal of Ichthyology, **8**, 92 (2021)
7. A.S. Khukhrin, O.I. Bundin, N.P. Tolmacheva, I.Yu. Agnaeva, T.F. Yanina, *Concept of the Scientific and Innovative Educational Cluster of the Agro-Industrial Complex*, M., All-Russian Research Institute for Organization of Production, Labor and Management in Agriculture (2015)
8. N.V. Smorodinskaya, *Globalized Economy: From Hierarchies to a Networked Treasure*, M., IE RAS (2015)

Role of clusters in development of the digital ecosystem of the agro-industrial complex

Elena Dorzhieva*, *Evdokia* Dugina, *Sayan* Alexeev, and *Nadezhda* Bulatova

East Siberia State Technical University, Klyuchevskaya st., 40 v, 670013 Ulan-Ude, Russia

Abstract. The main trends in agriculture over the past few years have been export orientation and digitalization. Since 2019, the departmental project "Digital Agriculture" has been implemented. In 2021, a draft order of the government of the Russian Federation "On the Strategic Direction in the Field of Digital Transformation of the Agro-Industrial and Fishery Complex of the Russian Federation until 2030" was developed, which provides for development of a national online platform for promoting the agricultural products and the launch of a modeling and forecasting system for the agro-industrial complex. Currently, the beneficiaries of digitalization are large agricultural holdings and the IT industry (manufacturers of sensors and software developers for machinery and equipment, software companies, fertilizer producers and telecom operators). In order for small and medium-sized farms that do not have sufficient resources for end-to-end digitalization of production and business processes to be able to join the innovation race, it is necessary to look for a format of interaction in which competition and cooperation are organically combined. The article discusses one of such forms of association – clusters; explores their potential role in development of platform network architectures; the similarities and differences of cluster and digital industrial systems are analyzed. It is concluded that the flexibility and openness of cluster structures allows actors to find their functional areas and niches and organically fit into the created digital ecosystem of the agro-industrial complex.

1 Introduction

In the digital economy, the list of cluster policy tasks is expanding: clusters become both subjects of a single digital space and objects of use of digital technologies. As a result of the digitalization of global supply chains, transformation of traditional business models, formation of global networks of highly adaptive integrated companies, implementing "smart" distributed production, self-optimizing on the basis of autonomous "communication" between digital devices are taking place.

Clusters are distinguished by high adaptability, flexibility, allowing them to integrate into network structures and industrial systems, an organic combination of competition and cooperation, developing the ability to work in collaboration, informal relationships between

* Corresponding author: elendorg@mail.ru

participants that contribute to formation of a culture of cooperation. Their role in development of Industry 4.0, which is fundamentally changing the paradigm of industrial development and blurring the boundaries between companies, sectors and regions, has not yet been fully defined. A number of researchers believe that clusters can become the core of platform network architectures [1, 2, 3].

2 Materials and Methods

The theoretical and methodological basis of the study was the papers of domestic and foreign scientists studying the peculiarities of the activity of clusters in the agro-industrial complex and the issues of managing the implementation and use of the ICT.

The instrumental and methodological apparatus is based on the basic methods of general scientific knowledge, theoretical, scientometric economic analysis, observation, generalization, abstraction, system analysis and synthesis, and expert assessment.

The studies used legislative and regulatory acts of the Russian Federation, data from ministries and departments, study results of the Higher School of Economics, institutes of the Russian Academy of Sciences, analytical reviews devoted to assessing the level of digitalization and clustering the Russian agro-industrial complex.

3. Results and Discussion

Blurring the spatial, temporal, industrial and sectoral boundaries leads to the fact that the staged value chain is gradually giving way to networks and further platforms [4]. Clusters, on the one hand, are geographically localized and have a key specialization, while digital industrial systems, due to remote interaction through Internet communications, carry out production everywhere and are focused on the diversification of activities. On the other hand, clusters and digital industrial systems have common characteristic properties (Table 1) [5].

Table 1. General properties of cluster and digital industrial systems.

Cluster systems	Digital industrial systems
Cooperation and competition between participants	Concept of "connected enterprise", in accordance with which the participants in the value chain interact with each other, while actively practicing outsourcing and diversification of suppliers
Exchange of knowledge, advanced technological solutions, diffusion of innovations	High level of communication technologies, growth of network density and intensity of network interconnections
Flexibility and adaptability of cluster structures to changes in external and internal factors	Mandatory requirement for participants is the ability to flexibly integrate into existing networks of enterprises and timely adjust strategies to achieve competitive advantages
Integration of production processes opens up access to resources and production factors (components, equipment, qualified personnel)	"Manufacturing as a service" model use is transforming manufacturers into service providers seeking to personalize customer experiences and diversify their businesses. This blurs the boundaries between sectors and makes it possible to take advantage of the resources and factors of the network participants on terms of lease, subcontracting, arrangement of combined production, and etc.

Clusters can become a testing ground for digital solutions because:

– the cluster environment favors the transfer of knowledge, development of cooperation between science and business, expansion of personnel competencies (within the framework of Industry 4.0, reformatting of the labor market is inevitable, as a result of which a number of professions will disappear and the demand for highly qualified workers with a wide range of competencies will increase);

– due to the factor of spatial proximity, informal contacts with contractors arise, which are practically impossible to maintain in a remote format. Clusters can become a platform for implementation of large-scale projects that require cooperation with actors from other regions, entry into global value chains and transformation into geographic agglomerations [6, 7].

According to the IAS (Information and Analytical System) of the Russian Cluster Observatory and GISIP (Geographic Information System of Industrial Parks, Technoparks and Clusters of the Russian Federation), 10 clusters in Russia can enter the digital ecosystems of the agro-industrial complex: the agro-industrial cluster of the Novgorod region; aquaculture and fisheries of the Astrakhan region; cluster for production and processing the dairy products "Donskoe dairy products" of the Rostov region; dairy cluster of the Vologda region; food cluster of the Republic of Tatarstan; agro-industrial cluster "Kuban" in the Krasnodar Territory; wine territorial cluster "Don Valley" in the Rostov region; biotechnology (a cluster for deep processing of grain in the Millerovsky district of the Rostov region); agrobiotechnological industrial cluster of the Omsk region; territorial and sectoral cluster AGROPOLIS "ALKIAGROBIOPROM" in the Republic of Tatarstan.

The analysis showed that among ten Russian clusters producing agricultural products (with a total number of participating companies - 201 and the number of employees in these companies - 31,663 people):

– only 2 have reached the average level of organizational development (agro-industrial cluster "Kuban" and territorial-sectoral cluster AGROPOLIS "ALKIAGROBIOPROM");

– half (50 %) are regional ones; their participants do not include enterprises and organizations from other constituent entities of the Russian Federation;

– 4 clusters (40 %) diversified their activities, covering a number of areas, which will be an advantage in formation of the national digital ecosystem of the agro-industrial complex.

The state agri-food policy is currently focused on the digital transformation of the agro-industrial complex with minimal participation of workers in production and business processes. The fourth industrial revolution affected all industries of the economy, including traditionally conservative agriculture. Use of innovative technologies, such as IoT (Internet of Things), AI (artificial intelligence), blockchain, makes it possible to create open digital platforms that cover all links and sectors of the agro-industrial complex and provide market players with a huge scope of options for making business decisions that provide a synergistic effect from expanding the interaction of industries and establishing a system of agri-food relations. The list of digital technologies that can be implemented in the agro-industrial complex is very extensive: from GIS (Geographic Information Systems) to the Internet of Things.

According to Order of the Ministry of Agriculture of the Russian Federation dated February 25, 2020 No. 84, a national platform "Digital Agriculture" shall be created by 2024, which will become the basis for building a whole ecosystem of additional services for the agro-industrial complex. In October 2021, the draft order of the government of the Russian Federation "On the Strategic Direction in the Field of Digital Transformation of the Agro-Industrial and Fishery Complex of the Russian Federation until 2030" was submitted for approval to the Ministry of Digital Development, Communications and Mass Media of the Russian Federation. The project provides for introduction of digital twin technologies, AI, IoT, drones, robotics, predictive analytics, remote sensing of the Earth; development of an online platform for promotion of Russian agricultural products and launch of modeling and forecasting systems.

However, directions of digitalization of the agro-industrial complex specified by the Ministry of Agriculture in the departmental project "Digital Agriculture", which is being implemented until 2024, raise many questions from practitioners. The criticisms point out that the digital technologies and solutions being introduced run counter to real optimization and savings, bringing benefits not to farmers, but to representatives of the IT industry. So, by 2023, the government plans to ensure the digitalization of reindeer herding farms ("Roadmap on the issues of indigenous peoples of the North, Siberia and the Far East of the Russian Federation in terms of development of animal husbandry"): 30 % of the reindeer population owned by farms receiving state subsidies will be equipped with collars with radio beacons; at least 30 % of reindeer farms will be equipped with quadrocopters; it is also planned to introduce support measures for operators providing satellite communication services to villagers and reindeer herders to ensure that nomadic and semi-nomadic people receive government services. According to preliminary estimates of experts, equipping a herd with a set of equipment costs about 2 million rubles, that is, in total, it will take about 1 billion rubles to arrange control and accounting. Meanwhile, the chairman of the board of the Association of Reindeer Herders M. Madejski expresses doubts about the technical feasibility and economic efficiency of this project, pointing out the more fundamental problems of the industry: transport and logistics, lack of specialists in reindeer farms, lack of mass demand for reindeer products, the need for protection livestock from predators [9].

In addition to ineffective public administration, barriers to digital transformation of the agro-industrial complex are as follows:

– low level of ICT penetration in rural areas and high tariffs for communication;

– low degree of processing and margins of the a-business, which makes the industry unattractive for investors;

– dependence on foreign equipment and technologies; non-customized foreign IT solutions;

– lack of funding and reluctance of agricultural companies to introduce digital technologies, lack of highly qualified personnel;

– heterogeneity and abundance of data in agriculture in the absence of effective tools for their processing;

– a lot of heterogeneous external factors that affect the yield and are often uncontrollable, which requires development of new digital forecasting methods;

– presence on the market of various platforms in absence of cross-platform solutions and industry standards; lack of information about existing solutions, manufacturers, implementation cases, achieved effects [10].

The most active implementation of digital solutions is carried out by agro-industrial holdings that have sufficient financial resources. Nevertheless, even large agricultural enterprises implement IT in fragments, there is no end-to-end digitalization of processes.

In terms of industries, the most in-demand technologies include:

– in agriculture - ERP (resource planning systems), satellite positioning technologies, equipment monitoring systems, field digitization, unmanned aerial vehicles;

– in the dairy industry — elements of a "smart farm", intelligent production management systems and RFID systems (a method for automatic identification of objects);

– in the meat industry — robotization of processing.

At the moment, innovations in the agricultural sector are chaotic, the level of awareness and penetration of the offered products among farmers is not high enough. The average score of digitalization in Russian regions is 6.4 out of 10, while the part of the regions that provide a large share of agricultural products is below the average score.

The digital divide is growing between agricultural holdings and small and medium-sized farms, which, due to a number of reasons (significant implementation costs, increased investment risks, lack of qualified personnel), cannot afford the large-scale use of digital technologies. Nevertheless, the potential demand for them from the agro-industrial complex is quite high: access to digital technologies offers advantages in terms of forming strategic partnerships, establishing relationships with suppliers, finding competent workers, openness of information, availability of support services, entering markets and arranging direct sales to consumers.

The digitalization vector in the agro-industrial complex can be represented as three steps for agricultural holdings and two steps (without the last one) for small and medium-sized agricultural producers (Table 2).

Table 2. Stages of digitalization in the agro-industrial complex.

Stages/steps	Characteristics
I. Agricultural producer with efficient business processes	Full integration of internal accounting systems, a single digital back office, use of dashboarding, creation of a production database.
II. Digital technology company	Use of innovative technologies such as precision farming, artificial intelligence, computer vision (CV), machine learning (ML), and etc. Online supply chains, omnichannel sales (both offline and online).
III. Ecosystem of the agricultural holding	Formation of infrastructure for innovation. Integration of market participants into a single system based on the digital platform of the agricultural holding. Sale of digital solutions/services in the agro-industrial complex is a separate area of activity.

To increase profitability, agricultural enterprises need to reach the second stage, and in 7-10 years, leadership in the industry will require the third stage of development. Currently, the overwhelming majority of Russian agricultural companies are faced with the tasks of the first two stages.

Note that the cluster creates conditions for development of all forms of farms within the most suitable market niche for them from the point of view of agroeconomic science and practice, using such a mechanism of interaction between small, medium and large enterprises as contracting.

Let us turn to foreign experience: agricultural holdings in the USA and Western European countries practice the so-called "incomplete" integration, in which the primary stage of agricultural production is an independent business process implemented by farmers (either through service cooperation or through contracting). The integrator, in turn, provides farmers with advice, maintenance, animals, and etc.

The advantages of "contract farming" are as follows:

– farmers become production partners of agricultural holdings, receive guarantees for sale of products and the possibility of technological renewal;

– agricultural holdings may not invest additional funds in production assets and use high-quality, environmentally friendly raw materials.

In our opinion, cooperation and integration of large, medium and small businesses in the countryside (in the form of the same cluster associations) is a prerequisite for transition to a competitive agricultural model and sustainable agriculture in the context of digital transformation.

4. Conclusion

For the effective implementation of the main tasks - ensuring the country's food security, preserving rural areas, saving ecosystems - agriculture shall eliminate the digital divide between it and related industries. Currently, digital technologies in agriculture are being introduced pointwise, mainly by large agricultural holdings, which expect that digitalization will allow them to enter a new level of competition.

Industrial large and advanced medium-sized farms are the main industrial environment for implementation of the IT innovations, while small peasants and farms prevailing in Russia do not have sufficient financial capabilities to digitalize agricultural production. However, opposition of family commodity farms and large entrepreneurial structures is counterproductive: each organizational form occupies its own functional space, and mobilization of all resources and tools in order to develop one of them when causing collateral damage to the other will lead to a distortion of the multi-structured agricultural economy that threatens its stability. It is necessary to create conditions for development of all forms of farms, which is possible when arranging production and business processes using the cluster model. Clusters in the agro-industrial complex are hybrid structures, whose members include both agricultural holdings and farms, the effective interaction of which can be carried out through contracting. The flexibility and openness of cluster structures allows actors to find their functional areas and niches and fit organically into the digital ecosystem of the agro-industrial complex, which implies a transition from a linear type of business organization to a "platform" one by ensuring network interaction between manufacturers, suppliers and retailers, and includes a full life cycle from production, movement, consumption, to utilization of agricultural products.

Acknowledgments

This work was made within the grant "Young Researchers of the East Siberia State University of Technology and Management".

References

1. M. Brettel, N. Friederichsen, M. Keller, M. Rosenberg, International Journal of Mechanical, Aerospace, Industrial, Mechatronic and Manufacturing Engineering, **8**, 37–44 (2014)
2. J. Alcácer, J. Cantwell, L. Piscitello, Journal of International Business Studies, **47**, 499–512 (2016)
3. R. Strange, A. Zucchella, Multinational Business Review, **25**, 174–184, 3 (2017)
4. M. Götz, B. Jankowska, European Planning Studies, **25**, 1633–1653 (2017)
5. M. Götz, Foresight and STI Governance, **13**, 72–83 (2019)

6. F. Belussi, European Urban and Regional Studies, **12**, 247–268 (2005)
7. M. Delgado, M.E. Porter, S. Stern, Defining clusters of related industries (NBER Working Paper 20375), Cambridge, MA, National Bureau of Economic Research (2014) http://www.nber.org/papers/w20375.
8. A. Konyavsky, Yu. M. Akatkin, E. D. Yasinovskaya, Information Security, **2**, 14-16 (2018)
9. N. Korolev, A. Polukhin, Kommersant, **174/P**, 7 (2021)
10. S. Vodianova, ICT Analytics and Digital Media. https://json.tv/.

Information and image support of the activities of the Agrarian University

Marina Alemasova[1,*], *Dmitry* Galkin[1], *Ludmila* Kurbatova[2], *Zhoia* Olovyanikova[3] and *Svetlana* Popova[1]

[1]Michurinsk State Agrarian University, Internationalnaya St., 101, 393760 Michurinsk, Russia
[2]Perm National Research Polytechnic University, Akademika Koroleva Street, 15, Perm, Russia
[3]Pskov State University, Lenin Square, 2, Pskov, Russia

Abstract. The article examines the activities of public relations specialists of Michurinsk State Agrarian University in its information and image support. The paper analyzes communication technologies aimed at promoting the higher educational institution of an agrarian profile, agricultural education and agrarian science in general. It is argued that the formation of a positive image of Michurinsk State Agrarian University is facilitated by the organization of effective interaction with the media, positioning of the activities of the university, its scientists, students; regular media monitoring; formation and strengthening of corporate culture; development of corporate identity, mission, slogan, logo; organizing and holding special events; preparation and participation in exhibitions and fairs; maintaining a corporate website; production of a corporate publication, advertising activities. Specific examples of the use of communication technologies in an educational institution are given. The conclusion is made that with the creation of a department for public relations and management of public relations, press and office work, the communicative activity of the agrarian university began to be carried out professionally and in a coordinated manner. The existing communication system ensures the formation of an effective positive information field of the university. The communicative policy of the educational institution is focused on the long-term prospect of building a serious positive dialogue with target audiences, on the prospect of trust.

1 Introduction

Nowadays, managing your own information flows becomes a necessity for any organization that wants to function in society for a long time and effectively. The higher education system is no exception. In conditions of intense competition "universities have to look for and find non-trivial ways to attract a future contingent of students" [1, 83]. Without the efforts of advertising and public relations specialists with the necessary set of knowledge and personal qualities, including creativity, courage, initiative, resistance to stress [2-3] the successful implementation of this task is not possible.

* Corresponding author: mal_2112@mail.ru

2 Materials and Methods

When writing this article, we used works devoted to the peculiarities of advertising and PR-activities in a higher educational institution; annual reports of the department of public relations and department of public relations, press and office work of Michurinsk State Agrarian University; publications and broadcasts in the media (newspapers, magazines, almanacs, radio, television) about events, projects, discoveries of university scientists; souvenir and advertising products. Methods of observation, analysis and description were used.

3 Results and Discussion

One of the main tasks of public relations in a higher education institution is to form a positive public opinion about the institution; control of the information field around the positioning object. For the effective promotion of agrarian universities, agricultural education and agrarian science in general, various communication technologies are used, combined into a system of thought-out strategically built actions.

PR activities have been carried out since 2004 in Michurinsk State Agrarian University. First, the position of a public relations specialist was introduced and then a PR department was formed with the aim of activating and increasing the effectiveness of work in the field of positioning the agricultural university in the external environment, expanding the scale of work to popularize agricultural education. Currently, there is a public relations department, the purpose of which is to provide information and image support for the university's activities or, using the language of professional terms, to increase the publicity capital of the basic subject of PR [4].

The department interacts with all structures of the university; mass media (newspapers, television, radio stations and Internet news portals) and information agencies, press services and public relations departments of the city of Michurinsk-Science City of the Russian Federation, the administration of the Tambov region and agrarian universities of the country.

A key area of public relations activities is the organization of interaction with the media – media relations. PR specialists build trusting positive relationships with journalists of newspapers, magazines, radio, television serving as a base and resource for the use of other communication technologies. Constant work with the press, includes the distribution of press releases, the publication of image articles about the personalities of the university, the holding of special events for the press (briefings, press conferences). Preparation of answers to their official inquiries contributes to the formation of a positive image of the university [5].

In the process of carrying out public relations activities at the university, fruitful cooperation has been established with the mass media of the city, region and country. Thanks to the comprehensive and regular coverage of the life of the university in the media, public confidence in the university has increased; the activities of Michurinsk State Agrarian University have become transparent in the training of personnel and the implementation of research activities for the domestic agricultural industry. In the press and on the radio there are constant headings "News of Michurinsk SAU". Journalists of "Tochka", "Vesti Tambov" and "Novy Vek" TV companies regularly cover the activities of the university in their news and feature stories which are initiated by PR specialists. The materials are mainly devoted to the scientific and innovative activities of the university, the discoveries and inventions of Michurinsk scientists, which contribute to the formation of an attractive image of agricultural science. News materials are also being prepared for the website of the Ministry of Agriculture of the Russian Federation.

Management specialists actively cooperate and regularly interact with the editors of professional journals, on the pages of which materials about the university and its scientific developments are published. These are the almanac "Universities of Russia", the magazines "Adviser in the field of education", "Adviser to the accountant", "Education: goals and prospects", etc.

In the summer of 2021 Michurinsk was visited by representatives of the Federation Council TV channel "TOGETHER - RF", who filmed for the cycle of programs "Science Cities". This program is about the past, present and future of Russian cities and towns with high scientific and technical potential. A big story about the only agricultural science city and towns in Russia was broadcast on TV, and a separate part of it was devoted to Michurinsk State Agrarian University (from 19 minutes) [6].

Since PR-activity is informational, the indicator of its effectiveness is the number of prepared information products, the number of addressees who received information products and the number of changes that occurred as a result of receiving information products [7]. According to these criteria, the information support of the activities of Michurinsk State Agricultural University is carried out competently and fruitfully.

During a difficult situation for the university related to the integration of educational institutions of Michurinsk on the basis of Michurinsk State Agrarian University a lot of information materials were published highlighting the need and the process of combining four higher educational institutions and colleges which made it possible to correctly adjust the public to this process and not cause resonant sentiments.

The university also publishes its own corporate publication "The Bulletin of Michurinsk State Agrarian University", the editor-in-chief of which is the head of the public relations department. The professional management of the newspaper contributed to its dramatic change. It has become more interesting and modern by changing its content and using full color printing in the production of mass printed product.

One of the most important components of the communication system of an educational institution is the formation and strengthening of corporate culture. It unites teachers, employees, students of the university into a team of like-minded people for the effective implementation of its mission, which was developed with the active participation of public relations specialists. Corporate culture defines the strategic goals of the educational institution. The text of the mission is aimed at both teachers and staff of the agrarian university and its students. Its main provisions determine the principles of educational, scientific activities; emphasize the university's focus on spiritual and moral education of the personality of students, the disclosure of the creative abilities of employees and students, the desire of the university to integrate into the international educational space.

The most important component of corporate culture is the corporate identity – a set of graphic forms and principles of building visual communication, united by one idea, the main task of which is to highlight the organization and create a recognizable image in the eyes of target audiences: teachers, employees, students, applicants, employers, partners, authorities, media representatives. Corporate identity includes a name, a coat of arms, an anthem, a flag, a uniform, a logo, a slogan.

The corporate identity of the Michurinsk State Agrarian University, which was developed by the university's PR specialists, is made taking into account the specifics of the organization, the agrarian profile of many specialties. Green and white were chosen as corporate colours. Green is a symbol of unity with nature, white is a sign of purity and harmony.

The university logo is an apple framed in a semicircle with a stylized map of the Russian Federation inside. Since ancient times the circle has been considered a sign of unity and perfection. Apple orchards are a "visiting card" of Michurinsk. Apple varieties bred by I.V. Michurin and his followers including scientists from Michurinsk State Agrarian

University are known throughout Russia, which is demonstrated by the map inside the apple (fig. 1).

Fig 1. The university logo

Each institution of the university has its own logo that reflects the specifics of its activities. Despite the characteristic differences, they all contain unifying elements, which are ears of wheat (a sign of wealth), a master's hat and an open book (symbols of knowledge).

The university created and patented the brand name of university eco-products "101 VITAMIN".

Corporate symbols on the clothes of representatives of the youth movements of the educational institution began to distinguish the students of the agrarian university of the Tambov region from others during mass events.

The slogan of Michurinsk State Agrarian University is "Together – to success". The phrase is laconic, easy to remember. It is featured on all advertising and souvenir products produced by the university.

Michurinsk State Agrarian University has its own anthem (words by V.N.Zmeevsky, music by N.P. Ermilova –teachers of the university), the performance of which opens all official events held at the university. The flag consists of two horizontal stripes: white and green. At the top of the flag there is a logo with the full name of the university. When designing business letters and various documents, the university uses the proprietary font - "Bangkok Cyr". Elements of branding are used in the design of corporate publications, the official website, souvenir and printing products, outdoor advertising and exhibition stands.

Organization and holding of special events is a significant component of the university's communication policy and a tool for strengthening its corporate culture. Various events are regularly held: concert programs dedicated to calendar holidays, anniversaries in the history of the university; contests and sports competitions in which students, teachers and university employees participate. For the student audience are organized competitive programs "Student Spring", "Freshman's Debut", "Student Family", "Mister and Miss of Michurinsk SAU", "Best Leader of Michurinsk SAU ". They "are designed to bring variety and revitalization into the daily work of the team, contribute to the formation and strengthening of a positive image of the university" [8, 5].

The competition "Mister of Michurinsk SAU", invented by students of PR specialists, is a kind of meeting "without ties "for the leadership of the university. The organizers managed to go beyond the standard perception of the leadership, to show in an unusual perspective those who are more often used to see in situations of strict officialdom, to “bring them closer” to students, teachers, and employees. Representatives of the administration recalled their student years, told funny stories that happened to them, peeled potatoes, demonstrated the ability to replace, if necessary, a spouse in the kitchen, made solemn speeches in honor of their wives, and performed lyric songs. The rector and his team were assessed by a student jury, very loyal to all the participants, apparently counting on a similar attitude towards themselves in the future. The competition revealed unexpected, previously unknown to the majority of those present, talent of the university administration, helped to raise the level of awareness of each participant as a person. Such contests, in addition to a good mood and sincere fun, set the stability of the moral and psychological climate of the team, help support the activities of the management, and form a single corporate spirit [8, 6].

Exhibition activities have become a widespread means of promoting the university. Participation in exhibitions and fairs supports the image of the educational institution both in the eyes of wide public groups and target audiences. The exhibition allows attracting the attention of potential applicants and employers to the activities of the university. Thanks to the PR-specialists of the university, the direction related to exhibition activities has reached a new level. Participating in major events of the all-Russian level ("Golden Autumn", "Russia, Looking to the Future!" , "Gardener's Day"), Michurinsk State Agrarian University positions itself as a higher educational institution of an agro-social profile, implementing training directions for the integrated development of rural areas.

The beginning of the second decade of the new millennium has finally confirmed the importance of the global information environment in society. The World Wide Web has created a new - informational - paradigm that promotes the availability of information, its mobility and lack of attachment to certain time parameters.

The Internet has become a part of the life of every citizen of Russia: people work there, communicate, and arrange their personal life. People are spending more and more time online. This is also related to the aspiration of domestic organizations to the Runet sphere. Michurinsk State Agrarian University is no exception [9, 141].

The representation of an educational institution on the Internet is an essential element of communication.

"The news feed of the site of Michurinsk State Agrarian University is popular and is included in the list of topical news sources. More and more Internet resources refer to articles from www.mgau.ru/about/news/ "[9, 142].

Users receive information about the know-how in the educational institution; the system of continuous agribusiness education; find out for which developments young scientists receive millions of grants; why eighty-year-old applicants enter the university, what form fruits grow in the "Green Valley" of the university, how employees help to grow vegetables in the Far North.

The development of telecommunication and digital technologies, mobile devices, computer technology, the Internet creates the prerequisites for the use of digital mail - a modern way to promote a brand. Considering that young people - applicants, students, graduates (key segments of the target audience of the university) - are most susceptible to the communicative impact of modern gadgets, the university widely uses direct-mail, forming and using personal electronic address databases for sending advertising and information messages university. Potential students can receive the necessary information promptly and directly.

Advertising technologies are used to position the activities of the university. Advertising is the most traditional channel of promotion, allowing applicants to be informed about the specialties and areas of training of the university. Advertising professionals use all types of traditional advertising. Advertising messages are published in print media, they sound on radio, television, they are present in brochures, reference books, calendars, postcards, leaflets, leaflets, booklets, memos. Outdoor advertising is placed on the facades of buildings, transport, pylons, and guy lines. The advertising carriers of Michurinsk SAU are billboards, posters, stickers, light boxes, banners. Particular attention is paid to the placement of advertising messages on the Internet - contextual advertising, flash banners, targeted advertising in social networks, etc.

Advertising of Michurinsk State Agrarian University changes conceptually every year.

Developers of advertising messages always strive to get away from traditional ideas in the field of higher education, sometimes to combine the incompatible. For example, one advertising concept was based on the idea of speed. The emphasis was placed on the combination of the concepts of "speed" and "knowledge". The result is an original advertising phrase - "Education at Michurinsk SAU - at the ultimate speed of knowledge!" The billboard showed a speedometer (as if through a magnifying glass), its arrow went off scale for the last division of 160 km / h, an additional mark was placed behind it, but instead of a number there was a university logo: studying at the university will allow you to quickly move in step with the times with the acquired luggage of modern knowledge [10].

During the advertising campaign "Choosing Michurinsk SAU: Yesterday, Today, Tomorrow" two main ideas are combined: the theme of time, expressed in the categories of the past, present and future (yesterday, today, tomorrow) and the theme of choice. Key message: The right choice an applicant made yesterday is a happy student today and a successful graduate tomorrow. Key messages: choosing Michurinsk SAU for training, the students did not regret their decision: they found their favorite work here, opportunities for creative self-realization, found new friends; students of the agrarian university are confident in their future; the university today gives its students such knowledge that will help them achieve success tomorrow.

Another concept of the advertising campaign is based on the idea of the success and professional competence of university graduates.

Names and photographs were used in advertising materials, positions of forty-five well-known specialists in the field of politics, culture, business, agriculture (heads of agricultural enterprises, heads of districts, rectors and directors of scientific research institutes), who at various times studied at Michurinsk SAU, were indicated. Promotional phrase: "Applicant! These people have achieved a lot by being educated at Michurinsk State Agrarian University! If you want to be successful go to Michurinsk SAU!" motivated potential applicants to study at the agrarian university. On the billboard and printed products in the center of the advertising material a free space was left for a photograph with the inscription "In 10 years your photograph will be here." The authors of the idea pursued the goal of encouraging applicants to present their photograph at this place and, accordingly, project it in the minds of themselves as a university graduate.

In the advertising campaign "Become Successful with Michurinsk SAU" the concept was implemented, according to which Michurinsk SAU was positioned as a university capable of giving its students the opportunity to achieve professional success and maximize self-realization. The key word "successful" focused on the following key messages: the university occupies a leading position in the region; it develops using innovative approaches to learning, strives to maintain and strengthen its leading position among the universities of the region; gives its students such knowledge that will help them to achieve success in the future.

University students (two girls and three boys) became the faces of the advertising campaign. The text of the commercial (audio and video) is “Success is always remembered! They trust him! He is always chosen because he makes others successful! Successful means the first! Successful means brave! Successful means the best! Get an education at Michurinsk SAU and become successful! Michurinsk Agrarian University - Together to Success! "

During advertising campaigns, advertising materials are produced: banners, posters for schools, brochures for applicants, video and audio clips, original layouts of publications in the newspapers "Moskovsky Komsomolets", "AIF-Tambov" and "Tambovskaya Zhizn", a banner for the official website university. Videos about the university are broadcast on the "Russia-2 TV channel" (Tambov), "STS", "TNT", "REN" and "Zvezda". The audio clips are broadcast on the radio channels "Hit FM", "Avtoradio-Michurinsk", "Avtoradio-Lipetsk", "Main Radio", "Russkoe Radio", "Road Radio - Tambov".

Advertising activity involves the production of souvenirs that are regularly updated at the university. It includes branded notebooks, key rings, pens, paper and plastic bags, chargers, badges, caps, T-shirts.

Thus, "Advertising provides information, informs / attracts attention, ensures the presence of the university in the advertising and information sphere. Today advertising technologies are a necessary resource and a means of ensuring the successful promotion of an educational institution" [11, 184].

4 Conclusion

Advertising and PR technologies are widely used by Michurinsk State Agrarian University to position its activities. Their complex use allows solving the problems of increasing the popularity of the university and strengthening its reputation, building the trust of target audiences, creating a positive information field. The communicative policy of the university is focused on the long-term prospect of building a serious confidential dialogue with target audiences.

With the introduction of the position of a PR specialist at Michurinsk State Agrarian University and later with the creation of a department for public relations and administration of public relations, press and office work, the communicative activity of the agrarian university began to be carried out professionally and in a coordinated manner. Relations have been established with the media and authorities. The work of the corporate website has been adjusted. High-quality advertising products are being produced.

References

1. E. Pervushina, Higher educ. in Russia, **11**, 83-89 (2008)
2. Professional competencies of a communication specialist, ed. Krivonosov, 227 (2021)
3. L Semenova, A. Krivonosov, E. Karmalova, N. Danilova, Comprehensive program of professionally oriented training of specialists in communication, European Proceedings of Social and Behavioral Sciences, EpSBS, Grozny, Russia (2021)
4. E, Bykova, Medialinguistics in terms and concepts, 305-306 (2018)
5. I. Savelyeva, Modern media and the media market, 70-73 (2020)
6. Vmeste RF, https: //vmeste-rf.tv.
7. E. Bykova, D Gavra, Bulletin of St. Petersburg University. Language and Literature, 14-2, 275-284 (2017)
8. M. Alemasova, MediaVector, 1, 4-7 (2021)
9. D. Galkin, PR and advertising in a changing world: Regional aspect, **10**, 141-144 (2012)
10. D. Galkin, Modern advertising technologies: theory and practice, 41-46 (2012)
11. E. Malikova, Communication management problems: from theory to practice, 180-184 (2013)

New organizational mechanisms of infrastructure development of small agricultural businesses in Russia

M. M. Voytyuk[1], *T. E.* Marinchenko[2,*], and *V. A.* Voityuk[2]

[1]Moscow branch of "Rosinformagrotech" SPC "Giproniselkhoz", Stromynka st., 11, 107014 Moscow, Russia
[2]Rosinformagrotech, Lesnaya, 60, 141261 Moscow region, Russia

Abstract. The developed infrastructure of a small agricultural business (ISAB) ensures the innovative development of production of small enterprises of the agro-industrial complex (AIC), solves social problems of rural areas and provides employment for the rural population, performing a number of functions, including: property, financial, consulting, information, legal, innovative , social and export support for small agricultural businesses (SAB). The aim of the study is to identify the problems of the infrastructure development of the SAB and search for their solutions. The methods of generalization, economic and mathematical modeling, functional analysis were used, the data of the Ministry of Economic Development of Russia, the reporting of the Ministry of Agriculture of Russia, publications of foreign and domestic authors and other sources served as the material for the study. It was proposed to use the institution of "commercial procurement", to functionally expand the "financial" service, to organize a Single Industry Center for Infrastructure Development (Center), which will ensure effective interaction of all participants through the use of a digital platform based on a service-oriented architecture (SOA). The platform needs to be integrated with government information systems that support the infrastructure development of the SAB. The cluster organization of infrastructure facilities will increase efficiency, will contribute to the concentration of resources and their greater purposefulness. The main result of the implementation of new organizational mechanisms should be the effective development of the IMAB.

1 Introduction

By ISAB we mean a set of state, non-state, public, educational and commercial organizations of production (transport, power lines, water supply, etc.) and social (trade, housing and communal services, health care and education, consumer services, etc.) plans that provide a supportive environment for the operation of the SAB. In this context, the following infrastructure objects are significant: public organizations that unite SAB to solve

* Corresponding author: 9419428@mail.ru

specific problems; commercial and industrial facilities; educational organizations that provide personnel training and professional development; information, consulting organizations and mass media; funds to support SAB, microfinance organizations and other financial services for SAB, etc.

The development of ISAB has been actively studied in the Russian Federation for almost 30 years, due to the fact that it not only ensures the innovative development of production of small enterprises AIC, but also solves the social problems of rural areas and, first of all, the life support and employment of the rural population [1, 2]. From an economic point of view, the development of ISAB will accelerate the creation of a competitive environment in the food market and satisfy the needs of buyers with new products and goods [3]. With flexibility and mobility, ISAB is able to fill empty economic niches and provide efficient infrastructure services not only to small, but also to large businesses and the population.

The main problem in the development of ISAB is the lack of effective organizational mechanisms for its formation. This is due to the fact that the global crisis that occurs in the world every 3-4 years negatively affects the functioning of the SAB and does not allow it to form an effective infrastructure for its activities [4]. Internal problems arise due to the fact that the functions of infrastructural development are distributed among various ministries and departments that pursue their goals, lead to ineffective development mechanisms and require significant adjustments at different levels of government. This is confirmed by the goals of the national project "Small and Medium Enterprises and Support for Individual Entrepreneurial Initiatives" and its tasks related to the search and development of effective organizational mechanisms for the development of ISAB.

Despite numerous Russian and foreign studies of the issue [5-8], the national policy in this area is still ineffective and requires changes to the ISAB development paradigm, including clarification of the functional features of infrastructure development and the formation of new organizational mechanisms, taking into account modern trends of changes in institutional environment and national strategic policy in the field of support and development of small business [9].

2 Materials and methods

The research methodology is based on a systematic approach and includes a set of methods, including: the method of generalization, economic and mathematical modeling, and statistical, functional analysis. The empirical basis of the analysis was the data of the Ministry of Economic Development of Russia, RosStat, reports of the Ministry of Agriculture of Russia, publications of domestic and foreign researchers.

3 Results

In order to develop new mechanisms for the development of infrastructure, it is necessary to determine its essence. There are several concepts of the essence of infrastructure. In the traditional sense, the SAB infrastructure is a subsystem of agricultural facilities that provide services of a production and social nature. Infrastructure itself does not create added value, but provides the production processes where it is formed, develops favorable conditions for SAB workers and the entire rural community. Infrastructure provides various services to create favorable conditions for the operation and effective reproduction of SAB, including in terms of the implementation of the functions of property, financial, consulting, information, legal, innovation, social and export support, which are carried out at the request of enterprises within the authority of infrastructure facilities [10].

At present, as practice proves, it is advisable to supplement the list of forms of infrastructure services with the institution of "commercial procurement", to functionally expand the financial service to "financial guarantee", since for SAB, in the face of a shortage of its own funds, it becomes important to obtain guarantees for loans and borrowings. Cluster organization of infrastructure facilities increases the efficiency of infrastructure services, contributing to the concentration of resources and their focus on solving complex SAB problems [4, 8].

Expanding the number and improving the quality of infrastructure services (support) ensures the development of the SAB infrastructure, which has enormous potential and has good reason to play an important socio-economic role in the development of the agricultural sector and rural community. Despite the positive dynamics in the development of SAB over the past decade, there are still many problems with the development of infrastructure, limiting the rate of its growth [1, 3, 9, 11]. To identify these problems, a functional-integration approach was used, which allowed integrating ISAB functions into groups in order to concentrate resources on solving large tasks and effectively performing infrastructure services tasks. As for the tasks of infrastructural development, the formation and assembly of organizational mechanisms is carried out in the main categories: financial and economic, organizational and technical, regulatory and legal, information and consulting.

Today it is difficult to imagine a SAB organization without infrastructure support for its production activities, providing information to consumers without consulting services. The knowledge gained in the course of personnel training is converted into information resources, therefore it is recommended to include them in the functions of information and consulting support. Norms and laws define the limits of the duties, rights and responsibilities of the SAB, which are in fact so-called formal institutions. The economic process includes the financial aspects of infrastructure services, which are inseparable from each other. In this regard, the functional-integration approach forms the basis for the development of infrastructure, increasing the number and improving the quality of infrastructure services for SAB, expands the scope of this approach.

It should be noted that due to the fact that a comprehensive infrastructure development system has not been created in the agricultural sector, which should include a network of facilities necessary for SAB's activities, state programs for the development of agribusiness are not fully implemented and budget funds are not used effectively enough. Another problem of infrastructure development is associated with the fact that often infrastructure facilities are organized formally, this leads to fictitious provision of services, confusion of concepts and functions of various facilities.

Such disunity and imbalance in the formation of infrastructure, the ineffectiveness of the provision of infrastructure services is explained by the lack of a clear state policy for the development of ISAB, the absence of a single center for infrastructure development - a coordinator of formation and development, territorial heterogeneity of infrastructure services and a general lack of infrastructure facilities, especially in remote rural areas.

The solution to these problems is possible with the creation of a single Center, with subsidiaries at various levels of management of the agricultural industry, with the formation of a digital platform for the provision of SAB infrastructure services using SOA (service-oriented architecture) technologies.

4 Discussions

Currently, AIC does not have a single body that carries out the coordinating functions of infrastructure development. Such functions are imputed to various ministries and departments, which hinders the coordination of their actions in terms of infrastructure development. The problem is aggravated by different goals and objectives of these structures, which does not contribute to the unification of strategic and tactical goals for infrastructure development and the provision of ISAB services [12, 13]. It is possible to solve the above problems by creating a Center, the main purpose of which will be the coordination of infrastructure development, and its functions: analysis of the existing infrastructure and identification of problems in its development; identification of the optimal structure and composition of infrastructure facilities; development of strategic development plans and support for public-private partnerships in the construction of infrastructure facilities; updating of normative documents on infrastructure development; methodological support for the standardization of infrastructure services; information and consulting service of SAB on solving problems of infrastructure development.

The creation of such a Center is an important part of the organizational development mechanism of ISAB. The experience of small agricultural businesses in the USA, Japan, Great Britain, EU countries and other foreign countries proves that business structures managed to achieve great success in the development of infrastructure when creating special institutions for infrastructure development [14-18]. This is reflected in the developed roadmap "Sustainable growth of the non-resource sector of the Russian economy." To implement its functions, it is advisable for the Center to include structures in the main areas of infrastructure development and infrastructure support: a department for organizational and technical development, regulatory and financial and economic support, information and consulting support [8, 10, 15].

The Business Development Department will provide an analysis of the existing SAB infrastructure, based on which it determines the need for new infrastructure, interaction with the existing infrastructure and other stakeholders, including industry authorities and the public.

The Regulatory Support Department will analyze the existing regulatory framework for infrastructure development and will develop regulations, standards and rules for the provision of infrastructure services.

The Department of Financial and Economic Support will analyze the financial and economic activities of the infrastructure, the effectiveness of the provision of financial support, the assessment of investment projects, including in public-private partnerships, in the creation of new infrastructure facilities and reconstruction of existing ones, etc.

The department of information and consulting support will collect and statistically analyze information, transfer information reports to SAB structures, provide advisory support, as well as train and retrain specialists in infrastructure development. For the interaction of all structures of the Center and small businesses, it is advisable to form a single digital platform for the development of ISAB.

In the context of AIC's transition to the digital economy, the formation of the core and architecture of a digital infrastructure development platform is a prerequisite for the implementation of the infrastructure development concept. The core of the digital platform should be the main elements that ensure the operation of the platform, its functions and interface required for its development and creation of applications for various groups of participants [13].

The main purpose of a single digital platform is to provide services and technical support for existing information systems for the development of ISAB and the development of new services, including user authorization, centralized data entry, providing reference information, bringing data to the status of legal significance. Among the services, the following should be highlighted: interface platform, data storage, network verification and security systems, analytical support service. Such an arrangement of elements of a single digital platform will ensure the uniformity of data entry, increase the reliability of primary information, and optimize the cost structure.

A prerequisite for the creation of the Center is the implementation of primary operations, including: analysis of the legal basis for the interaction of all participants in the digital space and preparation of recommendations for their implementation; study of the functional features of potential participants in the digital space and digitization of their data; formation of information exchange and coordination of interaction between participants; development of technical agreements and standards for information exchange. The platform should provide the ability to use the user's work computer and the browser on the mobile device to work in the network. It is advisable to organize the operation of the platform using redundant communication channels and load balancers to support the operation of the accumulated software and hardware for protection against natural disasters. Technological modification of PaaS (Platform as a Service) is considered as a way of providing all platform participants with a ready-made software environment and tools for its customization. The elements of PaaS are hardware, operating systems, a database management system (DBMS), then testing and development tools, including digital services that ensure the implementation of all functions. PaaS technology is based on SOA, which is able to integrate various applications with maximum flexibility and allows you to build the IT infrastructure of a digital platform in such a way that loosely coupled components are linked into a single whole, equipped with interfaces for interaction using standard protocols. A software package developed in accordance with SOA, as a rule, is implemented as a set of web services interacting with well-known standard protocols (SOAP, WSDL, etc.), and each web service provides a business function. Thus, the proposed digital platform will ensure interoperability, including rational and open relationships for all participants in their activities and in the IT sphere.

The main functions of the digital platform include:

- creation and implementation of management decisions for the development of the SAB infrastructure through the formation of a new environment of trust;
- ensuring the coordinated actions of the platform participants in the implementation of investment projects for the development of infrastructure;
- increasing the effect of management decisions in the development of infrastructure as a result of reducing time and financial costs, providing equipment, testing and developing applications;
- implementation of the integration of the digital platform and information systems of the participants, providing communication at the level of all business processes, data structures and services;
- centralization of data entry, promotion of information from reference books and classifiers;
- identification of participants, authentication and authorization of their information systems;
- use of extended digital signatures to impart legal significance to information and ensure electronic document flow;
- expanding the capabilities of a single point of data entry for the digital platform and information systems of the participants;
- providing access to the existing state information resources and systems of scientific and technical information, including monitoring the development of the infrastructure of a small agricultural business;
- provision of disaster recovery and network security;
- provision of continuous professional training.

The ISAB digital development platform must be combined with the Unified Identification and Authentication System in the infrastructure that ensures the information and technological interaction of information systems used to provide state and municipal services in electronic form (ESIA), the Unified Interdepartmental Electronic Interaction

System, the National Data Management System (NDSM) and other information systems that support the infrastructure development of SAB in order to provide access to statistical data, information and analytical materials, ready-made business solutions and business plans.

It is advisable to develop and operate the ISAB digital development platform using the principles of openness and integrability, scalability and flexibility of settings, efficiency and modifiability, security of the entire system and durability. As the expected results from the use of a single digital platform, SAB will receive: effective interaction with all participants through a single access point, shorten the time for the provision of infrastructure services, reduce the cost of acquiring information technologies, increase cyber-security and the ability to expand and improve the quality of analytics data through big-data technologies.

The formation of the organizational mechanism should be carried out in several stages. The first stage includes the development of methodological support for the process of creating a Center at the federal and subsidiaries at the regional levels of AIC management, including methodological recommendations on the activities of all centers and regulations for the interaction of all participants. The second stage provides for consulting activities for the creation of subsidiary centers, based on the holding of educational and methodological meetings with stakeholders on their creation and functioning. At the third stage, subsidiary centers and digital platforms are created on their basis. At the fourth stage, the implementation of information systems, resources and platforms of all participants is carried out. The fifth stage provides for the large-scale deployment of the functions of all subsidiary centers of digital platforms.

New in the organizational mechanism is the interaction of all participants in the ISAB development process on the basis of a single Center, a digital platform and information technologies adapted to the specifics of SAB's infrastructural activities. It is also proposed to conduct the management of the Center and its regional network under the Ministry of Agriculture of Russia, which will allow specialists to more fully implement the tasks of the industry for the development of ISAB and develop rural areas. A positive result of the implementation of new organizational mechanisms for the development of ISAB should be such an infrastructure development that fully satisfies the needs of SAB, will contribute to improving the quality of infrastructure services, creating a single digital platform for organizations and developing infrastructure through the interaction of stakeholders in the process under the control of a single Center that implements complex tasks facing before AIC.

5 Conclusions

The study of the organizational development mechanism of ISAB made it possible to identify the problems of infrastructure development and outline ways to solve them, taking into account the essence of infrastructure facilities and the characteristics of the services they provide. Based on the functional-integration approach, new organizational mechanisms for the development of the SAB infrastructure have been identified, including the creation of a new institution - a single body for infrastructure development in the agricultural industry (Center), where the effective interaction of all participants in the SAB infrastructure development process is provided by a digital platform based on SOA. Its organization will allow integrating various applications with maximum flexibility, using distributed, loosely coupled, replaceable components equipped with standardized interfaces for interaction of all participants in infrastructure development using standardized protocols. The main result of the implementation of new organizational mechanisms should be the effective development of ISAB, capable of providing favorable conditions for SAB employees and the entire rural community.

References

1. S. V. Frolova, et al., BIO Web of Conferences, **17,** 00004 (2020)
2. V.V. Bezpalov, et al., Management, economic engineering and rural development, **184,** 431 (2018)
3. E.Yu. Cherkesova, et al., European Research Studies Journal, **XXI(1)**, 287 (2018)
4. V. Kovshov, et al., Montenegrin Journal of Economics, **153**, 179-188 (2019)
5. A. Caloffi, M. Mariani, Environment and Planning, Politics and Space, **361,** 28 (2018)
6. G.A. Vaquero, et al., European Research on Management and Business Economics, 26 (3), 127-133 (2020)
7. E. Cherkesova, et al., IOP Conf. Ser.: Earth Environ. Sci. 12th International Scientific Conference on Agricultural Machinery Industry, 012040 (2019)
8. M. Voytyuk, et al., IOP Conf. Ser.: Earth Environ. Sci., **677,** 022083 (2021)
9. R.G. Yanbykh, Economy of agriculture of Russia, **11**, 80-83 (2017)
10. M.M. Voityuk, *Formation of a regional strategy for the development of infrastructure in rural areas*, 268 (2011)
11. N.A. Kuznetsova, et al., IOP Conf. Ser.: Earth Environ. Sci., 677, 022043 (2021)
12. O. F. Pyatova, et al., BIO Web of Conferences, 17 (2020)
13. M.M. Voityuk, *Implementation of investment projects for infrastructure development in rural areas*, 168 (2015)
14. M.G. Martinez, *Open Innovation in the Food and Beverage Industry* (2013)
15. V.Yu. Dianova, A.V. Tkach, O.I. Zhukova, Fundamental and applied research of the cooperative sector of the economy, **3**, 14-20 (2018)
16. R. Krishnan, et al., Resources, Conservation and Recycling, **168**, 105253 (2021)
17. M.F. Abu Hussin, et al., IOP Conf. Ser.: Earth Environ. Sci., 175(1), 012166 (2018)
18. J. Overall, World Review of Entrepreneurship, Management and Sustainable Development, **13(1)**, 69 (2017)

Issues of agriculture digitalization in the Russian Federation (legal aspect)

Alexander Savoskin, *V.V.* Kalitskaya, *O.A.* Rykalina, *O.V.* Mustafina, and *I.M.* Perminova

Ural State University of Economics, 8 March/NarodnayaVolya Str., 62/45, 620144 Ekaterinburg, Russia

Abstract. The President of the Russian Federation has set the task of achieving "digital maturity" of the key economic sectors, including agriculture, by 2030. At the same time, the digitalization of the agro-industrial complex is clearly lagging behind the pace of digital transformation in other sectors of the national economy. There are several reasons for this: the regulatory standards of the agricultural crops cultivation and the farm animals breeding are not perfect; the legal regulation of digitalization in the field of both production and agriculture is fragmented; the use of digital tools and information technologies in agricultural activities appears to be quite complicated. The article not only analyzes modern acts affecting the digital transformation of agriculture in the Russian Federation (including acts of the Ministry of Agriculture of the Russian Federation) and the problems of their application, but also suggests measures aimed to stimulate agricultural producers and food-processing companies, introducing innovative digital technologies, as well as provides recommendations for the use of special legal regimes that entered into force in 2021.

1 Introduction

The President of the Russian Federation, in Decree No. 474 dated July 21, 2020 "On the National Development Goals of the Russian Federation for the period up to 2030", set the task to "ensure digital transformation" by 2030, an integral part of which is to achieve the "digital maturity" of agriculture as one of the key sectors of the domestic economy. The stated national development goal of Russia has replaced the previously established goal of "ensuring the accelerated introduction of digital technologies in the economy" (Decree of the President of the Russian Federation dated 07.05.2018 No. 204 "On national goals and strategic objectives of the development of the Russian Federation for the period up to 2024"). It is obvious that only the digitalization course outlined in the Message of the President of the Russian Federation to the Federal Assembly of the Russian Federation in 2018 has now not only transformed into a national goal, but also continues to expand and be filled with new components, covering virtually all spheres of life. At the same time, a special emphasis on the digitalization of economic relations remains, as clearly evidenced by the number of program documents and achievements in the digitalization of the economic sphere. At the same time, an analysis of the acts and statistics of the authorities shows that the digitalization of the agro-industrial complex in the Russian Federation is significantly lagging behind the pace of digital transformation of other sectors of the economy.

Digitalization of all areas of modern life is a natural and inevitable process, but the process is developing largely spontaneously, under the influence of market laws and the achievements of the scientific and technical process. Given the insufficient pace of the introduction of digital technologies in agriculture and assuming prospects from their use, the Russian state in recent years has been actively trying to regulate the relevant sphere of public relations, first of all defining the status, rights and obligations of participants in special relations arising on the Internet and to direct the process of digitalization of agriculture in a certain direction, and in some cases accelerate it.

In such conditions, agricultural producers and processors of agricultural products are involved in this process. At the same time, digitalization may entail for them both additional responsibilities, related to the introduction of appropriate technologies or requirements, and new opportunities, due to the favorable, primarily infrastructural conditions created by the Russian Federation (for example, the expansion of the zone of sustainable mobile Internet coverage outside settlements).

At the same time, the modern legal regulation of digitalization of agriculture in Russia faces several problems at once. The current regulatory regulation of the agro-industrial complex is extremely fragmented and does not have sufficient depth, i.e. it does not take into account the specifics of agricultural producers and processors of agricultural products (in comparison with other types of production and processing), which in turn significantly complicates the development of acts in the field of digitalization of agriculture.

Today, the main acts ensuring the digitalization of agriculture are the Decree of the President of the Russian Federation dated 21.07.2016 No. 350 "On measures to implement the state scientific and Technical policy in the interests of agricultural development" and the Food Security Doctrine (approved by the President of the Russian Federation on January 21, 2020). At the same time, the legal regulation of the agriculture digitalization is actual entrusted to the Government of the Russian Federation, which, as an executive authority, is called upon to implement the agricultural policy carried out in the country, rather than generate it.

This is clearly evident from the abundance and content of the acts of the Government of the Russian Federation in the area we are analyzing, and above all, from the materials of the State Program for the Development of Agriculture and Regulation of Agricultural Products, Raw Materials and Food Markets (approved by the Decree of the Government of the Russian Federation No. 717 of July 14, 2012), the Development Strategy of the agro-industrial and fisheries complexes of the Russian Federation for the period up to 2030 (approved by the Government of the Russian Federation on April 12, 2020), Strategies for the Development of the Food and Processing Industry of the Russian Federation for the period up to 2020 (approved by the Government of the Russian Federation on April 17, 2012), Strategies for Sustainable Development of Rural Territories of the Russian Federation for the period up to 2030 (approved by the Government of the Russian Federation on February 2, 2015), a long-term strategy for the development of the grain complex of the Russian Federation until 2035 (approved by the Government of the Russian Federation on August 10, 2019), etc. It is noteworthy that many program documents were adopted in the form of orders of the Government of the Russian Federation, i.e. they are not normative legal acts, which directly contradicts their purpose and content as program acts, which can negatively affect the introduction of innovative information technologies in agroecosystems.

The acts of the Ministry of Agriculture of the Russian Federation are at the next level of legal regulation. The attention is drawn to the fact that among the acts of the Federal Ministry there are practically no documents on digitalization in particular. The only significant exception is the departmental project "Digital Agriculture" with the implementation period of 2019 - 2024.

The analysis shows that documents in the field of digitalization of agriculture in the Russian Federation are accepted not only by state authorities, but also by various committees, councils or bureaus attached to them, which are not among the power institutions, which raises a big question about the legal force of the documents approved by them and their ability to have a regulatory impact and ensure the digital transformation of agriculture.

2 Materials and Methods

The publication is based on the analysis of regulations of the President of the Russian Federation, the Government of the Russian Federation, the Ministry of Agriculture of the Russian Federation and official statistical data (including reports of authorities and reports of scientific organizations prepared by orders of authorities), presented on the website of the Ministry of Agriculture of the Russian Federation (https://mcx.gov.ru), the official website of the Federal Scientific and Technical Program for the Development of Agriculture for 2017 - 2025 (https://fntp-mcx.ru/documents.html), as well as on materials from open data of federal executive bodies posted in accordance with the Federal Law "On providing access to information on the activities of state bodies and local self-government bodies."

The article proposes a model of legal regulation of the digitalization of agriculture in the Russian Federation on the basis of the dialectical unity of its legal security and general scientific methods (analysis, synthesis, deduction and induction, abstraction, structural and functional method) and special (formal legal, method of legal design, formally -logical, systemic, technical and legal analysis, statistical method). After that, the optimal variant of legal regulation of digitalization of enterprises of the agro-industrial complex in Russia is proposed.

3 Results and Discussion

The first, and from a formal legal point of view, the key problem of legal regulation of the digitalization of agriculture in the Russian Federation is the problem of the lack of a generally recognized conceptual apparatus. The word "digitalization" has become firmly established in the scientific and normative circulation, but it is understood in completely different ways. As the normative acts have instilled and most researchers use this term, describing the modern conditions of social development, without explaining or specifying its content [1, 12]. The same authors who reveal the content of the term digitalization sometimes take opposite positions. For example, S.V. Lipen. combines digitalization with informatization [3], and, for example, in a collective monograph edited by Yu.A. Tikhomirov and S. B. Nanba the concepts of "digitalization", "informatization" and "robotization" are differentiated [11].

Another problem of legal regulation of digitalization of agriculture is its cross-sectoral nature. According to the apt remark of O.V. Tanimova digitalization is a natural phenomenon that has arisen on the path of development of the legal system, since it erases the boundaries between branches of law, because information and technologies are already present in every branch [6]. Digitization, penetrating into all spheres of human life, society, business and the state, cannot always be regulated within the framework of codified documents, but rather, on the contrary, is dispersed across various acts that refer not only to different branches of law, but also to different levels of legal regulation. At the same time, the norms on digitalization are gradually becoming a common denominator and, in the future, are able to determine a single logic of law.

Actual digitalization inevitably involves new and new social relations into the sphere of law, which previously either did not exist in principle, or did not require legal regulation, or could not be regulated by law. Researchers refer to these relationships as relationships, the subjects of which are virtual "personalities", relationships associated with legally significant identification of a person in virtual space, as well as relationships arising in connection with the implementation of human rights in virtual space (for example, the right to access the Internet) , relationships focused on the application of robotics [9] and many others.

At the same time, Russia does not have a comprehensive federal law that would determine the directions of digitalization in the Russian Federation, its stages, responsible persons, sources of funding, or at least consolidate the conceptual framework used. On the contrary, digitalization issues today are regulated by an extremely large number of disparate acts, many of which are not at all included in the traditional system of sources of law for Russia. For example, a very large array of provisions related to the digitalization of agriculture is contained in the national program "Digital Economy of the Russian Federation". This document was approved by the minutes of the meeting of the Presidium of the Council under the President of the Russian Federation for Strategic Development and National Projects No. 7 dated June 4, 2019.

Taking into account that the term "digitalization" is one of the key ones for this study and in recent years it has been used not only in special literature, but also in regulations of the Government of the Russian Federation and the Ministry of Agriculture of the Russian Federation, it is extremely important to determine its content. Moreover, its technical interpretation is at odds with the existing regulatory framework.

In a purely technical sense, digitization is the conversion of information into digital form, that is, digital transmission of data encoded into discrete signal pulses [7]. Apparently this is why, for example, A.A. Saurin, interpreting the term digitalization, defines it as a complex of relations for the generation, processing and transmission of information on the principles of a binary code [4]. This approach is directly based on the normative definition of informatization as relations associated with the search, receipt, transmission, production and dissemination of information using information technology, given in subparagraph 1 of paragraph 1 of Art. 12 Federal Law of July 27, 2006 No. 149-FZ "On Information, Information Technologies and Information Protection".

It is obvious that digitalization is associated with the use of digital technologies, where the term "digital" is used to denote a special way of storing and (or) transmitting data using a binary code, where the binary code implies use of pulses (value 1) or their absence (value 0). The translation of analog and other information into digital form constitutes the technical basis for digitalization [2].

At the same time, modern digitalization of agriculture is clearly not limited to the use of only digital technologies, but presupposes a much broader approach, namely the active introduction of information technologies, which in their content encompass digital technologies. In other words, when speaking of digitalization (or digital transformation), regulatory acts imply the introduction of a whole range of the latest information technologies.

This approach is not accidental, since the essence of technical means and the content of technologies that form the basis of these processes are the starting points for understanding the term digitalization. If we turn to recent history, then the mechanization processes were launched first (through the introduction of mechanical devices and aggregates). Then, with the advent of computers, the era of automation began. These processes have largely become the basis of modern transformations. The emergence of the Internet and the massive introduction of computer technology gave rise to the emergence of information and communication technologies that launched the processes of informatization. That is why, while agreeing with the position of S.M. Zubarev believe that "informatization has not only become the initial stage of digitalization, but also continues to be its integral part, since the information sphere and information technologies are immanently linked to the digital sphere and digital technologies "[2]. At the same time, like any later phenomenon, based on the achievements of earlier ones, digitalization allows solving more voluminous and complex problems that are not available at the level of informatization alone.

In this context, V.G. Khalina and G.V. Chernov in their statement that "digitalization is the main modern trend in the development of the economy and society, based on the transition to a digital format for presenting information, which is aimed at increasing the efficiency of the economy and improving the quality of life" [10].

The difficulty of establishing the content of the term "digitalization" and its subsequent clear normative consolidation are predetermined by the fact that it is not strictly legal from the outset. On the contrary, initially digitalization is of a socio-economic and technical nature, which in turn actively influences the sphere of public administration and the development of legislation.

In this respect, we should quote the latest statements of legal scholars on the subject. Thus, O.V. Tanimov defines the concept of digitalization as the process of transforming a certain sphere of life through the introduction of digital technologies or innovative technological solutions [5]. A.A. Kapina reveals the essence of economy digitalization through the emergence of new digital means and their active implementation in economic relations [8]. S.M. Zubarev defines digitalization as "the process of applying digital technologies in various fields" [2], V.G. Khalin and G.V. Chernova call digitalization a transition to a digital format of presenting information, aimed at improving the quality of life and increasing the efficiency of the economy [10].

As we have mentioned above, an important problem of agriculture digitalization is the lack and inconsistency of legal regulation. Therefore, one of the goals of this paper is to clarify the legal field of digital transformation of the agro-industrial complex, that is, those legal acts that determine the field of digital transformation of agriculture.

The first of the six projects of the "Digital Economy of the Russian Federation" national program, namely the "Normative Regulation of Digital Environment" project, is aimed to solve this problem and improve the legal field of digitalization. The key objective of this project is to create a flexible and differentiated system of legal regulation of the digital economy for each industry. The federal project "Normative Regulation of Digital Environment" provides for the development and adoption of a number of regulatory legal acts aimed at removing the priority barriers that hinder the development of digitalization in the economy, and primarily in the areas of civil turnover, financial technologies, telecommunications, identification, intellectual property, standardization and others. Also, work is underway on conceptual acts, which are to create opportunities for developing a new, more effective change management system, including through the development of platforms for technological and organizational piloting of new digital technologies. It is this very path - the path of creating digital platforms – that those few projects regulating the digitalization of agriculture choose to follow.

A digital platform is a comprehensive solution that provides automation of the most important processes in the agricultural sector. It enables to monitor the fields' development and condition on an interactive map by means of satellite images; receive reliable information from reference books about soils, fertilizers, crops, diseases and pests; receive the most accurate weather data from the autonomous weather stations; keep electronic journals about the activities carried out in the fields.

Digital platforms enable to analyze satellite DVI images in real time. The data from the images are processed given the expected development phases of each crop, calculated by the sum of the accumulated temperatures and the entered sowing dates. The platform takes into account what phase of development is expected from a given crop, and reflects in color those areas where the actual level differs from the calculated or average.

In a word, a digital platform provides sustainable farm management and informed decision-making based on operational data and forecasts. And alas, it is the only measure to be taken for agriculture digitalization, planned by the acts of the Minister of Agriculture of the Russian Federation.

4 Conclusions

The current regulatory acts, as well as digital information products presented on the market, do not guarantee the digitalization of agriculture, especially in the situation of a rather passive role of the Federal Ministry of Agriculture.

The Federal Law of July 31, 2020 No. 258-FZ "On Experimental Legal Regimes in the Sphere of Digital Innovations in the Russian Federation" is intended to stimulate this process, allowing enterprises involved in the implementation of digital innovations to practically apply them while the restrictions established by regulatory legal acts are being removed.

According to the forecasts of the Ministry of Economic Development and Trade, in 2021 it is planned to launch 9 experiments that will provide citizens and organizations with better and cheaper services, 4 projects related to the introduction of artificial intelligence. At the same time, the field of agriculture is not represented by any of these projects. More important, the Ministry of Agriculture did not react in any way not only to the adoption of this law in 2020, but also to its entry into force in February 2021. This all happens despite the law explicitly states its applicability in the field of agriculture.

In connection with the abovementioned, it is the activity of agricultural producers and agricultural products processors, introducing innovative digital solutions, that comes to the fore, since the law "On Experimental Legal Regimes in the Sphere of Digital Innovations in the Russian Federation" does not yet contain benefits for agricultural enterprises. It specifically assumes the possibility of introducing a special (respectively, preferential) legal regime and will determine the conditions for its introduction, while the regimes already enable enterprises to reduce the time and costs for the development, testing and implementation of new technologies, as well as reduce legal risks.

The experimental legal regime implies the temporary application of a special regulation instead of a general one. This is about those cases where the latter significantly complicates the implementation of digital innovations or makes it impossible. The regime is applied only to its participants in the areas of development, testing and implementation of digital innovations.

Given the jurisdiction of the Russian Federation and the subjects of the Russian Federation, it is permissible to establish experimental legal regimes both at the federal and regional levels, which enables to take into account the interests of not only federal companies, but also regional enterprises, which is especially important for the agricultural sector, that traditionally focuses on the level of the subjects of the Russian Federation.

The introduction of an experimental legal regime is not an automatic process and implies a declarative procedure for its development and establishment. For this, an enterprise (or a group of initiating enterprises) has the right to submit an initiative proposal to the authorized body. The proposal should describe the planned special regulation, measures to reduce the risks of harm to human life, health or property, etc., attach a draft program of the experimental legal regime and a package of accompanying documents.

Naturally, most agricultural enterprises will not be able to independently develop the entire package of documents. Therefore, it is extremely important to empower the regional ministries of agriculture with the authority to provide methodological and other legal support to agricultural enterprises in order to accompany the latter's initiatives to introduce special legal regimes.

References

1. M.S. Arabyan, K.M. Gil'manova, Customs Matter, **4**, 17-21 (2019)
2. S.M. Zubarev, Actual problems of Russian law, **6**, 23-26 (2020)
3. S.V. Lipen', Actual problems of Russian law, **8**, 22-33 (2019)
4. A.A. Saurin, Constitutional and municipal parvo, **8**, 26-27 (2019)
5. O.V. Tanimov, Russian justice, **7**, 2-5 (2019)
6. O.V. Tanimov, Actual problems of Russian law, **2**, 11-18 (2020)
7. Explanatory Dictionary of the Information Society and the New Economy, http://www.xn--80aacc4bir7b.xn-D0%B2%D0%B0%D1%80%D0%B8/%D1%82%D0%BE8F-digitization
8. K.T. Anisina, B.G. Badmayev, I.V. Bit-Shabo, *Financial law in the context of the development of the digital economy*, 320 (2019)
9. T.YA. Khabriyeva, Journal of Russian Law, **9**, 10-11 (2018)
10. V.G. Khalin, G.V. Chernova, Management Consulting, **10**, 46-63 (2018)
11. N.V. Antonova, S.B. Bal'khayeva, ZH.A. Gaunova, *Legal concept of robotization*, 10 (2019)
12. I.D. Yagofarova, State power and local government, **11**, 11-15 (2019)

Insurance as a risk management system in green construction

Kheda Murtazova[1,*], and *Salambek* Aliyev[2]

[1]Chechen State University named after A.A. Kadyrov, Grozny, Russian Federation
[2]Grozny State Oil Technical University named after Academician M.D. Millionshchikova, Grozny, Russian Federation

Abstract. The study of the problems of the impact of climate change on economic development has become in recent years one of the main directions of economic research. At the same time, along with the development of a global macroeconomic policy in the field of climate and green building, more and more attention is paid to the analysis of corporate strategies to reduce risks and adapt to the consequences of climate change. Without large-scale business investments in green innovative technologies and the introduction of corporate standards for reducing greenhouse gas emissions, it is impossible to achieve long-term targets for reducing global climate risks.

1 Introduction

Many major companies participate in the discussion of the problem of climate change, inform about plans to reduce greenhouse gas emissions in volumes that exceed the established limits, as well as achieve the status of a "carbon neutral company", talk about the priority of introducing "green" technologies in their products and business processes. More than 1,500 of the largest companies include a section on greenhouse gas emissions in their annual sustainability reports [3].

Despite the lack of an unambiguous definition and classification in the specialized literature, green technologies include a wide class of innovations aimed at reducing the negative impact on the environment in such areas as renewable energy, construction of new and reconstruction of existing buildings and other industries.

The growing economic attractiveness of new green technologies is evidenced by the growth of investments in this sector.

Buildings account for a significant part of the world's consumption of resources: 40% of energy, 30% of raw materials and materials, 25% of water and 12% of land resources, about 30% of global greenhouse gas emissions. Therefore, the use of green technologies in the construction of new and modernization of existing buildings is most often viewed by companies as a tool for managing climate risks, energy and resource conservation.

Construction is one of the most capital-intensive and organizationally complex sectors of the economy. This causes a high degree of risk for all participants in the construction

* Corresponding author: fu.ggni@mail.ru

services market, who may face not only objectively determined risks associated with the complexity and duration of the production cycle, but also with the dishonesty of contractors and potential partners [1-2].

The risks accompanying each stage of construction from the project to the commissioning of facilities are varied and are usually measured in significant amounts. In addition, recently from the side of financial institutions and banks to ensure the protection of the financial resources they provide, the requirements for compulsory insurance of construction risks have been put forward, which requires building a system of relationships that takes into account the interests of all participants in the construction process.

In Russia, insurance of construction and assembly risks is not compulsory, and despite the emerging trend of growth in insurance contracts in this industry, it lags far behind world practice.

Since it is completely impossible to avoid risks, a scientifically grounded approach to risk management in the implementation of investment projects is required, which will allow, on the basis of the developed algorithm, to identify the level of uncertainty and minimize the negative impact of risk on the activities of the organization. All this necessitates ensuring economic protection of all participants involved in the construction and investment process from possible risks.

2 Materials and Methods

Investment activity is a prerequisite for the development and competitiveness of Russian construction enterprises, since construction is the most capital-intensive sector of the economy. It traditionally involves design and engineering organizations, uses expensive technologies and unique equipment, and concludes large supply contracts. It is obvious that the risks inherent in each stage of the project are varied and significant, and the possible losses during its implementation can be measured in huge amounts. Insurance of construction risks as part of the construction investment process at all its stages from the development of a feasibility study to the commissioning of the facility allows participants to compensate for unforeseen losses incurred during construction and provides protection for investment in construction [4].

The construction area in all countries is considered a high-risk area where there is a high probability of unfavorable and unpredictable situations during the implementation of an investment project (an increase in investment costs, a decrease in profits, etc.). In investment and construction activities (ISD), research, identification and identification of risk factors underlies the effective implementation of the economic interests and needs of all stakeholders. The study of risk management problems in the context of political, economic and regulatory transformations of the basic rules of doing business in the construction industry is one of the most difficult economic problems and provides the following main approaches: avoidance, prevention, insurance, transfer and acceptance of risk [5].

Unfortunately, as world and domestic practice shows, despite significant efforts to reduce risk (introduction of new standards, use of innovative technologies), when carrying out construction and installation works (CMP), the developer often faces the problem of situations that are difficult to predict and can lead to to an increase in construction time due to damage or loss of building materials, construction equipment, equipment during installation and installation, property of third parties, damage to life and health of people, and there is also a risk of a significant increase in the cost of objects from the initial investment amount [6].

Analysis of existing classifications of risks in relation to the investment and construction sector and an attempt to create on the basis of this classification a unified long-term methodology for increasing the efficiency of investment resources of construction organizations remains unrealized. As a rule, these techniques are aimed at solving specific tactical problems. The reasons for the occurrence and the degree of influence on investment activities for each risk is individual.

The most optimal solution in the context of analyzing the objectively evolving situation and taking into account the economic interests of IUD participants is the insurance of construction and installation risks. In this case, there is no need to accumulate funds on the accounts of the reserve fund to cover possible damage and the contractor can use these funds as working capital, in addition, the terms of restoration of the object can be significantly reduced by compensating the insurer for additional costs associated with the urgent delivery of building materials, equipment and overtime [7].

Construction and installation insurance is a real guarantee of ensuring economic protection of project participants from losses caused as a result of various types of dangerous situations. Unlike other types of property insurance, insurance of construction risks has a specific format of insurance coverage provided, the peculiarity of which is that, due to the multiplicity of possible risks of a construction object, the insurance contract for a construction object does not contain a list of them, but on the contrary contains a list of cases, the damage from which is not is subject to reimbursement by the insurer.

In international practice, insurance coverage does not cover losses [8]:

- caused by defects and errors arising in the design and manufacture of materials;
- as a result of gross negligence of the policyholder and deliberate damage caused by him or his representative;
- as a result of extraordinary and military events, the arrest of property by a court decision;
- as a result of strikes or mutiny;
- damage caused by exposure to nuclear energy.

Thus, insurance can cover all civil and industrial construction projects, with the exception of those listed in the list.

3 Results and Discussions

Currently, providing effective insurance coverage against all risks that may occur during construction and installation work at a construction site, in the Russian and international practice of insurance of construction risks, the insurance standard provides for the use of two types of insurance - CAR (Contractor`s All Risks) and EAR (Engineering All Risks) [9].

CAR insurance provides effective protection against all risks of all participants in the investment and construction process from possible losses. This type of insurance also allows you to reduce the construction estimate, since the provided insurance premium is always lower than the risk premium included in the construction estimate. This type of insurance is most in demand all over the world. As a rule, for the construction of state and municipal facilities, the requirement for the conclusion of CAR insurance is mandatory. More and more often, for obvious reasons, developers, credit organizations, architects, and consulting firms insist on the conclusion of this type of insurance contract [10].

EAR construction and erection risks insurance provides for the insurance of all possible risks during the installation of machines, mechanisms, and the erection of steel structures. In this case, the policyholders may be: the manufacturer or supplier of the installation object; firms directly performing installation; customer; credit organizations. The insurance coverage provided by the insurance contract applies to the territory specified in the insurance contract, usually a construction site.

The period of validity of the insurance contract is determined by the participants and covers the time interval from the moment the equipment (building materials) is unloaded on the construction site specified in the contract and until the acceptance or commissioning of the construction facility. The most common drawback is the designation by the beginning of activities in the insurance contract of the moment the contract enters into force, the agreement is valid from the date of its conclusion or the transfer of the insurance premium. The validity of insurance for construction and installation works is determined from the moment of commencement of work and is valid until the delivery of the object. If the construction contract provides for post-launch warranty service, the insurance coverage can be extended for this period, for example, one year, after the acceptance of the object. Maintenance risk insurance can be standard - it covers only damage caused during the maintenance of equipment, or extended, in addition to the standard risk coverage, insurance for losses caused by errors in the process of construction work [11].

Risk management primarily involves preventing the possibility of their occurrence (avoiding risk) and mitigating the consequences throughout the entire life cycle of the project (insurance against risk, transferring risk to the partner who is able to manage it more effectively). As practice shows, it is impossible to keep all risk factors under control. One of the approaches to managing and limiting the impact of risks in construction, caused by external and internal factors, is "the creation of information systems to monitor various risk factors - from changes in federal legislation to incorrect performance of the task" [5].

Reducing risks during the implementation of an investment project can be ensured:

- creation of a reserve fund for unforeseen expenses associated with design errors, non-fulfillment of financial obligations on the part of partners, etc .;
- the creation of a reserve of time in case of an increase in the timing of certain types of construction work;
- a decrease in construction volumes due to production and non-production losses, etc.

The size of the reserve fund for unforeseen expenses is determined by the customer in accordance with the norms stipulated by the Urban Planning Codes of the Russian Federation: “no more than 2% - for non-production capital construction projects; no more than 3% percent - for capital construction facilities for industrial purposes and linear facilities; no more than 10% - for unique, especially dangerous and technically complex capital construction projects ”. [12]

In addition, the amount of the insured amount, by agreement of the parties, can be paid separately for construction and installation works, construction and installation equipment and the cost of purchased or refurbished technical equipment and machinery in accordance with the estimated cost of the construction object. Construction companies increasingly include in the field of insurance coverage expensive high-performance equipment, the repair and maintenance of which require considerable financial costs.

4 Conclusions

The main part of modern green buildings are office buildings, the greatest contribution to their economic efficiency is made by an increase in labor productivity, which, according to G. Katz's calculations, accounts for, on average, 75% of the total economic effect obtained during the operation of the "green" production he studied. buildings. the main factor in the growth of labor productivity is the decrease in the number of days missed due to illness and without a valid reason. The difference in geographic and economic conditions does not allow extrapolating foreign estimates of the effectiveness of "green" buildings on the Russian construction market. Other reasons for the slow introduction of green technologies into the practice of Russian construction companies include the conservatism of the participants in the real estate market, insufficient information about available technologies, as well as the lack of specialized specialists at all stages, including the development of the concept and design of "green" buildings.

References

1. Civil Code of the Russian Federation, 742
2. Town planning code of the Russian Federation, 481
3. Tax Code of the Russian Federation, 25(4), 263
4. Federal Law No. 39-FZ dated 25.02. 1999 "On investment activities in the Russian Federation carried out in the form of capital investments"
5. J. Pinkse, A. Kolk, International Business and Global Climate Change, Routledge, (2009)
6. Ernst & Young Business Risks Report (2010)
7. Common Carbon Metric for measuring Energy Use & Reporting Greenhouse Gas Emissions from building operations, UNEP SBCI (2009)
8. Green IT: Why Mid-size Companies are investing now, Info-Tech Research Group (2009)
9. Energy and Small and Medium Enterprise, US Agency for International Development (2001)
10. C. Mines, How Tech Will Be Critical to Corporate Sustainability in 2011 (2011)
11. The Business Value of “Green” Document Management Solutions, White paper, IDC (2010)
12. Climate change: global study 2010, Synovate Research Group (2010)

Model of state support for organic farming of Kazakhstan in the context of creation of international regulatory framework for the industry

Dilara S. Samenbetova[1], and *Oleg Yu.* Patlasov[2,3,*]

[1] Omsk Humanitarian Academy, 4th Cheluskintsev st., 2a, 644144 Omsk, Russia
[2] Pushchino State Institute of Natural science, 3 Prospekt nauki, Pushchino, Russia
[3] K.G. Razumovsky Moscow State University of Technologies and Management (the First Cossack University), 73, Zemlyanoy Val street, Moscow, Russia

Abstract. The paper evaluates the status and prospects for a wider market capacity of organic foods and items. The paper analyzes regulatory framework for organic market in Kazakhstan. The paper describes the conditions for the development of the organic market. The transformation of global food market structures has influenced the organic farming. Demand for organic products has been growing rapidly. Whereas the European Union has put in place at the time an effective system for regulating organic production, the EAEU, CIS, and SCO countries can see this production taking shape only now. It is argued that Kazakhstan is harnessing its agricultural potential only partially, including organic food production. The paper proves that the development of green farming entrepreneurship in Kazakhstan will enable agricultural organizations, aided by the measures of state support, to form a new niche in the market for organic products and expand the supply of organic products in the world market. New approaches are proposed for the first time in regard to formulating regulations, model laws, and recommendations of international unions.

1 Introduction

In 2013 Kazakhstan gained new opportunities in the development of the organic market following the approval of the concept for the transition of the Republic of Kazakhstan to green economy for 2013-2020 [1, p.10]. Within the shortest possible time, the Government developed standards for agricultural products.

Pursuant to the concept and based on international standards, the Government adopted the Law "On Production of Organic Goods" in 2015. This law governs organic production processes, which, in turn, will enable Kazakhstan to join the international organic market and, thus, the state will be able to regulate the import and export of domestic products. Kazakhstan possesses huge lands, which are traditionally farmed without the use of

* Corresponding author: opatlasov@mail.ru

synthetic fertilizers and pesticides [5]. This aspect carries a huge privilege for the EAEU country. Production and sales of organic farm produce is an objective national competitive advantage for Kazakhstan's agro-industrial complex.

The law is in line with the rules and regulations of the international law of IFOAM and, thus, complies with international quality and control standards for organic products [21]. The adoption of this law allows Kazakhstan to participate in the world organic trade market and has positive implications for the development of farming in Kazakhstan. Along with the adoption of the main law regulating the organic foods market of the Republic of Kazakhstan, the Government has also introduced new standards and rules for organic farming.

Currently, many countries have put in place laws and standards regulating the production and labeling of organic products, with special international organizations established in order to monitor compliance. Relative to other CIS countries, Kazakhstan adopted the Law on Organic Agriculture quite recently. Well, yet the standards and rules have been adopted relatively recently. I would like to note that the executive orders for the law administration are still at the stage of development. Kazakhstan has not yet designed a national overarching plan for the development of an organic production system.

The organic market of local products should be considered from a perspective of the world organic market. If we compare the experience and indicators of other foreign countries in setting up organic production, it can be noted that Kazakhstan is only at the initial stage of the organic market development.

2 Materials and methods

The paper evaluates the status and prospects for a wider market capacity of organic food products. The research presents the model for enhancing the ways in which state support is offered to the organic farming market in Kazakhstan and has revealed the linkage between organic farming in Kazakhstan and environmental, social and economic indicators.

As the study has showed, Kazakhstan, like other EAEU states, faces similar challenges when it comes to organic agriculture: underdeveloped internal market for organic food products, weak financial institutions and the lack of comprehensive government support aimed at encouraging the production of organic food; lack of national statistics on the production and export of organic products, and inadequate regulatory framework for both the internal and external organic markets.

The research has applied general scientific methods for getting to know socio-economic phenomena and processes: logical and analytical methods, classification, systemic and integrated approaches, synthesis and comparative analysis. Scientific novelty of the research outcome is to develop proposals on how to improve the institutional and legal foundations of the organic market, where such proposals are about putting in place a mechanism for the state supervisory body to accredit private and public certification organizations, setting up country's own certification authority that would comply with international generally accepted standards, harmonizing national laws with international laws and introducing standards for subsidizing organic agriculture into the legal field of Kazakhstan.

The hypothesis of the study is that governmental measures to support the agricultural organic market – as a link between production and consumption – are a prerequisite for ensuring and implementing country's green entrepreneurship, improving the quality of food security, and whether it is permissible to use organic mineral fertilizers in building organic farming. However, the organization and infrastructure of the above-mentioned market are imperfect, requiring appropriate theoretical and practical applications, reliable justifications, and the formation of a new market segment.

3 Results and discussion

Over the past 30 years, more than 187 countries out of 230 countries have been actively developing organic farming. Primary consumers of the organic foods market are the USA, Canada, the European Union countries – such as Austria, Germany, Great Britain, France – and Japan. It should be noted that about 78% of organic products are consumed by highly developed industrial countries of Europe and the United States. According to review by FiBL & IFOAM - Organic International (2020), 72.3 million hectares of farming land are eco-clean (including uncultivated areas) [15].

There is no official statistics on organic foods production in Kazakhstan, with no public register of organic producers available either. According to the National Accreditation Center, in 2019 Kazakhstan ranked 9th in the exports of organic foods to the EU countries, having increased supplies to 85,675 tons, although in 2018 exports amounted to 50,250 tons. That being the case, Kazakhstan ranks 6th in Asia in terms of the total area of organically certified land, but in the future the country may become Asia's third country after China (3,135,000 ha) and India (1,938,221 ha). The following countries have become the main exporters: Great Britain, Italy, Germany, France, Belgium, the Netherlands, Poland, Russia, Ukraine and other countries [1, p. 32]. According to the annual statistics by FiBL & IFOAM - Organic International (2020) - FiBL survey 2021, Kazakhstan – as at year-end of 2019 – has joined top 10 countries with a high growth in the use of organic lands (Figure 1).

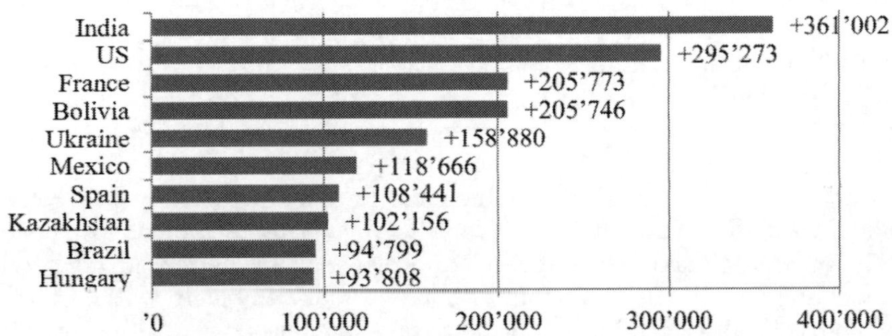

Fig. 1. Top ten countries with the highest gain in organic land in 2019 (hectares) [21].

Likewise, according to statistics at year-end 2018, Kazakhstan has used 192,134 hectares of land for organic production purposes [15]. This roughly makes 0.1% of the share of the total organic land. If we analyze this figure for 2017, we can see that the size of organic lands has decreased. In 2017, the size of organic land was 256,741 hectares, which is 64,607 hectares less than in 2018. But in 2019, area of agricultural organic land increased by almost 102,156 hectares. In the context of 10 years, Kazakhstan has certainly achieved a gain of almost 42.5%. There is no official statistics nationwide on producers, but according to the data provided by the source presented above, 63 producers make organic food products in Kazakhstan, which includes 22 businesses engaged in the storage and processing of organic products, 7 businesses doing imports, and 14 in exports. According to the annual statistics by FiBL & IFOAM - Organic International (2020) - FiBL survey 2021, at year-end 2019 Kazakhstan joined DAC countries (Development Assistance Committee) list with the largest areas of organic agricultural land in 2019 (Figure 2).

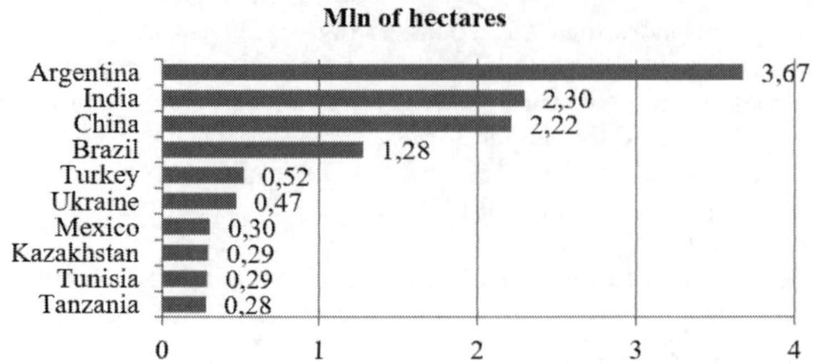

Fig. 2. Top 10 countries from DAC list with the largest areas of organic agricultural land in 2019 [21].

For farmers of Kazakhstan, organic products are still an unknown variable in the large realm of agriculture. Apart from more obvious advantages of natural production, organic production is more complicated due to an extensive number of requirements for the growing process. Currently Kazakhstan has more than 63 such producers, and the real organic pioneers are the farmers of the Kostanay Region. Over 15 types of farm produce are exported from Kazakhstan to Europe - peas, Sudan grass, duck wheat, lentils, flax, sainfoin, soybeans, chickpea, oats, spring wheat, sunflower, millet, false flax, wheatgrass and spring barley. Globally known flax – which is actively purchased by Germany, Switzerland and Italy – is also exported from Kazakhstan. Organic lentils are also very popular abroad, purchased in large volumes by Italy and Turkey. Exports are also made up of cereals, but not in such large volumes. There is a farm in the Almaty Region which deserves special presentation - the farm is the only one in Kazakhstan certified for the wild collection of medicinal herbs, the main buyer of which is located in Germany.

In recent years, Kazakhstan has become one of the international leaders in the export of flax. Thus, in 2020, local producers sold almost 15 thousand tons of certified organic flax worth about 9 million euros. In 2019, they sold about 7 thousand tons for about 4 million euros. In general, Kazakhstan is the 10th top player in the European organic market of oilseeds. Also, Kazakhstan sold about 6 thousand tons of organic wheat worth about 1.5 million euros. Well, here we face a paradoxical situation in which although more organic grain is produced in Kazakhstan, it is exported less than flax. One reason is that in recent years Europe has seen the emergence of more local farmers growing organic wheat. Russia is also a strategic export partner. In the Russian Federation, the domestic organic market in 2018 was worth $183 million euros. Russia's own production supplies only 20% of the demand for organic foods, with 80% coming from imports. Therefore, there are no artificial barriers in Russia that would prevent organic products from Kazakhstan from being on the shelves. In addition, it is convenient for our farmers to be in a single market within the EAEU, as well as to enjoy simple logistics. Russia buys organic foods from Kazakhstan, but so far there have been fewer orders from Russia than from Europe.

One of the main aspects for the development of the organic foods market in Kazakhstan is a matter of organic certification. In order to secure international certification, farmers and processing companies request certification from certification organizations. They are certified depending on the target market: under the terms of Council Regulation (EC) 834/07 for exports of products to the European market or under the terms of the NOP for exports of products to the North American market [9, p. 73]. The Certificate enables producers to label their products as organic product abroad. The following international companies currently operate as certifiers in Kazakhstan: Austria Bio Garantie GmbH;

CERES Certification of Environmental Standards GmbH; Ecocert SA; Institute for Marketecology (IMO); Istituto Certificazione Etica e Ambientale; LACON GmbH; SGS Austria Controll-Co. GmbH; Ecoglobe; Organic standard.

Kazakhstan has large enough areas for organic farming, which make up roughly 300 thousand hectares. But in order to certify land as organic, large financial investments are needed, which makes it impractical for farmers to request certification for less than 1000 hectares of arable land. Well, here the issue of reducing certification costs comes into play. For example, decreasing the price of certification by 2.5 times will respectively reduce the threshold for organic farming to 400 hectares. This measure will enable even small farmers to start organic farming. The south of the country is able to grow vegetables and fruits according to organic standards. Organic farming requires at least 100 hectares of land. But the south doesn't have such areas, as all the land plots are fragmented [12]. So the South does not do organics. Only in the Almaty Region there is a farm that exports organic soybeans, so reducing the minimum certifiable area to 40 hectares will change the situation and make the south the driver of organic production [11].

Quite high export figures have no impact on the internal market whatsoever. Farmers are not willing to develop this segment, with many not even aware of how to establish a supply chain to retail stores. Thus, producers and buyers have to be reached out in order to build awareness, which needs to be supported at government level. At the moment, the Union of Organic Producers of Kazakhstan is actively engaging with farmers in Kazakhstan in order to raise their awareness, offering pre-audit consultations. The opportunities and risks of organic farming are explained to producers, letting the farmer further decide whether to send products for export or sell the finished product in the domestic market. Therefore, the Union of Organic Producers needs to keep raising farmers' awareness across each aspect. Focus should be placed not only on the production, but also on processing, drying, primary treatment, land allotment, storage, and networking for sales systems. So, an important step in this respect will be to set up an export-oriented trade and processing cooperative (or a system of cooperatives) that would unite several regional producers. This will make it possible to make organic products and move away from the raw feeds model. In accordance with the standards, an organic producer must store and transport in a proper manner, i.e. there has to be a certified transport mode. This makes it relevant to consider creating in the future a northern regional hub in Kazakhstan to promote local organic products. It is necessary to focus on this issue. There is also a need to consider whether it is possible to establish logistics centers in Nur-Sultan and Almaty with appropriate storage facilities for the export of eco-friendly products to the EAEU countries and farther abroad.

Another strategic area of green entrepreneurship in Kazakhstan is the promotion of the umbrella brands of Qazaq Organic and Halal Qazaqstan. Organic items are already produced under the Qazaq Organic Food logo, and they are also exported to Europe. Thus, the goal behind establishing umbrella brands is to create a trendy character and branding that will unite all isolated brands according to certain identity principles. In the future, Qazaq Organic Food plans for additional export opportunities through recognition both in the domestic and foreign markets among producers and consumers. Therefore, additional coverage is needed in the local market for the Qazaq Organic Food brand. Country branding is becoming more and more popular, with more and more countries around the world committing considerable resources to the development of their national brands [10, p.100]. Of course, the issue of a single umbrella identifier applies to all umbrella brands. Consumers of organic foods are already familiar with the following brands: Zharkol-007 LLP, Bio Tau Zher LLP, Honey KZ SEC, and Karzhau Farm Business. For them, these products associate with domestic organic production. The Federation of Organic Agriculture Movements – KAZFOAM, the Kazakh Research Institute of the Agricultural Economics and Rural Development, and the Union of Organic Producers of Kazakhstan are the main groups that support and implement organic agriculture in Kazakhstan. Having done the analysis of the strategic areas for the development of green entrepreneurship in Kazakhstan, it can be argued that the state does not provide direct support. There are indirect measures of regulation in the form of a legislative framework and institutions, but they also need to be improved.

As part of the study of the domestic retail market for organic products in Kazakhstan, the authors conducted a survey of Internet users with a poll called "Analysis of the Organic Market for Products of Kazakhstan". About 500 respondents over the age of 20 were interviewed, involving the working population from governmental structures and businesses. Based on a social study, the authors made conclusions about the attitude of Internet users to the organic food market. There is no solid awareness yet. Residents of Kazakhstan learn about most of the benefits of organic products from the Internet, with the consumer not really aware of the difference between the labels such as "organic", "organic products", "BIO", "GMO-free", "eco-products", "natural product", not knowing about strict requirements for the production and labeling of organic products. For some, all organic products are just a sales pitch. This gives rise to a relevance of issue of certification and labeling. After all, only the "organic" label and a conformity mark next to it make any consumer 100% confident about the natural origin of raw materials and high quality without the use of any chemicals, preservatives, GMOs, and so on. Moreover, a large share of consumers in Kazakhstan is unaware of the labeling of organic products using a special conformity mark. In most cases, a consumer will read the information on the package that this food product complies with the established GOST. In Kazakhstan, there is no unified certification system in accordance with an international certificate, while getting a certificate from international bodies is costly. Therefore, it is utterly unrealistic for a producer to enter the international market without state support.

4 Conclusions

Based on the above conclusions and findings, the authors have made the following recommendations. First, domestic organic producers in Kazakhstan should raise public awareness of the benefits of environmentally friendly products. Consumers should be able to distinguish between pseudo-labels and special marks (approved at the legislative level). Secondly, there is no register of organic producers in Kazakhstan and, in fact, the state does not provide adequate support to them. With that being said, according to the survey, many respondents trust domestic producers for the most part due to the proven benefits of natural products. The "Qazaq Organic Food" State Program is gathering pace, so the communications and outreach policy to promote organic products must be arranged in such a way as to allow the local public to become familiar with domestic producers and trust them just as they would trust foreign producers. Thirdly, when assessing the capacity of the organic foods market, the authors have revealed that primary consumers are upper income families with children. Hence, the awareness-raising policy must emphasize the benefits of consuming organic products on a daily basis. Family shopping is not a spontaneous act, so the sales points need to be conveniently located and be on everyone's lips. Availability of online stores and delivery channels increases the probability of purchase. In their marketing strategies, organic producers should place the right emphasis, accommodating the preferences of married couples with children. Thus, we believe that these results of the sociological survey and the measures will increase the share of organic food, and can be used for further research.

Today, the issue of improving a number of measures of state support for organic production has become relevant. If Kazakhstan wants to develop organic farming and agriculture, then it is essential that a whole range of measures be delivered. Regulation and establishment of new national standards is considered to be the most important task of the country's government in the near future. In order to increase the share of organic exports, it is necessary to harmonize national standards with those of the exporting countries. One of the drivers of this process will be the accreditation of Kazakhstan's certification body in the exporting country.

Currently the local situation in Kazakhstan with the certification of organic products is that it takes at least 3 years to get certified in the EU. A total of 19 companies have been accredited by the EU to grant certification to producers of organic items in Kazakhstan. The nearest laboratory that meets international standards is located in Chisinau and Kiev, with 90% of organic exports from Kazakhstan certified by Ukrainian partners. Therefore, establishing our own certification authority is a key issue for the development of organic agriculture in Kazakhstan. It is necessary to set up laboratories that will test products for hazardous substances. The estimated cost of this would be approximately 10 million euros. So, at the moment, a farmer incurs enormous costs of international certification procedure: a lot of money is spent to invite a specialist, send lab samples for verification, as well as to implement transportation and storage. With a farmer spending at the moment about 10 and more thousand euros to obtain a certificate of conformity, this will cost producers 3-4 thousand euros once there is a local certification authority. So, the price of organic products in the market will be only 20% more expensive than traditional items. Thus, once there is an in-house certification authority and laboratory in the country, this will reduce the unit cost of organic products, which will have a beneficial effect on the development of the domestic market. With an in-house lab, Kazakhstan will be able to offer certification to the nearest countries, which also lack their own organics certification resources: Kyrgyzstan, Uzbekistan, Tajikistan, Russia (border areas). Due to its territorial advantage, Kazakhstan will be able to become a strategic partner for these countries. Certification of organic products will make it possible to finally deal with labeling and pseudo-labeling issue, when the "bio" and "eco" marks will be considered only a sales pitch, while the "organic" label will inspire confidence. Consequently, there will be an opportunity to promote the national brand of environmentally friendly products – Qazaq Organic Food – in the organic market of the Republic of Kazakhstan.

The next measure in the development of organic products is to improve the Law of the Republic of Kazakhstan "On the Production of Organic Goods" adopted in 2015. Along with this, it is necessary to update the regulations and standards governing all players in the organic market: farmers, traders, and certification authorities.

We believe it is practical and important to enact the following legislative innovations. We would like to emphasize a few points in this respect

1. The Law of the Republic of Kazakhstan "On the Production of Organic Goods" does not contain a definition on self-certification, i.e. most of the small organic producers are not actually certified by a third party, but are self-certified under a system developed by IFOAM and called the PGS – Participatory Guarantee System (or Community Guarantee System) [12]. But according to the law, such products cannot be called organic. Some countries also do not deal with self-certification; however, for small producers – that start their organic production – this measure will help to join the domestic market, and then export.

2. In Kazakhstan, there is no mandatory official statistics on organic products, no register of organic suppliers as well as no database of organic seeds. In the EU, this sort of database operates at the legislative level, but the EU exporters do not require it from other countries. But maybe in the future, once the harmonization of standards takes place, lack of such a database on Kazakhstan will affect the opportunities to sell locally made products to the European markets.

3. The Law of the Republic of Kazakhstan "On the Production of Organic Goods" does not have an article on mandatory certification of importers in Kazakhstan. This measure will help the Ministry of Agriculture to analyze the market of organic imports.

4. Introduction of the requirement that foreign certification bodies operating in the country shall be mandatorily registered in Kazakhstan. Since Kazakhstan is now on the EU gray lists for organic products, this measure will allow government agencies to rehabilitate the country's position on the list of faithful suppliers. As already noted, Kazakhstan is in the group of risk countries, which makes certification 2.5 times more expensive than for countries that are considered well-to-do by the EU.

5. Harmonization of domestic and international certification. At the moment, farmers with certificates from foreign countries of the EU, the US, Canada, and China cannot sell their products in the domestic market as organic, and they need to obtain local certification. This is why farmers – who have already incurred financial costs due to receiving international certification – no longer want to request domestic certification, which has a bad effect on the development of the domestic market.

6. The Law of the Republic of Kazakhstan "On the Production of Organic Goods" does not define the issue of subsidies. That is, the Law does not describe either interest rates for the purchase of fertilizers for organic agricultural production, or any preferential loans and grants for beginning entrepreneurs. However, we would like to note that Order of the Minister of Agriculture of the Republic of Kazakhstan No. 107 dated March 30, 2020 approved the rules for subsidizing gains in yields and quality of crops. The rules clearly state that there are several subsidizing areas in accordance with the Law, including subsidies for mineral fertilizers (except for organic fertilizers). This factor does not contribute to organic farming in Kazakhstan. So-called per hectare support is also needed, especially during conversion period, when a business is moving from classic to organic production, and this subsidy should be 3-5 times higher than classic support.

At the moment, the law is more of a framework rather than a delivery plan, with a lot of work needed in the preparation of by-laws aimed at putting in place a delivery mechanism. The National Accreditation Center deals with internal certification. NAC specialists noted that there are no applications for local certification, which shows that the domestic market is developing very slowly in Kazakhstan. Thus, it is necessary to avoid double certification, which will help increase the share of the domestic organic market by reducing production costs. The process of harmonization of European and domestic standards requires strong focus.

Relating to the development of organic farming in the CIS countries, the main issue that needs to be addressed is the synchronization of the countries' laws and shaping the general law on organic agriculture, which would be harmonized with all national laws. There are no systematized national strategies in the EAEU states, with the regulatory measures often posing additional barriers and not encouraging the production of organic foods and items. Thus, a prospective dimension for integration between EAEU states in developing an organic market is the following: working on national aware-raising and consultative systems, harmonization of systems of public registration of bio-drugs, crop nutrients and protection, international cooperation, introduction of systems for keeping track of organic products. The first step in this direction has already been made - the roadmap approved on August 23, 2021. Entering into an international treaty aimed at ensuring the free movement of organic farm produce across the EAEU will simplify certification procedures, and will contribute to fair competition, and more importantly, cooperation. The next step should be to create a common legislative model for all countries.

It is important for domestic agricultural organic producers to make it into the "List of Competitive Products" of the CIS countries, which currently includes 330 producers of Kazakhstan in the "Agricultural and Food Products" section.

It is also necessary to consider the international cooperation of the SCO. Measures to enhance regulatory and legal framework as part of the updated Program for Multilateral Trade and Economic Development of the SCO member states, as well as practical interchange and implementing innovative solutions in the agricultural sector particularly concerning organic agriculture will make it possible to establish a single legal environment for international cooperation in organic farming. Particular emphasis shall be placed on the infrastructure for marketing organic products, including to international markets, specifically relating to establishing a national lab in the Republic of Kazakhstan certified in the EU. The state has not yet allocated roles between state authorities, ministries and market players such as producers, consumers, suppliers, and traders. There is no system for collecting information on the number of producers, certified land, market size and other components, which would serve as a basis for further development forecasts. Improving the legislative framework, legal and institutional structures in the future will be able to produce a synergizing effect in the development of both the internal and external markets for organic products.

References

1. State Program for the Development of the Agro-Industrial Complex of the Republic of Kazakhstan for 2017-2021. http://extwprlegs1.fao.org.
2. Law on Organic Products and Amendments to Certain Legislative Acts of the Russian Federation. Adopted by State Duma (2018) https://rosorganic.ru
3. Law of the Republic of Kazakhstan No. 423-V dated NOV27 (2015) "On Production of Organic Goods", http://adilet.zan.kz.
4. Draft Concept of Development of Measures to Support Domestic Producers and Processors of Agricultural Goods based on Domestic Food Aid Arrangements within WTO Green Box, http://www.mcx.ru.
5. On Organic Products and Amendments to Certain Legislative Acts of the Russian Federation. Federal Law (2018) No. 280-FZ.
6. Order of the Ministry of Health of Russia dated January 15, (2020) N. 8 approved the Strategy for Promotion of Public Healthy Lifestyle, Prevention and Control of Non-Communicable Diseases until 2025.
7. Sanitary Regulations and Standards 2.3.2.2354-08 "Additions and amendments No. 8 to Sanitary and Epidemiological Rules of SRG 2.3.2.1078-01 "Hygienic Requirements for Food Safety and Nutritional Value of Foods" (registered with Russia's MoJ on March 22, (2002), reference No. 3326)
8. Ye. Varavin, M. Kozlova. Economy of the Region, **14(4),** 1282 (2018)
9. V. Grigoruk, Y. Klimov, *Development of Organic Farming in the World and Kazakhstan,* 153 (2016) http://www.fao.org.
10. I. Degtyarev, G. Tokareva, O. Shalina, Bulletin of Perm National Research Polytechnic University. Social and Economic Sciences, **3,** 180 (2016)
11. N. Krasovskaya, A. Deyev, S. Chernomorenko, Science Journal, **6** (2018) https://esj.today.
12. V. Kruchinina, VGUIT Bulletin (Voronezh State University of Engineering Technologies), **79(2),** 296 (2017)
13. O. Mironenko. Russia's Organic Market. Results of 2017. Prospects for 2018, Mironenko Analitika (2017-2018)

14. D. Samenbetova, O. Patlasov, Human Science: Humanitarian Research, **2(36)** 210 (2019)
15. FiBL& IFOAM -The World of Organic Agriculture - Statistics and Emerging Trends (2019) https://ciaorganico.net
16. L.M. Jespersen, D.L. Baggesen, E. Fog, Organic Agriculture, 243 (2017)
17. L. Marsh, E. Zoumenou, C. Cotton, F. Hashem, Agriculture, 125 (2017)
18. E. Roos, A. Mie, M. Wivstad, Agronomy for Sustainable Development, 14 (2018)
19. F.A. Shafi, D. Rennie, Procedia Social and Behavioral Sciences, **49,** 360 (2012) https://www.sciencedirect.com
20. S. Mann, Socioeconomics of Agriculture, 41 (2018) https://www.researchgate.net
21. The Food and Agriculture Organization (FAO), http://www.fao.org.
22. The World of Organic Agriculture - Statistics and Emerging Trends (2019), https://ciaorganico.net

AUTHOR INDEX

AUTHOR INDEX

AUTHOR INDEX

EDP Sciences – Web of Conferences
17, Avenue du Hoggar
Parc d'Activité de Courtabouf
BP 112
F-91944 Les Ulis Cedex A
France

ISBN 978-1-7138-4241-5

CURRAN ASSOCIATES INC.
proceedings.com